DYNAMIC SOIL-STRUCTURE INTERACTION

Prentice-Hall International Series in Civil Engineering and Engineering Mechanics

William J. Hall, Editor

DYNAMIC SOIL-STRUCTURE INTERACTION

JOHN P. WOLF

Electrowatt Engineering Services Ltd.

PRENTICE-HALL, INC., Englewood Cliffs, N.J. 07632

Library of Congress Cataloguing in Publication Data

WOLF, JOHN P. (date)

Dynamic soil-structure interaction.

(Prentice-Hall international series in civil engineering and engineering mechanics)
Bibliography: p.
Includes index.

1. Soil dynamics. 2. Structural dynamics. 3. Foundations. 4. Earthquake engineering. I. Title. II. Series: Prentice-Hall civil engineering and engineering mechanics series.

TA710.W59 1985 624.1′5136 83-25477
ISBN 0-13-221565-9

Editorial/production supervision
and interior design: *Theresa A. Soler*
Manufacturing buyer: *Anthony Caruso*

Printed in the United States of America

10 9 8 7 6 5 4 3 2 1

ISBN 0-13-221565-9 01

Prentice-Hall International, Inc., *London*
Prentice-Hall of Australia Pty. Limited, *Sydney*
Editora Prentice-Hall do Brasil, Ltda., *Rio de Janeiro*
Prentice-Hall Canada, Inc., *Toronto*
Prentice-Hall of India Private Limited, *New Delhi*
Prentice-Hall of Japan, Inc., *Tokyo*
Prentice-Hall of Southeast Asia Pte. Ltd., *Singapore*
Whitehall Books Limited, *Wellington, New Zealand*

CONTENTS

PREFACE

In the seismic analysis of a structure founded on rock, the motion experienced by the base is essentially identical to that occurring in the same point before the structure is built. The calculation can thus be restricted to the structure excited by this specified motion. In the case of a soft site, two important modifications arise for the same incident seismic waves from the source. First, the free-field motion at the site in the absence of the structure is strongly affected. Second, the presence of the structure in the soil will change the dynamic system from the fixed-base condition. The structure will interact with the surrounding soil, leading to a further change of the seismic motion at the base. This text on soil–structure interaction deals with both the free-field response and the actual interaction analysis. Soil–structure interaction is also important for other loading cases (e.g., arising from unbalanced mass in rotating machinery).

The effect of soil–structure interaction is recognized to be important and cannot, in general, be neglected. Even the seismic-design provisions applicable to everyday building structures permit a significant reduction of the equivalent static lateral load compared to that applicable for the fixed-base structure. For the design of critical facilities, especially nuclear-power plants, very complex analyses are required which are based on recent research results, some of which have not been fully evaluated. This has led to a situation where the analysis of soil–structure interaction has become a highly controversial matter.

A uniform approach to analyze both the free-field response as well as the actual interaction analysis is presented in this text. This rigorous procedure, based on wave propagation, makes use of such familiar concepts as the direct-stiffness method of structural analysis and also applies such recent developments as the boundary-element method. The results of vast parametric studies are included, which can be used directly by the analyst. Examples from actual practice demonstrate that these methods are being applied. Besides the rigorous procedure, simple approximate methods are developed which nevertheless capture the essential features of soil–structure interaction. This allows the analyst to perform preliminary calculations with simplified models to determine the key parameters before starting with complicated computations.

Many aspects of this field are so well established that it is difficult to assign credit for them. Credit references have thus been restricted to those necessary for copyright reasons. Sincere apologies are offered to anybody who might feel offended by being left out. The author has been influenced over the years by the research published by many authorities; to name just a few (in alphabetical order): Professors E. Kausel, J. E. Luco, J. Lysmer, J. M. Roesset, and A. S. Veletsos.

At the end of each chapter a summary is included. Then problems are formulated which not only allow the student to corroborate full understanding of the analytical techniques, but which lead to new insights into the various aspects. The problems thus form an important part of the text. The reader who has no intention of solving the problems in detail will find it advantageous to glance through them. As a rule, a detailed solution procedure is provided and the results, in many cases quite important on their own merit, are presented in the form of figures.

The course on which this text is based has been taught for the past few years at the Swiss Federal Institute of Technology in Zurich. It is offered for advanced undergraduate and graduate students in civil engineering. As a prerequisite some knowledge of structural dynamics is essential which can be acquired in a one-term course.

The important contributions of the author's colleagues and students are gratefully acknowledged. In particular, the author would sincerely like to thank Messrs. G. von Arx, K. Bucher, G. Darbre, P. Obernhuber, P. Skrikerud, D. Somaini, and B. Weber for their dedicated efforts. Finally, the author is indebted to Electrowatt Engineering Services Ltd. for its financial support.

John P. Wolf

1

INTRODUCTION

1.1 OBJECTIVE OF SOIL–STRUCTURE INTERACTION ANALYSIS

Structural dynamics deals with methods to determine the stresses and displacements of a structure subjected to dynamic loads. The dimensions of the structure are finite. It is thus rather straightforward to determine a dynamic model with a finite number of degrees of freedom. The corresponding dynamic equations of motion of the discretized structure are then formulated, and highly developed methods for solving them are readily available. In general, however, the structure will interact with the surrounding soil. It is thus not permissible to analyze only the structure. It must also be considered that in many important cases (e.g., earthquake excitation) the loading is applied to the soil region around the structure; this means that the former has to be modeled anyway. The soil is a semi-infinite medium, an unbounded domain. For static loading, a fictitious boundary at a sufficient distance from the structure, where the response is expected to have died out from a practical point of view, can be introduced. This leads to a finite domain for the soil which can be modeled similarly to the structure. The total discretized system, consisting of the structure and the soil, can then be analyzed straightforwardly. However, for dynamic loading, this procedure cannot be used. The fictitious boundary would reflect waves originating from the vibrating structure back into the discretized soil region instead of letting them pass through and propagate toward infinity. This need to model the unbounded foundation medium properly distinguishes soil dynamics from structural dynamics.

The fundamental objective of the analysis of soil–structure interaction is

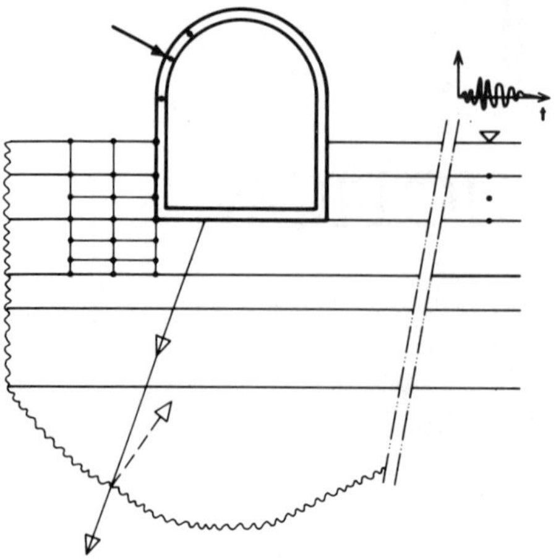

Figure 1-1 Fundamental objective of analysis of soil–structure interaction.

illustrated in Fig. 1-1. A specified time-varying load acts on a structure embedded in layered soil. The dynamic response of the structure and, to a lesser extent, of the soil is to be calculated, taking into account the radiation of energy of the waves propagating into the soil region not included in the model. The discretized model of the structure and of the soil domain is shown schematically. The semi-infinite soil domain, represented by a layered half-space, represents an energy sink.

1.2 TYPES OF PRESCRIBED LOADINGS, IN PARTICULAR SEISMIC EXCITATION

Many types of time-varying loads acting directly on the structure can arise: periodic loads originating from rotating machinery in buildings, impact loads [e.g., the crash of an aircraft onto a nuclear power plant (which can govern the design, although the probability of occurrence is small)], blast loadings, and so on. Probably the most important loading, and definitely the most complicated to analyze, is earthquake excitation, which acts primarily on the soil.

Earthquakes are caused by a sudden energy release in a volume of rock lying on a fault. This source is normally located a large distance away and at a significant depth from the site. Even if all details of how the source mechanism works and the data of the travel path of the seismic waves to the site were available (which is, of course, not the case), it would still be impossible to model all aspects because of the size of the dimensions compared to those of the structure. In any event, the many uncertainties involved make it meaningless to analyze the complete earthquake-excitation problem.

Today's state of the art of earthquake engineering allows only the influence of the local site conditions on the seismic input motion to be taken into account. The procedure normally followed can be characterized as follows: In the control point located at the surface of the so-called free field (site prior to construction, i.e., without the excavation and without the structure), the earthquake motion (e.g., the acceleration as a function of time) is specified (Fig. 1-1). To be able to do this, the seismic hazard of the region is assessed. The structural engineer or the licensing authority specifies the acceptable probability that the earthquake used for design will be exceeded during the life of the structure. In this evaluation the type of the structure will play an important role. For a potentially hazardous structure such as a nuclear power plant this probability will be selected to be very small. It will be chosen to be somewhat larger but will still have a dominant effect for a structure which has to remain fully operational during an earthquake (e.g., a hospital or a fire station). This then allows the determination of the most important parameter, that which is assumed to characterize the motion (e.g., the peak ground acceleration). The other parameters, such as the duration of the motion and the frequency content, are selected—mostly empirically—by the engineering seismologist, based on past earthquakes of the region. All this should allow the source mechanism, the transmission path, the local geology, and the soil conditions at the site to be taken into account very approximately. These activities leading to the definition of a design motion in a selected control point precede the actual soil–structure interaction analysis. The largest uncertainties arise in these preliminary phases. Quite arbitrary assumptions with far-reaching consequences have to be made. It is outside the scope of this work to discuss these very important aspects of earthquake engineering in any depth.

To analyze soil–structure interaction, it is sufficient to think of the prescribed motion as being derived from an observed record at this very site, or at least at a similar site. Starting from this control motion in one point, the earthquake motion throughout the free field (characterized by its spatial and temporal variation) is calculated. As will become apparent when discussing the site response, this can again be achieved only by making quite arbitrary and stringent assumptions regarding the wave pattern in the control point. These assumptions, which will, of course, also be influenced by the opinion of the engineering seismologist as described above, will affect the characteristics of the free-field response as far as the amplitudes and the frequency content away from the control point are concerned. However, all solutions are compatible with the local soil conditions. If, for example, the motion in the control point at the surface is assumed to arise from vertically propagating waves, the variation of the free-field motion with depth can be calculated (Fig. 1-1). In this case, the motion in the horizontal direction is uniform. Besides influencing the free-field response, the local soil conditions will also affect the actual interaction of the soil–structure system. Excavating the soil and building the structure cause changes in the dynamic system. The structure will interact with the soil along the embedment, thus modifying the seismic motion of the free field. The seismic motion acting on the structure is thus not known before the soil–structure interaction analysis is performed. In summary, for seismic excitation, analyzing soil–structure interaction consists of two distinct parts: first, determining the free-field response of the site, and second, calculating the modification of the seismic motion, the actual interaction, when the structure is inserted into the seismic environment of the free field. In the following, mostly seismic excitation will be examined. The loadings applied directly to the structure are, however, contained as special cases in the general formulation.

1.3 EFFECTS OF SOIL–STRUCTURE INTERACTION

To illustrate the salient features of soil–structure interaction, the dynamic response of a structure founded on rock is compared to that of the same construction, but embedded in soil. Only qualitative statements are, of course, possible at this stage. The two identical structures with a rigid base (consisting of the basemat and the side walls) are shown in Fig. 1-2a. The soil layer rests on top of the rock. As the distance between the two structures is small, it can be assumed that in the rock, the incident seismic waves arriving from the source of the earthquake are the same for both structures. For the sake of simplicity, a horizontal motion that propagates vertically is selected. The particle motion is shown in Fig. 1-2 as solid arrows having lengths proportional to the earthquake excitation. The control point is chosen at the free surface of the rock (point A).

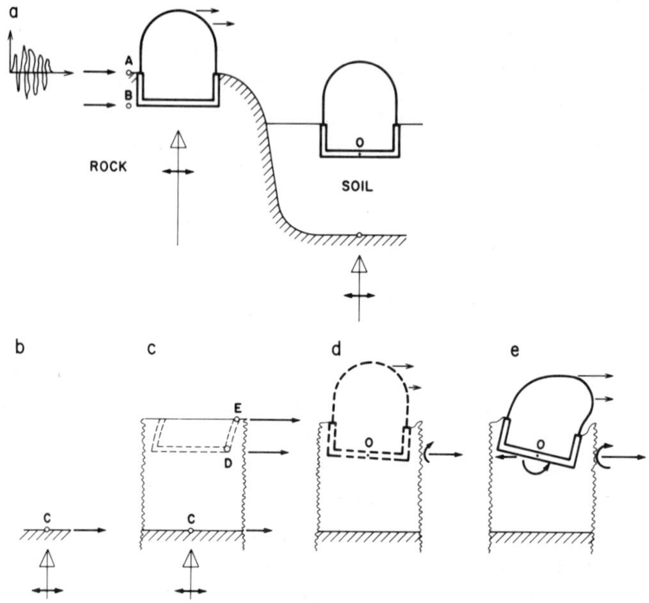

Figure 1-2 Seismic response of structure founded on rock and on soil. (a) Sites; (b) outcropping rock; (c) free field; (d) kinematic interaction; (e) inertial interaction.

From a practical point of view, the motion throughout the rock (e.g., in point *B*) will be the same.

For the structure on the rock, this (horizontal) motion can be applied directly to the base of the structure. The input acceleration resulting in the applied horizontal inertial loads will be constant over the height of the structure. During the earthquake, an overturning moment and a transverse shear acting at the base will develop. As the rock is very stiff, these two stress resultants will not lead to any (additional) deformation at the base. The resulting horizontal displacement of the base is thus equal to the control motion; no rocking motion arises at the base. For a given control motion, the seismic response of the structure depends only on the properties of the structure.

For the structure founded on soft soil, the motion of the base of the structure in point *O* (Fig. 1-2a) will be different from the control motion in the control point *A* because of the coupling of the structure–soil system. To gain insight into how the soil affects the dynamic response of the structure, it is convenient to distinguish between the three following effects. First, the motion of the site in the absence of the structure and of any excavation, what is called the free-field response, is modified (Fig. 1-2c). If there were no soil on top of the rock in point *C* of Fig. 1-2c, the motion in this fictitious rock outcrop shown in Fig. 1-2b would hardly differ from the control motion of the rock in point *A*. The presence of the soil layer will reduce the motion in point *C* (Fig. 1-2c). This wave will propagate vertically through the soil layer, resulting in motions in points *D* and *E* which differ from that in *C*. Points *D* and *E* are nodes in the free field which will subsequently lie on the structure–soil interface (the base) when the structure has been built. In general, the motion is amplified, but not always (depending on its frequency content), thus resulting in horizontal displacements that increase toward the free surface of the site. Second, excavating and inserting the rigid base into the site will modify the motion (Fig. 1-2d). The rigid base will experience some average horizontal displacement and a rocking component. This rigid-body motion will result in accelerations (leading to inertial loads) which will vary over the height of the structure, in contrast to the applied accelerations in the case of a structure founded on rock. This geometric averaging of the seismic input motion will be the result of the so-called kinematic-interaction part of the analysis, as discussed later in great detail. Third, the inertial loads applied to the structure will lead to an overturning moment and a transverse shear acting at point *O* (Fig. 1-2e). These will cause deformations in the soil and thus, once again, modify the motion at the base. This part of the analysis will henceforth be referred to as "inertial interaction."

Figure 1-2 also illustrates the main effects of taking soil–structure interaction into consideration. First, the seismic-input motion acting on the structure–soil system will change (Fig. 1-2d). Because of the amplification of the site (free-field response), the translational component will in many cases be larger than the control motion and, in addition, a significant rocking component will arise for an embedded structure. Each frequency component of the motion is affected differently, resulting in, for example, an acceleration time history which is quite different from the control motion. This amplification of the seismic motion is held responsible for the fact that structures founded on a deep soft-soil site have been damaged more severely in actual earthquakes than have neighboring structures founded on rock. Second, the presence of the soil in the final dynamic model (Fig. 1-2e) will make the system more flexible, decreasing the fundamental frequency to a value which will, in general, be significantly below that applicable for the fixed-base structure. The implication of this reduction will depend on the frequency content of the seismic-input motion. In certain cases, the fundamental frequency will be moved below the range of high seismic excitation, resulting in a significantly smaller seismic input "felt" by the structure. The shape of the vibrational mode will also be changed. The introduced rocking of the base will affect the response, especially at the top of a tall structure. Third, the radiation of energy of the propagating waves away from the structure will result in an increase of the damping of the final dynamic system (Fig. 1-2e). For a soil site approaching an elastic half-space, this increase will be significant, leading to a strongly reduced response. For a soil site consisting of a shallow layer, it is possible that no waves propagate away from the structure. In this case, only the material damping of the soil will act, and no beneficial effect on the seismic response is to be expected. In any soil–structure analysis, it is very important to determine for a specific site whether the loss of energy by radiation of waves can actually take place. It is intuitively plausible that the soil–structure interaction increases the more flexible the soil is and the stiffer the structure is. On the other hand, it will be negligible for a flexible structure founded on firm soil.

It is obvious from the many opposing effects that it is, in general, impossible to determine a priori whether the interaction effects will increase the seismic response. If, however, the first effect discussed above (the change of the seismic-input motion, Fig. 1-2d) is neglected when the interaction analysis is performed, the response will in many circumstances be smaller than the fixed-base response obtained from an analysis that neglects all interaction effects. For this approximate interaction analysis, the control motion is directly used as input motion in the final dynamic system (Fig. 1-2e). The fixed-base analysis leads to larger values of the global response, as, for example, the total overturning moment and the total transverse shear, and thus to a conservative design. The displacement at the top of the structure relative to its base may be larger because of the foundation rocking if one takes soil–structure interaction effects into account. This could influence the spacings between adjacent buildings and in very extreme cases also increase the second-order effects. Economic considerations normally dictate that, when designing structures, that reduction in seismic forces which results from considering the approximate soil–structure interaction analysis be used. There are also some exceptional cases where the simplified interaction effects will govern the design. It is important to stress that this approximate method of calculating the interaction, a method that neglects the free-field site analysis (Fig. 1-2c) and the geometric averaging due to the embedment effect (Fig. 1-2d), is inconsistent. For special structures (e.g., nuclear power plants) which cannot be designed on the basis of the global response alone, soil–structure interaction is always analyzed, considering all effects.

It is true that taking the flexibility of the underlying soil into account when calculating the seismic response of a structure does complicate the analysis considerably. It also makes it necessary to estimate additional key parameters, which are difficult to determine, such as the properties of the soil or the conditions of the embedment. An exact deterministic solution is indeed impossible. The important aspects of soil–structure interaction can, however, be modeled approximately in an appropriate model. The results will be physically logical, but will require engineering judgment to evaluate them. Varying the important parameters in the analysis should lead to response predictions which will consequently result in a safe design. Improved procedures should be used even if uncertainties do exist.

1.4 ASSUMED LINEARITY

Soils behave strongly nonlinearly when excited by the quite high earthquake levels of interest to structural engineers. True nonlinear analyses are, however, not yet possible, although quite promising nonlinear constitutive models for soil do exist. The parameters appearing in these relationships can be selected for a particular site. The cost of performing such an analysis, although higher than for a linear calculation, can be justified for an important structure. The reason is actually conceptional. As explained above, soil–structure interaction for seismic excitation implicitly assumes the law of superposition to be valid. The total solution equals the sum of the responses of the free field and of the interaction part when the structure is inserted into the seismic environment of the free field. But already when determining the free-field site response by starting from the motion in the control point, superposition is used. This means that a linear system is actually calculated. It is, of course, possible to select the material properties so that they are compatible with the strains reached during the excitation. This can be achieved by iteration, calculating a series of linear systems, adjusting the material properties at the end of each iteration. The fact that only linear or such quasi-linear analyses are possible shows how primitive today's state of the art of soil–structure interaction for seismic excitation really is. Within the framework of linear analyses, it is, however, possible to model the radiation characteristics of the unbounded foundation domain with the same accuracy as the finite dynamic model of the structure and of the soil represented in the model. It should also be remembered that linear soil–structure interaction has to be fully understood before nonlinear methods, which will eventually allow permanent settlements to be calculated, are developed. This simplification of performing only a linear analysis should also be seen in connection with the unsatisfactory way of specifying the seismic motion in the control point, which forms the input to the soil–structure interaction analysis. In performing the analysis, the engineer will vary parametrically the soil properties and study the sensitivity of the response of the dynamic system for extreme cases. It is to be expected that the resulting design will be sufficiently conservative.

1.5 TYPES OF PROBLEMS ENCOMPASSED

Although it is not possible to model all details of geometry and material of a practical problem, the following essential features can be captured. The soil consists of horizontal layers resting on a half-space, both consisting of isotropic viscoelastic material with hysteretic damping (Fig. 1-3). The properties vary

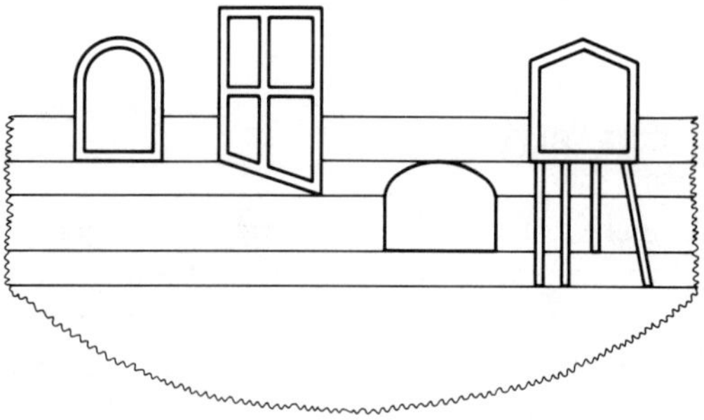

Figure 1-3 Soil–structure interaction system.

with depth but remain constant within the individual layers. This configuration is called a layered half-space. The structures can be embedded. The shape of the soil–structure interface is quite general: the basemat and the walls can, for example, be inclined and can vary from flexible to rigid. Pile foundations are permissible. The number and the location of the battered and vertical piles of different lengths and different dimensions of the cross section can be freely chosen. Buried structures (caissons, tunnels, pipe systems) can be analyzed. The linear structures can exhibit hysteretic damping. The three-dimensional nature of the problem is properly represented. More than one structure can be analyzed at the same time, thus taking the through-soil coupling effects of nearby structures into account.

As will be shown later, computationally efficient procedures do exist for calculating all aspects of soil–structure interaction when the soil can be modeled as a horizontally layered half space (Fig. 1-3). If the soil in the vicinity of the structure–soil interface is irregular, the analysis can still be performed. This finite soil domain can be regarded as part of the structure.

1.6 DIRECT AND SUBSTRUCTURE METHODS

Conceptionally, probably the easiest way to analyze soil–structure interaction for seismic excitation is to model a significant part of the soil around the embedded structure and to apply the free-field motion at the fictitious boundary discussed in connection with Fig. 1-1. (This direct procedure would even allow certain nonlinear material laws of the soil to be considered.) However, the number of dynamic degrees of freedom in the soil region is high, resulting in a large computer-storage requirement and in significant running time. As the law of superposition has to be implicitly assumed to be valid in a soil–structure interaction analysis anyway, it is computationally more efficient to use the substructure method.

The various aspects of the substructure method are explained in great depth in other chapters of this text. In this introduction, only a rough idea is needed to understand the advantages. No equations are developed here. For the embedded structure shown in Fig. 1-4, the substructure method proceeds as follows. The nodes of the dynamic model of the structure are shown, whereby those located along the embedment on the soil–structure interface are indicated as circles. The free-field response of the site (without any embedment) is calculated first. It will be shown that these motions are needed only in those nodes at which the structure is subsequently inserted (shown as circles). The interaction part which then follows has two steps. In the first, the unbounded soil is analyzed as a dynamic subsystem. The force–displacement relationship of the degrees of free-

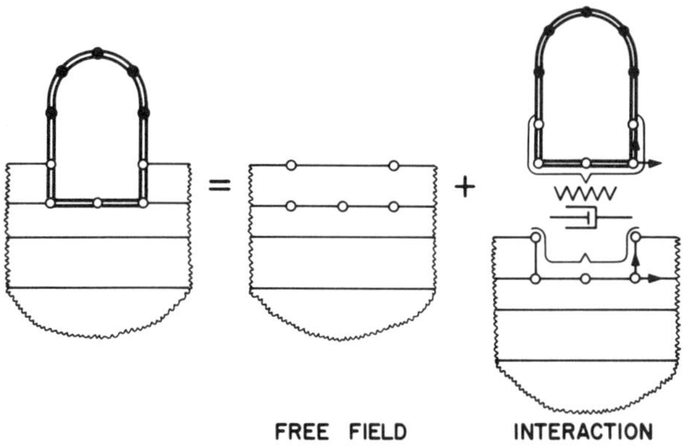

Figure 1-4 Seismic soil–structure interaction with substructure method.

dom of these same nodes, which will be in contact with the structure, is determined. These so-called dynamic-stiffness coefficients of the soil can physically be interpreted as a generalized spring, a spring–dashpot system. In the second step, the structure supported on this spring–dashpot system is analyzed for a loading case which depends on the free-field motion. Using the substructure method allows the complicated soil–structure system to be broken down into more manageable parts which can be more easily checked. The intermediate results can be interpreted and "smoothed" to take uncertainties into account. In some cases, parametric studies can be restricted to one part. The important parameters are more easily identified and their influence on the response more readily understood. If the free-field motion is changed, the dynamic stiffness of the soil does not have to be recalculated. The same applies if the structure is modified during the design process. The two subsystems, the soil and the structure, can also be calculated by completely different methods. In simple cases, results applicable to one of the subsystems can be used without any computation. It is possible to analyze the structure with much more detail than the soil system. For instance, an axisymmetric solution can be used for the soil in connection with a full three-dimensional analysis of the nonaxisymmetric structure to determine the torsional response. In the direct method, in which the responses of the structure and the soil are determined simultaneously, the structure can be modeled only very coarsely. It is then often necessary to model the structure more precisely in a second stage, whereby one applies the resulting motion of the base of the structure, in order to determine the detailed response. The models of the structures used in the two stages, however, have to be consistent. This means that the properties of the coarser model have to be determined by a systematic reduction procedure from those of the finer one.

The unbounded soil medium can best be described for harmonic excitation. As will be shown later, the dynamic stiffness taking the radiation of energy into account will be frequency dependent and complex. Material damping can also easily be introduced for harmonic excitation. This means that linear soil–structure interaction analysis is best handled in the frequency domain, using the so-called complex-response method. The advantages of working in the frequency domain are important. The transfer functions allow the first few natural frequencies and the corresponding approximate damping ratios to be determined. The accuracy in the intermediate and higher-frequency ranges can be checked easily. Once the transfer functions have been determined, it is also possible to calculate efficiently the final response for another seismic environment.

(Quasi-)linear behavior of the structure–soil system is postulated. The formulation can, however, be expanded straightforwardly to apply also to the case in which the nonlinearity is restricted to the structure. A thorough treatment of this procedure, which works in the time domain using convolution operators, is beyond the scope of this text.

1.7 ORGANIZATION OF TEXT

The fundamentals of a discrete dynamic system are summarized in Chapter 2. This then allows the basic equation of motion for the analysis of soil–structure interaction by the substructure method to be derived in Chapter 3. Using a simple introductory example, the important effects of soil–structure interaction are evaluated approximately. The various possibilities of modeling the structure are discussed in Chapter 4. In Chapter 5 the fundamentals of wave propagation in Cartesian and cylindrical coordinates are established. This leads to the dynamic-stiffness matrix of the layered half-space assembled from those of the individual soil layers and of the rock. The seismic free-field response of the site, which has to be determined before the actual interaction analysis can be performed, is discussed in Chapter 6. The various types of waves, the vertically incident and inclined body waves as well as the surface waves, which will make up the seismic motion, are addressed. Vast parametric studies, which also include the analysis of actual sites, are evaluated to be able to demonstrate the importance of the free-field response of a site. Chapter 7 deals with the modeling of the other substructure, the soil. After reviewing the various possibilities, the boundary-element method is used to calculate the dynamic-stiffness coefficients of surface and embedded foundations. Alternative formulations of the equation of motion are derived in Chapter 8. An approximate method, which works directly in the time domain, is also included. Chapter 9 contains engineering applications. The interaction effects, including those arising from horizontally propagating waves, are evaluated. Examples from actual practice follow. Finally, the (limited) recorded field performance is compared to the results of the analysis, which are found to be in good agreement.

SUMMARY

1. A structure excited by a dynamic load (e.g., a seismic motion) interacts with the surrounding soil. In contrast to the structure, the soil is an unbounded domain whose radiation condition has to be taken into account in the dynamic model.
2. Analyzing soil–structure interaction for seismic excitation consists of two distinct parts: First, determining the free-field response of the site and second, calculating that modification of the seismic motion which results from the actual interaction when the structure is inserted into the seismic environment of the free field. The law of superposition is assumed, and thus linearity as well.
3. When comparing a structure embedded in soil with the same structure founded on rock, the seismic motion at the base of the structure is affected in

three ways: First, the free-field response is modified, usually amplified. Second, excavation of the soil and insertion of the base of the (massless) structure into it will alter the seismic input, resulting in an averaging of the translation and in an additional rotational component (kinematic interaction). Third, the inertial loads determined from this seismic input will further change the seismic motion along the base (inertial interaction). Besides affecting the seismic input, the presence of the soil also makes the dynamic system more flexible and the radiation of energy of the propagating waves away from the structure increases the damping of the total dynamic system.

4. (Quasi-)linear three-dimensional analysis of soil–structure interaction is possible for one or more structures with flexible or rigid bases of arbitrary shape embedded in a horizontally layered half-space with material damping.

5. In the substructure method, the unbounded soil is first examined and can be represented by a generalized spring, on which the structure is then supported for the interaction analysis.

2

FUNDAMENTALS OF DISCRETE DYNAMIC SYSTEM

The fundamental equations of a discrete dynamic system are summarized in this chapter. Only those relations that are actually used in this text are specified. For a thorough treatment of this subject, including all derivations and a discussion of the assumptions and limitations, the established textbooks in structural dynamics and in the theory of vibrations should be studied. The main objective is to define the nomenclature that is used in this text.

2.1 EQUATIONS OF MOTION IN THE TIME DOMAIN

The elastic continuum is divided into elements interconnected only at a finite number of nodes. Using the finite-element method to perform the spatial discretization, the equations of motion are established by assembling the element matrices. This leads to

$$[M]\{\ddot{r}\} + [C]\{\dot{r}\} + [K]\{r\} = \{R\} \tag{2.1}$$

The vector $\{r\}$, which is a function of time, contains the displacements of all unconstrained degrees of freedom of all nodes. A derivative with respect to time is denoted as a dot. The matrices $[M]$, $[C]$, and $[K]$ represent the mass matrix, the damping matrix, and the static-stiffness matrix, which are constant for a linear system. The vector $\{R\}$ denotes the prescribed loads, which are a function of time, acting in the direction of the displacements in all nodes. If the loading case consists partly of prescribed motions as a function of time at support nodes, the equations are also formulated for these degrees of freedom. In this case the elements in $\{R\}$ corresponding to these displacements represent the unknown reaction forces. Viscous damping is assumed in Eq. 2.1. The radiation of energy due to the propagation of waves away from the region of interest belongs to this type of damping (see Section 5.1).

2.2 TRANSFORMATION TO MODAL AMPLITUDES

The free-vibration mode shapes $\{\phi_j\}$ and the corresponding natural frequencies ω_j of the undamped system follow from the solution of the eigenvalue problem

$$[K]\{\phi_j\} = \omega_j^2[M]\{\phi_j\} \tag{2.2}$$

The vector $\{\phi_j\}$ and ω_j^2 represent the jth eigenvector and eigenvalue, respectively. Assembling all suitably scaled $\{\phi_j\}$ in $[\Phi]$ and all ω_j^2 as the elements on the diagonal matrix $[\Omega]$, the orthogonality conditions are as follows

$$[\Phi]^T[M][\Phi] = [I] \tag{2.3a}$$

$$[\Phi]^T[K][\Phi] = [\Omega] \tag{2.3b}$$

where $[I]$ represents the unit matrix. The number of selected mode shapes can be selected smaller than the number of degrees of freedom.

Introducing the transformation

$$\{r\} = [\Phi]\{y\} \tag{2.4}$$

where $\{y\}$ denotes the vector of the modal amplitudes (coordinates), the equations of motion (Eq. 2.1) are formulated as

$$\{\ddot{y}\} + [\Phi]^T[C][\Phi]\{\dot{y}\} + [\Omega]\{y\} = [\Phi]^T\{R\} \tag{2.5}$$

The vector $[\Phi]^T\{R\}$ is the generalized modal load vector. In general, Eq. 2.5 still represents a coupled system. If it is assumed that the orthogonality condition also applies to the damping matrix, Eq. 2.5 is rewritten as

$$\{\ddot{y}\} + 2[\zeta][\Omega]^{1/2}\{\dot{y}\} + [\Omega]\{y\} = [\Phi]^T\{R\} \tag{2.6a}$$

or formulated for each modal amplitude y_j:

$$\ddot{y}_j + 2\zeta_j\omega_j\dot{y}_j + \omega_j^2 y_j = \{\phi_j\}^T\{R\} \tag{2.6b}$$

The diagonal matrix $[\zeta]$ contains as elements the modal damping ratios ζ_j.

As will be demonstrated later (Section 8.4), the orthogonality condition for damping is (approximately) satisfied for the structure but not for the complete structure–soil system. Classical modes thus do not exist.

2.3 EQUATIONS OF MOTION FOR HARMONIC EXCITATION

For a harmonically varying load with the excitation frequency ω,

$$\{R\} = \{P\} \exp(i\omega t) \tag{2.7}$$

the response (after the initial transient) will be of a similar form:

$$\{r\} = \{u\} \exp(i\omega t) \tag{2.8}$$

The vectors $\{P\}$ and $\{u\}$ represent the (complex) amplitudes of the loads and of the displacements, respectively. The equations of motion (Eq. 2.1) are formulated as

$$[S]\{u\} = \{P\} \tag{2.9}$$

where the dynamic-stiffness matrix $[S]$ is specified as

$$[S] = [K] + i\omega[C] - \omega^2[M] \tag{2.10}$$

The amplitudes $\{u\}$ follow from Eq. 2.9 by direct solution. As the total system will always have some damping, $[S]$ is not singular for any ω.

The transformation to modal amplitudes can also be performed. With

$$\{u\} = [\Phi]\{z\} \tag{2.11}$$

Eq. 2.9 is transformed to

$$([\Omega] + i\omega[\Phi]^T[C][\Phi] - \omega^2[I])\{z\} = [\Phi]^T\{P\} \tag{2.12}$$

or if the orthogonality condition is also valid for $[C]$,

$$(\omega_j^2 + 2i\omega\zeta_j\omega_j - \omega^2)z_j = \{\phi_j\}^T\{P\} \tag{2.13}$$

In principle, different letters are used in this text to distinguish a variable in the time domain (e.g., r for the displacement) from the corresponding amplitude in the frequency domain (u). However, to be able to simplify the nomenclature in a specific example, this distinction is not enforced if no confusion can arise. For instance, if in a parametric study the displacement amplitudes u as a function of frequency are discussed in depth for many cases, and a maximum value is also shown for a transient excitation, u_{max} (and not r_{max}) will be used to designate this quantity.

2.4 CORRESPONDENCE PRINCIPLE

The material damping occurring in the soil and the structure involves a frictional loss of energy. This so-called (linear) hysteretic damping is independent of frequency. It can be introduced into the solution, when working in the frequency domain, by using the correspondence principle. The latter states that the damped solution is obtained from the elastic one by replacing the elastic constants by the corresponding complex ones. This means that for the static-stiffness matrix

$$[K^*] = [K](1 + 2\zeta i) \tag{2.14}$$

applies, where ζ is the damping ratio. Different damping ratios for the two types of body waves are discussed in Section 5.2. The viscous-damping matrix $[C]$ representing the radiation of energy in the soil will also be a (much more complicated) function of the elastic constants (see Section 5.1). If no viscous damping is present, $[S]$ (Eq. 2.10) can be expressed by

$$[S] = [K](1 + 2\zeta i) - \omega^2[M] \tag{2.15}$$

After transformation to modal amplitudes, this case is specified as

$$[\omega_j^2(1 + 2\zeta i) - \omega^2]z_j = \{\phi_j\}^T\{P\} \tag{2.16}$$

2.5 DISCRETE FOURIER TRANSFORM

The Fourier integral of the excitation represents a nonperiodic load which can be regarded as a periodic one of infinite duration. This integral is an extension of the Fourier series to the case of a load having continuous frequency components. The Fourier transform pair formulated for one element of the load vector is specified as

$$P(\omega) = \int_{-\infty}^{+\infty} R(t) \exp(-i\omega t)\, dt \tag{2.17a}$$

$$R(t) = \frac{1}{2\pi}\int_{-\infty}^{+\infty} P(\omega) \exp(i\omega t)\, d\omega \tag{2.17b}$$

The normalization factor $1/(2\pi)$ is introduced into the equation for $R(t)$. Some derivations use this factor in the other equation, that is, in the definition of the amplitude per unit ω of the load component at frequency ω $[= P(\omega)/(2\pi)]$. In other formulas for the Fourier transform pair, the signs of the arguments of the exponential functions are reversed.

In the numerical-analysis procedure, the load has to be assumed to be periodic with a finite period T. To minimize errors the load period is extended by including an interval of zero load. This so-called quiet zone will allow the free vibration components of the response of the system to be damped out when the new period of the load starts. The period T is divided into n equal increments of Δt each, where for the efficiency of the calculation n is selected as a power of 2. The frequency span is also divided into the same number of increments $\Delta\omega$, where the frequency increment (and the lowest frequency) $\Delta\omega = 2\pi/T$. With $t_j = j\,\Delta t = jT/n$ and $\omega_l = l\,\Delta\omega = l2\pi/T$, Eq. 2.17a formulated as a finite sum of discrete terms equals

$$P(\omega_l) = \Delta t \sum_{j=0}^{n-1} R(t_j) \exp\left(-2\pi i \frac{lj}{n}\right) \qquad l = 0, 1, \ldots, n-1 \tag{2.18a}$$

and the inverse transform (Eq. 2.17b) is specified as

$$R(t_j) = \frac{\Delta\omega}{2\pi} \sum_{l=0}^{n-1} P(\omega_l) \exp\left(2\pi i \frac{lj}{n}\right) \qquad j = 0, 1, \ldots, n-1 \tag{2.18b}$$

As the nonperiodic load is actually replaced by a suitably chosen periodic one, this numerical-analysis procedure can also be used to calculate periodic loads as, for example, arising from machine vibrations. The term $\Delta\omega\, P(\omega_l)/(2\pi)$ represents the Fourier coefficient for the frequency ω_l.

When evaluating the discrete Fourier transforms, a lot of duplication will occur when evaluating the exponential functions. Taking full advantage of this fact, using a binary representation and applying the results of one step immediately in the next, a very efficient scheme results, which is called the Fast Fourier Transform.

The highest frequency ω_{max} contained in the load equals

$$\omega_{max} = \frac{n}{2}\Delta\omega = \frac{\pi}{\Delta t} \tag{2.19}$$

2.6 METHOD OF COMPLEX RESPONSE

By applying the discrete Fourier transform to the loads discretized at discrete points $\{R(t_j)\}$, the amplitudes of the loads in the frequency domain $\{P(\omega_l)\}$ calculated at all discrete frequencies ω_l are determined (Eq. 2.18a). In analogy to Eqs. 2.9 and 2.10, the equations of motion in the frequency domain are equal to

$$[S(\omega_l)]\{u(\omega_l)\} = \{P(\omega_l)\} \tag{2.20}$$

with

$$[S(\omega_l)] = [K] + i\omega_l[C] - \omega_l^2[M] \tag{2.21}$$

The soil's contribution to $[K]$ and $[C]$ will in general depend on ω_l (Sections 5.1 and 9.1). The material damping of the structure can be included by using $[K^*]$ instead of $[K]$ (Eq. 2.14). Solving Eq. 2.20 for $\{u(\omega_l)\}$ results in the amplitudes of the displacements in the frequency domain. Using the inverse transform of Eq. 2.18b, but formulated for the displacements

$$\{r(t_j)\} = \frac{\Delta\omega}{2\pi}\sum_{l=0}^{n-1} u(\omega_l)\exp\left(2\pi i\frac{lj}{n}\right) \tag{2.22}$$

the displacements in the time domain at discrete points $\{r(t_j)\}$ follow.

The equations of motion in the frequency domain are never solved for all points of the Fourier transform. The frequency increment to solve Eq. 2.20 is selected as quite large, possibly smaller in the neighborhood of the first few peaks of the response occurring at the natural frequencies of the system. Intermediate values follow by interpolation, whereby the analytically available solutions for systems having one or two degrees of freedom (anchored at the calculated frequencies) can be used.

3

BASIC EQUATION OF MOTION

3.1 FORMULATION IN TOTAL DISPLACEMENTS

3.1.1 Flexible Base

To derive the basic equations of motion, it is sufficient to examine a single structure embedded in soil for earthquake excitation (Fig. 3-1). The structure's base consisting of the basemat and of the adjacent walls is assumed at first to be flexible. The structure is discretized schematically as shown. Subscripts are used to denote the nodes of the discretized system. The nodes located on the structure–soil interface are denoted by b (for base), the remaining nodes of the structure by s. In the substructure method no nodes are introduced in the interior of the soil.

The dynamic system consists of two substructures, the actual structure and the soil with excavation. To differentiate between the various subsystems, superscripts are used when necessary. The structure is indicated by s (when used with a property matrix), the other substructure, the soil with excavation, by g (for ground). In the following it is appropriate to work also with other subsystems for the soil (Fig. 3-2). The soil without excavation, the so-called free field, is denoted by f, and e is used to designate the excavated soil.

The dynamic equations of the motion are formulated in the frequency domain. The amplitudes of the total displacements are denoted by $\{u^t\}$, which is a function of the discrete value of the frequency ω. The subscript used to indicate a discrete value of ω in Eqs. 2.20 and 2.21 is deleted. The word "total" (superscript t) expresses that the motion is referred to an origin that does not

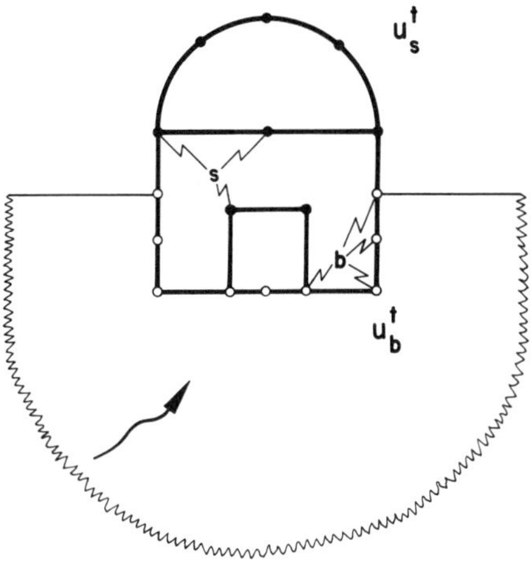

Figure 3-1 Structure–soil system.

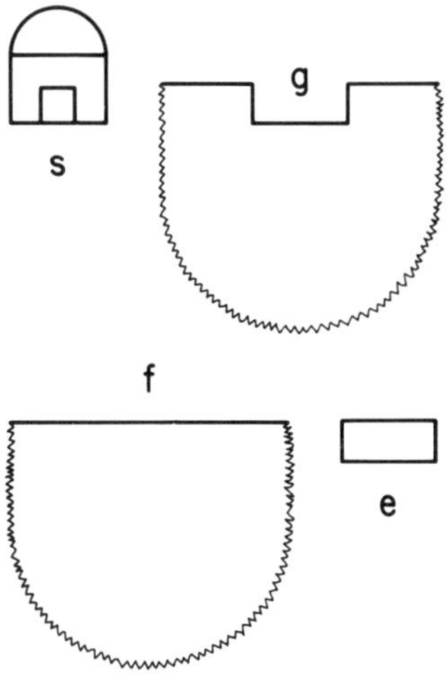

Figure 3-2 Reference subsystems.

move. The order of this vector equals the number of dynamic degrees of freedom of the total discretized system. The vector $\{u^t\}$ can be decomposed into the subvectors $\{u_s^t\}$ and $\{u_b^t\}$. The vector symbols are deleted in Fig. 3-1 for the sake of conciseness. The dynamic-stiffness matrix $[S]$ of the structure, which is a bounded system, is calculated as (Eq. 2.15)

$$[S] = [K](1 + 2\zeta i) - \omega^2[M] \tag{3.1}$$

where $[K]$ and $[M]$ are the static-stiffness and mass matrices, respectively. The hysteretic-damping ratio ζ, which is independent of frequency, is assumed to be constant throughout the structure. The formulation can straightforwardly be expanded to the case of nonuniform damping. Some formulas, however, cannot be written so concisely. $[S]$ can also be decomposed into the submatrices $[S_{ss}]$,

$[S_{sb}]$, and $[S^s_{bb}]$. To avoid using unnecessary symbols, the superscript s (for structure) is used only when confusion would otherwise arise. The equations of motion of the structure are formulated as

$$\begin{bmatrix} [S_{ss}] & [S_{sb}] \\ [S_{bs}] & [S^s_{bb}] \end{bmatrix} \begin{Bmatrix} \{u^t_s\} \\ \{u^t_b\} \end{Bmatrix} = \begin{Bmatrix} \{P_s\} \\ \{P_b\} \end{Bmatrix} \tag{3.2}$$

where $\{P_s\}$ and $\{P_b\}$ denote the amplitudes of the loads and of the interaction forces with the soil, respectively.

The dynamic-stiffness matrix of the soil $[S^g_{bb}]$ (Fig. 3-3) is not as easy to determine as that of the structure, as the soil is an unbounded domain. Conceptionally, $[S^g_{bb}]$ could be determined by eliminating all degrees of freedom not lying on the structure–soil interface of a mesh of the soil extending to infinity. The vector $\{u^g_b\}$ denotes the displacement amplitudes of the soil with excavation for the earthquake excitation. For the reference system of the free field, $[S^f_{bb}]$ and $\{u^f_b\}$ are the dynamic-stiffness matrix and the vector of displacement amplitudes, respectively. The dynamic-stiffness matrix $[S^e_{bb}]$ of the excavated soil, a bounded domain, follows analogously from Eq. 3.1, using the properties of the soil.

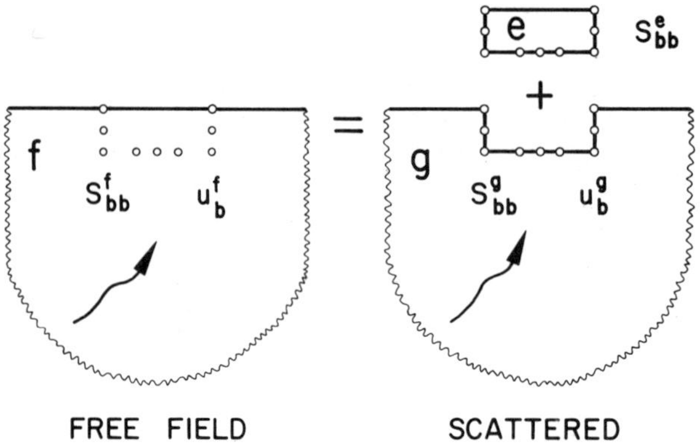

Figure 3-3 Dynamic-stiffness matrix and earthquake excitation referred to different reference systems of soil.

For earthquake excitation, the nodes not in contact with the soil (the subscript s stands for the structure) are not loaded. Setting $\{P_s\} = \{0\}$ in Eq. 3.2 leads to

$$[[S_{ss}][S_{sb}]] \begin{Bmatrix} \{u^t_s\} \\ \{u^t_b\} \end{Bmatrix} = \{0\} \tag{3.3}$$

Both substructures contribute to the dynamic equilibrium equations of the nodes b lying on the structure–soil interface. The contribution of the soil is discussed first. For the displacement amplitudes $\{u^g_b\}$, the interaction forces acting in the nodes b, arising from the soil with excavation, vanish, as for this loading state, the line that will subsequently form the structure–soil interface is a free surface (Figs. 3-2 and 3-3). The interaction forces of the soil will thus depend on the motion relative to $\{u_b^g\}$. They are equal to $[S_{bb}^g](\{u_b^t\} - \{u_b^g\})$. Including the contribution of the structure ($\{P_b\}$ in Eq. 3.2), the equations of motion for the nodes in contact with the soil (subscript b) are formulated as

$$[S_{bs}]\{u_s^t\} + [S_{bb}^s]\{u_b^t\} + [S_{bb}^g](\{u_b^t\} - \{u_b^g\}) = \{0\} \tag{3.4}$$

Combining Eqs. 3.3 and 3.4, the equations of motion of the total structure-soil system are

$$\begin{bmatrix} [S_{ss}] & [S_{sb}] \\ [S_{bs}] & [S_{bb}^s] + [S_{bb}^g] \end{bmatrix} \begin{Bmatrix} \{u_s^t\} \\ \{u_b^t\} \end{Bmatrix} = \begin{Bmatrix} \{0\} \\ [S_{bb}^g]\{u_b^g\} \end{Bmatrix} \tag{3.5}$$

In this formulation, the earthquake excitation is characterized by $\{u_b^g\}$, that is, the motion in the nodes (which will subsequently lie on the structure–soil interface) of the ground with the excavation. This so-called "scattered" motion is not easy to determine. It is desirable to replace $\{u_b^g\}$ by $\{u_b^f\}$, the free-field motion, which does not depend on the excavation (with the exception of the location of the nodes in which it is to be calculated).

The free-field system results when the excavated part of the soil is added to the soil with excavation (Fig. 3-3). This also holds for the assembly process of the dynamic-stiffness matrices

$$[S_{bb}^e] + [S_{bb}^g] = [S_{bb}^f] \tag{3.6}$$

By stipulating that the "structure" consist of the excavated part of the soil only, Eq. 3.4 can be formulated for this special case. With $[S_{bs}] = [0]$, $[S_{bb}^s] = [S_{bb}^e]$, and $\{u_b^t\} = \{u_b^f\}$,

$$([S_{bb}^e] + [S_{bb}^g])\{u_b^f\} = [S_{bb}^g]\{u_b^g\} \tag{3.7}$$

results. Introducing Eq. 3.6 in Eq. 3.7 leads to

$$[S_{bb}^f]\{u_b^f\} = [S_{bb}^g]\{u_b^g\} \tag{3.8}$$

This equality of forces is quite a remarkable result in its own right. Although for the substructure of the soil with excavation, the line with the nodes b (where the motion is equal to $\{u_b^g\}$) is a free surface, as discussed above, the forces $[S_{bb}^g]\{u_b^g\}$ are not zero. The influence of an exterior boundary with an applied earthquake motion also has to be taken into account when calculating the forces in nodes b. This is explained further in Section 8.2, where the basic equations of motion are rederived starting from those of the direct formulation of soil–structure interaction.

Substituting Eq. 3.8 in Eq. 3.5 results in the discretized equations of motion

$$\begin{bmatrix} [S_{ss}] & [S_{sb}] \\ [S_{bs}] & [S_{bb}^s] + [S_{bb}^g] \end{bmatrix} \begin{Bmatrix} \{u_s^t\} \\ \{u_b^t\} \end{Bmatrix} = \begin{Bmatrix} \{0\} \\ [S_{bb}^f]\{u_b^f\} \end{Bmatrix} \tag{3.9}$$

In this formulation in total displacements, the load vector is expressed as the product of the dynamic-stiffness matrix of the free field $[S_{bb}^f]$ (discretized in the nodes at which the structure is subsequently inserted) and of the free-field motion $\{u_b^f\}$ in the same nodes. This physical interpretation is remarkable. The load acts only on the nodes b at the base of the structure. The interaction part of soil–structure analysis is thus formulated as a so-called source problem, with the source being located on the structure–soil interface. Only outgoing waves that propagate toward infinity arise (see Section 7.1). As expected for seismic excitation, the nodes s not connected with the soil are unloaded. It should be emphasized that in the system f for which $[S_{bb}^f]$ is calculated, the soil is not excavated (Fig. 3-3). As the boundary of the free-field region is regular, $[S_{bb}^f]$ should be easier than $[S_{bb}^g]$ to calculate (see Section 7.5.3). The vector $\{u_b^f\}$ has only to be calculated in those nodes b which subsequently will lie on the structure–soil interface. Procedures to determine $[S_{bb}^g]$, $[S_{bb}^f]$ and $\{u_b^f\}$ are discussed in depth in Chapters 7 and 6, respectively. The derivation of the basic equations of motion is based on substructuring with replacement. By adding the excavated part of the soil (system e) to the irregular system g, a regular system f is formed on which the load vector depends. See Problems 3.2, 3.3, and 3.4 for further applications of this concept, which is also valid for applied loads and not only prescribed motions.

The amount of interaction of the embedded part described by the left-hand side of Eq. 3.9 depends on $[S_{bb}^s] + [S_{bb}^g]$, which, after making use of Eq. 3.6, is equal to

$$[S_{bb}^s] + [S_{bb}^g] = [S_{bb}^s] - [S_{bb}^e] + [S_{bb}^f] \tag{3.10}$$

The difference of the property matrices of the structure and of the soil in the embedded region $[S_{bb}^s] - [S_{bb}^e]$ is thus of importance. Using Eq. 3.10, Eq. 3.9 is reformulated as

$$\begin{bmatrix} [S_{ss}] & [S_{sb}] \\ [S_{bs}] & [S_{bb}^s] - [S_{bb}^e] + [S_{bb}^f] \end{bmatrix} \begin{Bmatrix} \{u_s^t\} \\ \{u_b^t\} \end{Bmatrix} = \begin{Bmatrix} \{0\} \\ [S_{bb}^f]\{u_b^f\} \end{Bmatrix} \tag{3.11}$$

This represents the equations of motion of a discretized system (consisting of the structure and in the embedded region of the difference of the structure and of the soil) supported on a generalized spring described by $[S_{bb}^f]$. The excitation consists of a prescribed support motion $\{u_b^f\}$ acting at the end of the generalized spring that is not connected to the structure. This is shown schematically in Fig. 3-4, where the matrix and vector symbols as well as the subscripts have been omitted. It is important to note that the support motion to be applied at the end of the generalized spring is the free-field response $\{u_b^f\}$ at the structure–soil interface and not at some (fictitious) boundary at a depth where the underlying medium can be regarded as very stiff. For instance, consider a structure supported on the surface of a layer of soil resting on rigid bedrock. The generalized spring represents the stiffness and damping of the layer built in at its base at the top of the bedrock. The applied support motion is the free-field response at the free surface of the layer and not at its base. For a better understanding of the foregoing, it is helpful to think of the generalized spring as having a length approaching zero.

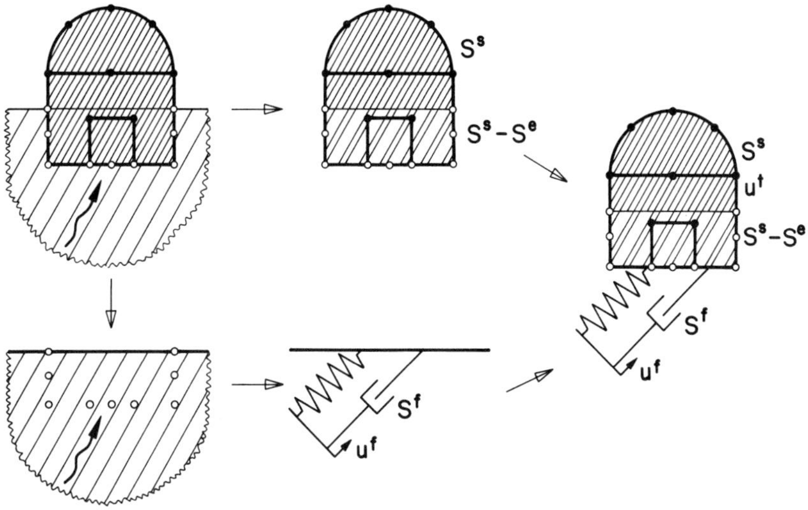

Figure 3-4 Physical interpretation of basic equation of motion in total displacements (flexible base).

Equation 3.9 describes the basic equations of motion, formulated for a structure with a flexible base. It represents a simple, but general procedure to calculate even the most general case of soil–structure interaction. All other formulations which are derived in Sections 3.2, 8.2, and 8.3 are not more powerful. They are discussed only because valuable physical insight can be gained from the equations, which work with relative displacements.

Equation 3.8 could be used to calculate the "scattered" earthquake excitation $\{u_b^g\}$:

$$\{u_b^g\} = [S_{bb}^g]^{-1}[S_{bb}^f]\{u_b^f\} \tag{3.12}$$

It is, however, unnecessary to determine this seismic motion of the soil modified by the excavation $\{u_b^g\}$ (Fig. 3-3), as Eq. 3.9 can be used. It is worth mentioning that $\{u_b^g\}$ has no real existence (i.e., it does not occur in the real soil–structure system).

3.1.2 Rigid Base

The base, consisting of the basemat and the adjacent walls, can be assumed to be rigid for many practical applications (see Section 4.3). This compatibility constraint on the structure–soil interface leads to a slight modification of the formulation. At the same time the physical significance of the terms appearing in the equations can be discussed.

The same structure–soil system of Fig. 3-1 is shown, but with a rigid base in Fig. 3-5. In this case, the total motion at the base $\{u_b^t\}$ can be expressed as a function of the total rigid-body motions of a point O $\{u_o^t\}$ as

$$\{u_b^t\} = [A]\{u_o^t\} \tag{3.13}$$

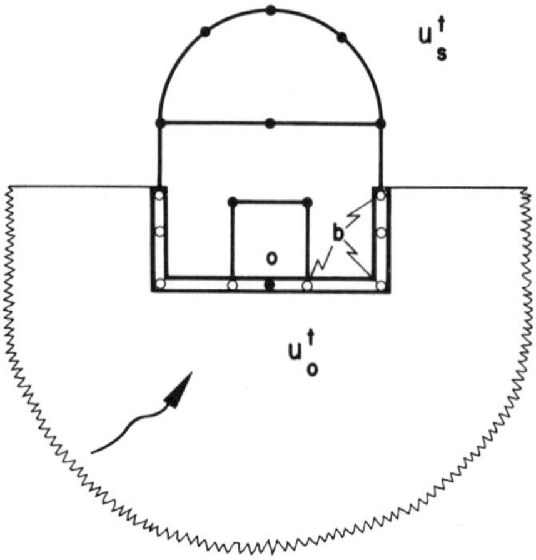

Figure 3-5 Structure–soil system with rigid base.

For a three-dimensional base, $\{u_o^t\}$ contains the amplitudes of three displacements and three rotations. The matrix $[A]$ represents the kinematic transformation with geometric quantities only. In Problem 3.1 the $[A]$-matrix for a two-dimensional case is presented.

For a rigid base, the motion of the structure–soil interface which represents the boundary between the two subsystems thus depends only on $\{u_o^t\}$. Compared to a flexible base, the number of degrees of freedom is reduced. In the substructure of the soil with excavation (system g), the compatibility constraints of the rigid base are enforced (Fig. 3-6). The base is massless (dashed line). Obviously, the free-field motion is unchanged.

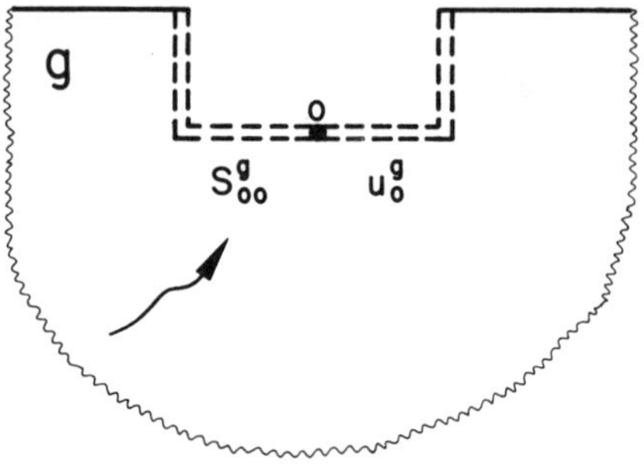

Figure 3-6 Reference soil system with excavation and rigid structure–soil interface.

Introducing this transformation of variables,

$$\begin{Bmatrix} \{u_s^t\} \\ \{u_b^t\} \end{Bmatrix} = \begin{bmatrix} [I] & \\ & [A] \end{bmatrix} \begin{Bmatrix} \{u_s^t\} \\ \{u_o^t\} \end{Bmatrix} \tag{3.14}$$

in Eq. 3.9 and premultiplying by the transposed transformation matrix defined by Eq. 3.14 leads to

$$\left[\begin{array}{c:c} [S_{ss}] & [S_{so}] \\ [S_{os}] & [S_{oo}^s] + [S_{oo}^g] \end{array}\right] \begin{Bmatrix} \{u_s^t\} \\ \{u_o^t\} \end{Bmatrix} = \begin{Bmatrix} \{0\} \\ [A]^T[S_{bb}^f]\{u_b^f\} \end{Bmatrix} \tag{3.15}$$

where

$$[S_{so}] = [S_{sb}][A] \tag{3.16a}$$

$$[S^s_{oo}] = [A]^T[S^s_{bb}][A] \tag{3.16b}$$

$$[S^g_{oo}] = [A]^T[S^g_{bb}][A] \tag{3.16c}$$

$[I]$ denotes the unit matrix. $[S^s_{oo}]$, $[S_{so}]$, and $[S_{os}] = [S_{so}]^T$ are the dynamic-stiffness submatrices of the structure with a rigid base. They are normally directly established when discretizing the structure by selecting models that take account of the geometric constraints of the rigid base. Equations 3.16a and b are thus not explicitly used. $[S^g_{oo}]$ represents the dynamic-stiffness matrix of the soil with excavation for a rigid structure–soil interface (Fig. 3-6). In a general three-dimensional case, $[S^g_{oo}]$ describes the amplitudes of the three forces and moments acting in point O which lead to unit amplitudes of displacements and rotations in the same point of the rigid base connected to the soil.

Equation 3.15 represents the equations expressed in total motion. As in the case of a flexible basemat, the load vector depends on the free-field motion $\{u^f_b\}$ in those nodes that will subsequently lie on the structure–soil interface. The load vector $[A]^T[S^f_{bb}]\{u^f_b\}$, which in a three-dimensional case consists of three forces and three moments, acts in point O of the rigid base. No need really exists to calculate the motion of any other reference soil system.

To gain insight into the physical significance of the load vector, it is meaningful to calculate the seismic motion of the ground system (soil accounting for the excavation) with the compatibility constraints of the rigid base enforced. The seismic motion of this reference subsystem shown in Fig. 3-6 is denoted as $\{u^g_o\}$ and represents a scattered wave motion. The compatibility constraints of the rigid base for the soil–subsystem g are formulated analogously as in Eq. 3.13:

$$\{u^g_b\} = [A]\{u^g_o\} \tag{3.17}$$

Substituting this equation in Eq. 3.8, premultiplying by $[A]^T$, and using Eq. 3.16c results in

$$[S^g_{oo}]\{u^g_o\} = [A]^T[S^f_{bb}]\{u^f_b\} \tag{3.18}$$

or

$$\{u^g_o\} = [S^g_{oo}]^{-1}[A]^T[S^f_{bb}]\{u^f_b\} \tag{3.19}$$

This equation can, of course, be derived directly from Eq. 3.15, deleting all matrices associated with the structure, and setting $\{u^t_o\} = \{u^g_o\}$. As $\{u^f_b\}$ along the walls of the embedded structure varies with depth, a rotational component is also present in $\{u^g_o\}$ even for vertically propagating shear waves.

Substituting Eq. 3.18 in Eq. 3.15 leads to the equivalent of Eq. 3.5 for a structure with a rigid base.

$$\begin{bmatrix} [S_{ss}] & [S_{so}] \\ [S_{os}] & [S^s_{oo}] + [S^g_{oo}] \end{bmatrix} \begin{Bmatrix} \{u^t_s\} \\ \{u^t_o\} \end{Bmatrix} = \begin{Bmatrix} \{0\} \\ [S^g_{oo}]\{u^g_o\} \end{Bmatrix} \tag{3.20}$$

This equation in total motion can be physically interpreted as illustrated in Fig. 3-7. The discretized structure specified by $[S^s]$ is supported on a generalized

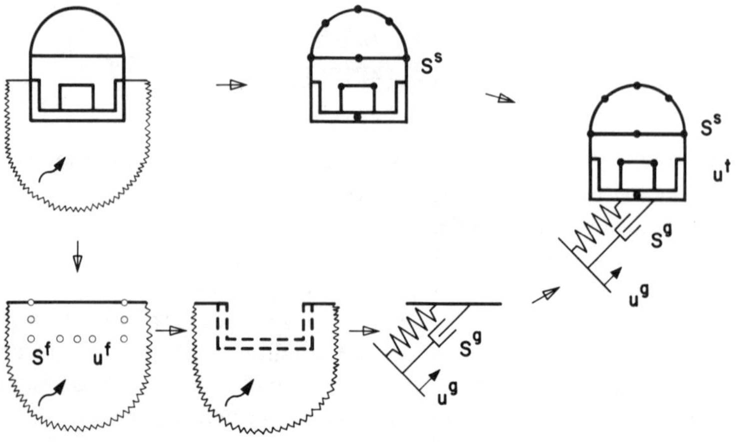

Figure 3-7 Physical interpretation of basic equation of motion in total displacements (scattered waves).

spring characterized by $[S^g_{oo}]$. The end of the generalized spring not connected to the structure is excited by $\{u^g_o\}$, which is calculated from $\{u^f_b\}$ and $[S^f_{bb}]$. This interpretation is also valid with minor adjustments for the structure with a flexible base (Eq. 3.5). The vector and matrix symbols, as well as the subscripts, have been deleted in Fig. 3-7. By setting $\{u^t_o\} = \{0\}$ in Eq. 3-20, it can be deduced that $[S^g_{oo}]\{u^g_o\}$ represents the amplitudes of the forces exerted on the rigid base in point O by the seismic motion when the base is kept fixed. They are sometimes called driving forces. Equation 3.5 can be interpreted analogously for a structure with a flexible base.

3.1.3 Special Foundations

It is assumed that the soil on the exterior of the line with nodes b consists of a layered half-space, as discussed in Chapter 1. This allows efficient formulations to be used to calculate $[S^g_{bb}]$ and $\{u^f_b\}$. Normally, this line will coincide with the base of the structure and thus form the structure–soil interface. If, however, an irregular soil region around the structure exists, as shown in Fig. 3-8, the basic equation (Eq. 3.9) can still be used. In this case, the bounded irregular soil region is regarded as part of an expanded structure. The nodes b lie on a line separating the irregular soil region and possibly the base of the structure from the regular horizontally layered soil region.

The basic equations of motion (Eqs. 3.9 and 3.15) not only apply to the single embedded structure shown in Figs. 3-1 and 3-5, but can also be used to analyze the more general structural configurations discussed in Chapter 1 in connection with Fig. 1-3. In this case, $\{u^t_b\}$ contains the degrees of freedom of the nodes lying on the structure–soil interfaces of all structures, for a pile founda-

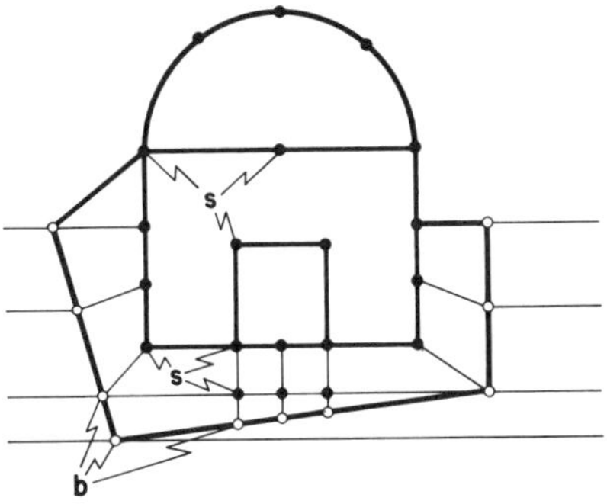

Figure 3-8 Structure–soil system with irregular soil region.

tion of the nodes along all piles. The matrices $[S^f_{bb}]$ and $[S^g_{bb}]$ are full. Analogously, the other degrees of freedom of all structures are assembled in $\{u^t_s\}$. Each structure contributes to the dynamic-stiffness matrix of the total structural configuration. Part of the structure–soil interface can be flexible, and part of it rigid. This case is illustrated schematically in Fig. 3-9 for an embedded structure with a rigid basemat and adjacent walls supported on piles. Nodes $\bar{b}$ lying on the line which subsequently will form the rigid base and nodes b are introduced. The dynamic degrees of freedom are $\{u^t_s\}$, $\{u^t_o\}$, and $\{u^t_b\}$. The motion of the nodes $\bar{b}$ can be expressed as a function of $\{u^t_o\}$, defining $[A]$. The load vector corresponding to the dynamic-equilibrium equations of point O equals $[A]^T([S^f_{\bar{b}\bar{b}}]\{u^f_{\bar{b}}\} + [S^f_{\bar{b}b}]\{u^f_b\})$; that of nodes b equals $[S^f_{b\bar{b}}][u^f_{\bar{b}}] + [S^f_{bb}]\{u^f_b\}$.

For the special case of rigid soil the formulation is also valid. In this case, no soil–structure interaction occurs; the structure is built in at its base with the prescribed motions $\{u^t_b\}$. This also follows from the equation of motion at the base (from Eq. 3.5: $\{u^t_b\} = \{u^g_b\}$). The equations of motion of the built-in structure with prescribed motions at its base $\{u^t_b\}$ are derived from the equations

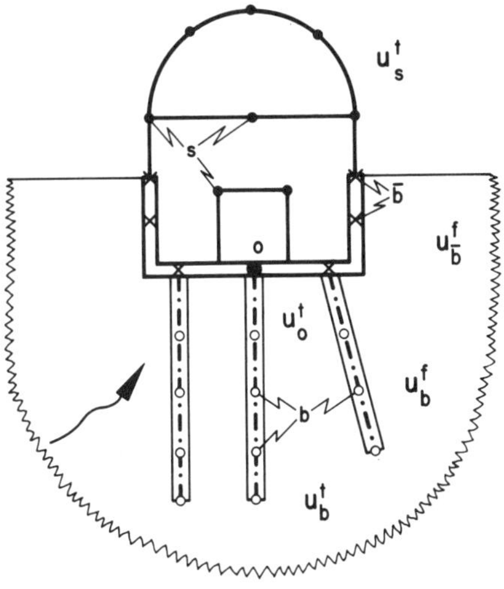

Figure 3-9 Structure supported on piles.

corresponding to the nodes of the structure not lying on the base. The equations expressed in the total motion $\{u_s^t\}$ follow from Eq. 3.5 as

$$[S_{ss}]\{u_s^t\} = -[S_{sb}]\{u_b^t\} \tag{3.21}$$

As expected, only the dynamic-stiffness matrix of the structure appears in the equations.

3.2 KINEMATIC AND INERTIAL INTERACTIONS

In the equations of motion formulated in total displacements (Eq. 3.9), the loads for earthquake excitation are applied only at the nodes located on the structure–soil interface. The structural engineer is, however, accustomed to work with seismic inertial loads (calculated as the product of the mass of the structure and some suitably chosen earthquake excitation) which are applied in all nodes of the structure. In general, this excitation is not equal to the free-field motion, but has to be determined prior to the actual dynamic analysis. The analysis is thus performed in the following two steps. The total displacements can be split into those caused by the so-called kinematic interaction (superscript k) and by the inertial one (superscript i).

3.2.1 Flexible Base

For a structure with a flexible base (Fig. 3-1), this procedure results in

$$\{u_s^t\} = \{u_s^k\} + \{u_s^i\} \tag{3.22a}$$

$$\{u_b^t\} = \{u_b^k\} + \{u_b^i\} \tag{3.22b}$$

In the kinematic-interaction part, the mass of the structure is set equal to zero by definition. Introducing this condition in Eq. 3.5 and using Eq. 3.1 results in

$$\begin{bmatrix} (1 + 2\zeta i)[K_{ss}] & (1 + 2\zeta i)[K_{sb}] \\ (1 + 2\zeta i)[K_{bs}] & (1 + 2\zeta i)[K_{bb}^s] + [S_{bb}^g] \end{bmatrix} \begin{Bmatrix} \{u_s^k\} \\ \{u_b^k\} \end{Bmatrix} = \begin{Bmatrix} \{0\} \\ [S_{bb}^g]\{u_b^g\} \end{Bmatrix} \tag{3.23}$$

Using Eq. 3.8, the nonzero load vector could also be formulated as $[S_{bb}^f]\{u_b^f\}$.

From the set of equations associated with nodes s, it follows that

$$\{u_s^k\} = -[K_{ss}]^{-1}[K_{sb}]\{u_b^k\} = [T_{sb}]\{u_b^k\} \tag{3.24}$$

The matrix $[T_{sb}]$ represents the quasi-static transformation, which is a function of the static-stiffness matrix of the structure. Each column of $[T_{sb}]$ can be visualized as the static displacements of the nodes s of the structure when a unit displacement is imposed at a specific node b. All other displacements at the nodes b are zero, and no loads are applied at the nodes s. In general, geometry alone is not sufficient to calculate $[T_{sb}]$.

Substituting Eq. 3.24 in Eq. 3.23 leads to

$$((1 + 2\zeta i)([K_{bb}^s] - [K_{bs}][K_{ss}]^{-1}[K_{sb}]) + [S_{bb}^g])\{u_b^k\} = [S_{bb}^g]\{u_b^g\} = [S_{bb}^f]\{u_b^f\} \tag{3.25}$$

For a general configuration, the displacement amplitudes for kinematic interaction of the nodes located on the structure–soil interface $\{u_b^k\}$ differ from those of the scattered motion $\{u_b^g\}$ (and also from the free-field $\{u_b^f\}$). For a structure supported on the surface (Fig. 3-10), $[S_{bb}^g] = [S_{bb}^f]$, as $[S_{bb}^e] = [0]$ (Eq. 3.6 and Fig. 3-3). For vertically incident body waves, the elements in $\{u_b^f\}$ are constant for the nodes b located on the free surface. For a horizontal and vertical uniform earthquake excitation, $\{u_b^f\}$ contains the constant amplitudes of this motion denoted as u^h and u^v, respectively, shown in Fig. 3-10. Multiplying ($[K_{bb}^s] - [K_{bs}][K_{ss}]^{-1}[K_{sb}]$) with this vector results in zero forces, as the structure moves as a rigid body. Thus for a surface structure excited by vertically incident waves, $\{u_b^k\}$ equals $\{u_b^f\}$. In this case $[T_{sb}]$ can be established from rigid body kinematics (horizontal and vertical translations u^h and u^v). For the sake of conciseness, the vector and matrix symbols as well as subscripts have been deleted in Fig. 3-10.

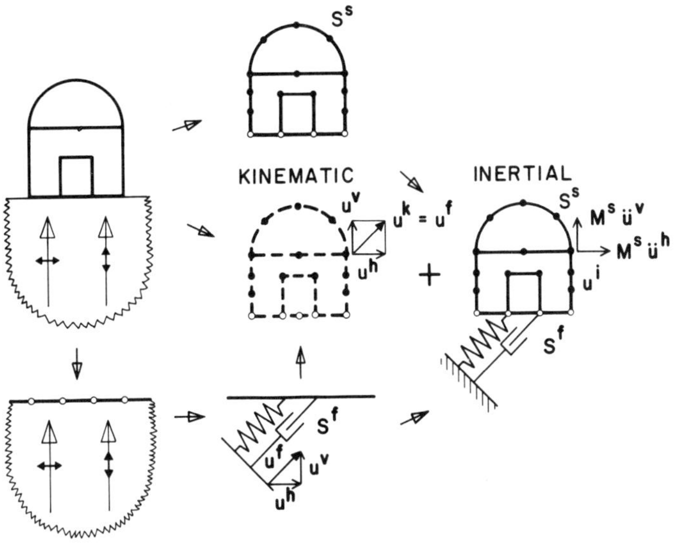

Figure 3-10 Physical interpretation of kinematic and inertial interactions (surface structure with vertically incident waves).

Proceeding further with the derivation of the formulation, Eq. 3.22 is substituted in Eq. 3.5, and making use of Eq. 3.23 leads to the equations of motion for the inertial interaction part

$$\begin{bmatrix} [S_{ss}] & [S_{sb}] \\ [S_{bs}] & [S_{bb}^s] + [S_{bb}^g] \end{bmatrix} \begin{Bmatrix} \{u_s^i\} \\ \{u_b^i\} \end{Bmatrix} = \omega^2 \begin{bmatrix} [M_{ss}][M_{sb}] \\ [M_{bs}][M_{bb}^s] \end{bmatrix} \begin{Bmatrix} \{u_s^k\} \\ \{u_b^k\} \end{Bmatrix} \quad (3.26)$$

The term $\omega^2\{u^k\}$ represents the negative-acceleration amplitude of the kinematic motion. For the inertial-interaction part of the dynamic response, the load vector thus consists of the negative inertia loads (mass of structure times acceleration) determined from $\{u^k\}$ (and not from $\{u_b^f\}$). The presence of this familiar load term represents the advantage of this method.

Summarizing, by omitting the mass of the structure and subjecting the dynamic system to the same load vector (which depends on the free-field motion), the kinematic motion is first calculated. The latter determines the loading to be applied in the actual dynamic analysis, the inertial interaction part. The procedure is illustrated in Fig. 3-11. As, however, for a flexible basemat, the equations of motion for the kinematic-interaction part (Eq. 3.23) are just as complicated as those for the total motion (Eq. 3.9), this procedure should not be used. For a rigid basemat, this method is appropriate, as is demonstrated in the following.

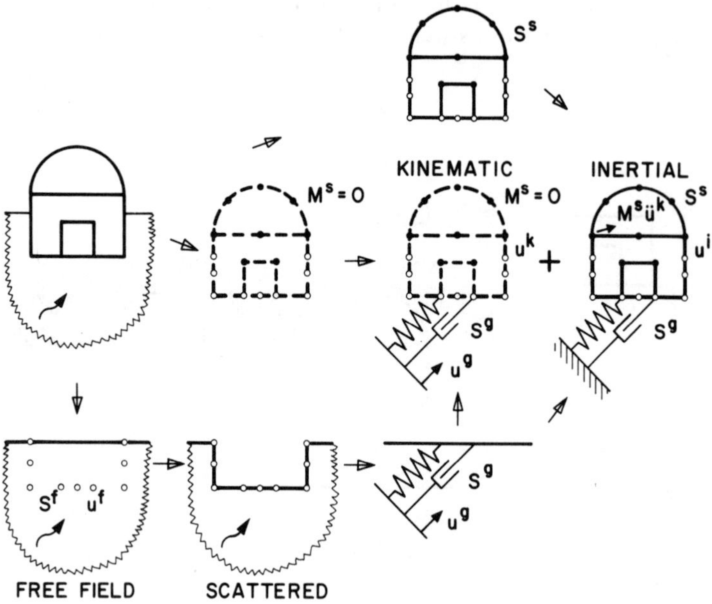

Figure 3-11 Physical interpretation of kinematic and inertial interactions (flexible base).

3.2.2 Rigid Base

For a structure with a rigid base (Fig. 3-5) the total motion is split up into those caused by kinematic and inertial interactions. Analogously to Eq. 3.22,

$$\{u_s^t\} = \{u_s^k\} + \{u_s^i\} \tag{3.27a}$$

$$\{u_o^t\} = \{u_o^k\} + \{u_o^i\} \tag{3.27b}$$

can be formulated.

Setting the mass of the structure equal to zero for the kinematic-interaction part of the analysis transforms Eq. 3.20 to

$$\left[\begin{array}{c|c} (1+2\zeta i)[K_{ss}] & (1+2\zeta i)[K_{so}] \\ (1+2\zeta i)[K_{os}] & (1+2\zeta i)[K_{oo}^s] + [S_{oo}^g] \end{array}\right] \begin{Bmatrix} \{u_s^k\} \\ \{u_o^k\} \end{Bmatrix} = \begin{Bmatrix} \{0\} \\ [S_{oo}^g]\{u_o^g\} \end{Bmatrix} \tag{3.28}$$

Solving this equation leads to

$$\{u_s^k\} = [T_{so}]\{u_o^k\} \tag{3.29a}$$

$$\{u_o^k\} = \{u_o^g\} \tag{3.29b}$$

with

$$[T_{so}] = -[K_{ss}]^{-1}[K_{so}] \tag{3.30}$$

The quasi-static transformation matrix $[T_{so}]$ is determined formally from the static part of the equilibrium equations of nodes s in Eq. 3.28. This matrix is not actually calculated using Eq. 3.30, but follows directly from rigid-body kinematics, analogously to the matrix $[A]$. The matrix $[T_{so}]$ depends on geometric quantities only and not on the stiffness properties of the structure. This is illustrated in Fig. 3-12. For the sake of clarity, the inner part of the structure shown in Fig. 3-5 is deleted. The undeformed structure is shown as a thin line, the displaced structure for kinematic interaction as a dashed line, and the deformed structure exhibiting the total motion as a solid line. The symbols for vectors and matrices are omitted.

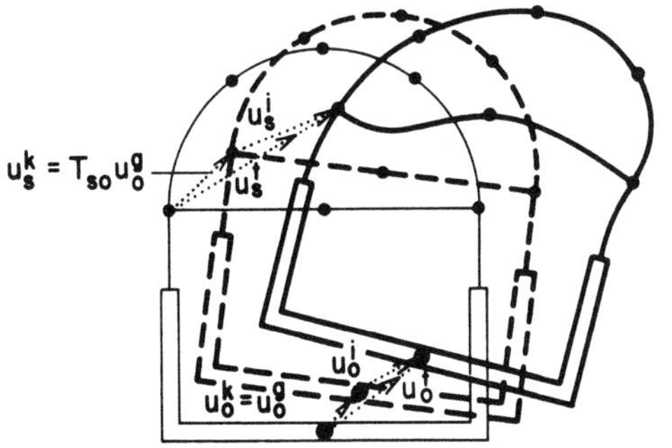

Figure 3-12 Kinematic and inertial motions.

In deriving Eq. 3.29b, use is made of

$$[K_{oo}^s] - [K_{os}][K_{ss}]^{-1}[K_{so}] = [0] \tag{3.31}$$

which expresses static equilibrium of the structure.

It follows from Eq. 3.29b that the motion of the rigid base in the kinematic-interaction part of the analysis (structure massless) is the same as that of the ground enforcing the rigid-body kinematics along the base. This is obvious, as the presence of a massless structure does not change the response of the base. With $\{u_o^g\}$, the kinematic interaction part of the motion follows throughout the structure using rigid-body kinematics (Eq. 3.29a).

Substituting Eq. 3.27 in Eq. 3.20 and using Eq. 3.28 leads to the equations of motion of the inertial-interaction part:

$$\begin{bmatrix} [S_{ss}] & [S_{so}] \\ [S_{os}] & [S_{oo}^s] + [S_{oo}^g] \end{bmatrix} \begin{Bmatrix} \{u_s^i\} \\ \{u_o^i\} \end{Bmatrix} = \omega^2 \begin{bmatrix} [M_{ss}][M_{so}] \\ [M_{os}][M_{oo}^s] \end{bmatrix} \begin{Bmatrix} \{u_s^k\} \\ \{u_o^k\} \end{Bmatrix} \tag{3.32}$$

The load vector is equal to the negative product of the mass of the structure and of the seismic-acceleration amplitude, the latter being determined by applying the acceleration amplitude $\{\ddot{u}_o^g\}(= -\omega^2\{u_o^g\})$ in point O at the base and applying rigid-body kinematics.

The kinematic and inertial motions are illustrated in Figs. 3-12 and 3-13. Summarizing, from the knowledge of the free-field response $\{u_b^f\}$, the motion of the kinematic interaction $\{u_o^k\}(= \{u_o^g\})$ follows from Eq. 3.19. Using rigid-body kinematics, the inertia loading to be applied throughout the structure in the inertial-interaction part of the analysis is determined. This physical interpretation of $\{u_o^g\}$ is important. The structure is supported on the generalized spring; no support motion is applied for the inertial-interaction part. Figure 3-13 should be compared to the corresponding one illustrating the formulation in total motions using scattered waves (Fig. 3-7). Reference should also be made to the discussion in Chapter 1 in connection with Fig. 1-2c, d, and e.

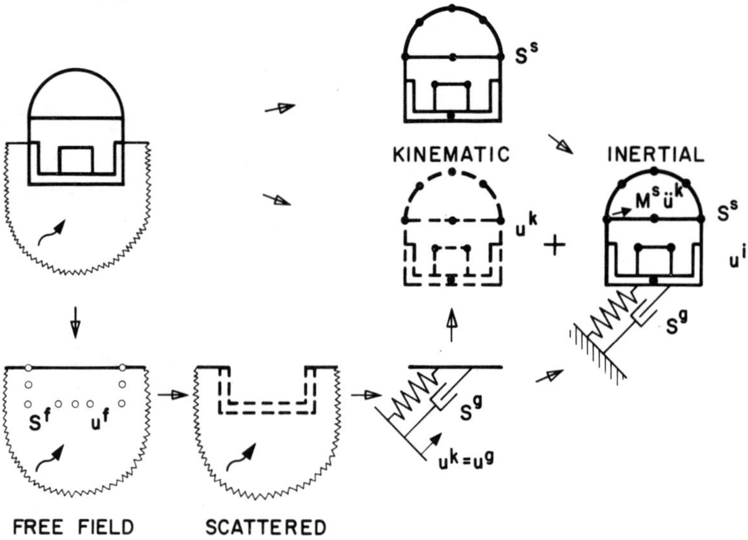

Figure 3-13 Physical interpretation of kinematic and inertial interactions (rigid base).

In the formulations derived above, for the substructure of the soil with excavation (system g), the compatibility constraints of the rigid base are enforced. The base is assumed to be massless. This is indicated in the figures by showing the base as a dashed line. Alternatively, it is possible to develop formulations where the mass of the base is included in the subsystem of the soil. This not only affects $[S_{oo}^g]$ and $\{u_o^g\}$ but also the modeling of the actual structure, which does not include the mass of the base. The inertial-interaction part of the dynamic analysis is modified as well. This scheme is not further investigated.

Other possibilities exist to split the total motion into two parts. For instance, the free-field response can be used to define one part. This method of the so-called quasi-static transmission of the free-field input motion and other procedures are discussed in Sections 8.2 and 8.3.

3.3 APPLIED LOADS AND THEIR CHARACTERIZATION

3.3.1 Rotating Machinery

Unbalanced mass in rotating machinery installed in buildings will lead to periodic loads acting on the structure, as over a very large number of cycles virtually the same variation of the load with time will occur. In general, loads and moments acting in all three directions will be applied to the structure. As an example, the vertical component $R(t)$ determined as the resultant of the measured vertical reactions of the four supports of a weaver's loom (Sulzer-Ruti Machinery Works Ltd. 8630 Ruti, Switzerland) is plotted over a complete cycle of 360° in Fig. 3-14. The weight of the machine operating at 9 Hz equals 27.46 kN, which is shown as a dashed line. The amplitudes of the load in the frequency domain $P(\omega) = \text{Re}\,(P) + i\,\text{Im}\,(P)$ follow from Eq. 2.18a [multiplied by $\Delta\omega/(2\pi)$]. The absolute value $\sqrt{[\text{Re}\,(P)]^2 + [\text{Im}\,(P)]^2}$ and the phase angle $\arctan\,[\text{Im}\,(P)/\text{Re}\,(P)]$ of the first harmonic (at 9 Hz) and of the higher ones up to the twentieth (at 180 Hz) are shown in Table 3-1. Transients also have to be investigated, resulting, for example, from the starting-up procedure of the machine or from extraordinary conditions, as a short circuit.

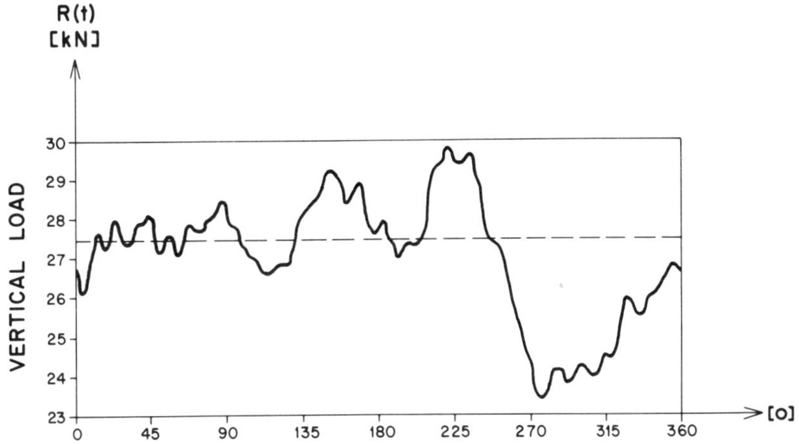

Figure 3-14 Vertical periodic load over one cycle.

3.3.2 Impact

As an example of a short-duration nonperiodic load, the impact forces are discussed. If the target (i.e., the structure) has a large mass and is quite stiff compared to the missile, the interaction occurring between the missile and the

TABLE 3-1 Fourier Amplitudes of the Vertical Periodic Load

	Amplitude	
Harmonic Number	Absolute Value (kN)	Phase Angle (deg)
1	1.518	138
2	1.223	59
3	0.555	− 16
4	0.695	−126
5	0.580	74
6	0.232	−142
7	0.144	−129
8	0.052	− 43
9	0.122	131
10	0.107	38
11	0.063	− 81
12	0.061	−103
13	0.080	−144
14	0.048	−165
15	0.132	− 86
16	0.177	−137
17	0.052	73
18	0.029	−154
19	0.012	0
20	0.016	0

target during impact can be neglected. This applies, for example, for the postulated aircraft crash on certain structures of nuclear-power plants. In this case of a soft missile, a load–time relationship for impact on a rigid target can be established which depends on the initial velocity v_0 of the missile, and on the distribution of its mass and of its buckling load. The force–time diagram for a Boeing 707-320 crashing onto a rigid target is shown in Fig. 3-15.

3.3.3 Earthquake

Earthquake time histories recorded on firm ground are quite irregular. This is caused by the complexities of the source mechanism, the reflections and refractions at the irregular interfaces, and the dispersion of the waves along the travel path.

For a ground motion $r_g(t)$ specified in the time domain, the Fourier transform $u_g(\omega)$ (Eq. 2.17a) is a complete measure, as its (inverse) transform recovers $r_g(t)$. For engineering-design aspects, the response spectrum that specifies the maximum response of a one-degree-of-freedom dynamic system excited by $r_g(t)$

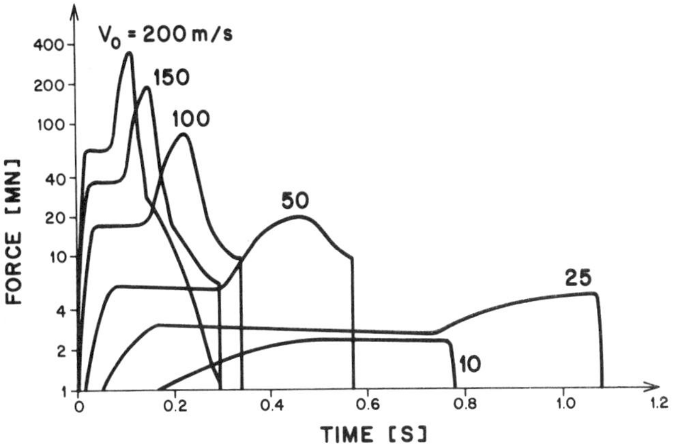

Figure 3-15 Force–time diagrams for Boeing 707-320.

represents a more important characterization (Fig. 3-16a). The maximum relative displacement $r_{\max}$ is calculated by varying the natural frequency $\omega = \sqrt{k/m}$ and the damping ratio $\zeta = c/(2\sqrt{km})$ of the dynamic system. This so-called spectral displacement $S_d = r_{\max}$ is used to define two other related measures: the spectral pseudo-velocity $S_v = \omega S_d$, which is statistically very close to the maximum relative velocity $\dot{r}_{\max}$ (with the exception of very low frequencies), and the spectral pseudo-acceleration $S_a = \omega^2 S_d$, which from a practical point of view is equal to the maximum total acceleration $\ddot{r}^t_{\max}$. Because the three spectral values are related by powers of the natural frequency ω, the response spectrum can be displayed on a so-called tripartite plot, from which all three spectral values can be read. A total of four logarithmic scales are present. As an example, the response spectra of an acceleration time history of the 1971 San Fernando earthquake, the H 115 record with the N 11E component with $r_{g\,\max} = 0.134$ m, $\dot{r}_{g\,\max} = 0.282$ m/s, and $\ddot{r}_{g\,\max} = 0.224$ g (Fig. 3-16b) are plotted for the indicated damping ratios ζ in Fig. 3-16c. The shape is jagged, reflecting the incidentals of a particular ground motion.

To determine the shape of the design-response spectrum, the response spectra of comparable excitations which can be expected at the site are processed statistically. This processing results in the peaks and dips being smoothed out. Different frequency ranges will be governed by different sizes and locations of earthquakes. In particular, by including specific records of potential earthquakes that could occur, the engineering seismologist can contribute significantly to a realistic shape of the design-response spectrum applicable to a specific site. The duration of the earthquake does affect the shape, as the longer the shaking, the larger the chance is of a large response arising. As the shape will be a smooth curve, small deviations of the dynamic properties of the system will have only a negligible influence on the structural response.

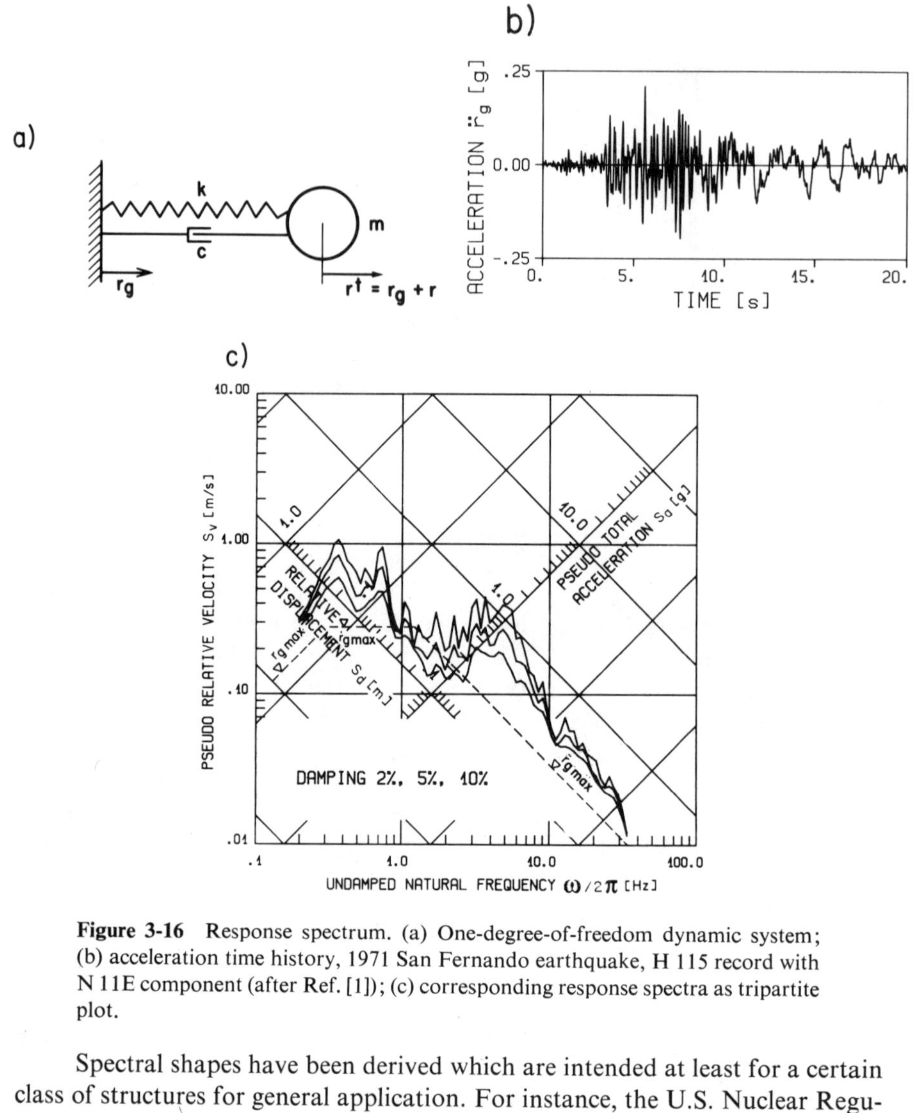

Figure 3-16 Response spectrum. (a) One-degree-of-freedom dynamic system; (b) acceleration time history, 1971 San Fernando earthquake, H 115 record with N 11E component (after Ref. [1]); (c) corresponding response spectra as tripartite plot.

Spectral shapes have been derived which are intended at least for a certain class of structures for general application. For instance, the U.S. Nuclear Regulatory Commission (NRC) published in December 1973 the revised Regulatory Guide 1.60, "Design Response Spectra for Seismic Design of Nuclear-Power Plants," which is applicable for a wide range of sites, excluding unusually soft ones.

The shapes of the spectra for the horizontal direction are shown as dashed lines for $\zeta = 0.02$, 0.05, and 0.10 in Fig. 3-17b. Possibly, for very flexible structures (which are hardly to be encountered in a nuclear-power plant) with a natural frequency below 1 Hz, the specified values are not conservative enough. For the vertical direction, the design spectrum can be selected as two-thirds of

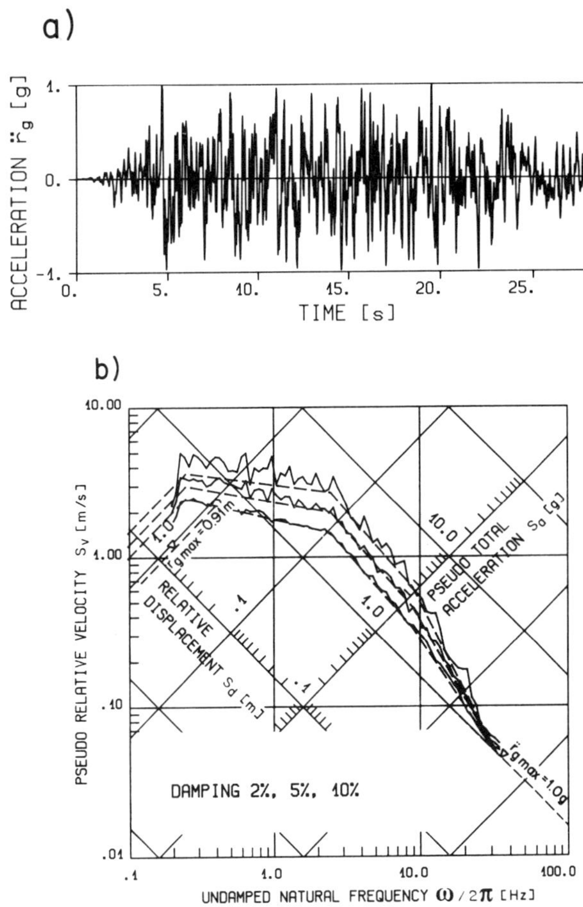

Figure 3-17 Design motion. (a) Artificial acceleration time history; (b) horizontal U.S. NRC design response spectrum, normalized to 1.0*g*.

the value in the horizontal one throughout the frequency range of interest, or, for a near-field earthquake, can even be selected as equal.

Based on a seismic hazard study and after selecting the acceptable probability that the earthquake used for design will be exceeded, the anchor value follows which is needed to scale the response spectrum. The anchor value is normally chosen as the maximum (effective) acceleration of the ground motion. This high-frequency acceleration of the response spectrum directly determines the total acceleration loads for structures which are quite stiff. In other cases, the maximum relative velocity, which is a measure of the kinetic energy, could possibly be used.

Once the design-response spectrum is known, the maximum force can easily be calculated for a single-degree-of-freedom system. It is determined as

(Fig. 3-17b) $kr_{max} = kS_d$, or, alternatively, as mS_a. For a multi-degree-of-freedom system with classical modes, the maximum response also follows directly from the design-response spectrum, assuming a certain combination of the contributions of the individual modes. For a typical soil–structure interaction system, classical modes, however, do not exist. In this case, so-called artificial time histories, which are compatible with the design-response spectrum, have to be calculated. This also applies in other circumstances, for example, when a subsystem is decoupled for analysis (Section 4.1) or for a nonlinear analysis. The calculation of artificial time histories, which is based on the theory of random processes, considers the general statistical characteristics of reasonable earthquakes. The amplitudes of the various frequencies are adjusted so that the artificial earthquake's response spectrum just envelops the specified design response spectrum for each damping value of interest. This is the case for the artificial time history shown in Fig. 3-17a, with its response spectra plotted as solid lines in Fig. 3-17b. Obviously, a unique time history does not exist. In general at least two horizontal time histories and one vertical (scaled to two-thirds) which are statistically independent are needed for analysis. The other horizontal and the vertical time histories used for practical applications in this text are not shown.

This motion of 30-s duration can be digitized at increments $\Delta t = 0.01$ s. If 4096 points are selected, about 10 s of zero motion can be added. The highest frequency contained in the motion equals 50 Hz (Eq. 2.19). Its contribution is, however, not accurately reproduced.

This procedure is consistent with today's state of knowledge, contributing to a design criterium which is insensitive to the exact values of parameters difficult to estimate. It is to be expected that in the future, sufficiently reliable earthquake records which fit the situation encountered will be available, eliminating the need of using artificial earthquakes.

As discussed in Section 6.1, the actual design motion is not sufficient to define the seismic loading case. In addition, the location (i.e., the control point) in which the motion acts and the wave pattern have to be specified. It is also worth mentioning that seismic design criteria will also address other aspects (e.g., damping ratios, methods of analysis, etc.).

3.4 INTRODUCTORY EXAMPLE

3.4.1 Statement of Problem

The different aspects of considering soil–structure interaction in the seismic analysis are discussed qualitatively in Chapter 1 (Fig. 1-2). The following examination is restricted to the actual interaction analysis illustrated in Fig. 1-4. For a surface structure excited by vertically incident waves, only the inertial interaction part of the analysis really has to be analyzed (Fig. 3-10). The latter can be examined qualitatively for a horizontal excitation with an amplitude u_g

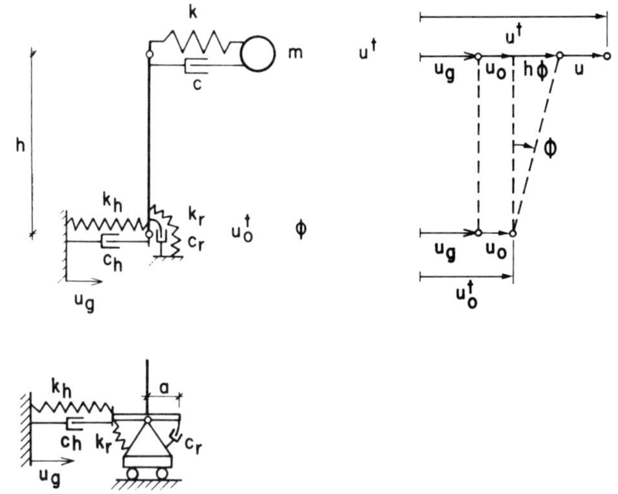

Figure 3-18 Model with one dynamic degree of freedom.

of frequency ω, based on the simple model shown in Fig. 3-18. The structure is modeled with a mass m, a lateral stiffness with a spring coefficient k, and a damper with a coefficient c, which are connected to a rigid bar of height h. This is the straightforward structural model of a single-story building frame or of a bridge whose girder can be regarded as rigid longitudinally compared to the (hinged) columns (Fig. 3-19). The values m, k, c, and h as well as a characteristic length a of the rigid base are easily determined. In addition, the idealized structure can also be interpreted as the model for a multistory structure which responds essentially in the fixed-base condition as if it were a single-degree-of-freedom system. The effective values m, k, and c are associated with the fundamental mode of vibration of the structure built in at its base; h is the distance from the base to the centroid of the inertial forces. Methods to calculate these values are specified in Chapter 4 (see Problem 4.5). The fixed-base frequency of the structure is denoted as ω_s; ζ represents the hysteretic damping ratio of the structure.

$$\omega_s^2 = \frac{k}{m} \tag{3.33}$$

$$c = \frac{2k\zeta}{\omega} \tag{3.34}$$

The damping coefficient c is thus frequency dependent. At the other end of the bar, at the base, the soil's dynamic stiffness is attached. The latter consists of a spring and, as energy is radiated and dissipated in the soil, of a damper. The corresponding coefficients are denoted as k_h and c_h in the horizontal direction and as k_r and c_r in the rotational (rocking) direction. No mass exists at the base of the structure. In the lower part of Fig. 3-18, the mechanism of the

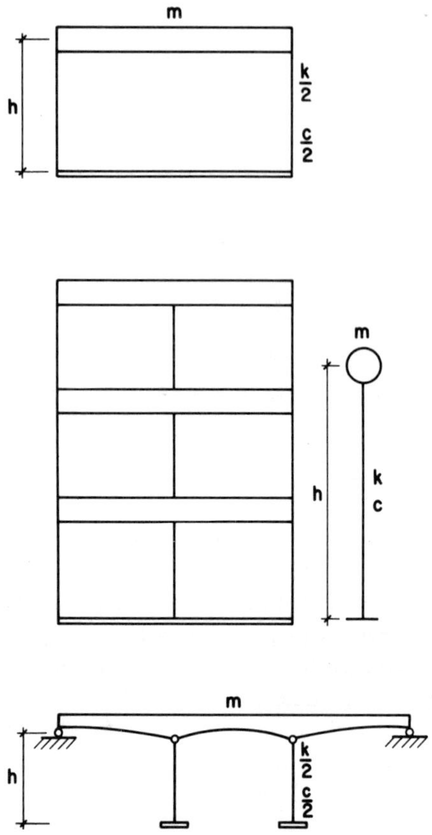

Figure 3-19 Structures that can be discretized as an equivalent model with one dynamic degree of freedom.

dynamic stiffness is shown in detail. All springs and dampers have a length approaching zero. Although the system has three degrees of freedom, the total lateral displacements of the mass with amplitude u^t, of the basemat with amplitude u_o^t and the rocking amplitude ϕ, only one of them is dynamic, as there is only one mass. This is the same number as that of the structure built in at its base (i.e., not considering soil–structure interaction). The model shown in Fig. 3-18 is thus well suited to identify the key parameters affecting soil–structure interaction and to study their effects. It is advantageous to split the total displacement amplitudes into their components

$$u^t = u_g + u_o + h\phi + u \tag{3.35a}$$

$$u_o^t = u_g + u_o \tag{3.35b}$$

where u_0 is the amplitude of the base relative to the free-field motion denoted as u_g (g for ground) and u represents the amplitude of the relative displacement of the mass referred to a moving frame of reference attached to the rigid base and which is equal to the structural distorsion (Fig. 3-18).

The modeling of the soil is treated rigorously in Chapter 7. At this stage, approximations are introduced, which nevertheless lead to valuable engineering-type procedures. The horizontal-force amplitude of the soil P_h is formulated as

$$P_h = k_h u_o + c_h \dot{u}_o \tag{3.36}$$

where, as will be shown later on (Section 5.1), k_h and c_h depend on the frequency of excitation ω. As discussed in connection with Eq. 3.4, only the motion of the base relative to that of the ground results in forces. For a soil without material damping, the corresponding equation is written as

$$P_x = k_x u_o + c_x \dot{u}_o \tag{3.37}$$

where the subscript x denotes the horizontal direction for a purely elastic soil. For a harmonic excitation

$$\dot{u}_o = i\omega u_o \tag{3.38}$$

applies. Substituting Eq. 3.38 in Eq. 3.37 leads to

$$P_x = k_x\left(1 + i\omega\frac{c_x}{k_x}\right)u_0 = k_x(1 + 2\zeta_x i)u_0 \tag{3.39}$$

where ζ_x represents the ratio of the viscous radiation damping in the horizontal direction

$$\zeta_x = \frac{\omega c_x}{2k_x} \tag{3.40}$$

The material damping of the soil is introduced, in an approximate manner, by multiplying the spring coefficient k_x (for frequency ω) with the factor $(1 + 2\zeta_g i)$, where ζ_g is the hysteretic damping ratio. This results in

$$P_h = k_x(1 + 2\zeta_x i + 2\zeta_g i)u_o \tag{3.41}$$

where the subscript h is used again, as a damped soil is addressed. Comparing Eq. 3.41 with Eq. 3.36 and using Eqs. 3.38 and 3.40,

$$k_h = k_x \tag{3.42a}$$

$$c_h = c_x + \frac{2}{\omega}\zeta_g k_x \tag{3.42b}$$

follow. Assuming the structure to be rigid ($k = \infty$) and that the foundation cannot rock ($k_r = \infty$), the corresponding natural frequency ω_h is specified by

$$\omega_h^2 = \frac{k_h}{m} \tag{3.43}$$

Analogously, for the rocking degree of freedom, the moment amplitude of the soil M_r is formulated as

$$M_r = k_r\phi + c_r\dot{\phi} = k_\phi(1 + 2\zeta_\phi i + 2\zeta_g i)\phi \tag{3.44}$$

The subscript ϕ denotes the rocking direction of the undamped soil. The ratio ζ_ϕ of the viscous radiation damping of the elastic soil in the rocking direction is defined as

$$\zeta_\phi = \frac{\omega c_\phi}{2k_\phi} \tag{3.45}$$

Equations 3.44 and 3.45 lead to

$$k_r = k_\phi \tag{3.46a}$$

$$c_r = c_\phi + \frac{2}{\omega}\zeta_g k_\phi \tag{3.46b}$$

Similarly, for a rigid structure ($k = \infty$) whose foundation is only allowed to rock ($k_h = \infty$), the natural frequency ω_r follows from

$$\omega_r^2 = \frac{k_r}{mh^2} \tag{3.47}$$

3.4.2 Equations of Motion of Coupled System

The equations of motion for this structure with a rigid basemat can be established either by applying Eq. 3.20 or Eq. 3.32 or, starting from scratch, by formulating the dynamic equilibrium of the mass point and the horizontal and rotational equilibrium equations of the total system (Fig. 3-18). Using Eqs. 3.41, 3.42a, 3.44, and 3.46a leads to

$$-m\omega^2(u + u_0 + h\phi) + k(1 + 2\zeta i)u = m\omega^2 u_g \tag{3.48a}$$

$$-m\omega^2(u + u_0 + h\phi) + k_h(1 + 2\zeta_x i + 2\zeta_g i)u_0 = m\omega^2 u_g \tag{3.48b}$$

$$-mh\omega^2(u + u_0 + h\phi) + k_r(1 + 2\zeta_\phi i + 2\zeta_g i)\phi = mh\omega^2 u_g \tag{3.48c}$$

Dividing Eqs. 3.48a and b by $m\omega^2$, Eq. 3.48c by $mh\omega^2$ and using Eqs. 3.33, 3.43, and 3.47 results in the following symmetric equations of motion of the coupled system:

$$\begin{bmatrix} \frac{\omega_s^2}{\omega^2}(1 + 2\zeta i) - 1 & -1 & -1 \\ -1 & \frac{\omega_h^2}{\omega^2}(1 + 2\zeta_x i + 2\zeta_g i) - 1 & -1 \\ -1 & -1 & \frac{\omega_r^2}{\omega^2}(1 + 2\zeta_\phi i + 2\zeta_g i) - 1 \end{bmatrix} \cdot \begin{Bmatrix} u \\ u_o \\ h\phi \end{Bmatrix} = \begin{Bmatrix} 1 \\ 1 \\ 1 \end{Bmatrix} u_g \tag{3.49}$$

For a specified ω with the corresponding u_g, the response determined from solving this equation depends on ω_s, ω_h, ω_r, ζ, ζ_x, ζ_ϕ, and ζ_g. It is worth pointing out that (even for constant spring and damping coefficients) ζ_x (Eq. 3.40) and ζ_ϕ (Eq. 3.45) depend on ω.

Eliminating u_o and $h\phi$ from Eq. 3.49 leads to

$$u_0 = \frac{\omega_s^2}{\omega_h^2} \frac{1 + 2\zeta i}{1 + 2\zeta_x i + 2\zeta_g i} u \tag{3.50a}$$

$$h\phi = \frac{\omega_s^2}{\omega_r^2} \frac{1 + 2\zeta i}{1 + 2\zeta_\phi i + 2\zeta_g i} u \tag{3.50b}$$

u is expressed as

$$\left(1 + 2\zeta i - \frac{\omega^2}{\omega_s^2} - \frac{\omega^2}{\omega_h^2} \frac{1 + 2\zeta i}{1 + 2\zeta_x i + 2\zeta_g i} - \frac{\omega^2}{\omega_r^2} \frac{1 + 2\zeta i}{1 + 2\zeta_\phi i + 2\zeta_g i}\right) u = \frac{\omega^2}{\omega_s^2} u_g \tag{3.51}$$

3.4.3 Equivalent One-Degree-of-Freedom System

To gain further physical insight, an equivalent one-degree-of-freedom system is introduced. Its properties (natural frequency $\tilde{\omega}$, ratio of hysteretic damping $\tilde{\zeta}$) are selected such that when excited by the replacement excitation $\tilde{u}_g$, essentially the same response u results as for the coupled system described by Eq. 3.51. A tilde (~) is used to denote the properties of this replacement oscillator. For harmonic motion, the equation of motion of the equivalent one-degree-of-freedom system is equal to

$$(-m\omega^2 + i\omega\tilde{c} + \tilde{k})u = m\omega^2 \tilde{u}_g \tag{3.52}$$

or with

$$\tilde{\omega}^2 = \frac{\tilde{k}}{m} \tag{3.53}$$

$$\tilde{c} = \frac{2\tilde{k}\tilde{\zeta}}{\omega} \tag{3.54}$$

$$\left(1 + 2\tilde{\zeta} i - \frac{\omega^2}{\tilde{\omega}^2}\right) u = \frac{\omega^2}{\tilde{\omega}^2} \tilde{u}_g \tag{3.55}$$

results. Note that the mass m is the same in the two dynamic systems.

The equivalent frequency $\tilde{\omega}$ is determined by noting that in an undamped system, the response is infinite at this natural frequency. Setting the coefficient of u equal to zero with $\zeta = \zeta_x = \zeta_\phi = \zeta_g = 0$ and with $\omega = \tilde{\omega}$ in Eq. 3.51 leads to

$$\frac{1}{\tilde{\omega}^2} = \frac{1}{\omega_s^2} + \frac{1}{\omega_h^2} + \frac{1}{\omega_r^2} \tag{3.56}$$

or substituting Eqs. 3.33, 3.43, and 3.47 yields

$$\tilde{\omega}^2 = \frac{\omega_s^2}{1 + k/k_h + kh^2/k_r} \tag{3.57}$$

It follows that the fundamental frequency $\tilde{\omega}$ of the soil–structure system is always smaller than the fixed-based frequency of the structure ω_s.

To establish the equivalent damping ratio $\tilde{\zeta}$, products of ζ, ζ_x, ζ_ϕ, and ζ_g are neglected compared to unity. This operation transforms Eq. 3.51 to

$$\left[1 + 2\zeta i - \frac{\omega^2}{\omega_s^2} - \frac{\omega^2}{\omega_h^2}(1 + 2\zeta i - 2\zeta_x i - 2\zeta_g i) - \frac{\omega^2}{\omega_r^2}(1 + 2\zeta i - 2\zeta_\phi i - 2\zeta_g i)\right]u = \frac{\omega^2}{\omega_s^2}u_g \tag{3.58}$$

Substituting Eq. 3.56 into the left-hand side of Eq. 3.55, the corresponding relationship for the equivalent system results:

$$\left(1 + 2\tilde{\zeta} i - \frac{\omega^2}{\omega_s^2} - \frac{\omega^2}{\omega_h^2} - \frac{\omega^2}{\omega_r^2}\right)u = \frac{\omega^2}{\tilde{\omega}^2}\tilde{u}_g \tag{3.59}$$

Equating the left-hand sides of Eqs. 3.58 and 3.59 (which have the same real parts) leads to

$$\tilde{\zeta} = \zeta\left(1 - \frac{\omega^2}{\omega_h^2} - \frac{\omega^2}{\omega_r^2}\right) + \zeta_x\frac{\omega^2}{\omega_h^2} + \zeta_\phi\frac{\omega^2}{\omega_r^2} + \zeta_g\left(\frac{\omega^2}{\omega_h^2} + \frac{\omega^2}{\omega_r^2}\right) \tag{3.60}$$

For resonance, $\omega = \tilde{\omega}$, this equation can be further simplified, using Eq. 3.56, to read

$$\tilde{\zeta} = \frac{\tilde{\omega}^2}{\omega_s^2}\zeta + \left(1 - \frac{\tilde{\omega}^2}{\omega_s^2}\right)\zeta_g + \frac{\tilde{\omega}^2}{\omega_h^2}\zeta_x + \frac{\tilde{\omega}^2}{\omega_r^2}\zeta_\phi \tag{3.61}$$

The equivalent damping ratio $\tilde{\zeta}$ is evaluated at resonance and then used over the whole range of frequency. For consistency ζ_x and ζ_ϕ should also be evaluated for $\omega = \tilde{\omega}$. The four different parts contributing to $\tilde{\zeta}$ are clearly separated in this equation. If no radiation damping occurs in the horizontal and rocking directions ($\zeta_x = \zeta_\phi = 0$) and if the material-damping ratio of the structure is equal to that of the soil ($\zeta = \zeta_g$), $\tilde{\zeta} = \zeta$ results, as is to be expected. As under normal circumstances ζ_g will not be smaller than ζ, the equivalent damping ratio $\tilde{\zeta}$ will, even in the absence of radiation damping, be somewhat larger than the damping ratio ζ of the structure, which would apply if soil–structure interaction were not taken into account. An alternative derivation based on energy considerations is addressed in Problem 3.10.

Finally, comparing the right-hand sides of Eqs. 3.51 and 3.55, it follows that

$$\tilde{u}_g = \frac{\tilde{\omega}^2}{\omega_s^2}u_g \tag{3.62}$$

The equivalent effective seismic input $\tilde{u}_g$ will thus always be smaller than u_g.

Summarizing, the seismic response of the coupled system shown in Fig. 3-18 is, from a practical point of view, the same as that of the equivalent one-degree-of-freedom system resting on rigid ground. This replacement oscillator is defined by the natural frequency $\tilde{\omega}$ (Eq. 3.57) and by the hysteretic-damping ratio $\tilde{\zeta}$ (Eq. 3.61) and is subjected to an effective ground-displacement amplitude $\tilde{u}_g$ (Eq. 3.62). The displacement amplitude u of this equivalent system is equal to that of the structural distortion. The corresponding transverse-shear force in the structure equals $k|u|$. Alternatively, the pseudo-acceleration of the equivalent oscillator $(\tilde{\omega}, \tilde{\zeta})$ can be calculated based on the original seismic input u_g, resulting in $\tilde{\omega}^2|u|(\omega_s^2/\tilde{\omega}^2)$. (The last factor reflects the different seismic excitation to be applied.) Multiplying this value with m leads to $m|u|\omega_s^2 = k|u|$, which is of course again the transverse-shear force. This represents a simple procedure applicable to design which takes soil–structure interaction into account. The amplitudes of the horizontal displacement of the base u_o relative to the free-field motion and of the rotation ϕ follow from Eq. 3.50. Again, neglecting squares of damping terms compared to unity results in

$$u_o = \frac{\omega_s^2}{\omega_h^2}(1 + 2\zeta i - 2\zeta_x i - 2\zeta_g i)u \tag{3.63a}$$

$$h\phi = \frac{\omega_s^2}{\omega_r^2}(1 + 2\zeta i - 2\zeta_\phi i - 2\zeta_g i)u \tag{3.63b}$$

$$u + u_o + h\phi = \omega_s^2\left(\frac{1}{\tilde{\omega}^2} + 2(\zeta - \zeta_g)i\left(\frac{1}{\tilde{\omega}^2} - \frac{1}{\omega_s^2}\right) - \frac{2\zeta_x i}{\omega_h^2} - \frac{2\zeta_\phi i}{\omega_r^2}\right)u \tag{3.63c}$$

The last equation describes the motion of the mass point relative to that of the free field. It has to be calculated to design, for example, the gap between neighboring buildings or the movement of a bridge abutment.

3.4.4 Dimensionless Parameters

For a specific excitation, the response of the dynamic system will depend on the properties of the structure compared to those of the soil. For the model illustrated in Fig. 3-18, the following dimensionless parameters are introduced:

1. The ratio of the stiffness of the structure to that of the soil:

$$\bar{s} = \frac{\omega_s h}{c_s} \tag{3.64a}$$

where c_s is the shear-wave velocity of the soil. As for certain tall buildings ω_s is approximately inversely proportional to h, $\omega_s h$ will be constant for this type of structure. For decreasing stiffness of the soil, $\bar{s}$ increases.

2. The slenderness ratio

$$\bar{h} = \frac{h}{a} \tag{3.64b}$$

where a represents a characteristic length of the rigid base (e.g., the radius for a circular basemat).

3. The mass ratio

$$\bar{m} = \frac{m}{\rho a^3} \tag{3.64c}$$

where ρ represents the mass density of the soil (shear modulus $G = \rho c_s^2$).

4. Poisson's ratio ν of the soil.
5. Hysteretic-damping ratios of the structure ζ and of the soil ζ_g.

To introduce the frequency of the excitation ω, the ratio ω/ω_s is defined.

These dimensionless parameters are sufficient to characterize the response of the structure founded on a half-space. The methods discussed in Section 7.2 allow the corresponding coefficients of the soil k_h, c_h, k_r, and c_r to be calculated, which depend on the frequency of excitation. As a crude approximation, the following expressions, which are frequency independent, are used for the undamped soil.

$$k_x = \frac{8Ga}{2-\nu} \tag{3.65a}$$

$$c_x = \frac{4.6}{2-\nu}\rho c_s a^2 \tag{3.65b}$$

$$k_\phi = \frac{8Ga^3}{3(1-\nu)} \tag{3.65c}$$

$$c_\phi = \frac{0.4}{1-\nu}\rho c_s a^4 \tag{3.65d}$$

The values k_x and k_ϕ are the static-stiffness coefficients for a rigid circular basemat of radius a. Expressing the properties $\tilde{\omega}$ and $\tilde{\zeta}$ of the replacement oscillator (Eqs. 3.57 and 3.61) in terms of the dimensionless parameters leads to

$$\frac{\tilde{\omega}^2}{\omega_s^2} = \frac{1}{1 + \dfrac{\bar{m}\bar{s}^2}{8}\left[\dfrac{2-\nu}{\bar{h}^2} + 3(1-\nu)\right]} \tag{3.66a}$$

$$\tilde{\zeta} = \frac{\tilde{\omega}^2}{\omega_s^2}\zeta + \left(1 - \frac{\tilde{\omega}^2}{\tilde{\omega}_s^2}\right)\zeta_g + \frac{\tilde{\omega}^3}{\omega_s^3}\frac{\bar{s}^3\bar{m}}{\bar{h}}\left[0.036\frac{2-\nu}{\bar{h}^2} + 0.028(1-\nu)\right] \tag{3.66b}$$

3.4.5 Parametric Study

As expected, decreasing the stiffness of the soil ($\bar{s}$ increases) results in a decreasing of $\tilde{\omega}/\omega_s$. Augmenting the mass of the structure also leads to smaller values of $\tilde{\omega}/\omega_s$. The ratio $\tilde{\omega}/\omega_s$ can be regarded as characterizing the effect of soil–structure interaction. The latter is thus important for stiff structures with a large mass supported on flexible soil. In Fig. 3-20, $\tilde{\omega}/\omega_s$ and $\tilde{\zeta}$ are plotted as a function of $\bar{s}$, varying the slenderness ratio $\bar{h}$. The other parameters are specified in the caption. For squat structures ($\bar{h}$ small), whose mode shapes of the structure–soil system will predominantly consist of a translation, $\tilde{\zeta}$ is larger than for slender structures ($\bar{h}$ large), where the rocking motion is of paramount importance. In Fig. 3-21, the mass ratio $\bar{m}$ is varied. The variables correspond to a typical reactor building of a nuclear-power plant. Extreme cases are also included to emphasize the effects.

As will be explained in depth in further sections (Sections 5.1, 7.3, and 7.4), for certain sites no radiation damping exists below a specific frequency. In this

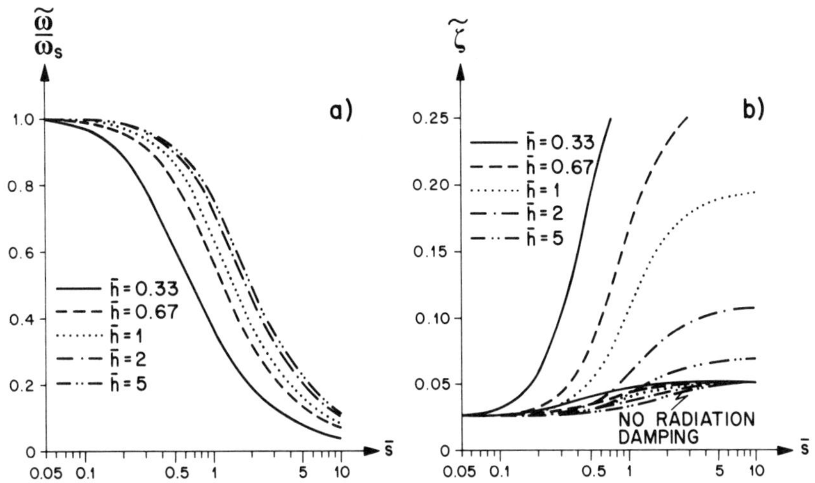

Figure 3-20 Properties of equivalent one-degree-of-freedom system ($\bar{m} = 3$, $\nu = 0.33$, $\zeta = 0.025$, $\zeta_g = 0.05$), varying slenderness ratio. (a) Natural frequency; (b) damping.

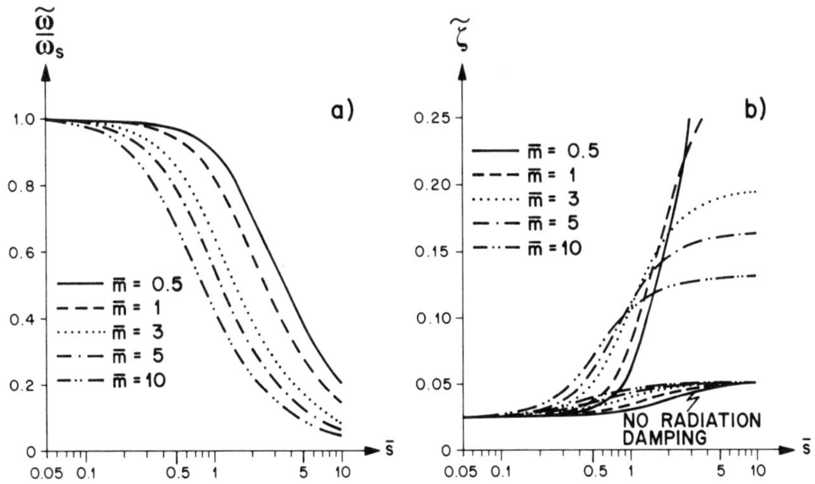

Figure 3-21 Properties of equivalent one-degree-of-freedom system ($\bar{h} = 1$, $\nu = 0.33$, $\zeta = 0.025$, $\zeta_g = 0.05$), varying mass ratio. (a) Natural frequency; (b) damping.

case, only the first two terms involving the material dampings are present in Eq. 3.66b. The corresponding effective damping ratio $\tilde{\zeta}$ is also shown in Figs. 3-20 and 3-21.

Although for the equivalent oscillator the damping ratio $\tilde{\zeta}$ is generally larger than ζ and the effective seismic input $\tilde{u}_g$ smaller than u_g, the structural response u for a specific ω can be larger or smaller than if soil–structure interaction is neglected. The latter follows from Eq. 3.55 for a stiffness ratio $\bar{s} = 0$ (soil rigid, $c_s = \infty$), that is, replacing $\tilde{\zeta}$ by ζ, $\tilde{\omega}$ by ω_s, and $\tilde{u}_g$ by u_g. If only the peak response is examined, which occurs at the natural frequency of the corresponding dynamic system ($\omega = \tilde{\omega}$, $\omega = \omega_s$), then that which takes soil–structure interaction into account $[= (u_g/2\tilde{\zeta})(\tilde{\omega}^2/\omega_s^2)]$ is practically always smaller than that for the same structure on rigid soil ($= u_g/2\zeta$). In Fig. 3-22a, the absolute value of the structural distortion $|u|$ (nondimensionalized with u_g) is plotted for $\bar{s} = 1$ and $\bar{s} = 0$ as a function of the dimensionless excitation frequency ω/ω_s. The parameters are again listed in the caption. In addition, selected values,

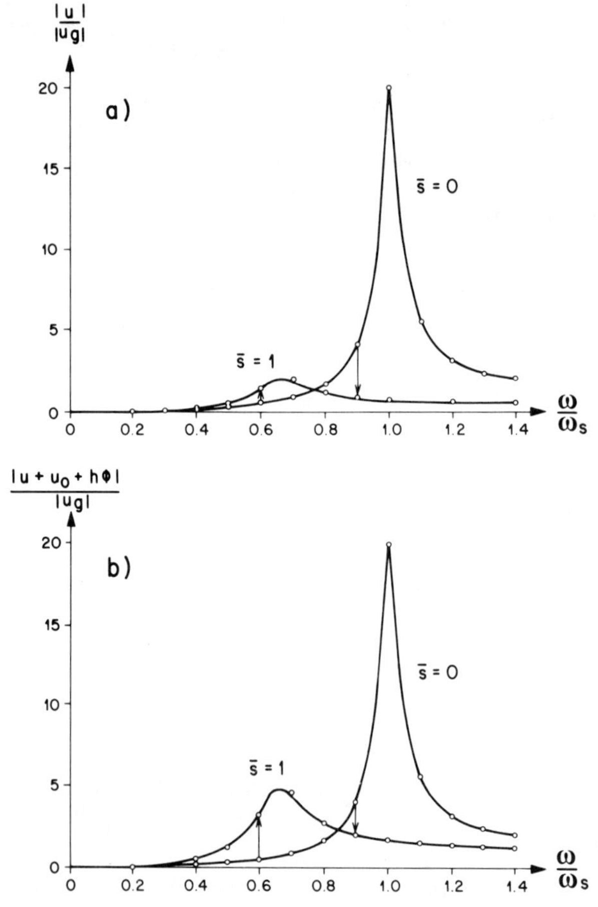

Figure 3-22 Influence of soil–structure interaction as a function of excitation frequency ($\bar{h} = 1$, $\bar{m} = 3$, $\nu = 0.33$, $\zeta = 0.025$, $\zeta_g = 0.05$). (a) Structural distortion; (b) displacement of mass relative to free field.

shown as dots, are plotted, determined from solving the coupled system (Eq. 3.49). The agreement with the results of the equivalent one-degree-of-freedom system is excellent. The absolute value of the displacement amplitude of the mass point relative to the free-field motion $|u + u_o + h\phi|$ (Eq. 3.63c) is plotted for the same parameters in Fig. 3-22b. Taking soil–structure interaction into account again reduces the peak response, although to a lesser extent. Disregarding soil–structure interaction, $u_o = \phi = 0$. Obviously, the interaction effect is negligible for very small and very large ratios of ω/ω_s. The motion at the base is examined in Problem 3.6.

Finally, the artificial time history described in Section 3.3 (Fig. 3-17a) is applied, normalized to 0.1 g. Its corresponding response spectrum closely follows the U.S. NRC Regulatory Guide 1.60. The maxima of the structural distortion u_{max} and of the displacement of the mass point relative to the free-field motion $(u + u_o + h\phi)_{max}$ are plotted as a function of the stiffness ratio $\bar{s}$ in Fig. 3-23a and b, respectively. The fixed-base frequency of the structure is varied. In contrast to the peak response for the harmonic motion, $(u + u_o + h\phi)_{max}$

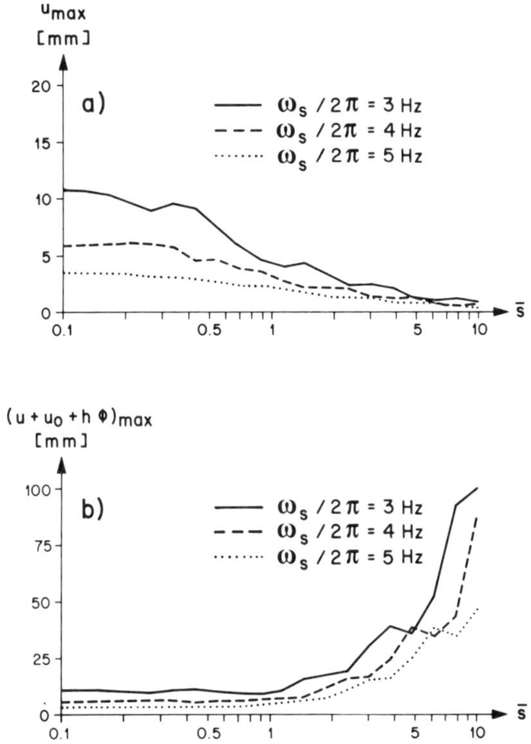

Figure 3-23 Maximum response, artificial time history ($\bar{h} = 1$, $\bar{m} = 3$, $\nu = 0.33$, $\zeta = 0.025$, $\zeta_g = 0.05$), varying fixed-base frequency. (a) Structural distortion; (b) displacement of mass relative to free field.

increases for (sufficiently large) increasing $\bar{s}$. The structural distortion u_{max} always decreases, when taking soil–structure interaction into account.

It is worth mentioning that the equation of motion of the equivalent one-degree-of-freedom system can also be solved directly in the time domain. Using frequency-independent spring and damping coefficients and assuming the damping whose ratio is $\tilde{\zeta}$ to be viscous, the following differential equation results:

$$\ddot{r} + 2\tilde{\zeta}\tilde{\omega}\dot{r} + \tilde{\omega}^2 r = -\frac{\tilde{\omega}^2}{\omega_s^2}\ddot{r}_g \tag{3.67}$$

The structural distortion as a function of time and the prescribed acceleration time history are denoted as r and $\ddot{r}_g$, respectively. See Problem 3.13 for an application.

An even simpler model can be selected to analyze the vertical earthquake excitation (see Problems 3.7, 3.8, and 3.9).

SUMMARY

1. By merging the two substructures, the actual structure and the soil with excavation, the basic equation of motion in the frequency domain is derived. The amplitudes of the total displacements are associated with nodes within the structure and on the structure–soil interface.
2. The corresponding coefficient matrix is formed by assembling the dynamic-stiffness matrices of the discretized structure and of the unbounded soil with excavation with degrees of freedom along the structure–soil interface (base). For a rigid base, the compatibility constraints are incorporated into the dynamic-stiffness matrices.
3. The corresponding load vector equals the product of the dynamic-stiffness matrix of the soil (free field) and the free-field-motion vector, both determined only in those nodes which subsequently will lie on the structure–soil interface. The earthquake excitation in other points of the free field is thus not required. The nodes of the structure not in contact with the soil are unloaded. Also for a structure with a rigid base, the load vector depends on this free-field motion. No need to determine the seismic-input motion of any other reference soil system exists.
4. The discretized system consists of the structure, and in the embedded region, of the difference of the structure and of the soil, and is supported on a generalized spring characterized by the dynamic-stiffness matrix of the soil (free field). That end of this spring not connected to the structure is excited by the free-field motion in those nodes which subsequently will lie on the structure–soil interface.
5. The basic equation expressed in total motion results in a simple procedure which can be used to calculate the most general case of soil–structure interaction straightforwardly and economically.

6. Alternatively, the load vector can also be determined as the product of the dynamic-stiffness matrix of the soil with excavation and the vector of the corresponding scattered seismic-input motion.
7. The total motion can be split up into that arising from kinematic and that from inertial interactions. An analysis with the same load vector as in the basic equation of motion and omitting the mass of the structure results in the kinematic motion. The latter forms the seismic acceleration which, when multiplied by the mass of the structure, leads to the inertial loads (acting in all nodes of the structure) of the inertial interaction. The coefficient matrix of the inertial interaction is the same as that of the basic equation of motion.
8. For a surface structure excited by vertically incident waves, the kinematic motion equals that of the free field. The corresponding acceleration (constant throughout the structure) multiplied by the mass of the structure leads to the loads of the inertial interaction.
9. For a general configuration with a flexible base, determining the kinematic motion is as complicated as calculating the total motion directly from the basic equation of motion and should thus not be performed.
10. For a rigid base, the kinematic motion at the base equals the scattered motion of the soil with excavation, whereby the compatibility constraints are enforced. Even when an embedded structure is subjected only to vertically incident waves, rotational components arise. The kinematic motion throughout the structure (and thus also the loads for the inertial interaction) follow from rigid-body kinematics. As the physical insight gained by splitting the procedure into two steps can be valuable, a viable alternative to solving the basic equation in total motion exists for a rigid base.
11. Unbalanced mass in rotating machines will lead to periodic loading which is characterized by the Fourier amplitudes of the first and higher harmonics.
12. The soft-missile impact on a rigid target results in a short-duration nonperiodic load which depends on the initial velocity of the missile and on the distribution of its mass and of its buckling load.
13. The design motion for earthquake excitation is specified as a design-response spectrum determined from a shape intended for general application and an anchor value. A corresponding artificial time history can be calculated whose response spectrum closely follows the design-response spectrum for the damping values of interest.
14. A simple system consisting of a vertical rigid bar with the translational and rocking springs and dashpots representing the soil attached at one end and at the other one, at a distance equal to the height, a spring and a dashpot connected to a mass, which models the structure, correctly captures the essential effects of soil–structure interaction.

15. The response of this system for a prescribed horizontal seismic excitation with a specified frequency is a function of the fixed-base frequency of the structure, of the frequencies for a rigid structure assuming in addition that either the rocking or the translational spring is rigid, the ratios for the viscous radiation damping in the horizontal and rocking directions and the ratios of the hysteretic damping of the structure and of the soil.

16. The coupled system can be replaced by an equivalent one-degree-of-freedom system. Its natural frequency, which is a function of the three frequencies defined above, is always smaller than the fixed-base frequency of the structure. Its damping ratio, which can be calculated by adding the contributions of the four damping ratios introduced above, will, in general, be larger than the hysteretic damping ratio of the structure. Its effective support motion will be smaller than that of the fixed-base structure, the factor being equal to the square of the ratio of the equivalent natural frequency to the fixed-base frequency.

17. The response is a function of the ratio of the stiffness of the structure to that of the soil, of the slenderness ratio, of the ratio of the mass of the structure to that of the soil, of Poisson's ratio of the soil, and of the ratios of hysteretic damping of the structure and of the soil.

18. Taking soil–structure interaction into account will reduce the peak structural distortion for harmonic excitation, while for a specific frequency of the excitation the result can be either smaller or larger than that of the fixed-base structure.

19. For an artificial earthquake time history, taking soil–structure interaction into account, will, in general, reduce the maximum structural distortion, while the maximum displacement of the structure relative to the free-field motion can be increased.

PROBLEMS

3.1. A mass m is supported by two massless truss bars of length l, modulus of elasticity E, area A, and damping ratio ζ, which are connected to a rigid massless basemat. This surface structure is shown in Fig. P3-1a. The dynamic-stiffness coefficients of the soil are denoted as S_h^g, S_v^g, and S_r^g in the horizontal, vertical, and rotational directions, respectively. The scattered motion (which is assumed to be specified) consists of the translational components with the amplitudes u_o^g and w_o^g and of a rotation with the amplitude β_o^g. Using the nomenclature indicated in Fig. P3-1a, formulate the equations of motion in total displacements (amplitudes with superscript t, Eq. 3.20) and in kinematic and inertial displacements (amplitudes with superscripts k and i, Eqs. 3.29 and 3.32).

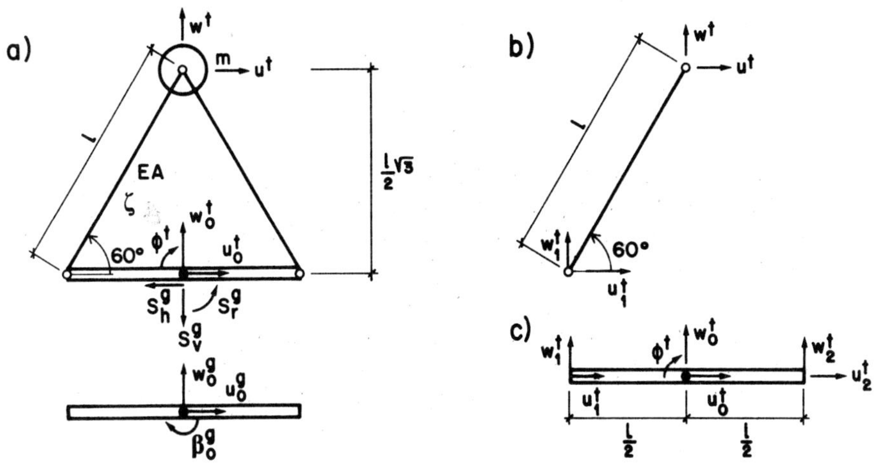

Figure P3-1 Mass supported by two truss bars on rigid basemat. (a) Structural system; (b) inclined bar; (c) rigid basemat.

Solution:

The dynamic-stiffness matrix (relating the amplitudes of the displacements u^t, w^t, u_1^t, and w_1^t to those of the loads) of the massless damped bar in Fig. P3-1b is equal to

$$\frac{EA}{l}(1 + 2\zeta i)\begin{bmatrix} \frac{1}{4} & \frac{\sqrt{3}}{4} & -\frac{1}{4} & -\frac{\sqrt{3}}{4} \\ \frac{\sqrt{3}}{4} & \frac{3}{4} & -\frac{\sqrt{3}}{4} & -\frac{3}{4} \\ -\frac{1}{4} & -\frac{\sqrt{3}}{4} & \frac{1}{4} & \frac{\sqrt{3}}{4} \\ -\frac{\sqrt{3}}{4} & -\frac{3}{4} & \frac{\sqrt{3}}{4} & \frac{3}{4} \end{bmatrix}$$

The kinematic-transformation matrix $[A]$ of the rigid basemat follows from Eq. 3.13 as (Fig. P3-1c)

$$\begin{Bmatrix} u_1^t \\ w_1^t \\ u_2^t \\ w_2^t \end{Bmatrix} = \begin{bmatrix} 1 & & \\ & 1 & \frac{l}{2} \\ 1 & & \\ & 1 & -\frac{l}{2} \end{bmatrix} \begin{Bmatrix} u_0^t \\ w_0^t \\ \phi^t \end{Bmatrix}$$

The matrices $[S_{so}]$ and $[S_{oo}^s]$ associated with the structure are calculated using Eqs. 3.16a and 3.16b, whereby in $[S_{bb}^s]$ the contribution of both bars is present. Adding the effect of the mass and of the soil, the equations of motion in total-displacement amplitudes are equal to (Eq. 3.20):

$$
\left(\begin{bmatrix} -\omega^2 m & & & & \\ & -\omega^2 m & & & \\ & & S_h^g & & \\ & & & S_v^g & \\ & & & & S_r^g \end{bmatrix}\right.
$$

$$
\left.+\frac{EA}{l}(1+2\zeta i)\begin{bmatrix} \frac{1}{2} & & -\frac{1}{2} & & -\frac{\sqrt{3}\,l}{4} \\ & \frac{3}{2} & & -\frac{3}{2} & \\ -\frac{1}{2} & & \frac{1}{2} & & \frac{\sqrt{3}\,l}{4} \\ & -\frac{3}{2} & & \frac{3}{2} & \\ -\frac{\sqrt{3}\,l}{4} & & \frac{\sqrt{3}\,l}{4} & & \frac{3l^2}{8} \end{bmatrix}\right)\begin{Bmatrix} u^t \\ w^t \\ u_o^t \\ w_o^t \\ \phi^t \end{Bmatrix} = \begin{Bmatrix} \\ \\ \\ \begin{bmatrix} S_h^g & & \\ & S_v^g & \\ & & S_r^g \end{bmatrix}\begin{Bmatrix} u_o^g \\ w_o^g \\ \beta_o^g \end{Bmatrix} \end{Bmatrix}
$$

The kinematic displacement amplitudes follow from Eq. 3.29 as

$$
\begin{Bmatrix} u^k \\ w^k \\ u_o^k \\ w_o^k \\ \phi^k \end{Bmatrix} = \begin{bmatrix} 1 & & \frac{l}{2}\sqrt{3} \\ & 1 & \\ 1 & & \\ & 1 & \\ & & 1 \end{bmatrix}\begin{Bmatrix} u_o^g \\ w_o^g \\ \beta_o^g \end{Bmatrix}
$$

The inertial-displacement amplitudes are determined using Eq. 3.32. The coefficient matrix of the left-hand side is the same as that in the case of total displacements but with the unknowns u^i, w^i, u_o, w_o^i, and ϕ^i. The right-hand side equals

$$
\omega^2\begin{bmatrix} m & & & & \\ & m & & & \\ & & & & \\ & & & & \\ & & & & \end{bmatrix}\begin{Bmatrix} u^k \\ w^k \\ u_o^k \\ w_o^k \\ \phi^k \end{Bmatrix} = \omega^2\begin{bmatrix} m & & \frac{l\sqrt{3}}{2}m \\ & m & \\ & & \\ & & \\ & & \end{bmatrix}\begin{Bmatrix} u_o^g \\ w_o^g \\ \beta_o^g \end{Bmatrix}
$$

3.2. The derivation of the basic equation of motion (Eq. 3.9) is based on substructuring with replacement. Adding the excavated part of the soil (the reference system e) to the irregular substructure of the soil with excavation (system g) results in a reference system with regular geometry, the free field f (Fig. 3-3). The latter's response to the seismic excitation $\{u_b^f\}$ is relatively easy to calculate, as will be shown in Chapter 6. The same applies to the dynamic-stiffness matrix $[S^f_{bb}]$ (Section 7.5). The load vector of the basic equation of motion is a function of the free-field reference system only.

This concept can be illustrated by solving the following simple static problem, whereby the loading is not a prescribed motion, but an exterior load. One span of a continuous beam with infinitely many spans of equal length l is loaded with an evenly distributed load p (Fig. P3-2). All spans have the same moment of inertia I with the exception of one of the spans adjacent to the loaded one, where it is denoted as I_s. The modulus of elasticity is E. Determine the moment distribution along this span, taking only the work of the moments into account. For the numerical calculation choose $I_s = 2I$.

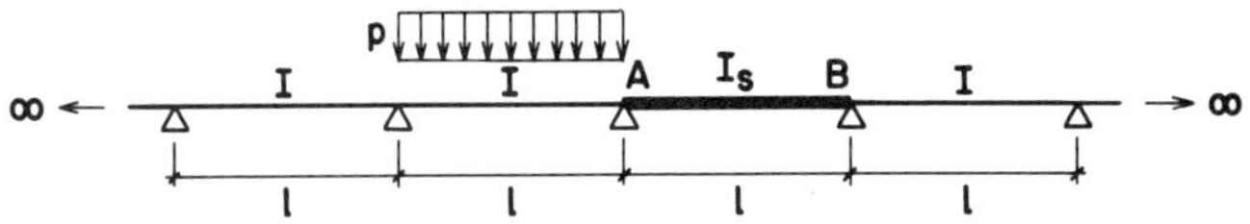

Figure P3-2 Continuous beam with infinitely many spans with one span having a different moment of inertia.

Hints:

The span A-B represents one of the substructures (denoted as s), the beams to the left of A and to the right of B, the other substructure being g. Replacing the span A-B with I_s by one with a moment of inertia I, the free-field system f results, which is a continuous beam with infinitely many identical spans, whose response is tabulated. In particular, the rotations at A and B for the distributed load follow, which correspond to the free-field response $\{u^f_b\}$. The moment that leads to a unit rotation at the end of the continuous beam with infinitely many identical spans is equal to the two elements of the diagonal matrix of the stiffness matrix $[S^g_{bb}]$. After adding to $[S^g_{bb}]$ the full 2×2 stiffness matrix of the span A-B with I (system e), i.e., $[S^e_{bb}]$, one obtains $[S^f_{bb}]$.

Results:

$$[S^s_{bb}] = \begin{bmatrix} 4\dfrac{EI_s}{l} & 2\dfrac{EI_s}{l} \\ 2\dfrac{EI_s}{l} & 4\dfrac{EI_s}{l} \end{bmatrix} \qquad [S^g_{bb}] = \begin{bmatrix} 3.46\dfrac{EI}{l} & \\ & 3.46\dfrac{EI}{l} \end{bmatrix}$$

$$[S^e_{bb}] = \begin{bmatrix} 4\dfrac{EI}{l} & 2\dfrac{EI}{l} \\ 2\dfrac{EI}{l} & 4\dfrac{EI}{l} \end{bmatrix}$$

$$[S^f_{bb}] = [S^g_{bb}] + [S^e_{bb}] = \begin{bmatrix} 7.46\dfrac{EI}{l} & 2\dfrac{EI}{l} \\ 2\dfrac{EI}{l} & 7.46\dfrac{EI}{l} \end{bmatrix}$$

$$\{u_b^f\} = \begin{Bmatrix} 0.0152\dfrac{pl^3}{EI} \\ -0.0041\dfrac{pl^3}{EI} \end{Bmatrix}$$

$$([S_{bb}^s] + [S_{bb}^g])\{u_b^t\} = [S_{bb}^f]\{u_b^f\}$$

with

$$I_s = 2I$$

$$\{u_b^t\} = \begin{Bmatrix} 0.0105\dfrac{pl^3}{EI} \\ -0.0037\dfrac{pl^3}{EI} \end{Bmatrix}$$

$$M_A = -0.54\frac{pl^2}{8}$$

$$M_B = 0.10\frac{pl^2}{8}$$

3.3. In a multistory building the vibrations at the top floor arising from periodic loads from a machine installed at another floor are unacceptably strong for the employees (Fig. P3-3). To reduce the dynamic response, a tuned-mass absorber consisting of a mass m, of a spring with a constant k, and of a damper with a constant c is installed at this top floor. Derive the equations of motion using substructure concepts.

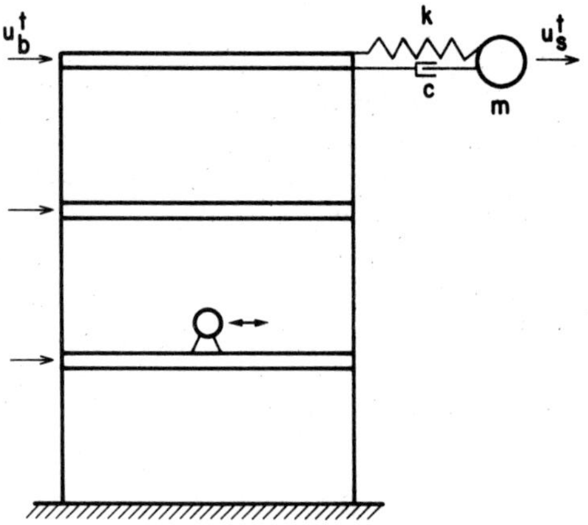

Figure P3-3 Structure excited by periodic load from machine with tuned-mass absorber.

Solution:

To establish the coupled equations of motion, the original structure without the absorber can be regarded as the free-field reference system f. Assuming the columns to be inextensional and lumping the mass of the structure on the level of the floors, the dynamic degrees of freedom of the (two-dimensional) structure are equal to the lateral displacements of the floors. The base of the structure is assumed to be fixed. The equations of motion (in the frequency domain) of the original structure, including the loads from the machine, can easily be formulated (Section 4.3). The corresponding displacement amplitudes at the top floor of the free field u_b^f then follow. Applying a harmonic load of unit amplitude at the top floor and calculating the inverse of the displacement amplitude at the same location leads to the dynamic-stiffness coefficient S_{bb}^f. In actual practice, u_b^f will often be measured and also S_{bb}^f will be determined experimentally with an eccentric vibrator varying the frequency to cover the range of interest. As the original structure is also one of the two substructures g, $S_{bb}^g = S_{bb}^f$. The other substructure consists of the absorber mechanism. The equations of motion of the coupled system then follow from Eq. 3.9 as

$$\begin{bmatrix} k + i\omega c - \omega^2 m & -k - i\omega c \\ -k - i\omega c & k + i\omega c + S_{bb}^f \end{bmatrix} \begin{Bmatrix} u_s^t \\ u_b^t \end{Bmatrix} = \begin{Bmatrix} 0 \\ S_{bb}^f u_b^f \end{Bmatrix}$$

3.4. If certain approximations are introduced, the method of substructuring with replacement can also be used to establish the differential equations of motion in the time domain. This is, for example, the case for a vertical pile supported on a Winkler type of foundation, that is, on individual springs whose coefficients $k(z)$ (per unit length) depend on the soil properties and on the radius a of the pile (Fig. P3-4). The seismic excitation of the free field consists of a vertically propagating shear wave with a horizontal displacement $r^f(z, t)$. The modulus of elasticity, the mass density, the area, and the moment of inertia are denoted as E, ρ, A, and I, respectively. The subscripts p and q are used to indicate the properties of the pile and of the soil, respectively. Derive the differential equation of motion, neglecting all damping terms, which is equal to

$$E_p I r^t_{,zzzz} + \rho_p A \ddot{r}^t + k r^t = k r^f + E_q I r^f_{,zzzz} + \rho_q A \ddot{r}^f$$

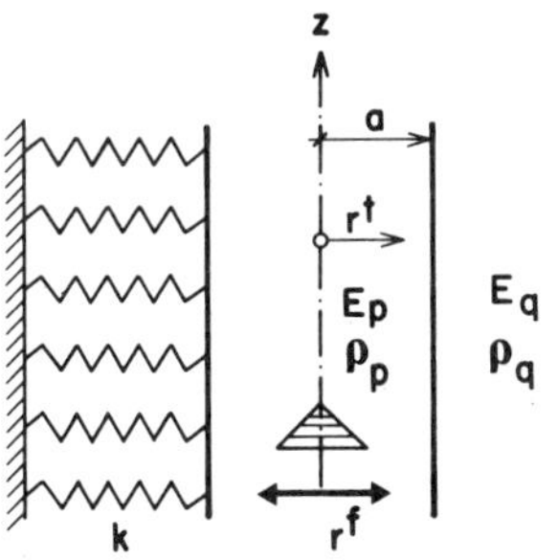

Figure P3-4 Pile supported on individual springs.

Solution:

The first two terms and the third one on the left-hand side correspond to the contributions of the pile [the "structure"(s)] and of the ground with excavation (system g), respectively, that is, of the two substructures. The right-hand side represents the product of the dynamic stiffness of the free field and its motion.

The last two terms on the right-hand side correspond to the contribution of the excavated soil, that is, the pile replaced by soil (system *e*). Introducing the wave equation of the soil (which is derived in Section 5.1 for harmonic excitation),

$$r^f_{,zz} = \frac{1}{c_s^2}\ddot{r}^f$$

the right-hand side can be rewritten, resulting in

$$E_p I r^t_{,zzzz} + \rho_p A \ddot{r}^t + k r^t = k r^f + \rho_q A \ddot{r}^f + \frac{E_q I}{c_s^4} \ddot{\ddot{r}}^f$$

The shear-wave velocity of the soil is denoted as c_s.

This example is only used to illustrate the method of substructuring with replacement. The influence of the last two terms on the right-hand side will, in general, be smaller than that of selecting isolated springs to model the soil, thus neglecting the coupling and damping effects.

3.5. Assume that the model of a structure consists of separate "sticks" (branches), which are not connected to each other and which are attached to the flexible base in one node only, this node in general being different for each stick. The stick could be a beam. Show that for such an embedded structure the motion of the kinematic interaction of the base $\{u_b^k\}$ is equal to that of the scattered motion $\{u_b^g\}$.

Solution:

The amplitudes $\{u_s^k\}$ of each stick can be eliminated independently of those of the other sticks (Eq. 3.23). The equilibrium equations of each stick result in

$$[K_{bb}^s] - [K_{bs}][K_{ss}]^{-1}[K_{sb}] = 0$$

Equation 3.25 leads to

$$[S_{bb}^g]\{u_b^k\} = [S_{bb}^g]\{u_b^g\}$$

that is, $\{u_b^k\} = \{u_b^g\}$.

3.6. For the coupled system used to analyze the horizontal excitation and which is examined in Section 3.4 (Fig. 3-18), plot the absolute value of the total-displacement amplitude of the base (nondimensionalized with $|u_g|$), $|u_o + u_g|/|u_g|$ versus the ratio of the excitation frequency to that of the fixed base ω/ω_s. Use the same parameters as in Fig. 3-22: $\bar{s} = 1$, $\bar{h} = 1$, $\bar{m} = 3$, $\nu = 0.33$, $\zeta = 0.025$, and $\zeta_g = 0.05$. For the range of frequencies where $|u_o + u_g|/|u_g|$ is larger than 1, the motion of the base with soil–structure interaction is larger than without. The curve exhibits a maximum and a minimum. Verify that the maximum and minimum occur approximately at $\tilde{\omega}/\omega_s$ and at $\omega_r/\sqrt{\omega_r^2 + \omega_s^2}$, respectively. The latter expression represents the natural frequency of the system with a rigid horizontal soil spring, divided by ω_s (Eq. 3.56 with $1/\omega_h^2 = 0$).

Results:

Solving the equations of motion (Eq. 3.49) for u_o the results plotted in Fig. P3-6 follow. Those based on examining the equivalent one-degree-of-freedom system (Eqs. 3.63a and 3.55) are very similar.

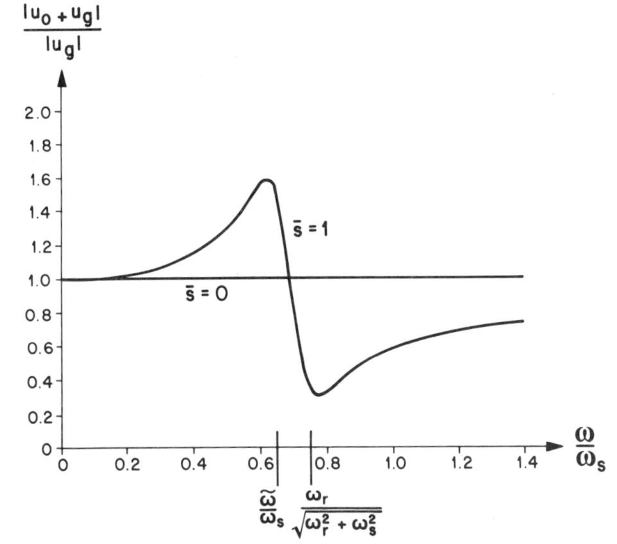

Figure P3-6 Total horizontal displacement at base as a function of excitation frequency ($\bar{h} = 1$, $\bar{m} = 3$, $\nu = 0.33$, $\zeta = 0.025$, $\zeta_g = 0.05$).

3.7. Analogously, as described in Section 3.4 for the horizontal excitation, the (even simpler) system shown in Fig. P3-7 can be used to analyze the vertical one with amplitude w_g. Two degrees of freedom, the amplitudes of the total vertical displacements of the mass w^t and of the massless base w_o^t, are introduced, whereby only one of them is dynamic. Derive the equations of motion using as unknowns the amplitudes of the relative displacements w and w_o and establish the properties of the equivalent one-degree-of-freedom system.

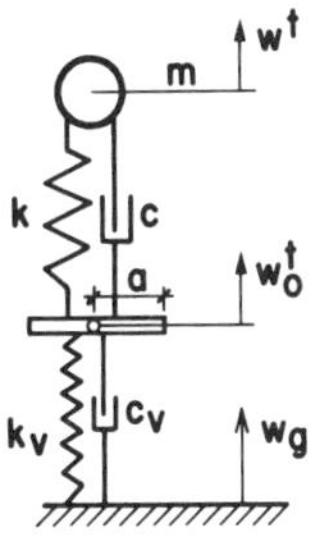

Figure P3-7 Model with one dynamic degree of freedom (vertical).

Solution:

$$w^t = w_g + w_o + w$$

$$w_o^t = w_g + w_o$$

The vertical-force amplitude of the damped soil P_v is formulated as

$$P_v = k_v w_o + c_v \dot{w}_o$$

where, as an approximation, the vertical-spring coefficient k_v is equal to that of the undamped case k_z. Defining the viscous-radiation damping ratio in the vertical direction ζ_z of the undamped soil as

$$\zeta_z = \frac{\omega c_z}{2k_z}$$

where c_z is the corresponding damping coefficient of the elastic soil and introducing the material damping ζ_g as

$$c_v = c_z + \frac{2}{\omega}\zeta_g k_z$$

leads to

$$P_v = k_z(1 + 2\zeta_z i + 2\zeta_g i)w_o$$

The following symmetric equations of motion result:

$$\begin{bmatrix} \frac{\omega_s^2}{\omega^2}(1 + 2\zeta i) - 1 & -1 \\ -1 & \frac{\omega_v^2}{\omega^2}(1 + 2\zeta_z i + 2\zeta_g i) - 1 \end{bmatrix} \begin{Bmatrix} w \\ w_o \end{Bmatrix} = \begin{Bmatrix} 1 \\ 1 \end{Bmatrix} w_g$$

where

$$\omega_v^2 = \frac{k_v}{m}$$

and ω_s and ζ are the fixed-base frequency and hysteretic-damping ratio of the structure, respectively.

The natural frequency $\tilde{\omega}$, the ratio of hysteretic damping $\tilde{\zeta}$, and the replacement excitation $\tilde{w}_g$ of the approximate equivalent one-degree-of-freedom system are equal to

$$\frac{1}{\tilde{\omega}^2} = \frac{1}{\omega_s^2} + \frac{1}{\omega_v^2}$$

$$\tilde{\zeta} = \frac{\tilde{\omega}^2}{\omega_s^2}\zeta + \left(1 - \frac{\tilde{\omega}^2}{\omega_s^2}\right)\zeta_g + \frac{\tilde{\omega}^2}{\omega_v^2}\zeta_z = \frac{\tilde{\omega}^2}{\omega_s^2}\zeta + \left(1 - \frac{\tilde{\omega}^2}{\omega_s^2}\right)(\zeta_g + \zeta_z)$$

$$\tilde{w}_g = \frac{\tilde{\omega}^2}{\omega_s^2} w_g$$

In the derivation of the properties of the equivalent one-degree-of-freedom system, it is assumed that products of the damping ratios can be neglected compared to unity. As the radiation damping is much larger in the vertical direction than for the other degrees of freedom, this assumption is questionable and introduces restrictions when using these formulas. See Problem 3.8 for an evaluation.

3.8. For the equivalent one-degree-of-freedom system for vertical excitation derived in connection with Fig. P3-7, the following dimensionless parameters can be defined:

1. The stiffness ratio of the structure and of the soil

$$\bar{s} = \frac{\omega_s a}{c_s}$$

2. The mass ratio

$$\bar{m} = \frac{m}{\rho a^3}$$

3. Poisson's ratio ν of the soil
4. The hysteretic-damping ratios of the structure ζ and of the soil ζ_g

As a crude approximation, the following (frequency-independent) formulas for the spring and damping coefficients can be used for a rigid circular basemat of radius a:

$$k_z = \frac{4Ga}{1-\nu}$$

$$c_z = \frac{4}{1-\nu}\rho c_s a^2$$

Express the equivalent properties of the one-degree-of-freedom system $\tilde{\omega}^2/\omega_s^2$ and $\tilde{\zeta}$ (see Problem 3.7) as a function of the dimensionless parameters. Plot $\tilde{\omega}/\omega_s$ and $\tilde{\zeta}$ as a function of $\bar{s}$ (in the range from 0.1 to 10) for $\bar{m} = 3$, selecting $\nu = 0.33$, $\zeta = 0.025$, and $\zeta_g = 0.05$. Also calculate the curve for $\tilde{\zeta}$, neglecting the influence of radiation damping. Compare the results with those shown in Fig. 3-20b for a small value of the slenderness ratio $\bar{h}$. Note that the radiation damping in the vertical direction is larger than that in the horizontal one. Derive a more accurate expression for $\tilde{\omega}$ and $\tilde{\zeta}$ (which is also more complicated) by equating the real and imaginary parts of the equation for w of the coupled system with those of the equivalent one-degree-of-freedom system.

Results:

$$\frac{\tilde{\omega}^2}{\omega_s^2} = \frac{1}{1 + \frac{\bar{m}\bar{s}^2}{4}(1-\nu)}$$

$$\tilde{\zeta} = \frac{\tilde{\omega}^2}{\omega_s^2}\zeta + \left(1 - \frac{\tilde{\omega}^2}{\omega_s^2}\right)\left(\zeta_g + \frac{\tilde{\omega}}{\omega_s}\frac{\bar{s}}{2}\right)$$

plotted in Fig. P3-8 as solid lines. The equation of motion for w of the coupled system equals

$$\left[1 - \frac{\omega^2}{\omega_s^2} - \frac{\omega^2}{\omega_v^2}\frac{1 + 4\zeta\zeta_z + 4\zeta\zeta_g}{1 + 4(\zeta_z + \zeta_g)^2} + \left(2\zeta + \frac{\omega^2}{\omega_v^2}\frac{-2\zeta + 2\zeta_z + 2\zeta_g}{1 + 4(\zeta_z + \zeta_g)^2}\right)i\right]w = \frac{\omega^2}{\omega_s^2}w_g$$

while that of the equivalent one-degree-of-freedom system equals

$$\left(1 - \frac{\omega^2}{\tilde{\omega}^2} + 2\tilde{\zeta}i\right)w = \frac{\omega^2}{\tilde{\omega}^2}\tilde{w}_g$$

which results in

$$\frac{1}{\tilde{\omega}^2} = \frac{1}{\omega_s^2} + \frac{1}{\omega_v^2}\frac{1 + 4\zeta(\zeta_z + \zeta_g)}{1 + 4(\zeta_z + \zeta_g)^2}$$

$$\tilde{\zeta} = \zeta + \frac{\tilde{\omega}^2}{\omega_v^2}\frac{\zeta_z + \zeta_g - \zeta}{1 + 4(\zeta_z + \zeta_g)^2}$$

These are plotted in Fig. P3-8 as dashed lines. These more accurate results should not be used in practice, as the direct solution of the two equations of motion represents a simpler approach.

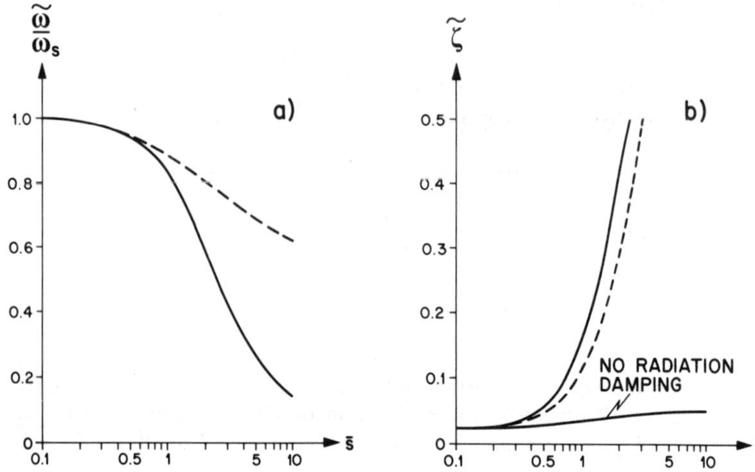

Figure P3-8 Properties of equivalent one-degree-of-freedom system (vertical) ($\bar{m} = 3$, $\nu = 0.33$, $\zeta = 0.025$, $\zeta_g = 0.05$). (a) Natural frequency; (b) damping.

3.9. For the coupled system analyzed in Problem 3.7, plot the absolute values of the vertical structural distortion (nondimensionalized with $|w_g|$), $|w|/|w_g|$ and of the displacement amplitude of the mass relative to the free field $|w + w_o|/|w_g|$ as a function of the ratio of the excitation frequency and that of the fixed-base frequency ω/ω_s. Use the spring and damping coefficients introduced in Problem 3.8. The following parameters apply: $\bar{s} = 1$, $\bar{m} = 3$, $\nu = 0.33$, $\zeta = 0.025$, and $\zeta_g = 0.05$. Compare the curves for $\bar{s} = 1$ with those for $\bar{s} = 0$, that is, not taking soil–structure interaction into account. For which range of ω/ω_s does neglecting soil–structure interaction lead to a smaller response? For the same set of parameters, plot the absolute value of the total vertical-displacement amplitude of the base $|w_o + w_g|/|w_g|$ as a function of ω/ω_s.

Results:

Solving the two equations of motion of the coupled system leads to the results plotted in Fig. P3-9.

3.10. An alternative derivation of the equivalent damping ratio $\tilde{\zeta}$ for the coupled system shown in Fig. 3-18 exists. It is based on energy considerations. For hysteretic damping, the dissipated energy (which is independent of the frequency of excitation) is proportional to the product of the damping ratio and of the strain energy. Adding the dissipated energies of the structure and of the soil in the horizontal and rocking directions, and setting this value equal to that of the equivalent one-degree-of-freedom system using the same total strain energy leads to the equation

$$\tilde{\zeta} = \frac{\zeta \dfrac{ku^2}{2} + \zeta_h \dfrac{k_h u_o^2}{2} + \zeta_r \dfrac{k_r \phi^2}{2}}{\dfrac{ku^2}{2} + \dfrac{k_h u_o^2}{2} + \dfrac{k_r \phi^2}{2}}$$

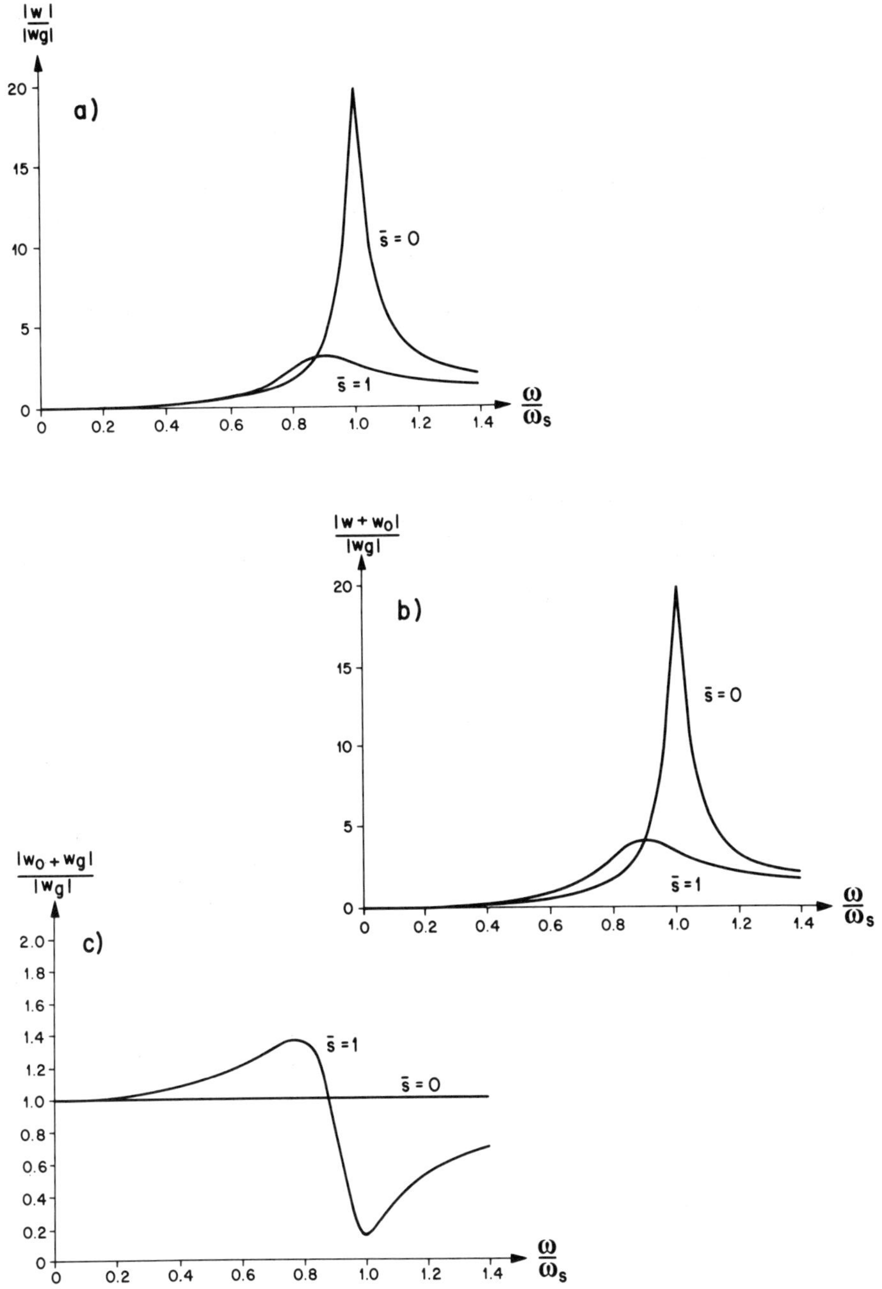

Figure P3-9 Influence of soil–structure interaction as a function of frequency of vertical excitation ($\bar{m} = 3$, $\nu = 0.33$, $\zeta = 0.025$, $\zeta_g = 0.05$). (a) Structural distortion; (b) displacement of mass relative to free field; (c) total displacement at base.

where ζ_h and ζ_r are the (hysteretic) damping ratios of the damped soil in the horizontal and rocking directions, respectively.

$$\zeta_h = \zeta_x + \zeta_g$$

$$\zeta_r = \zeta_\phi + \zeta_g$$

Neglecting the damping terms when formulating the equilibrium equations results in

$$u_o = \frac{k}{k_h} u$$

$$\phi = \frac{khu}{k_r}$$

Derive the equivalent damping ratio as specified in Eq. 3.61. Note that in this derivation, products of damping ratios compared to unity do not have to be formally neglected.

3.11. A rigid undamped structure of mass m and moment of inertia I (with respect to the center of mass, which is at a height h) rests on the surface of flexible soil (Fig. P3-11). For horizontal excitation with amplitude u_g, this system has two

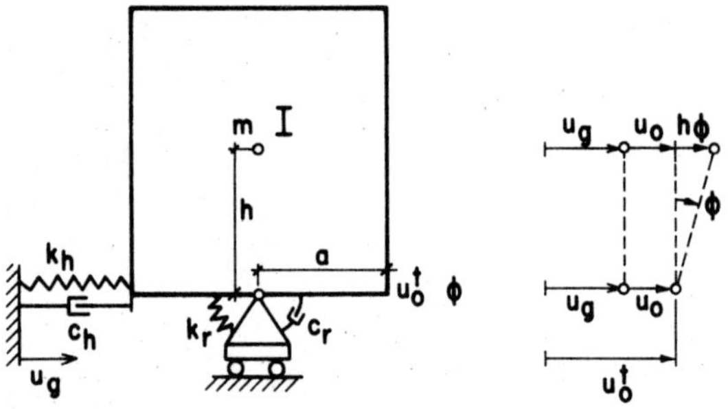

Figure P3-11 Rigid structure founded on flexible soil.

dynamic degrees of freedom, u_o and ϕ. Introducing the same approximation as discussed in connection with Fig. 3-18, the horizontal-force amplitude P_h of the damped soil is formulated as

$$P_h = k_h u_o + c_h \dot{u}_o = k_h(1 + 2\zeta_x i + 2\zeta_g i)u_o$$

where ζ_x and ζ_g are the ratios of the viscous radiation damping of the undamped soil in the horizontal direction and of the hysteretic-material damping, respectively. Analogously for the moment amplitude of the soil

$$M_r = k_r\phi + c_r\dot{\phi} = k_r(1 + 2\zeta_\phi i + 2\zeta_g i)\phi$$

applies. The two equations are the same as Eqs. 3.41 and 3.44.

(a) Derive the symmetric equations of motion for harmonic response using the unknowns u_o and $h\phi$. The coefficient matrix is a function of $\zeta_x, \zeta_\phi, \zeta_g, \omega$, and I/h^2m and of the frequencies $\omega_h(k_r = \infty)$ and of $\omega_r(k_h = \infty)$.

$$\omega_h^2 = \frac{k_h}{m}$$

$$\omega_r^2 = \frac{k_r}{I + h^2 m}$$

(b) Identify the stiffness matrix $[K]$ and mass matrix $[M]$ of the system and calculate the two natural frequencies ω_1 and ω_2 of the coupled system with the corresponding mode shapes $\{\phi_1\}$ and $\{\phi_2\}$.

(c) Under what condition can the fundamental frequency ω_1 be calculated approximately using the equivalent of Eq. 3.56?

$$\frac{1}{\omega_1^2} = \frac{1}{\omega_h^2} + \frac{1}{\omega_r^2}$$

(d) The structure–soil system does, in general, not have classical modes; that is, the transformation to modal coordinates does not decouple the damping terms. Identify the damping matrix $[C]$ and show that for $\omega = \omega_1$ the off-diagonal term of the transformed damping matrix $\{\phi_1\}^T[C]\{\phi_2\}$ does not vanish. Determine an approximate equation for the corresponding modal-damping ratio ζ_1 based on energy considerations, as in Problem 3.10.

Results:

(a)
$$\begin{bmatrix} \frac{\omega_h^2}{\omega^2}(1 + 2\zeta_x i + 2\zeta_g i) - 1 & -1 \\ -1 & \left[\frac{\omega_r^2}{\omega^2}(1 + 2\zeta_\phi i + 2\zeta_g i) - 1\right]\left(1 + \frac{I}{h^2 m}\right) \end{bmatrix} \begin{Bmatrix} u_o \\ h\phi \end{Bmatrix} = \begin{Bmatrix} 1 \\ 1 \end{Bmatrix} u_g$$

(b)
$$[K] = \begin{bmatrix} k_h & \\ & \frac{k_r}{h^2} \end{bmatrix} \qquad [M] = \begin{bmatrix} m & m \\ m & m + \frac{I}{h^2} \end{bmatrix}$$

with the unknowns u_o and $h\phi$.

$$|[K] - \omega_i^2[M]| = 0$$

$$\omega_{1,2}^2 = \tfrac{1}{2}\left(1 + \frac{h^2 m}{I}\right)(\omega_h^2 + \omega_r^2)\left[1 \mp \sqrt{1 - \frac{4}{(h^2m/I) + 1}\frac{\omega_h^2\omega_r^2}{(\omega_h^2 + \omega_r^2)^2}}\right]$$

$$([K] - \omega_i^2[M])\{\phi_i\} = 0$$

$$\{\phi_1\} = \begin{Bmatrix} u_{o1} \\ \left(\frac{\omega_h^2}{\omega_1^2} - 1\right)u_{o1} \end{Bmatrix} \qquad \{\phi_2\} = \begin{Bmatrix} u_{o2} \\ \left(\frac{\omega_h^2}{\omega_2^2} - 1\right)u_{o2} \end{Bmatrix}$$

(c) Setting the approximation

$$\omega_1^2 = \frac{\omega_h^2\omega_r^2}{\omega_h^2 + \omega_r^2}$$

equal to the equation specified above leads to

$$\frac{\omega_1^2}{(1 + h^2m/I)(\omega_h^2 + \omega_r^2)} = 0$$

which is approximately satisfied for

$$\omega_1^2 \ll \left(1 + \frac{h^2m}{I}\right)(\omega_h^2 + \omega_r^2)$$

(d) $$[C] = \frac{1}{\omega}\begin{bmatrix} k_h(2\zeta_x + 2\zeta_g) & \\ & \frac{k_r}{h^2}(2\zeta_\phi + 2\zeta_g) \end{bmatrix} = \begin{bmatrix} c_x + \frac{2\zeta_g k_h}{\omega} & \\ & \frac{1}{h^2}\left(c_\phi + \frac{2\zeta_g k_r}{\omega}\right) \end{bmatrix}$$

$$[\Phi] = \begin{bmatrix} 1 & 1 \\ \frac{\omega_h^2}{\omega_1^2} - 1 & \frac{\omega_h^2}{\omega_2^2} - 1 \end{bmatrix}$$

The off-diagonal term of $[\Phi]^T[C][\Phi]$ equals

$$c_x + 2\frac{\zeta_g k_h}{\omega} + \frac{1}{h^2}\left(\frac{\omega_h^2}{\omega_1^2} - 1\right)\left(\frac{\omega_h^2}{\omega_2^2} - 1\right)\left(c_\phi + 2\frac{\zeta_g k_r}{\omega}\right) \neq 0$$

The modal-damping ratio is determined on the basis of energy consideration:

$$\tilde{\zeta}_1 = \frac{\zeta_h(k_h u_o^2/2) + \zeta_r(k_r \phi^2/2)}{k_h u_o^2/2 + k_r \phi^2/2}$$

With

$$\zeta_h = \zeta_x + \zeta_g$$
$$\zeta_r = \zeta_\phi + \zeta_g$$

formulating equilibrium (neglecting damping terms)

$$\phi = \frac{h k_h u_o}{k_r}$$

and introducing the approximation discussed in part (c) leads to

$$\tilde{\zeta}_1 = \zeta_g + \frac{\omega_1^2}{\omega_h^2}\zeta_x + \frac{\omega_1^2}{\omega_r^2}\zeta_\phi$$

3.12. The rigid structure of density ρ_s of the shape of a triangle with a right angle (width $2b$, height $8b/3$) rests on the surface of flexible soil which is modeled as a homogeneous half-plane with shear modulus G, Poisson's ratio ν, density ρ, and material-damping ratio ζ_g. This could be regarded as a simplified model of a gravity dam supported on poor rock, whereby this two-dimensional case represents a cross section (Fig. P3-12a). As a crude approximation, the coefficients of the springs and dashpots of the undamped half-plane of unit depth are specified as

$$k_x = 0.35\pi\frac{G}{1-\nu}$$

$$c_x = 0.72\pi\rho c_s b$$

$$k_\phi = 0.5\pi\frac{Gb^2}{1-\nu}$$

$$c_\phi = 0.19\pi\frac{\rho c_s b^3}{1-0.8\nu}$$

To determine the dynamic-stiffness coefficients of the damped soil, the same approximations as discussed in Problem 3.11 are introduced.

(a) Making use of the results in Problem 3.11, formulate the symmetric equations of motion for harmonic response using the relative-displacement amplitudes u_o and $h\phi$ ($h = 8b/9$).

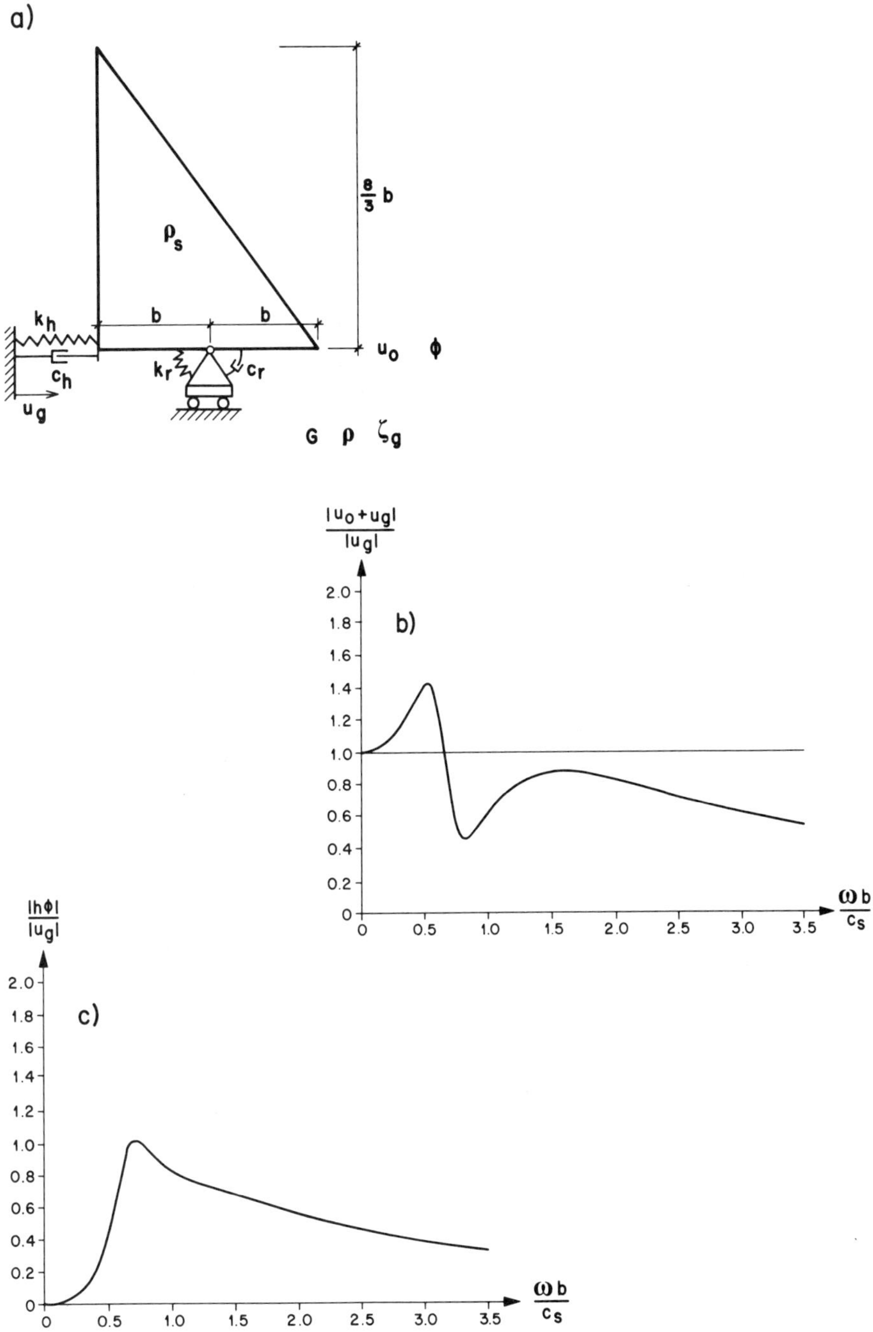

Figure P3-12 Rigid structure interacting with soil. (a) Triangular structure on half-plane; (b) total displacement at base; (c) rocking.

(b) As the structure is modeled as a rigid body, it is not possible to evaluate the effects of soil–structure interaction by comparing natural frequencies and damping ratios. Plot the ratios $|u_o + u_g|/|u_g|$ and $|h\phi|/|u_g|$ as a function of $\omega b/c_s$ for the following parameters: $\nu = 0.33$, $\rho_s/\rho = 1$, and $\zeta_g = 0.05$. Is soil–structure interaction important in this case?

Results:

(a) $$\frac{\omega_h^2}{\omega^2} = 0.41\frac{\rho}{\rho_s}\frac{c_s^2}{\omega^2 b^2}\frac{1}{1-\nu}$$

$$\frac{\omega_r^2}{\omega^2} = 0.42\frac{\rho}{\rho_s}\frac{c_s^2}{\omega^2 b^2}\frac{1}{1-\nu}$$

$$\zeta_x = 1.03\frac{\omega b}{c_s}(1-\nu)$$

$$\zeta_\phi = 0.19\frac{\omega b}{c_s}\frac{1-\nu}{1-0.8\nu}$$

(b) For the plots, see Fig. P3-12b and c.

3.13. A structure has the following properties: effective height $h = 20$ m, effective radius of basemat $a = 20$ m, effective mass $m = 50$ Gg, fixed-base frequency $\omega_s/2\pi = 4$ Hz, and material damping $\zeta = 0.05$. It is built on the surface of a half-space with Poisson's ratio $\nu = 0.33$, a mass density $\rho = 2.0$ Mg/m^3, a shear-wave velocity $c_s = 200$ m/s, and a damping ratio $\zeta_g = 0.05$. The design response spectrum of the U.S. NRC applies (Fig. 3-17b). Based on the approximations introduced in Section 3.4, calculate the ratio of the relative displacement (which is equal to the ratio of the total accelerations) of the structure founded on the soil described above to that of the one founded on rock. Use the response-spectrum method.

Results:

Dimensionless parameters: $\bar{s} = 2.5$, $\bar{h} = 1$, $\bar{m} = 3.12$

Equivalent one-degree-of-freedom system: $\dfrac{\tilde{\omega}}{\omega_s} = 0.32$, $\tilde{\zeta} = 0.17$

Response spectrum (Fig. 3-17b):

$$\omega_s/2\pi = 4 \text{ Hz}, \qquad \zeta = 0.05 \longrightarrow S_a = 2.9 \text{ g}$$

$$\tilde{\omega}/2\pi = 1.3 \text{ Hz}, \qquad \tilde{\zeta} = 0.17 \longrightarrow S_a < 1.45 \text{ g}$$

Ratio < 0.5

4

MODELING OF STRUCTURE

4.1 GENERAL CONSIDERATIONS

In contrast to the procedure often used in determining the static response, it is, in general, not permissible to perform the dynamic analysis in steps, calculating one part of the structure after the other (e.g., proceeding floorwise, starting at the top). The total dynamic system with a correct representation of the stiffness and the mass has to be modeled. This, however, does not mean that a very complicated detailed model is always necessary. It is well known that many dynamic loads excite only certain modes, mostly of low frequency. In many cases an approximate model capturing these essential dynamic properties is sufficient, also in view of the many uncertainties of the dynamic load, which are normally larger than for static ones. Selecting a crude model does not mean that the structural engineer is not familiar with sophisticated methods. Economical considerations in general also demand the use of a simple model for a dynamic analysis, as the computational effort of the latter can be an order of magnitude larger than that of a static analysis working with the same discretization. Procedures to establish such models in a direct way are discussed in Section 4.3. Alternatively, a large finite-element model can be first established and then reduced systematically by diminishing the number of dynamic degrees of freedom. Various methods to achieve this are outlined in Section 4.4. The structures and the applied loads vary widely, which makes modeling a difficult task. Fewer pitfalls are encountered in modeling the structure than the soil, as the dynamic properties of the structure are better known and the problem of representing the radiation of energy in the unbounded domain does not exist.

Lack of knowledge of the dynamic characteristics of the structure can in many cases be compensated by the choice of a more detailed model.

4.1.1 Frequency Content and Spatial Variation of Applied Loads

The applied loading influences the modeling of a structure significantly. The same structure excited by loads differing in frequency content and spatial variation has to be discretized differently. All modes with nonnegligible generalized modal loads in the range of frequency of the applied loading have to be represented. This means that impact loads with a high-frequency content demand a much finer model than the seismic excitation, for which only all modes with frequencies smaller than 30 Hz have to be included. As will be discussed in Section 4.2, an approximately axisymmetric surface structure (even with a flexible base) can be modeled very simply if excited by an earthquake arising from vertically incident waves. The complexity of the model of the same structure excited by a horizontally propagating wave for which the motions at the base of the structure differ from node to node increases by at least one order of magnitude. In both cases the frequency content of the earthquake, but not the spatial variation, is the same.

In addition to the frequency content and spatial variation of the applied loading, the type of result required plays an important role. If only global results are to be determined, the dynamic model will tend, in general, to be restricted to represent even fewer of the modes of low frequency. The shapes of the modes of high frequency with many nodes exhibit zones of positive and negative values. As the global values correspond to integrals of the mode shapes, the contribution of the higher modes will be small and can thus be neglected. For instance, the base shear of a building modeled as a vertical beam will be in equilibrium with the inertial loads. These are proportional to the product of the mass and the mode shape. The resultant of the inertial loads corresponding to the modes of high frequency will thus be small. Analogously, these modes will hardly contribute to the overturning moment at the base, another global result. In contrast, a mode of high frequency can significantly influence the local response, such as the displacement or the acceleration at a specific location (e.g., at the top of the structure) if the mode shape exhibits a large value at this point. Besides distinguishing between global and local results, it is essential to realize that modes of high frequency contribute significantly more to the maximum total accelerations than to the maximum relative displacements. As discussed in connection with the tripartite representation of a response spectrum (Section 3.3.3), the pseudo total acceleration (which is very close to the total acceleration) is equal to ω^2 times the relative displacement of a mode with frequency ω. Obviously, ω^2 is larger for modes with high frequencies than for those with low frequencies.

If the motion is expected to occur essentially in a plane, a two-dimensional model of the structure can be appropriate. This is in contrast to the modeling of the soil (see Section 7.4.2). Of course, results associated with the three-dimensionality of the structure (e.g., the torsional response) cannot be calculated in this way.

In many cases the base, which lies on the structure–soil interface, can be regarded as rigid, even if the basemat itself is not necessarily rigid, as cylindrical shells and neighboring shear walls can stiffen the base significantly. The assumption of a rigid base reduces the number of degrees of freedom of the structure–soil interface (see Sections 3.1 and 3.2) and allows a much simpler model of the structure to be used for earthquake excitation (Section 4.2).

The selected model of the structure determines the dynamic-stiffness matrices $[S_{ss}]$ and $[S_{sb}]$, $[S^s_{bb}]$ for the flexible base and $[S_{so}]$, $[S^s_{oo}]$ for the rigid base. These matrices appear in the corresponding basic equations of motion (Eqs. 3.9 and 3.15). The matrix $[S]$ of a bounded system depends on the static-stiffness matrix $[K]$, the damping ratio ζ, and the mass matrix $[M]$ (Eq. 3.1). The matrix $[K]$ depends on the (dynamic) material properties of the structure. The material-damping ratio ζ, which is actually a global representation of nonlinear effects, varies significantly with the stress level and the type of the loading.

4.1.2 Decoupling of Subsystem

The results of the crude dynamic model are in general not sufficient to design the structure directly. For instance, the stress resultant in an equivalent beam with a vertical axis representing all walls cut by a horizontal section between two adjacent floors of the dynamic model is not suitable for designing a specific portion of the wall. In most cases it will be necessary to analyze the structural element separately, using certain results of the crude model of the total structure–soil system (e.g., applying the maximum total acceleration as the loading).

Similarly, a subsystem with a small mass can be decoupled from the dynamic model. This could apply to a component fixed to the structure at one point. This procedure is illustrated in Fig. 4-1. For a dynamic load (e.g., an earthquake excitation specified at the surface of the free field denoted as u_f, or an impact load acting on the dome of the structure), the response u_c of the component with a static-stiffness matrix $[K_c]$, a mass matrix $[M_c]$, and a damping ratio ζ_c is to be determined. In the figure, the matrix symbols are deleted. In the first step, the dynamic model without the decoupled subsystem is analyzed. The acceleration or displacement time history u_s at the location where the subsystem is in reality connected to the structure is calculated. This motion will be influenced by the dynamic characteristics of the model of this first step. In the second step, this motion is applied to the support of the subsystem, resulting in its response u_c. In this method, the two systems are assumed to behave independently of one another with no feedback from the subsystem to the dynamic model of the first step. When implementing this procedure, the result of the first step, the time history of the motion at the support point of the

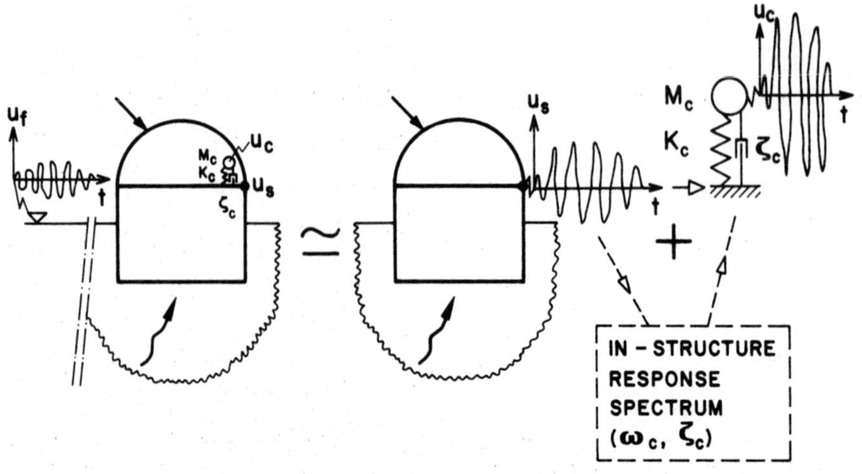

Figure 4-1 Uncoupling of subsystem with small mass.

subsystem, denoted as u_s in Fig. 4-1, is often displayed in the form of the so-called in-structure response spectrum. This represents the maximum response of a single-degree-of-freedom system characterized by its own natural frequency ω_c and damping ratio ζ_c. This concept is discussed in connection with the design response spectrum of the control motion in Section 3.3.3. Assuming the subsystem to have classical modes, its maximum response can be calculated straightforwardly with the response-spectrum method of dynamic analysis. Examples of such in-structure response spectra are contained in Section 4.3 (Fig. 4-9). Problem 4.1 addresses the uncoupling criterion.

4.2 SPATIAL VARIATION OF SEISMIC LOADS

4.2.1 Free-Field Motion

Before discussing the seismic load applied to the structure, the free-field response of the site has to be examined. It is convenient to distinguish between vertically incident waves and horizontally propagating waves. Chapter 6 contains a thorough discussion of the different types of waves. At this stage, the following simplified description illustrated in Fig. 4-2 is sufficient for an understanding of the concept. Vertically incident waves propagate vertically. The compressional P-wave for which the particle motion coincides with the direction of propagation causes the vertical component of the earthquake with amplitude u_z^f. The two shear waves SV and SH whose particle motion is perpendicular to the direction of propagation lead to the two horizontal components with amplitudes u_x^f and u_y^f. In a horizontal plane, the particle motion at a specific time is constant. The P-wave and the SV- and SH-waves propagate with the velocities c_p and c_s, respectively.

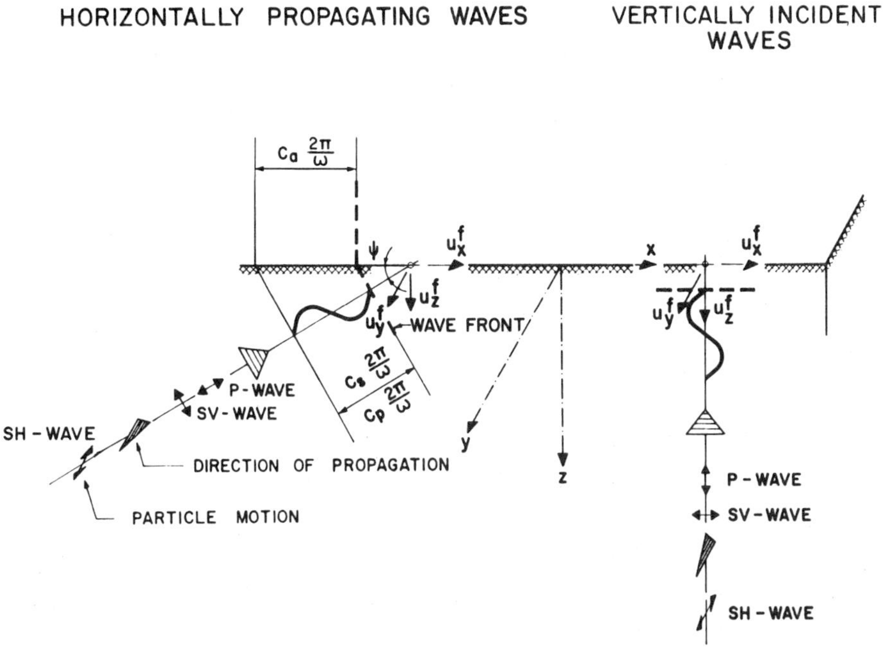

Figure 4-2 Free-field motion of horizontally propagating and vertically incident waves.

Selecting the angle of incidence ψ (measured from the horizontal x-axis in the vertical x-z plane) to be different from 90°, inclined body waves are generated. Both the P- and SV-waves will result in the horizontal and vertical free-field motions u_x^f and u_z^f. The other horizontal motion u_y^f is caused by the SH-wave. (The letters V and H denote that the particle motions take place in the vertical plane and in the horizontal direction, respectively.) As the wave front is inclined, the inclined body waves propagate horizontally across a horizontal plane in the x-direction with an apparent velocity c_a which is a function of the velocity of propagation of the body wave and of ψ. Projecting the wavelength measured along the direction of propagation onto the horizontal x-axis results in

$$c_a = \begin{cases} \dfrac{c_s}{\cos\psi} & \text{for SV- and SH-waves} \quad (4.1a) \\[2ex] \dfrac{c_p}{\cos\psi} & \text{for P-wave} \quad (4.1b) \end{cases}$$

This means that at a specific time, the particle motion at two points on the x-axis will differ. Besides inclined body waves, surface waves (so-called Rayleigh and Love waves) exist, which also propagate horizontally. These are not shown in Fig. 4-2.

It is appropriate to select the positive direction of the z-axis as pointing downward when addressing the soil, and upward when dealing with the structure. A simple transformation of coordinates is then needed when assembling the contributions of the two substructures.

4.2.2 Kinematic Motion of Surface Structure with Rigid Base

Turning to the spatial variation of the seismic loading applied to a structure, it is informative to split the total motion into components caused by the kinematic and inertial interactions. This procedure is presented in Section 3.2. A surface structure with a rigid basemat is first assumed. As discussed in connection with Fig. 3-10, the motion of kinematic interaction is equal to that of the free field for a surface structure with vertically incident waves. Since for a rigid basemat the scattered motion $\{u_o^g\}$ equals that of kinematic interaction (Eq. 3.29b), $u_o^g = u_x^f$, $v_o^g = u_y^f$, and $w_o^g = u_z^f$ apply. The six rigid-body components of $\{u_o^g\}$ are the three translations with amplitudes u_o^g, v_o^g, and w_o^g and the three rotations with amplitudes α_o^g, β_o^g, and γ_o^g, the latter being zero for this case. The displacements of kinematic interaction are constant throughout the structure for vertically incident waves, leading also to constant accelerations, which, multiplied with the mass of the structure, lead to the loads to be applied in the inertial-interaction part of the analysis (see Fig. 3-10).

Addressing the effect of horizontally propagating waves, the vertical component of the free-field motion u_z^f is examined first. The variation corresponding to u_z^f along the x-axis, which is the direction of (apparent) propagation, is shown in Fig. 4-3c. To determine $\{u_o^g\}$ accurately, Eq. 3.19 should be used, which requires the calculation of the dynamic-stiffness matrix of the soil $[S_{bb}^g]$ ($= [S_{bb}^f]$ for a surface structure). Because in this equation $[S_{bb}^f]$ and its inverse are present, the influence of the stiffness of the site on $\{u_o^g\}$ will be small. The vector $\{u_o^g\}$ can, for this discussion, be determined approximately, based on geometric considerations only (see Problems 4.2, 4.3, and 4.4). As the basemat is rigid, some "average" vertical displacement with an amplitude w_o^g and, in addition, a rocking component turning about the y-axis with an amplitude β_o^g arise. The corresponding motion of the kinematic interaction in the structure follows, using rigid-body kinematics. The amplitude β_o^g generates a horizontal displacement with an amplitude u_x^k, which increases linearly with height, and an additional vertical displacement proportional to the horizontal distance from the center of the basemat. The resulting motion is shown in the vertical x-z plane and in the horizontal x-y plane in Fig. 4-3c. The amplitude of the resulting vertical displacement is denoted as u_z^k. Analogously, the (tangential) horizontal component of the free-field motion u_y^f will result in a translational component with an amplitude v_o^g and a torsional component turning about the z-axis with an amplitude γ_o^g (Fig. 4-3b). In any horizontal plane of the structure, the motion

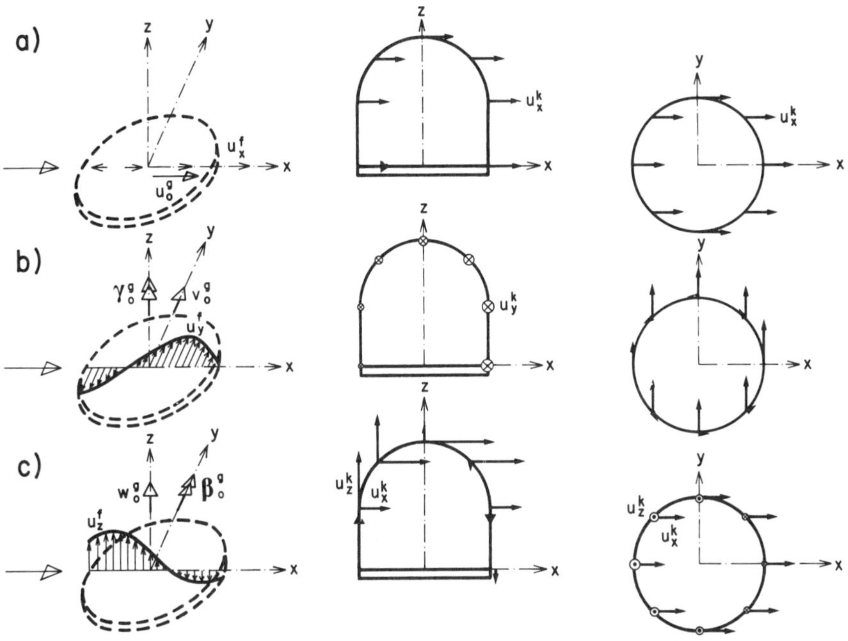

Figure 4-3 Kinematic motion of surface structure with rigid basemat for horizontally propagating wave. (a) Horizontal component coinciding with direction of propagation; (b) horizontal component perpendicular to direction of propagation; (c) vertical component.

of kinematic interaction consists of a constant displacement in the y-direction, onto which a constant tangential displacement is superimposed. Finally, the (radial) horizontal component of the free-field motion u_x^f leads only to a translational component with an amplitude u_o^g (Fig. 4-3a). The corresponding displacement amplitudes u_x^k are constant throughout the structure. Because of the averaging effect resulting in self-canceling, u_o^g, v_o^g, and w_o^g for horizontally propagating waves are smaller than the corresponding values, assuming the same wave as vertically incident.

The resulting kinematic motion of the rigid base as discussed above is only approximate. The important components of motion are properly represented. A rigorous analysis based on Eq. 3.19 would show that small additional terms can arise, assuming so-called welded contact (see Sections 7.3 and 7.4): For instance, for the free-field motion u_y^f, an additional rocking component with an amplitude α_o^g arises. These values can, however, be neglected. Even if this rigorous formulation is used, the kinematic motion of the rigid base will consist of six components at the most. The corresponding kinematic motion (displacements and accelerations) throughout the structure follows from rigid-body kinematics.

4.2.3 Kinematic Motion of Embedded Structure with Rigid Base

The embedded structure with a rigid base (basemat and side walls) is examined next. In Fig. 4-4 the kinematic motion resulting from the horizontal component of the free-field motion u_x^f assumed to arise from a vertically incident wave is shown. As u_x^f varies along the vertical wall of the embedment, an additional rocking component with an amplitude β_o^g is generated, leading to a linear variation of u_x^k with height and to a vertical component with an amplitude u_z^k. By the same reasoning, the horizontal component of the free-field motion u_y^f from a vertically incident wave leads to v_o^g and α_o^g. The amplitude u_z^f results in w_o^g and no additional rotational term.

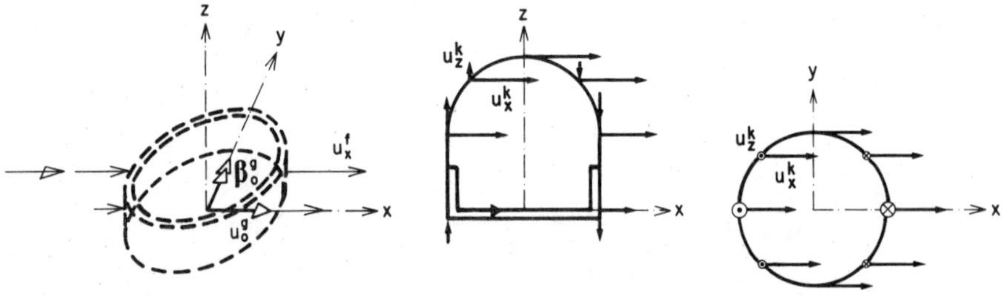

Figure 4-4 Kinematic motion of embedded structure with rigid base for vertically incident waves.

Using the results discussed in connection with Figs. 4-3 and 4-4, the response of the embedded structure with a rigid base for a horizontally propagating wave can be determined. For instance, the component of the free-field motion u_y^f propagating horizontally in the x-direction results in the translational component with an amplitude v_o^g, in the torsional component with an amplitude γ_o^g (Fig. 4-3), and in the rocking component with an amplitude α_o^g.

4.2.4 Kinematic Motion of Structure with Flexible Base

Examining the influence of a flexible basemat, the kinematic motion throughout such a structure is equal to that of the free field at the surface only if the structure is founded on the surface and only for vertically incident waves (Fig. 3-10). For an embedded structure or for horizontally propagating waves, the seismic loading case for a flexible base is complex. This will also affect the modeling. As discussed in Section 3.2, there is no advantage in splitting the total motion into components caused by kinematic and inertial interaction in this case.

4.2.5 Axisymmetric Structure

The spatial variation of the kinematic motion in an axisymmetric structure (e.g., a solid or a shell) is addressed next (Fig. 4-5). Only the case where the kinematic motion of the base can be described by the six components u_o^g, v_o^g, w_o^g,

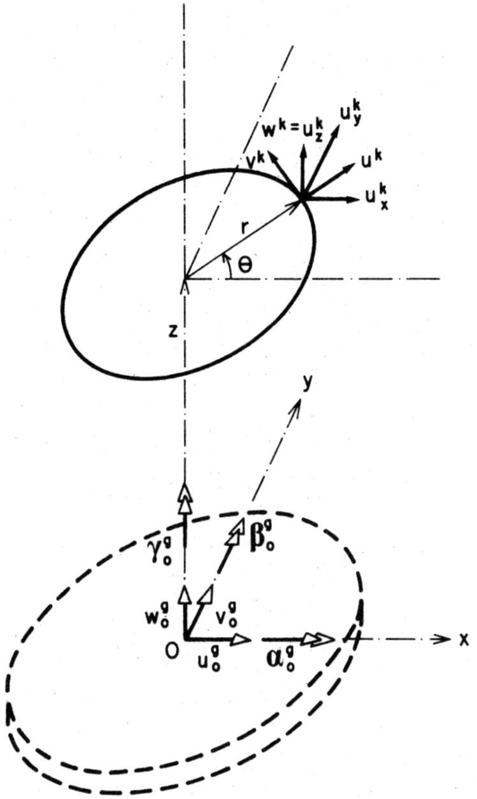

Figure 4-5 Spatial variation of kinematic motion in axisymmetric structure.

α_o^g, β_o^g, and γ_o^g at the center of the base O is examined. This applies, as discussed above, to a structure with a rigid base even for embedment and even for horizontally propagating waves and to a surface structure with a flexible basemat for vertically incident waves. For this latter case, the rotational components vanish. A horizontal section at the height z is shown in Fig. 4-5. In a point with the coordinates x, y, and z the components of the kinematic motion u_x^k, u_y^k, and u_z^k are

$$u_x^k = u_o^g + z\beta_o^g - y\gamma_o^g \tag{4.2a}$$

$$u_y^k = v_o^g - z\alpha_o^g + x\gamma_o^g \tag{4.2b}$$

$$u_z^k = w_o^g + y\alpha_o^g - x\beta_o^g \tag{4.2c}$$

Introducing cylindrical coordinates with the radius r and the circumferential angle θ,

$$x = r\cos\theta \tag{4.3a}$$

$$y = r\sin\theta \tag{4.3b}$$

and denoting the amplitudes of the components in this coordinate system as u^k and v^k leads to

$$u^k = \cos\theta\, u_x^k + \sin\theta\, u_y^k \tag{4.4a}$$

$$v^k = -\sin\theta\, u_x^k + \cos\theta\, u_y^k \tag{4.4b}$$

Substituting Eqs. 4.2 and 4.3 in Eq. 4.4 results in

$$u^k = \cos\theta\, u_o^g + \sin\theta\, v_o^g - \sin\theta\, z\alpha_o^g + \cos\theta\, z\beta_o^g \tag{4.5a}$$

$$v^k = -\sin\theta\, u_o^g + \cos\theta\, v_o^g - \cos\theta\, z\alpha_o^g - \sin\theta\, z\beta_o^g + r\gamma_o^g \tag{4.5b}$$

$$w^k = w_o^g + r\sin\theta\, \alpha_o^g - r\cos\theta\, \beta_o^g \tag{4.5c}$$

Circumferentially, the kinematic motion varies only with the zeroth and first harmonics of a Fourier series in θ. With respect to the plane $\theta = 0°$, the terms in u_o^g and β_o^g vary as the first symmetric harmonic, v_o^g and α_o^g as the first antimetric harmonic. The terms in w_o^g and γ_o^g vary as the symmetric and antimetric parts, respectively, of the zeroth harmonic. The inertial loads acting in the inertial-interaction part of the analysis will vary analogously, as they are calculated as the product of the mass and of the acceleration of the kinematic motion.

A finite-element discretization of the axisymmetric structure for inertial interaction (Eq. 3.32) can also be based on a Fourier expansion of the displacements in the circumferential direction.

$$u^i = \sum_n u_n^s \cos n\theta + \sum_n u_n^a \sin n\theta \tag{4.6a}$$

$$v^i = -\sum_n v_n^s \sin n\theta + \sum_n v_n^a \cos n\theta \tag{4.6b}$$

$$w^i = \sum_n w_n^s \cos n\theta + \sum_n w_n^a \sin n\theta \tag{4.6c}$$

For instance, u_n^s represents the nth amplitude of the radial displacement and is a function of r and z. The superscripts s and a denote the symmetric and antimetric parts, respectively, and i is used to indicate inertial interaction. As the trigonometric functions are orthogonal,

$$\int_0^{2\pi} \cos j\theta \cos l\theta\, d\theta = 0 \qquad \text{for } j \neq l \tag{4.7a}$$

$$\int_0^{2\pi} \cos j\theta \sin l\theta\, d\theta = 0 \tag{4.7b}$$

the dynamic-stiffness matrices for all harmonics of the symmetric and antimetric cases are independent from one another. The discretized loading applied in the finite-element model is proportional to the product of the kinematic and inertial motions in the same direction, integrated around the circumference $\left(\int_0^{2\pi} u^k u^i\, d\theta,\right.$ etc.). Because of Eq. 4.7 the loading cases also decouple. Thus the lth symmetric and antimetric loading terms only excite the corresponding lth displacement term.

This means that in the inertial-interaction analysis, only the zeroth and first harmonics are excited for the cases studied. The same applies to the total motion. For any axisymmetric surface structure, only the first harmonic (symmetric to the earthquake component) is nonzero for the horizontal component, and all harmonics, with the exception of the zeroth symmetric one, vanish for the vertical component, assuming vertically incident waves. For instance, for the case shown in Fig. 4-3b, only the zeroth antimetric harmonic and the first antimetric one are excited. The same applies if the structure with a rigid base is embedded.

In general, all three components of a specific harmonic are excited, even if the applied load acts only in one or two directions. For instance, w_o^g (arising from vertically incident w_z^f) leads to only a vertical loading w^k (Eq. 4.5) of the zeroth symmetric harmonic, and excites not only w_o^s, but also u_o^s and v_o^s. The amplitude u_o^g (resulting from u_x^f) leads to u^k and v^k, which give rise to nonzero u_1^s, v_1^s, and w_1^s.

The fact that for the earthquake excitations discussed above, only the zeroth and first harmonics are excited for axisymmetric structures is of paramount importance when modeling the structure. For many cases, beams with a corresponding annular section can be used as an approximation. In the theory of strength of materials, plane cross sections are postulated to remain plane. This means that, for example, for a beam with a vertical axis, the out-of-plane displacement u_z will vary as $\cos\theta$ around the circumference for an applied bending moment acting along the y-axis, corresponding to the term in w_1^s (Eq. 4.6c). For the section to translate in-plane as a rigid body, the displacement u_x,

$$u_x = u\cos\theta - v\sin\theta = u_1^s\cos^2\theta + v_1^s\sin^2\theta \tag{4.8}$$

has to be independent of θ. This can only be the case if $u_1^s = v_1^s$. Using beam theory, a constraint is thus introduced into the analysis which is not present in the theory of elasticity applied to axisymmetric solids and shells. Additional approximations thus apply when using beams. An applied normal force will lead to only an out-of-plane displacement u_z ($= w_o^s$). The corresponding in-plane displacements u_o^s and v_o^s are zero. Similarly, in a beam, an applied torsional moment will only result in v_o^a, setting $u_o^a = w_o^a = 0$.

4.2.6 Example

For the sake of illustration, the outer containment of a reactor building of a nuclear-power plant is examined. This so-called shield building of reinforced concrete built in on the level of the basemat (Fig. 4-6) consists of a cylindrical and a spherical part with a thickness of 1.20 m. The other dimensions are specified in Fig. 4-6. Two dynamic models are used for comparison: a shell model using 36 curved higher-order isoparametric frusta in the meridional direction with a Fourier series (zeroth and first harmonics) in the circumferential direction, and a beam model with 52 nodes. For the seismic loading case, a much coarser model with fewer nodes could have been selected (see Section 4.3.4). This discretization is also used for the calculation of impact loads, which excite much higher frequencies (see Section 4.3.2). Six cross-sectional values are determined for the beam. Consistent mass and mass moment of inertia are introduced in the beam model.

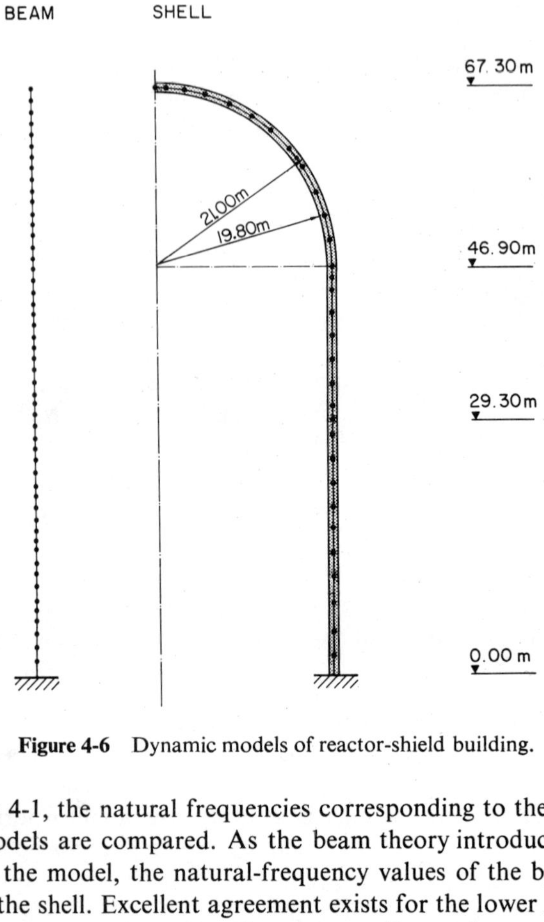

Figure 4-6 Dynamic models of reactor-shield building.

In Table 4-1, the natural frequencies corresponding to the first harmonic of the two models are compared. As the beam theory introduces a constraint, which stiffens the model, the natural-frequency values of the beam are higher than those of the shell. Excellent agreement exists for the lower modes. For the higher modes, discrepancies arise. The same applies to the mode shapes, as is shown in Fig. 4-7, where the elevations of the second and fifth are plotted. The lateral deflection of the beam is, in general, smaller than the radial displacement with the amplitude u_1^s of the shell (corresponds to the lateral displacement at $\theta = 0°$) and larger than the circumferential displacement with the amplitude v_1^s (corresponds to the lateral displacement of the shell at 90°). As $u_1^s \neq v_1^s$, a

TABLE 4-1 Horizontal Natural Frequencies (Hz) (First Harmonic)

	Mode Number				
Model	1	2	3	4	5
Beam	4.41	13.46	25.22	28.74	40.41
Shell	4.38	13.31	21.64	22.38	25.06

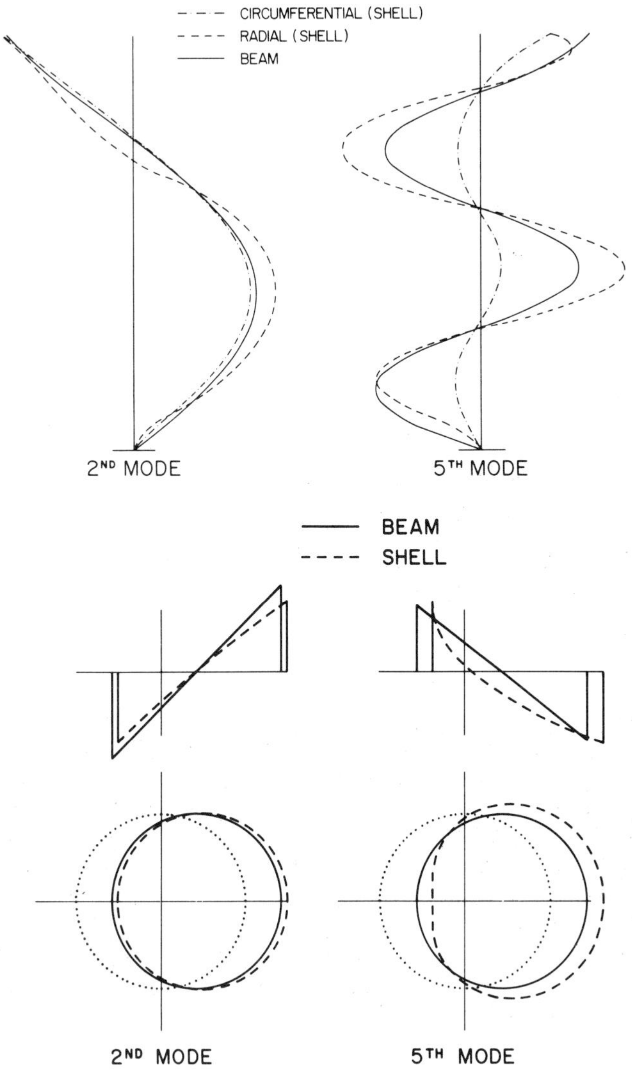

Figure 4-7 Horizontal mode shapes (Ref. [2]). (a) Elevation of mode shapes; (b) vertical and horizontal projections at level 29.30 m.

distortion of the cross section takes place, in contrast to the beam theory. This is demonstrated in Fig. 4-7b, where the horizontal and vertical projections at the indicated level are shown. As expected, the deviation from the results of the beam for the fifth mode is larger than for the second. For a horizontal earthquake associated with vertically propagating waves, the two models will lead to very similar results, as the generalized modal loads (participation factors) of the lower modes dominate. Even the in-structure response spectra, calculated, for example, at 0 and 90° around the circumference at this level of the shield building, can hardly be distinguished from that of the beam model. Any stiff floors will tend to force the shell model to behave more as a beam. The rigid basemat and, to a lesser extent, the spherical part have this effect in the shell model of the shield building.

In Table 4-2 the natural frequencies of the zeroth harmonic are compared.

TABLE 4-2 Natural Frequencies (Hz) (Zeroth Harmonic)

	Vertical		Torsional		
	Mode Number		Mode Number		
Model	1	2	1	2	3
Beam	13.21	39.63	9.65	28.56	45.87
Shell	12.72	23.10	9.65	28.28	44.15

The agreement is excellent for all indicated modes for the antimetric term, which corresponds to the torsional modes. For the vertical modes (symmetric term), the second mode already shows large discrepancies. This is caused by the "breathing" mode of the shell model (u_o^s, v_o^s) in the dome which leads to significant local bending behavior. The latter cannot be represented by a beam.

When modeling an axisymmetric structure as a beam, it is important to make use of all features of beam theory. For instance, the shield building is obviously not a slender structure. It is thus imperative that when calculating the static-stiffness matrix of a beam element, the contribution of the shear force to the strain energy be taken into account. Neglecting the contribution of the bending moment would be just as wrong, as shown in Table 4-3. It is interesting to note that for the higher modes with many nodes, the strain energy of the shear force becomes dominant. The mass moment of inertia in the beam model is of secondary importance. Neglecting its influence results in somewhat larger frequencies (results not shown).

TABLE 4-3 Horizontal Natural Frequencies of Beam (Hz)

	Mode Number				
Strain Energy	1	2	3	4	5
Moment only	5.99	28.07	58.45	58.45	85.54
Shear force only	6.05	18.18	29.86	29.86	41.48
Moment and shear force	4.41	13.46	25.22	28.74	40.41

4.3 DIRECT DISCRETIZATION

4.3.1 Finite-Element Model

In any discretization, the dynamic degrees of freedom in a structure have to be able to represent all significant inertial loads. To determine the dynamic-stiffness matrix, the displacements throughout the structure have to be expressed as a function of these degrees of freedom. The finite-element concept represents one approach to achieve this. It is so well established that it does not have to be treated in any detail. Each part of the structure with the same structural behavior (e.g., plate, shell, beam, solid) is divided into elements, varying the shapes, the dimensions, and the material properties of the elements, if necessary, but using as many identical elements as possible. The generalized displacements at the nodes located on the boundaries of the elements are the degrees of freedom. Their total number can be increased straightforwardly by selecting more elements. Interpolation functions, so-called shape functions, have been developed which express the displacement field of an element in terms of the nodal values of this element. Continuity across element boundaries is satisfied. The dynamic-stiffness matrix of an element is easily established, being the same for identical elements. Assembling the contributions of all elements results in the banded dynamic-stiffness matrix of the dynamic model, which allows efficient solution procedures to be applied.

In some instances the finite-element model of the total structure used for the static analysis can also be applied with no or only small modifications (e.g., by adding mass of nonstructural parts) for the dynamic one. This can be the case if the structure is of simple geometry (which does not mean that the structural behavior is necessarily easy to determine), or if beam elements are used. The number of degrees of freedom of the model using sophisticated elements, if appropriate, is relatively small. A few examples are doubly curved arch dams modeled with higher-order isoparametric elements, certain shell and plate structures, bridges modeled with three-dimensional curved beams, chimneys, and certain high-rise buildings. In other instances, the static finite-element model of a complex structure is so detailed that it would be uneconomical to use it in a dynamic analysis. To calculate results of acceptable accuracy, a coarse dynamic model is sufficient, which can be established either by reducing the degrees of freedom (Section 4.4) or by direct discretization. Many examples are contained in this section. Finally, either no model of the total structure is established for the static loading case, or it cannot capture the essential dynamic response. A detailed dynamic model then has to be created. The next example is such a case. In all circumstances, the structural behavior determined from the static analysis should be used when establishing a dynamic model. These results are especially valuable if the static loads act similarly to the applied dynamic loads (including the inertial ones). Calculating such additional loading cases in the static analysis can be advisable.

Some principles of the direct discretization are illustrated in the following. Real-world structures for which dynamic analyses have to be performed are used as far as possible, starting in this text with elaborate models and working toward approximate ones used for preliminary design.

4.3.2 Models for Impact Load

At first, the combined reactor-auxiliary and fuel-handling building, which form one structure of a nuclear-power plant, is analyzed for an impact force. The two buildings share the same basemat and surround the reactor building without any connections to it. The outer walls and some of the floors are also common to the two buildings. The box-type structure, which is assumed to be built in at its base, is shown with its most important dimensions in Fig. 4-8. These safety-relevant structures and the equipment located within have to be designed to resist the postulated crash of a Boeing 707-320 impacting at a velocity of 370 km/h. The force–time diagram for an impact on a rigid target is shown in Fig. 3-15. An impact area of 37 m^2 is assumed. Because of the content of high frequencies present in the force–time curve of the airplane crash and the small area of loading, high frequencies up to 100 Hz are excited. The modeling of the structure thus has to be rather detailed. The local behavior must be adequately represented, and all

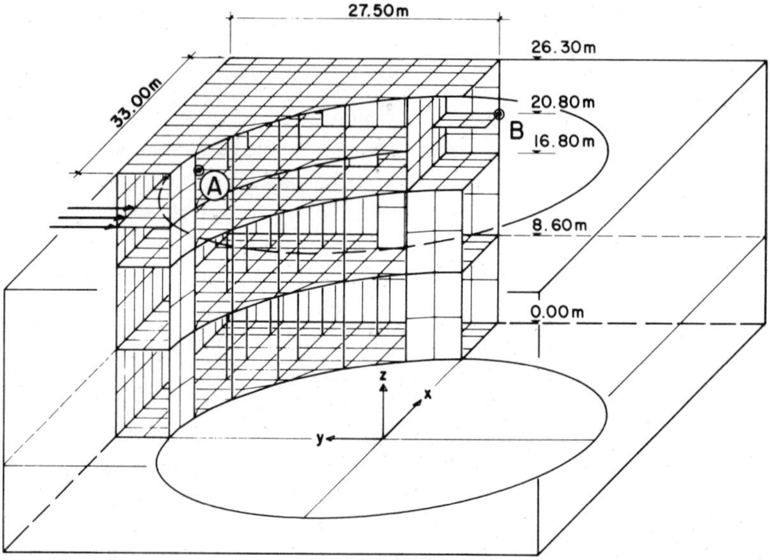

Figure 4-8 Dynamic model of reactor-auxiliary and fuel-handling buildings for impact analysis (Ref. [2]).

modes up to at least 100 Hz included. By assuming symmetry of the structure, only one-fourth has to be modeled (Fig. 4-8). For the horizontal impact force acting on the axis of symmetry as shown in Fig. 4-8, only two sets of boundary conditions have to be processed. A combination of plate elements with high-order shape functions (in bending and in stretching) for the walls and the floors and of three-dimensional beam elements for the columns is used, resulting in thousands of dynamic degrees of freedom for each set of boundary conditions. Material damping of 0.07 times the critical is selected. To design components located at points *A* and *B*, in-structure response spectra, as discussed in Section 4.1.2, are calculated. To be able to do this, the acceleration time histories in these two points have to be determined. Both points are located on the floor that is situated on the same level as the center of impact. The in-structure response spectra in the two horizontal directions are shown for 1% damping of the component in Fig. 4-9. For instance, a component with a fundamental frequency of 15 Hz located at point *A* will experience the following maximum response parallel to the impact force in the *y*-direction: a total acceleration of 4 g and a displacement relative to the floor of 5 mm. As is characteristic for impact loads, stiff components are subjected to high accelerations and small relative displacements. Moving from point *A* to point *B*, the response parallel to the direction of

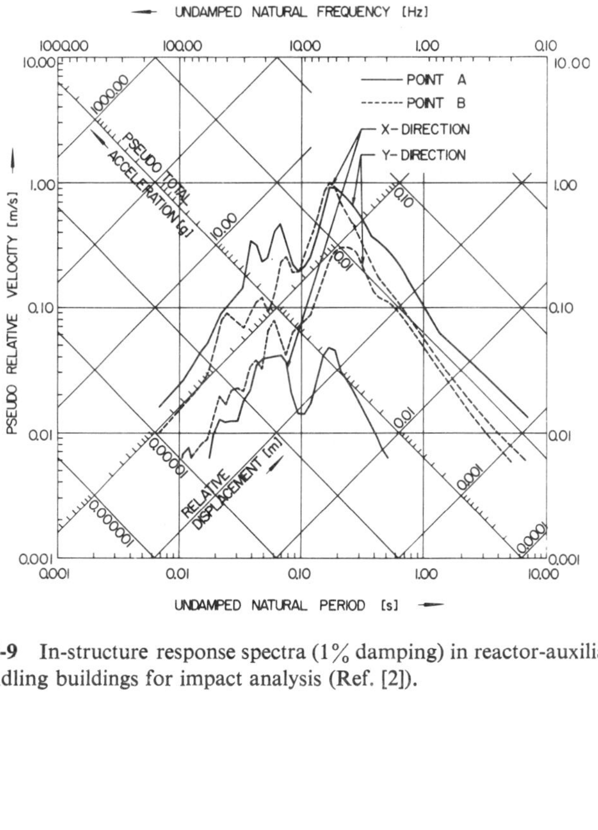

Figure 4-9 In-structure response spectra (1% damping) in reactor-auxiliary and fuel-handling buildings for impact analysis (Ref. [2]).

impact diminishes as expected, but increases in the other direction. Obviously, this effect could not be achieved using the vertical-beam model selected for the seismic excitation (Fig. 4-16).

Modeling axisymmetric structures by selecting a finite-element discretization in the meridional direction with a Fourier expansion in the circumferential direction is efficient not only for certain seismic loads, as discussed in Section 4.2, but also for impact loads. In this case, higher harmonics than the first must, however, be included. Each harmonic can be analyzed independently from the others. As an example, the same reactor-shield building whose dynamic-shell model is already treated in connection with Fig. 4-6 is analyzed, but subjected to the airplane crash (Fig. 4-10). The impact load versus time is specified in

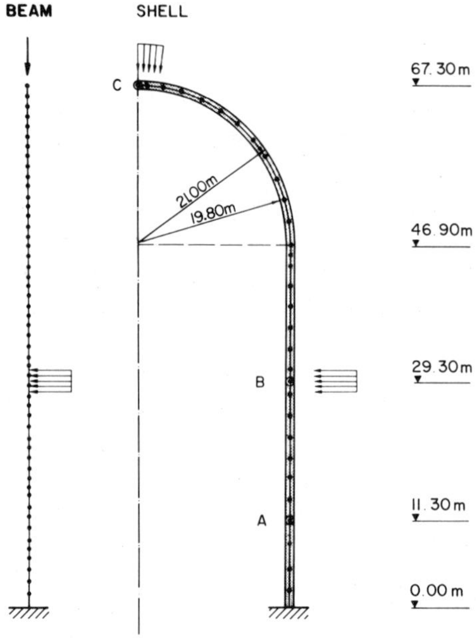

Figure 4-10 Dynamic model for reactor-shield building and impact locations (Ref. [2]).

Fig. 3-15. The damping ratio $\zeta = 0.07$ is selected. Two impact locations are examined. In the shell model, the load acting on the impact area of 3.10 m in the meridional times 12.00m in the circumferential directions for horizontal crash at point B is decomposed circumferentially into 16 Fourier terms. For a vertical crash at point C, a circular impact area of radius 3.44 m is selected, and hence only the zeroth harmonic is excited. For each harmonic, 333 dynamic degrees of freedom result. This allows all modes with natural frequencies up to approximately 200 Hz to be included in the shell model. The beam model of Fig. 4-10 is used for comparison.

The different structural behavior of the two models can best be illustrated by examining the in-structure response spectra. In Fig. 4-11, the horizontal-

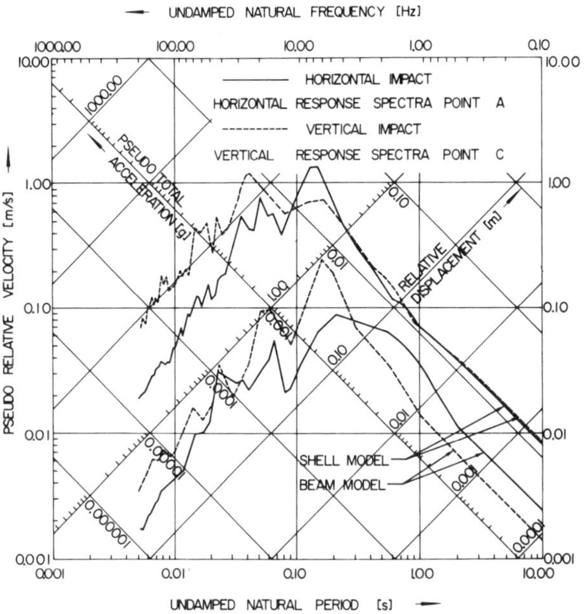

Figure 4-11 In-structure response spectra (1% damping) in reactor-shield building, beam as opposed to shell model (Ref. [2]).

response spectra at point *A* for horizontal impact and the vertical-response spectra at point *C* for vertical impact are compared for the two models. The beam solution underestimates the accelerations by one order of magnitude. Although the beam model can easily represent modes of up to 200 Hz, the assumption of no in-plane deformation turns out to be incompatible with the actual behavior of a loaded shell. This is further demonstrated in Fig. 4-12, which shows horizontal-response spectra calculated at selected points around the circumference at level *B* of the shell model. The aircraft impacts horizontally on the zero-degree meridian. It is worth mentioning that, for high frequencies, the response on the meridian opposite to that of impact (180° radial) is larger than the excitation on the meridian at 90° in the same direction (90° circumferential). The latter is, for a certain range of frequencies, even smaller than the response perpendicular to it (90° radial). These results emphasize the importance of the shell modes of higher harmonics. The results of the beam model, which are shown for comparison, again underestimate substantially the response in the frequency range of interest. The beam model fails so dramatically in representing the actual behavior of the shell structure because, first, it cannot simulate higher harmonics (from the second upward) and second, the assumption of the beam theory even constrains the first and zeroth harmonics. This is discussed in Section 4.2.5 (Fig. 4-7).

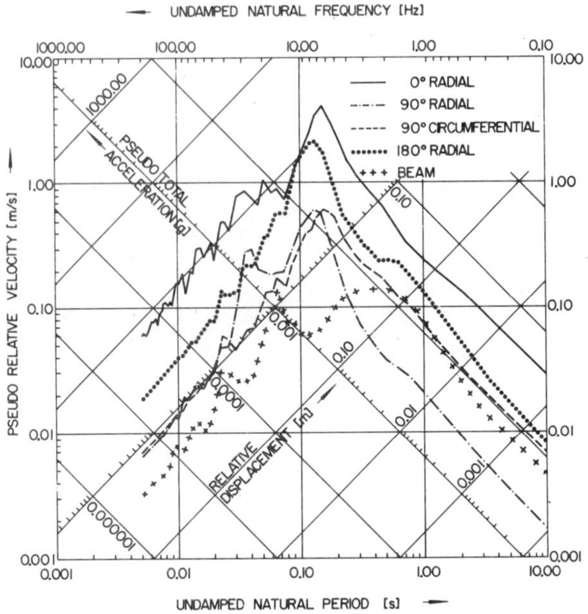

Figure 4-12 In-structure response spectra (1% damping) in reactor-shield building at level B, variation in points around circumference (Ref. [2]).

4.3.3 Hyperbolic Cooling Tower for Seismic Load

Turning to the seismic loading of axisymmetric structures, the hyperbolic cooling tower having variable thickness and an eccentric edge beam at the top is examined (Fig. 4-13). It is stiffened at the lower edge by tapering. Thirty-six pairs of columns form a V and rest on separate foundations without any connecting edge beam. This surface structure thus possesses a flexible base. As the diameter at the base is quite large, horizontally propagating wave effects (examined qualitatively in Section 4.2.1) can be important. For this loading case, the free-field motion at the separate foundations will be different. As discussed in Section 4.2, modes corresponding to harmonics higher than the first will be excited. Turning to the dynamic model, the axisymmetric shell is discretized with 30 higher-order finite-element frusta with an isoparametric expansion in the meridional and a Fourier expansion in the circumferential direction. (This quite large number is dictated by the loading-case wind, for which results are required in many points to be able to design the shell.) The stiffening-ring beam and the supporting columns are synthesized into a dynamic-stiffness matrix compatible with the axisymmetric shell element. All harmonics from the zeroth to the fifth are taken into account. A total of 167 modes (with nonnegligible generalized loads) which span the frequency range from 0.7 to 70 Hz are incorporated into the dynamic analysis. For vertically propagating waves, only the first and

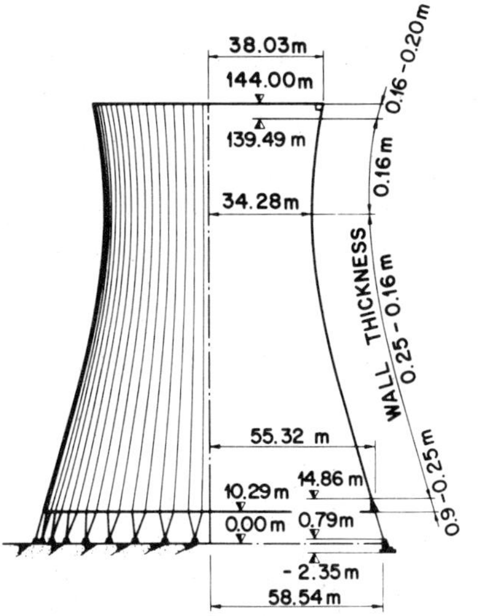

Figure 4-13 Hyperbolic cooling tower (Ref. [3]).

zeroth symmetric harmonics are excited for the horizontal and vertical components, respectively, of the earthquake. The first mode shape, with a natural frequency of 1.96 Hz of the first harmonic, is plotted along the meridional in Fig. 4-14. The amplitudes u_1^s, v_1^s, and w_1^s represent the radial, tangential, and vertical displacements of the first symmetric harmonic (see Eq. 4.6). Clearly visible is the large relative displacement between the upper and lower ends of the columns. This global shear distortion of the total tower leads to large seismic forces in the columns. Hence, for increasing seismic loads, this part of the tower is the first whose design is governed by the loading case of earthquake instead of by wind. The second mode, with a frequency = 2.93 Hz, of the first harmonic contributes less than 1% to the column forces and can thus be neglected. For a horizontal earthquake associated with vertically propagating waves, the tower can hence be adequately analyzed using only the first symmetric mode of the first harmonic, which leads to only one unknown (the amplitude of the mode shape shown in Fig. 4-14). While for the analysis of horizontally propagating waves, a standard finite-element discretiza-

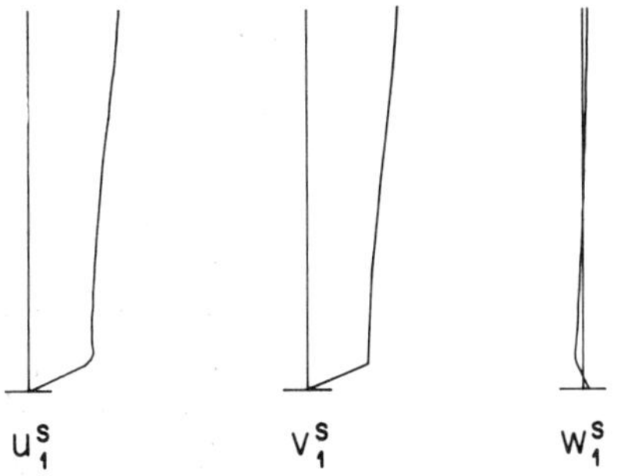

Figure 4-14 Shape of first mode of first symmetric harmonic (Ref. [3]).

tion covering the whole surface of the shell (with nodes also in the circumferential direction) can be used, this model would be very uneconomical for the calculation of vertically incident waves. In this type of model the mode with the overwhelming generalized load is not easily identified and separated from the others before the response analysis starts. Instead of using thin shells to model the tower, the discretization can also be based on beams (but only for vertically incident waves). The latter will approximate the zeroth and first harmonics extremely well. Only a few nodes along the vertical axis of the tower are needed, the number depending on how accurately the variation of the cross-sectional values of the beam elements lying between the nodes can be taken into account.

As developed in depth in Section 4.2, the simple spatial variation of the seismic inertial loading acting on an axisymmetric structure allows the latter to be modeled as a beam in the following cases: for a surface structure with a flexible basemat excited by vertically incident waves, and for all wave patterns for a structure with a rigid base (even embedded). This rule is also applied to structures which are not strictly axisymmetric. In practice, beam models are selected even in quite general cases. The number of nodes where the mass is concentrated is chosen as surprisingly small, their location coinciding with floors. Between two adjacent nodes, the most general type of a three-dimensional beam element is applied: The centers of gravity, of shear (= of twist), and of mass do not have to coincide and the corresponding principal axes need not be parallel. The mass properties are quite simple to calculate and the dynamic response is relatively insensitive to this modeling parameter. The real art lies in calculating the stiffness properties (i.e., the cross-sectional values). Concepts of folded-plate theory can sometimes be applied. At the nodes, eccentricities of the different vertical axes can arise, reflecting that stories will vary in plan.

4.3.4 Nuclear Structures for Seismic Load

As examples, the dynamic models of the reactor building and of the combined reactor auxiliary and fuel-handling structure are sketched. These models can be used for the final seismic analysis, which forms the basis for the design of the structures and for the calculation of the in-structure response spectra applicable to the components. The basemat, with a thickness of 4.5 m of the reactor building (Fig. 4-15), is assumed to be rigid. From its center four independent beams whose axes coincide are used to model the shield building, the free-standing steel containment, the so-called drywell, and the pressure vessel with its pedestal. Each beam has only three nodes (besides that at the basemat), resulting in a total of 13 nodes with 78 dynamic degrees of freedom in a three-dimensional model. The combined reactor auxiliary and fuel-handling structure (already introduced in Fig. 4-8), being a box-type structure, is modeled as a single vertical beam with 14 nodes leading to 84 dynamic degrees of freedom (Fig. 4-16). As the structure is not really symmetric over the total height, the ends of the beam elements are connected to the nodes, with horizontal eccentric-

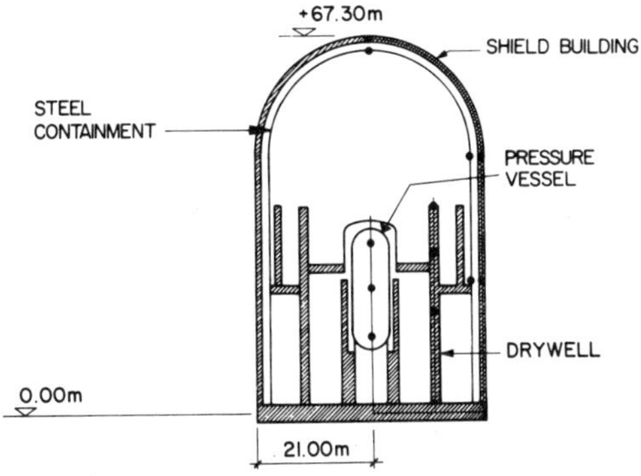

Figure 4-15 Dynamic model of reactor building for seismic analysis.

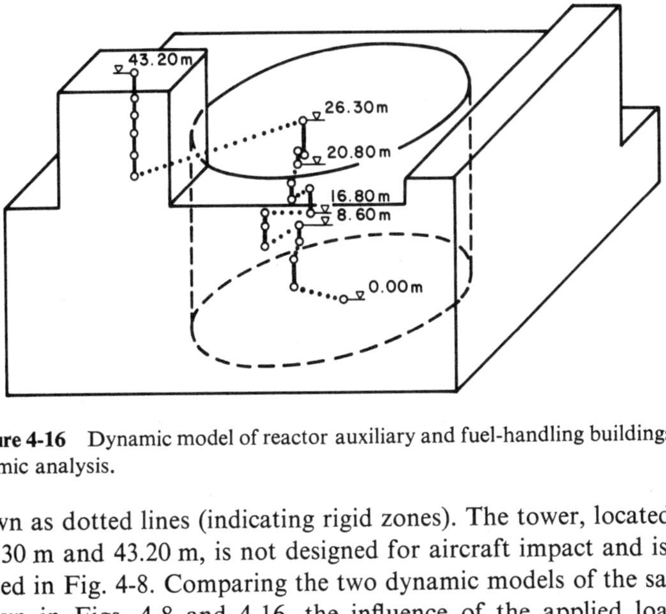

Figure 4-16 Dynamic model of reactor auxiliary and fuel-handling buildings for seismic analysis.

ities shown as dotted lines (indicating rigid zones). The tower, located between levels 26.30 m and 43.20 m, is not designed for aircraft impact and is thus not represented in Fig. 4-8. Comparing the two dynamic models of the same structure shown in Figs. 4-8 and 4-16, the influence of the applied load on the discretization becomes evident.

Modeling a structure as a vertical beam does not mean that the basemat has to be assumed to be rigid. For an axisymmetric structure, annular finite elements can be introduced to model the flexibility of the basemat. As an example, the axisymmetric reactor building is investigated again (Fig. 4-17). The dynamic model of the superstructure is the same as described above (Fig. 4-15). The flexible basemat is discretized with 11 annular finite elements in bending (left-hand side of Fig. 4-17). No mass is associated with the six circular nodes shown as thin lines. Their static degrees of freedom are eliminated. These nodes can thus be disregarded for the dynamic model. The beam representing the steel

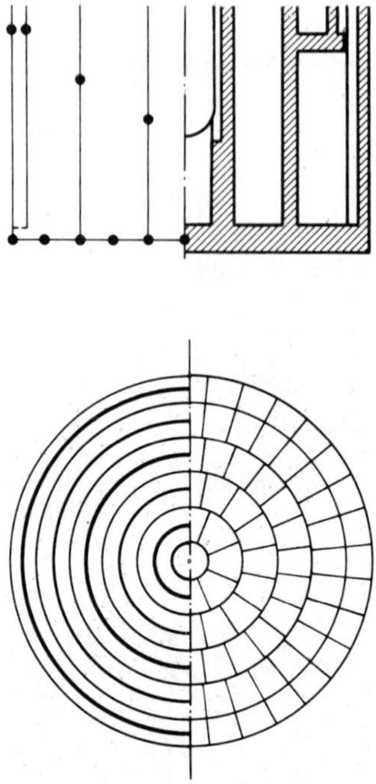

Figure 4-17 Dynamic model of flexible basemat of reactor building (after Ref. [4]).

containment is connected to the same node on the basemat as the shield building. The soil is discretized with 113 subdisks, as shown on the right-hand side of this figure. The modeling of the soil is treated in depth in Chapter 7. At this stage, it is sufficient to state that a resultant force having three components acts on each subdisk. The $[A]$-matrix defined in Eq. 3.13, which relates the (constrained) dynamic degrees of freedom of the basemat to those of the soil on the structure–soil interface, will thus have $3 \cdot 113 = 339$ rows. The basemat is assumed to be rigid in its plane, which leads to three dynamic degrees of freedom (two horizontal translations and twisting around the vertical axis). As the four branches of the superstructure are modeled as beams, the intersection lines for each beam on the level of the basemat have to deform as a rigid body (vertical translation and two rocking components). Each circle of subdisk centers which lies between two adjacent intersection lines with a beam is also constrained by an (independent) rigid-body motion. Thus five circular nodes (shown as heavy lines) with $5 \cdot 3 = 15$ dynamic degrees of freedom result. Together with the vertical displacement in the center of the basemat and the three components describing the rigid in-plane motion, a total of 19 dynamic degrees of freedom of the constrained basemat arise. This is also equal to the number of columns of the $[A]$-matrix. The total dynamic degrees of freedom of the structure–soil system equals $4 \cdot 3 \cdot 6 + 19 = 91$. A parametric study (whose results are not reproduced here) shows that for the thicknesses encountered in nuclear power plants (3 m and more), the basemat of the reactor building can be assumed to be rigid (see Section 9.3.1).

Modeling part of a building or a total structure as a vertical beam as in

Figs. 4-15 and 4-16 is possible only for a box-type structure. In addition, a rigid basemat, stiff floors, and a thick roof tend to force the horizontal cross section of the structure to act as a beam. Many such structures are encountered in nuclear power plants. The required radiation protection and the presence of heavy equipment and extraordinary load cases (e.g., aircraft impact) lead to such stiff buildings with large mass, for which the analysis of soil–structure interaction is important.

4.3.5 Frames with Shear Panels for Seismic Load

In frame-type structures with shear panels, the floors can be assumed rigid only in their own planes. The modeling procedure is explained using the three-dimensional building of Fig. 4-18 as illustration. The structure is made up of plane frames, located arbitrarily in plan. In the example there are three such frames, two of which share the same column. It is assumed that the frames, which are composed of beams and columns (without torsional stiffness) as well as walls and shear panels, resist forces only in their own plane. In each frame the elevations of the floors are the same. The floor diaphragms are assumed to be rigid in their planes. This allows the horizontal displacement of all frames at each floor level to be expressed in terms of three dynamic degrees of freedom (defined conveniently at the center of mass of the floor): two horizontal translations and a rotation around the vertical axis. Obviously, the horizontal beams are inextensional. The bending stiffness of the floor can be approximately represented in the beams of the frames. In each joint of the frame, a vertical displacement and an in-plane rotation are introduced as additional degrees of freedom. For the roof level, all degrees of freedom are indicated in Fig. 4-18. In the column that is common to the two frames, two vertical displacements are introduced, which

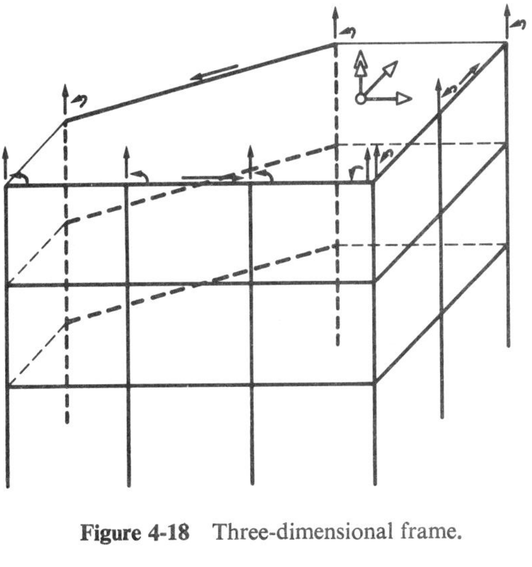

Figure 4-18 Three-dimensional frame.

will not turn out to be equal. As the two frames are assumed to be perpendicular, the rotations in this joint of the common column will be uncoupled. The dynamic-stiffness matrix of the building can be established using standard plane-frame programs.

Various approximations are possible. Also the columns can be assumed to be inextensional. The moment of inertia associated with the rotations of the joints of the frame can be neglected. This limits the dynamic degrees of freedom to three per floor. The contribution of a frame to the static stiffness of the system can be determined by applying sequentially horizontal unit forces at each floor, calculating the horizontal displacements at all levels, which leads to the flexibility matrix, which is then inverted. Alternatively, the number of dynamic degrees of freedom can also be reduced, using the procedure of static condensation described in Section 4.4.2.

This procedure of lumping mass in nodes where the dynamic degrees of freedom are defined works well if a significant part of the total mass is concentrated in small zones. This is, for example, the case for the frame structure shown in Fig. 4-18, for which most of the mass is present at the floor levels. The contribution of the columns of the frame is small and can easily be lumped in the same nodes.

4.3.6 Models Based on Generalized Displacements

For a structure with a distributed mass, the use of generalized displacements can be appropriate to determine the consistent mass matrix (and the static-stiffness matrix and the load vector). In the finite-element method, local generalized displacements (shape functions) are selected, leading to a banded dynamic-stiffness matrix. Global generalized displacements, which are defined throughout or over a large portion of the structure, can also be introduced. These deflection patterns must satisfy the geometric boundary conditions. As the formulation is analogous to that of the finite-element method, only the results are summarized, using a beam with coordinate x for illustration. The transverse displacement $w(x)$ is expressed as

$$w(x) = \sum_{i=1}^{n} z_i \phi_i(x) \tag{4.9}$$

where $\phi_i(x)$ and z_i are the generalized displacement and amplitude (generalized coordinate), respectively. The subscript i denotes the ith term and n the number of dynamic degrees of freedom. As the mass matrix $[M]$ and the static-stiffness matrix $[K]$ are the coefficient matrices of the quadratic forms corresponding to the kinetic and strain energies, respectively, the elements

$$m_{ij} = \int \phi_i(x)\phi_j(x)m(x)\, dx \tag{4.10}$$

$$k_{ij} = \int \phi_{i,xx}(x)\phi_{j,xx}(x)EI(x)\, dx \tag{4.11}$$

follow. The mass per unit length is $m(x)$, the moment of inertia $I(x)$, and the modulus of elasticity E. A comma denotes differentiation. In general, $[M]$ and $[K]$ are full matrices.

Further examples of structural models are contained in the engineering applications in Section 9.3. Simple models used to identify the key parameters of soil–structure interaction in a parameteric study are described in Sections 3.4, 9.1, and 9.2. Approximate models are described in Problems 4.5, 4.6, and 4.7.

4.4 REDUCTION OF NUMBER OF DYNAMIC DEGREES OF FREEDOM

Using the finite-element concept is the most straightforward and systematic method of discretization of a structure. When calculating the static-stiffness matrix, derivatives of the shape functions are required. As differentiating leads to a loss of accuracy, the model for a static analysis has to be quite detailed. As explained in Section 4.1, a much coarser model is in many cases sufficient for a dynamic analysis. This is also apparent from the fact that, to determine the mass matrix, the shape functions themselves are used. In a specific model, the inertial forces are thus represented much more accurately than the elastic ones. Before starting with the dynamic analysis, procedures should thus be applied which systematically reduce the large number of degrees of freedom used for the static analysis to the smaller number of the dynamic model. After performing the dynamic analysis, it must be possible to calculate the original degrees of freedom before determining elastic stresses. This method of establishing a dynamic model by reducing the number of degrees of freedom is attractive, as less experience of the analyst is needed, the procedure can more easily be automated, and the stresses should be more accurate than if the (reduced) model were derived directly. The results of the dynamic analysis are also consistent with those of the static one, as the same original model is used. The computational effort can, however, be high. It is worth stressing that reducing the number of degrees of freedom for the dynamic analysis results in an approximation. Various procedures, which can be modified further, exist.

4.4.1 Transformed Equations of Motion

The equations of motion with the original degrees of freedom are specified in total displacements for the flexible base in Eq. 3.9 and for the rigid base in Eq. 3.15. If the motion is decomposed into those of kinematic and inertial interactions, Eqs. 3.26 and 3.32 apply in the inertial interaction part of the analysis for the structure with a flexible and a rigid base, respectively. The equations of kinematic interaction do not have to be addressed in this context. As discussed in Section 3.2.1, there is no advantage in splitting the motion up into two parts for a structure with a flexible base, with the exception of a surface structure for vertically incident waves (for which case the kinematic motion equals that of the free field). For a structure with a rigid base, the kinematic-interaction analysis results in the scattered motion.

Consistent with the concept of substructuring, the degrees of freedom on the structure–soil interface (subscript *b* or *o*) are not modified. The contribution of the structure to the equations with the original degrees of freedom can be formulated as (Eq. 3.2)

$$[S]\{u\} = \{P\} \tag{4.12}$$

For a flexible base, $[S]$ is made up of $[S_{ss}]$, $[S_{sb}]$, and $[S^t_{bb}]$, and $\{u\}$ consists of $\{u_s\}$ and $\{u_b\}$ with the superscript *t* or *i*. The vector $\{P\}$ either is zero or, for a rigid base, is specified by the right-hand side of Eq. 3.32, which is a function of the properties of the structure.

All reduction procedures can be interpreted as a transformation from the original degrees of freedom $\{u\}$ to another, smaller set with amplitudes $\{u_r\}$. This is denoted as

$$\{u\} = [T]\{u_r\} \tag{4.13}$$

The transformation matrix is denoted as $[T]$, the index *r* standing for reduction. Substituting Eq. 4.13 in Eq. 4.12 and premultiplying the equation with $[T]^T$ results in

$$[S_r]\{u_r\} = \{P_r\} \tag{4.14}$$

where

$$[S_r] = [T]^T[S][T] \tag{4.15a}$$

$$\{P_r\} = [T]^T\{P\} \tag{4.15b}$$

Using the definition of $[S_r]$ (analogous to Eq. 3.1),

$$[S_r] = [K_r](1 + 2\zeta i) - \omega^2[M_r] \tag{4.16a}$$

$$[K_r] = [T]^T[K][T] \tag{4.16b}$$

$$[M_r] = [T]^T[M][T] \tag{4.16c}$$

follows. The various reduction methods differ in the choice of $\{u_r\}$ and of $[T]$, leading to different forms of the reduced property matrices $[K_r]$, $[M_r]$ and of the load vector $\{P_r\}$. The dynamic degrees of freedom on the structure–soil interface with amplitude $\{u_b\}$ will always be present in $\{u_r\}$. Besides displacement amplitudes of selected nodes of the structure, other variables, such as amplitudes of generalized displacements (Ritz vectors) or of mode shapes (generalized coordinates), can also be chosen in the set $\{u_r\}$.

4.4.2 Mass Lumping Followed by Static Condensation

The first procedure to be discussed is called mass lumping followed by static condensation. It is assumed that lumping the mass in specific nodes of the structure leads to an acceptable accuracy of the results. For instance, neglecting the mass moment of inertia in all joints of a three-dimensional frame reduces the dynamic degrees of freedom by one-half. Actually, the rotational inertia is not omitted, but it is concentrated at one node [e.g., at the center of the (rigid) base]. Assuming certain beams and columns to be inextensional allows a further reduction of the translational degrees of freedom. In addition, other translational degrees of freedom (e.g., every second one along the axis of a beam) can be chosen as free of mass. In all cases the total mass and the total mass moment of inertia must be properly modeled to be able to represent correctly the rigid-body motions of the structure. In this part of the analysis, which bears comparison with that of the direct discretization described in Section 4.3, the experience of the analyst is important. For instance, in the three-dimensional frame of Fig. 4-18 for a horizontal earthquake excitation, the inertial properties could be assigned only to the three dynamic degrees of freedom on each floor. The terms of the mass matrix corresponding to the vertical displacement and in-plane rotation of each column on every floor vanish. In many cases the number of dynamic degrees of freedom will be one order of magnitude smaller than in the original model.

After lumping and making use of Eq. 3.1, Eq. 4.12 with the original degrees of freedom is specified in partitioned form as

$$\left(\begin{bmatrix}[K_{rr}] & [K_{rm}]\\ [K_{mr}] & [K_{mm}]\end{bmatrix}(1+2\zeta i)-\omega^2\begin{bmatrix}[M_{rr}] & [0]\\ [0] & [0]\end{bmatrix}\right)\begin{Bmatrix}\{u_r\}\\ \{u_m\}\end{Bmatrix}=\begin{Bmatrix}\{P_r\}\\ \{0\}\end{Bmatrix} \tag{4.17}$$

The subscript r denotes all dynamic degrees of freedom with mass and m the massless nonessential variables. The former can be called the "master" and the latter the "slave" unknowns. Allocating the mass in discrete nodes not only affects the coefficient matrix, but also the right-hand side. The vector $\{P_m\}$ as a function of $[M_{mm}]$ (Eq. 3.32), which is zero, also vanishes. Using the lower partition of Eq. 4.17, $\{u_m\}$ can be eliminated:

$$\{u_m\} = -[K_{mm}]^{-1}[K_{mr}]\{u_r\} \tag{4.18}$$

This leads to the transformation matrix $[T]$ (see Eq. 4.13)

$$[T] = \begin{bmatrix}[I]\\ -[K_{mm}]^{-1}[K_{mr}]\end{bmatrix} \tag{4.19}$$

The unit matrix is denoted as $[I]$. Substituting in Eqs. 4.16b and 4.16c leads to

$$[K_r] = [K_{rr}] - [K_{rm}][K_{mm}]^{-1}[K_{mr}] \tag{4.20a}$$

$$[M_r] = [M_{rr}] \tag{4.20b}$$

The reduced equations of motion (Eq. 4.14) are

$$(([K_{rr}] - [K_{rm}][K_{mm}]^{-1}[K_{mr}])(1 + 2\zeta i) - \omega^2[M_{rr}])\{u_r\} = \{P_r\} \tag{4.21}$$

The band structure of the original equation (Eq. 4.12) is lost, as $[K_r]$ is, in general, a full matrix. If Eq. 4.12 already contains some zero-mass terms and no additional slave unknowns are introduced (i.e., no mass lumping is necessary), obviously the procedure is exact. Solving the original equation (Eq. 4.12) or the reduced one (Eq. 4.21) will in this case lead to identical results.

Various modifications are possible. The part of the procedure consisting of lumping of the mass can be skipped. In this case, $[M_{mm}]$ and $\{P_m\}$ (as well as

$[M_{rm}]$ for a consistent mass matrix) will be nonzero (Eq. 4.17). The master degrees of freedom with amplitudes $\{u_r\}$ are selected as those for which the ratios of the ith diagonal term m_{ii}/k_{ii} are the largest. This allows the reduction process to be automated. The amplitudes of the slave displacements $\{u_m\}$ are still assumed to follow from enforcing $\{u_r\}$ statically on the otherwise unloaded structure, leading to the same $[T]$ matrix (Eq. 4.19). The reduced equations are specified in Eq. 4.14 with Eqs. 4.15b and 4.16c, while $[K_r]$ still follows from Eq. 4.21. The reduced mass matrix $[M_r]$ is not diagonal any more. This modified procedure does not correspond to eliminating $\{u_m\}$ from the lower partition of the original equation of motion (Eq. 4.12). If this were done, the reduced coefficient matrix of $\{u_r\}$ would be a very complicated function of the excitation frequency ω. A frequency-independent static-stiffness matrix and a mass matrix, which is multiplied with ω^2, could not be identified. The same would apply to the right-hand side.

4.4.3 Substructure-Mode Synthesis

To derive the basic equation of motion in Section 3.1, the actual structure and the soil are interpreted as substructures. No approximation is introduced. For certain buildings, it can be appropriate to identify further substructures. This is, for example, the case for the reactor building shown in Fig. 4-15 with the shield building, the steel containment, the drywell, and the reactor-pressure vessel with its pedestal as substructures. A lot of design work is performed on each substructure independently of the others. For instance, for the pressure vessel with the pedestal built in on the level of the basemat, the first few vibrational modes will be known when the dynamic model of the total structure is established. Those of the shield building (Tables 4-1 and 4-2 and Fig. 4-7) are also given. Selected mode shapes and natural frequencies of the other subsystems, each built in, can easily be determined. When modeling the total structure, it is natural that the analyst wants to make use of this information. Selected vibrational modes of the substructures are used as shape functions. This second procedure to be examined is called substructure (or component)-mode synthesis. The method is approximate, as only certain vibrational modes of the substructures are normally included in the dynamic analysis of the total system. Inspection of the vibrational modes and of the natural frequencies of each subsystem built in and analyzed independently allows valuable physical insight. The advantages of the substructure method discussed in Section 1.6 apply to each individual substructure of the building. Judgment has to be applied when selecting the modes of the substructures. Those are included which will have a significant contribution to the important vibrational characteristics of the total structure.

The equations of motion are derived using for illustration Fig. 4-19a, which shows schematically the finite-element discretization with the original degrees of freedom with amplitudes $\{u\}$. Three substructures are identified.

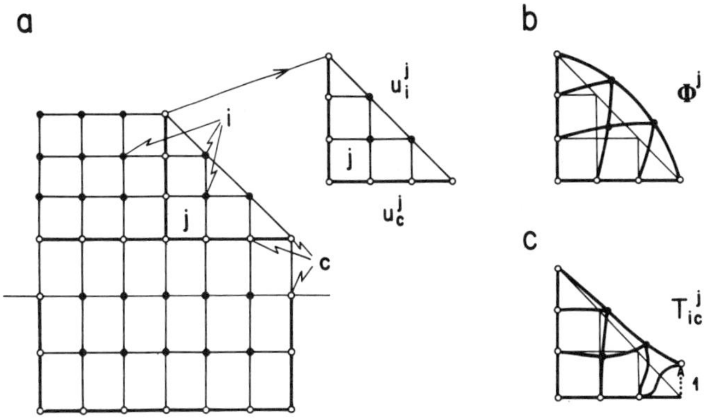

Figure 4-19 Substructure-mode synthesis. (a) Nomenclature; (b) mode shape; (c) quasi-static transmission of boundary motion.

The vector $\{u\}$ is partitioned into $\{u_i\}$, the displacement amplitudes at interior points, and into $\{u_c\}$, those at nodes on the common boundary of two adjacent substructures. The same partitioning applies to all property matrices and to the load vector. The vector $\{u_c\}$ has to be preserved in the reduced equation to be able to interconnect the substructures to form the total system. The vector $\{u_b\}$, with the displacement amplitudes on the structure–soil interface, is always contained in $\{u_c\}$. The vector $\{u_i\}$ will be expressed by the amplitudes $\{z\}$ of the vibrational modes of the substructures, assuming a fixed boundary ($\{u_c\} = \{0\}$). For each substructure with superscript j, the eigenvalue problem is formulated as (Eq. 2.2)

$$[K_{ii}^j][\Phi^j] = [M_{ii}^j][\Phi^j][\Omega^j] \tag{4.22}$$

$[\Omega^j]$ and $[\Phi^j]$ denote the matrices of the (lowest) eigenvalues (square of the natural frequencies) on the diagonal and of the corresponding eigenvectors (mode shapes), respectively, of the substructure j. The number of modes is selected smaller than the number of internal degrees of freedom. The orthogonality conditions of the suitably scaled $[\Phi^j]$ are as follows (Eq. 2.3):

$$[\Phi^j]^T[M_{ii}^j][\Phi^j] = [I] \tag{4.23a}$$

$$[\Phi^j]^T[K_{ii}^j][\Phi^j] = [\Omega^j] \tag{4.23b}$$

The following transformation is introduced:

$$\{u_i^j\} = [T_{ic}^j]\{u_c^j\} + [\Phi^j]\{z^j\} \tag{4.24}$$

$\{u_c^j\}$ denotes the displacement amplitudes on the boundary of substructure j.

Using the concept of static condensation, a column of $[T_{ic}^j]$ represents the static displacements at nodes i of the (otherwise unloaded) substructure j when a unit displacement component is enforced at one node on the boundary and all other boundary displacements vanish. The motion $\{u_c^j\}$ is thus quasi-statically transmitted to the interior nodes, resulting in the displacement amplitudes

$[T^j_{ic}]\{u^j_c\}$ in these points. In analogy to Eq. 4.18, $[T^j_{ic}]$ is specified as

$$[T^j_{ic}] = -[K^j_{ii}]^{-1}[K^j_{ic}] \tag{4.25}$$

The vector $\{z^j\}$ represents the amplitudes of the selected modeshapes of the substructure j. One column of $[\Phi^j]$ and $[T^j_{ic}]$ is shown in Fig. 4-19b and c. The vector symbols are omitted in this figure.

Equations 4.22 to 4.25 also apply to the ensemble of all substructures, assembling the contribution of each substructure. The superscript j is dropped when denoting the total structure.

The vector of the reduced displacement amplitudes $\{u_r\}$ consists of $\{z\}$ and $\{u_c\}$, leading to the following transformation:

$$\begin{Bmatrix} \{u_i\} \\ \{u_c\} \end{Bmatrix} = \begin{bmatrix} [\Phi] & [T_{ic}] \\ & [I] \end{bmatrix} \begin{Bmatrix} \{z\} \\ \{u_c\} \end{Bmatrix} \tag{4.26}$$

This equation corresponds to the transformation relating $\{u_r\}$ to $\{u\}$, as specified in Eq. 4.13. The coefficient matrix on the right-hand side represents $[T]$. Because of the special form of $[T]$ and of the orthogonality conditions (Eq. 4.23), $[K_r]$ (Eq. 4.16b), and $[M_r]$ (Eq. 4.16c) can be simplified.

Assuming a lumped-mass matrix ($[M_{ic}] = [0]$), the reduced equations of motion (Eq. 4.14), using Eqs. 4.15b and 4.16a, equal

$$\left[\begin{array}{c:c} [\Omega](1 + 2\zeta i) - \omega^2[I] & -\omega^2[\Phi]^T[M_{ii}][T_{ic}] \\ -\omega^2[T_{ic}]^T[M_{ii}][\Phi] & (1 + 2\zeta i)([K_{cc}] + [K_{ci}][T_{ic}]) \\ & \quad - \omega^2([M_{cc}] + [T_{ic}]^T[M_{ii}][T_{ic}]) \end{array}\right] \begin{Bmatrix} \{z\} \\ \{u_c\} \end{Bmatrix} = \begin{Bmatrix} [\Phi]^T\{P_i\} \\ [T_{ic}]^T\{P_i\} + \{P_c\} \end{Bmatrix} \tag{4.27}$$

Once again, the band structure present in the original equations (Eq. 4.12) is lost. If a consistent mass matrix is used, reference can be made to Section 8.2.5 (Eq. 8.39), where a special case of the substructure-mode synthesis is discussed in another context.

Instead of using the mode shapes, other generalized displacements (shape functions) could be used. They have to vanish on the boundary between two neighboring substructures. The deflected shape of the built-in substructure loaded by the expected dominant inertial loads could be selected. Assembling these shape functions of the substructure j in $[\Phi^j]$, Eq. 4.26 would still apply. As the orthogonality conditions of Eq. 4.23 are no longer valid, the first submatrix in Eq. 4.27 is changed to

$$(1 + 2\zeta i)[\Phi]^T[K_{ii}][\Phi] - \omega^2[\Phi]^T[M_{ii}][\Phi] \tag{4.28}$$

As the chosen generalized displacements will resemble the mode shapes, the off-diagonal terms are small. The other submatrices and the right-hand side are not affected.

SUMMARY

1. As the inertial forces depend on the displacements and are not a function of their derivatives as in the case of the elastic forces, a coarser model can be selected for the dynamic analysis than for the static one. In many cases only a few vibrational modes, generally the lowest, are excited significantly. The uncertainties of the dynamic load are generally larger than those of the static one. Economic considerations also demand a simple dynamic model, if at all possible.
2. The frequency content and the spatial variation of the applied load influence the dynamic model, in which all vibrational modes with nonnegligible generalized modal loads have to be represented. The type of result also affects the discretization: higher modes contribute more to local than to global results and are also more important for accelerations than for displacements.
3. A subsystem with a small mass can be decoupled from the structural model. The motion at the support point of the subsystem is calculated first and can be displayed in the form of an in-structure response spectrum. The subsystem is then analyzed independently for this support motion, neglecting the feedback from the subsystem to the structure.
4. For a structure with a rigid base, even embedded, the applied seismic inertial loads follow from using rigid-body kinematics from the three translational and three rotational components of the scattered motion of the massless base. This also applies for horizontally propagating waves. The resulting simple spatial variation of the inertial loads excites only the zeroth and first symmetric and antimetric modes of an axisymmetric structure. For a surface structure with a flexible base subjected to vertically incident waves, the applied inertial acceleration is constant throughout the structure. This excites, for the horizontal and vertical earthquakes in an axisymmetric structure, only the first and zeroth harmonics, respectively. For all these cases a simple dynamic model thus results. Constraining the excited harmonics somewhat allows, as an acceptable approximation, beams to be applied in the dynamic model for seismic excitation. The latter are also used in practice to model even box-type structures which deviate considerably from axisymmetry.
5. For the other cases of a structure with a flexible basemat, higher-order harmonics are excited for seismic loads. This is the case for an embedded structure or for a surface structure subjected to horizontally propagating waves. Beams can be used only in exceptional cases.
6. Dynamic models are normally established directly. For an impact load which is applied over a small area and whose load–time diagram exhibits significant high frequencies, a very detailed discretization is necessary to be able to model the higher modes accurately. For a box-type structure with a stiff base, a

vertical beam (whose center of gravity does not have to coincide with those of shear and mass and whose axis can exhibit discontinuities at the floor levels) can often be used as the dynamic model for seismic excitation. In frame-type structures with shear panels, the connecting floors can be assumed to be rigid only in their planes. The mass is either lumped at the nodes by inspection or by using generalized displacements.

7. Alternatively, the dynamic model can be derived from the detailed static finite-element discretization systematically by reducing the number of dynamic degrees of freedom. Two methods are described to achieve this transformation: mass lumping, followed by static condensation and substructure-mode synthesis. In the latter, the amplitudes of the vibrational modes of suitably chosen substructures are used as the dynamic degrees of freedom, together with the displacements on the boundaries separating the substructures.

PROBLEMS

4.1. Figure P4-1a shows symbolically a structure (subscript s) with a component (subscript c) attached to it. The criterion on uncoupling of the latter can be defined as a function of the mass ratio $\bar{m} = m_c/m_s$ and of the frequency ratio $\bar{\omega} = \sqrt{k_c/m_c}/\sqrt{k_s/m_s}$. Calculate the two natural frequencies ω_1 and ω_2 (nondimensionalized

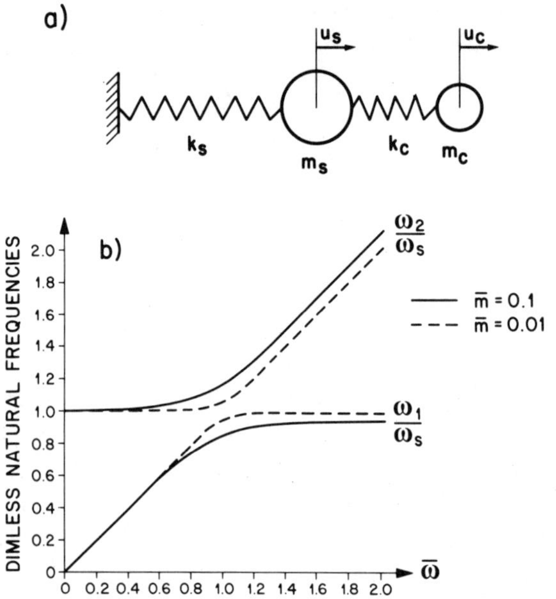

Figure P4-1 Uncoupling of subsystem. (a) Supporting system (structure) with supported subsystem (component); (b) natural frequencies of combined system;

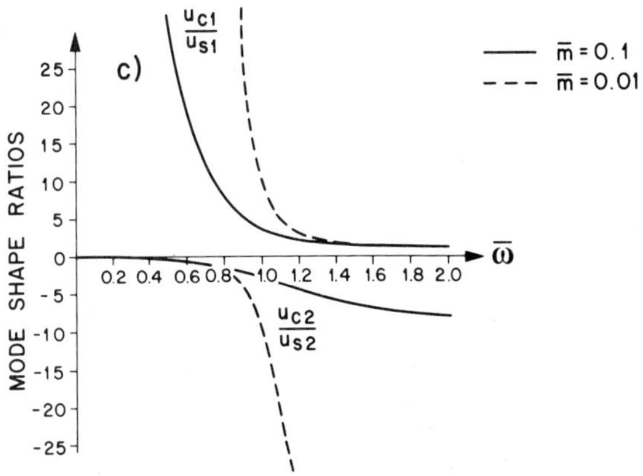

Figure P4-1 (Continued) (c) ratio of displacements.

with $\omega_s = \sqrt{k_s/m_s}$) and the corresponding ratios of the displacements u_c/u_s of the coupled system as a function of $\bar{\omega}$ and $\bar{m}$. Plot ω_1/ω_s and ω_2/ω_s versus $\bar{\omega}$ for $\bar{m} = 0.01$ and $= 0.1$ and, in addition, u_c/u_s versus $\bar{\omega}$ for the same two mass ratios.

Results:

$$\frac{\omega_{1,2}^2}{\omega_s^2} = \tfrac{1}{2}[1 + \bar{\omega}^2(1 + \bar{m}) \mp \sqrt{[1 + \bar{\omega}^2(1 + \bar{m})]^2 - 4\bar{\omega}^2}]$$

$$\frac{u_{c1}}{u_{s1}} = \frac{1}{1 - \omega_1^2/(\omega_s^2\bar{\omega}^2)}$$

$$\frac{u_{c2}}{u_{s2}} = \frac{1}{1 - \omega_2^2/(\omega_s^2\bar{\omega}^2)}$$

The results are plotted in Fig. P4-1b and c.

4.2. As discussed in Sections 3.1 and 3.2, the free-field motion acts at the lower end (i.e., the end that is not connected to the basemat of the structure) of a generalized spring–dashpot system. To evaluate approximately the scattered motion (which is equal to that of kinematic interaction) of a rigid surface structure subjected to horizontally propagating waves, the dashpots can be omitted, and continuously distributed springs with constant values can be selected.

For the quadratic rigid basemat of length $2a$ resting on distributed springs with a constant k (stiffness per area), determine the amplitude of the scattered motion u_o^g caused by the free-field motion u_x^f with a particle motion in the direction of propagation (x-axis), whereby the apparent velocity is denoted as c_a (Fig. P4-2a). Plot the ratio u_o^g/u_x^f as a function of $\omega a/c_a$.

Solution:

The harmonic displacement of frequency ω propagating in the positive x-direction can be written as $u_x^f \exp[i\omega(t - x/c_a)]$. The amplitude describing the spatial variation is thus equal to $u_x^f \exp(-i\omega x/c_a)$. This is equivalent to $u_x^f[\cos(\omega x/c_a + \phi)]$, where ϕ is the phase angle.

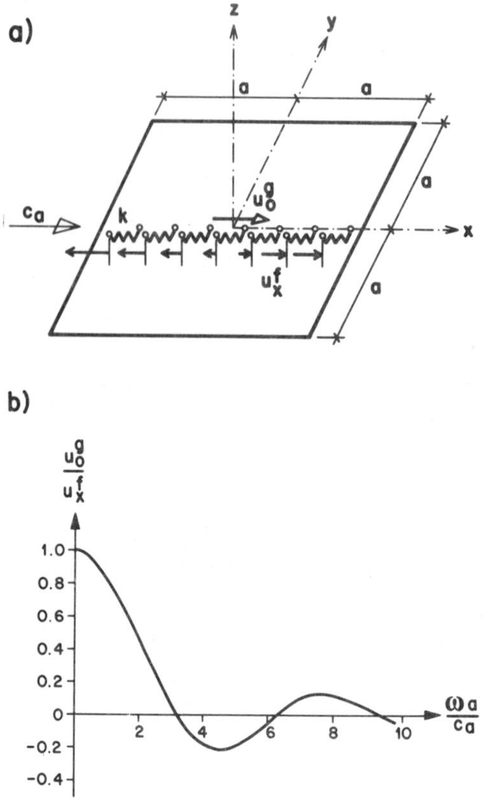

Figure P4-2 Kinematic motion for horizontally propagating wave with horizontal component coinciding with direction of propagation. (a) Rigid quadratic basemat with continuously distributed soil springs beneath; (b) translational component.

The amplitude u_o^g follows from Eq. 3.19, which, as a scalar relation, is equal to

$$u_o^g = S_{oo}^{g\,-1} A^T S_{bb}^f u_b^f$$

The term $A^T S_{bb}^f u_b^f$ is equal to the resultant arising from the forces of the springs subjected to the prescribed support motion $u_b^f = u_x^f$ and S_{oo}^g denotes the horizontal-stiffness coefficient. This leads to

$$u_o^g = \frac{1}{k4a^2} \int_{-a}^{+a} k2au_x^f \exp\left(\frac{-i\omega x}{c_a}\right) dx$$

$$\frac{u_o^g}{u_x^f} = \frac{1}{a} \int_o^a \cos \frac{\omega x}{c_a}\, dx = \frac{c_a}{\omega a} \sin \frac{\omega a}{c_a}$$

From the plot of u_o^g / u_x^f versus $\omega a / c_a$ (Fig. P4-2b) the self-canceling effect of the free motion propagating under the structure is clearly visible. Calculating an equivalent radius by equating the areas of the square and of the circle, the agreement with the exact solution in Fig. 7-30 for the circular basemat is shown to be good. This is the case because, in the equation for u_o^g, the dynamic stiffness and its inverse both occur. The details of how the soil is modeled will thus have only a small influence when calculating the scattered motion. Of course, the scattered motion could also be calculated directly for the circular basemat, which involves a Bessel function J_1 of order 1 and of the first kind:

$$u_o^g = \frac{1}{k\pi a^2}\int_{-a}^{+a} k2\sqrt{a^2 - x^2}\,u_x^f \exp\left(\frac{-i\omega x}{c_a}\right) dx$$

$$\frac{u_o^g}{u_x^f} = \frac{4}{\pi a}\int_0^a \sqrt{1-\left(\frac{x}{a}\right)^2}\cos\frac{\omega x}{c_a}\,dx = 2\frac{c_a}{\omega a}J_1\left(\frac{\omega a}{c_a}\right)$$

4.3. For the quadratic basemat with length $2a$ resting on distributed springs having a constant spring constant k (stiffness per area), determine the amplitudes of the scattered motion v_o^g and γ_o^g caused by the free-field motion u_y^f propagating with the apparent velocity c_a in the positive x-direction and whose horizontal particle motion is perpendicular to this direction (Fig. P4-3a). Plot the ratio $a\gamma_o^g/u_y^f$ as a function of $\omega a/c_a$ (see also Problem 4.2).

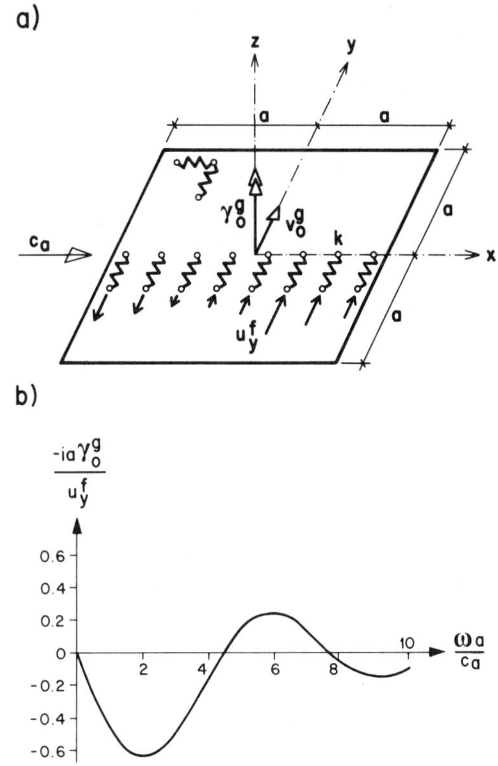

Figure P4-3 Kinematic motion for horizontally propagating wave with horizontal component perpendicular to direction of propagation. (a) Rigid quadratic basemat with continuously distributed soil springs beneath; (b) torsional component.

Solution:

$$v_o^g = \frac{1}{k4a^2} \int_{-a}^{+a} k2au_y^f \exp\left(\frac{-i\omega x}{c_a}\right) dx$$

analogous to equation for u_o^g (see Problem 4.2).

$$\gamma_o^g = \frac{1}{k8a^4/3} \int_{-a}^{+a} xk2au_y^f \exp\left(\frac{-i\omega x}{c_a}\right) dx$$

For the torsional component, S_{oo}^g is equal to the polar moment of inertia multiplied by k.

$$\frac{a\gamma_o^g}{u_y^f} = -i\frac{3}{2a^2} \int_0^a x \sin\frac{\omega x}{c_a} dx = -i\frac{3}{2}\frac{c_a}{\omega a}\left(\frac{c_a}{\omega a} \sin\frac{\omega a}{c_a} - \cos\frac{\omega a}{c_a}\right)$$

The factor $(-i)$ means that γ_o^g lags behind u_y^f (and v_o^g) by 90°. From the plot of $-ia\gamma_o^g/u_y^f$ versus $\omega a/c_a$ shown in Fig. P4-3b, the significant torsional motion with which the symmetric basemat is loaded is visible. The maximum occurs for $\omega a/c_a \sim 2$. After calculating an equivalent radius by equating the polar moments of inertia of the square and of the circle, the agreement with the exact values shown in Fig. 7-31 for the circular basemat is seen to be good.

4.4. For the square basemat with length $2a$ resting on distributed springs having a constant spring constant k (stiffness per area), determine the amplitudes of the scattered motion w_o^g and β_o^g caused by the free-field motion with a vertical component u_z^f propagating with the apparent velocity c_a in the positive x-direction (Fig. P4-4a). Plot the ratio $a\beta_o^g/u_z^f$ as a function of $\omega a/c_a$ (see also Problems 4.2 and 4.3).

Solution:

$$w_o^g = \frac{1}{k4a^2} \int_{-a}^{+a} k2au_z^f \exp\left(\frac{-i\omega x}{c_a}\right) dx$$

analogous to equations for u_o^g (see Problem 4.2) and for v_o^g (see Problem 4.3).

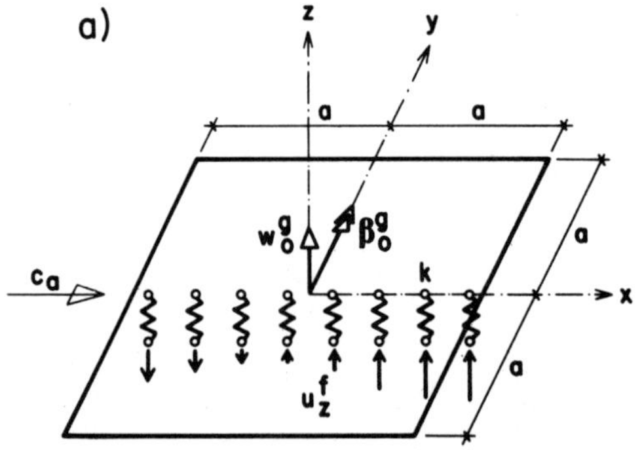

Figure P4-4 Kinematic motion for horizontally propagating wave with vertical component. (a) Rigid quadratic basemat with continuously distributed soil springs beneath;

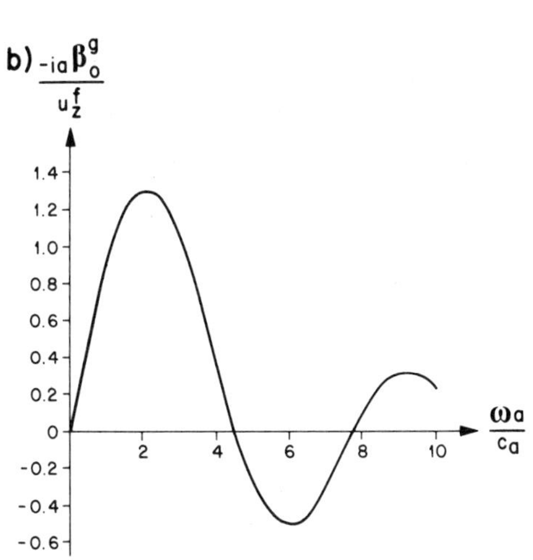

Figure P4-4 (Continued) (b) rocking component.

$$\beta_o^g = -\frac{1}{k4a^4/3}\int_{-a}^{+a} xk2au_z^f \exp\left(\frac{-i\omega x}{c_a}\right) dx$$

For the rocking component, S_{oo}^g is equal to the moment of inertia multiplied by k.

$$\frac{a\beta_o^g}{u_z^f} = i\frac{3}{a^2}\int_0^a x \sin\frac{\omega x}{c_a} dx = i3\frac{c_a}{\omega a}\left(\frac{c_a}{\omega a}\sin\frac{\omega a}{c_a} - \cos\frac{\omega a}{c_a}\right)$$

Apart from the sign, β_o^g is twice as large as γ_o^g. Comparing the plot of $-ia\beta_o^g/u_z^f$ versus $\omega a/c_a$ (Fig. P4-4b) with that of the exact solution after calculating an equivalent radius (Fig. 7-31), the good agreement is apparent.

4.5. As illustrated in Fig. 3-19, the simple coupled system with one dynamic degree of freedom used to analyze soil–structure interaction for seismic excitation (Fig. 3-18) can also be applied for a multistory building which responds predominantly in the fundamental mode. The system is described by the fixed-base frequency ω_s ($=\sqrt{k/m}$), the damping ratio ζ, the effective mass m, and the effective height h. Verify that the following equations apply:

$$m = \frac{(\sum_j m_j\phi_j)^2}{\sum_j m_j\phi_j^2}$$

$$h = \frac{\sum_j m_j\phi_j h_j}{\sum_j m_j\phi_j}$$

where m_j, h_j, and ϕ_j are the mass at story j, the distance of story j from the base, and the (fixed-base) fundamental mode shape at story j, respectively.

Solution:

To achieve the same interaction effects, the base shear H and the overturning moment M must be equal for the two systems built in at their bases. For the equivalent one-degree-of-freedom system,

$$H = kr = \omega_s^2 mr$$

$$M = hkr = \omega_s^2 hmr$$

apply, where r is the relative displacement. For the multistory building, $\{R\}$ in Eq. 2.6b is specified as

$$\{R\} = -[M]\{e\}\ddot{r}_g$$

where $\{e\}$ contains the value 1 for all horizontal degrees of freedom and zero for all others. Not enforcing the scaling specified in Eq. 2.3a,

$$y = \frac{\{\phi\}^T[M]\{e\}}{\{\phi\}^T[M]\{\phi\}} r$$

results for the modal amplitude y. The corresponding global values are as follows:

$$H = \omega_s^2 \{e\}^T[M]\{\phi\} y$$

$$M = \omega_s^2 \{h\}^T[M]\{\phi\} y$$

Equating the base shears leads to

$$m = \frac{\{e\}^T[M]\{\phi\}\{\phi\}^T[M]\{e\}}{\{\phi\}^T[M]\{\phi\}} = \frac{(\sum_j m_j \phi_j)^2}{\sum_j m_j \phi_j^2}$$

while h is determined by setting the overturning moments equal.

$$h = \frac{\{h\}^T[M]\{\phi\}\{\phi\}^T[M]\{e\}}{m\{\phi\}^T[M]\{\phi\}} = \frac{\{h\}^T[M]\{\phi\}}{\{e\}^T[M]\{\phi\}} = \frac{\sum_j m_j \phi_j h_j}{\sum_j m_j \phi_j}$$

4.6. The discretization of certain structures can lead to a model that is supported in a statically indeterminate way. As an example, the continuous bridge whose columns are built in at the level of the girder is addressed for horizontal excitation with an amplitude u_g (Fig. P4-6). [If the columns are hinged (Fig. 3-19) and not built in, the statically determinate model of Fig. 3-18 can be selected, which is examined in depth in Section 3.4.] Assuming the girder to be rigid compared to the stiffness of the columns results in the statically indeterminate model with one redundant shown in Fig. P4-6b. The mass m with the total displacement amplitude u^t can move only horizontally without exhibiting rotation. The base has two degrees of freedom, the horizontal translation with the amplitude u_o^t and the rocking with the amplitude ϕ. The column of length h (which for the sake of simplicity is assumed to be prismatic with the bending stiffness EI) has a static stiffness k ($= 12EI/h^3$) and hysteretic-damping ratio ζ. The fixed-base frequency is equal to

$$\omega_s^2 = \frac{k}{m}$$

It is advantageous to split the total displacement amplitudes into their components

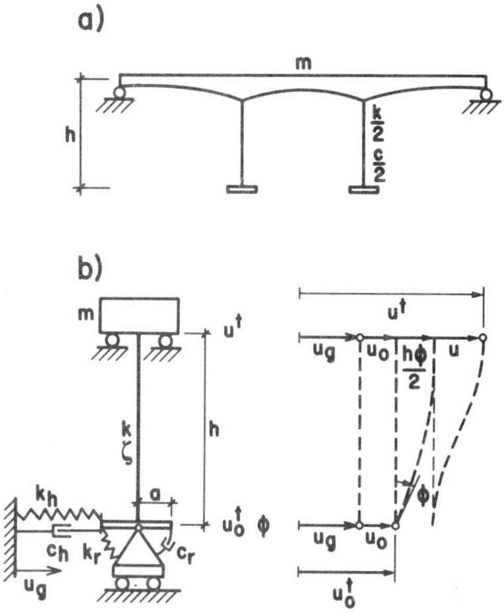

Figure P4-6 Continuous bridge girder with built-in columns modeled as redundant system with one dynamic degree of freedom. (a) Bridge; (b) dynamic model.

$$u^t = u_g + u_o + \frac{h\phi}{2} + u$$

$$u_o^t = u_g + u_o$$

where u_o is the amplitude of the base relative to that of the free field. The term $h\phi/2$ is the amplitude of the horizontal displacement caused by ϕ in the statically indeterminate system, which does not result in a shear force in the column (but in a constant moment), and u is the amplitude of the structural distortion. As discussed in depth in Section 3.4, the horizontal-force amplitude of the soil is approximated as

$$P_h = k_h u_o + c_h \dot{u}_o = k_h(1 + 2\zeta_x i + 2\zeta_g i)u_o$$

where k_h and c_h are the constants of the spring and the dashpot, respectively. The ratio of the radiation damping of the undamped soil is denoted as ζ_x, that of the material damping as ζ_g. Analogously, for the moment amplitude of the soil

$$M_r = k_r\phi + c_r\dot{\phi} = k_r(1 + 2\zeta_\phi i + 2\zeta_g i)\phi$$

applies.

(a) Derive the equations of motion for harmonic response using the unknowns u, u_o, and $h\phi$. Introduce the following natural frequencies:

$$\omega_h^2 = \frac{k_h}{m}$$

$$\omega_r^2 = \frac{k_r}{mh^2}$$

Express the coefficient matrix as a function of ω_s, ω_h, ω_r, ω, ζ_x, ζ_ϕ, ζ, and ζ_g.

(b) Proceeding as in Section 3.4, determine the properties of an equivalent one-degree-of-freedom system: its natural frequency $\tilde{\omega}$, its ratio of hysteretic damping $\tilde{\zeta}$, and its excitation $\tilde{u}_g$.

Results:

(a)

$$\begin{bmatrix} \frac{\omega_s^2}{\omega^2}(1+2\zeta i)-1 & -1 & -1 \\ -1 & \frac{\omega_h^2}{\omega^2}(1+2\zeta_x i+2\zeta_g i)-1 & -1 \\ \frac{\omega_s^2}{2\omega^2}(1+2\zeta i)-1 & -1 & 2\frac{\omega_r^2}{\omega^2}(1+2\zeta_\phi i+2\zeta_g i) + \frac{\omega_s^2}{6\omega^2}(1+2\zeta i)-1 \end{bmatrix} \cdot \begin{Bmatrix} u \\ u_o \\ \frac{h\phi}{2} \end{Bmatrix} = \begin{Bmatrix} 1 \\ 1 \\ 1 \end{Bmatrix} u_g$$

with

$$\omega_s^2 = \frac{12EI}{h^3 m}$$

(b) Neglecting products of the damping ratios compared to unity, the properties of the equivalent one-degree-of-freedom system are derived as

$$\frac{1}{\tilde{\omega}^2} = \frac{1}{\omega_s^2} + \frac{1}{\omega_h^2} + \frac{3}{\omega_s^2 + 12\omega_r^2}$$

$$\tilde{\zeta} = \zeta - \frac{\tilde{\omega}^2}{\omega_h^2}(\zeta - \zeta_g - \zeta_x) - \frac{36\tilde{\omega}^2\omega_r^2}{(\omega_s^2 + 12\omega_r^2)^2}(\zeta - \zeta_g - \zeta_\phi)$$

$$\tilde{u}_g = \frac{\tilde{\omega}^2}{\omega_s^2} u_g$$

4.7. Figure P3-11 shows the model of a rigid structure founded on flexible soil, which results in a simple system for analysis. This discretization can be of practical use in a preliminary analysis of a nuclear building which is very stiff and rests on relatively flexible soil. A rigid structure can, however, also be assumed in situations where, at a first glance, this would not be expected to be possible. It turns out that the shell of a hyperbolic cooling tower, although its wall thickness is very small compared to a characteristic length, can be modeled as a rigid body. This is possible because the supporting columns are extremely flexible in the horizontal direction. The shape of the first mode of the first symmetric harmonic (which dominates the response for a uniform horizontal earthquake excitation) shown in Fig. 4-14 of the tower of Fig. 4-13 demonstrates this fact. Thus the model shown in Fig. P4-7a can be used.

The columns of the tower rest on separate foundations, which it is possible to connect by a ring beam. As the separate foundations are quite far apart, they can be regarded as independent of one another when calculating the dynamic-stiffness coefficients of the total foundation. The horizontal-force amplitude P_h of the damped soil is expressed approximately as (see Section 3.4)

$$P_h = k_h u_o + c_h \dot{u}_o = k_h(1 + 2\zeta_x i + 2\zeta_g i)u_o$$

where ζ_x and ζ_g are the ratios of the viscous radiation damping of the undamped

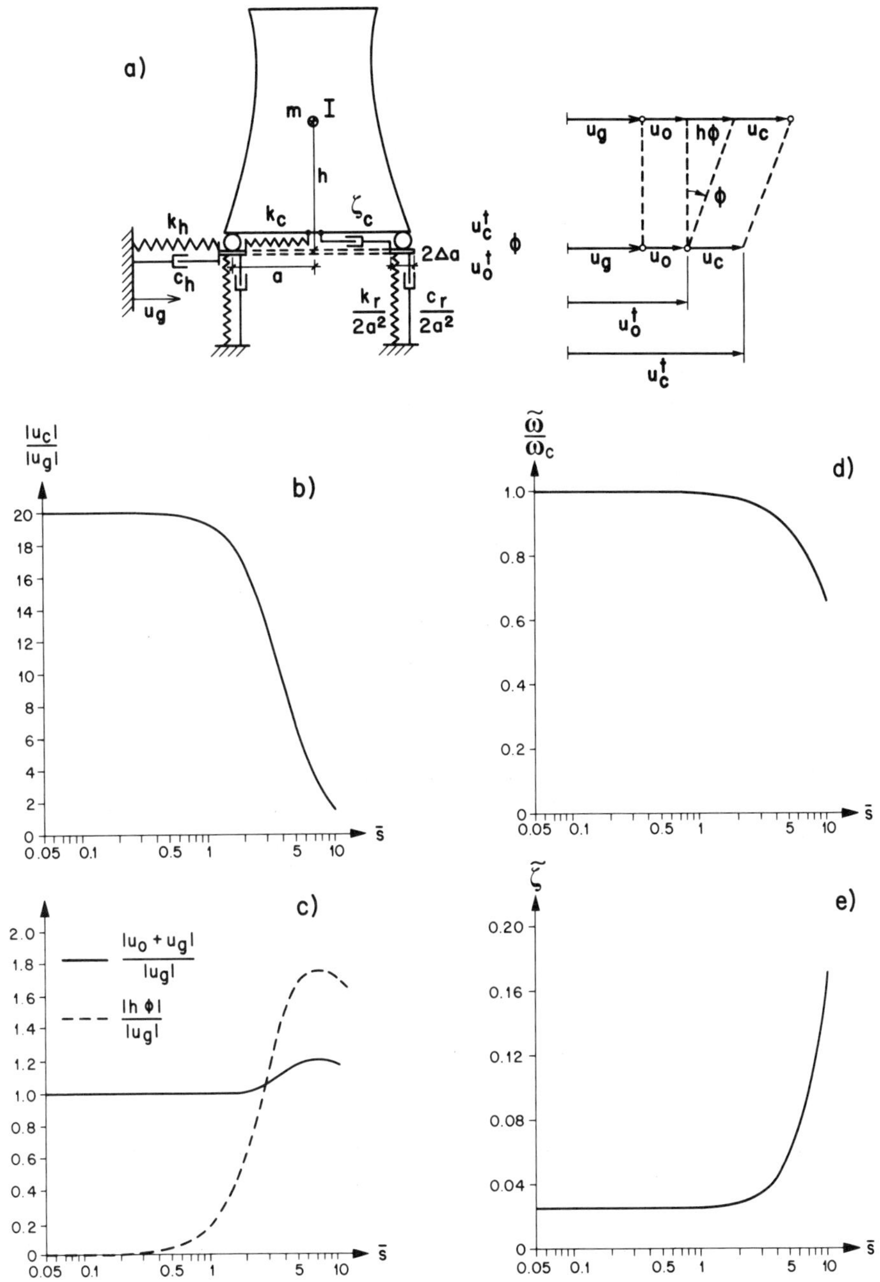

Figure P4-7 Hyperbolic cooling tower modeled as rigid shell on flexible columns founded on soil. (a) Dynamic model; (b) relative displacement of columns; (c) total displacement at base and rocking; (d) equivalent natural frequency; (e) equivalent damping ratio.

soil in the horizontal direction and of the hysteretic-material damping, respectively. Analogously, for the moment amplitude of the soil

$$M_r = k_r\phi + c_r\dot{\phi} = k_r(1 + 2\zeta_\phi i + 2\zeta_g i)\phi$$

applies. The amplitudes u_o and ϕ correspond to the displacement of the base relative to that of the free field and to the rocking, respectively. The columns are modeled in the horizontal direction as a spring with a constant k_c and a hysteretic-damping ratio ζ_c. The amplitude of the relative displacement between the upper and lower ends of the columns is denoted as u_c. In the vertical direction the columns are assumed to be rigid. The shell is represented as a rigid body with mass m and moment of inertia I. Its center of mass is located at a height h above the foundation.

(a) Modify the harmonic equations of motion derived in Problem 3.11 to take the flexibility of the columns into consideration. Use u_o, $h\phi$, and u_c as the unknowns. Besides ω_h and ω_r, introduce the natural frequency

$$\omega_c^2 = \frac{k_c}{m}$$

The coefficient matrix is a function of ω_c, ω_h, ω_r, ω, I/h^2m, ζ_c, ζ_x, ζ_ϕ, and ζ_g.

(b) Use the same crude approximation of Eqs. 3.65a and 3.65b and of Problem 3.8 to calculate the contribution of a separate foundation of radius Δa to the total dynamic stiffness of the undamped soil.

$$k_x = n\frac{8G\,\Delta a}{2-\nu}$$

$$c_x = n\frac{4.6}{2-\nu}\rho c_s\,\Delta a^2$$

$$k_\phi = \frac{4G\,\Delta a}{1-\nu}\sum_{i=1}^{n} x_i^2 = n\frac{4G\,\Delta a}{1-\nu}\frac{a^2}{2}$$

$$c_\phi = \frac{4}{1-\nu}\rho c_s\,\Delta a^2\sum_{i=1}^{n} x_i^2 = n\frac{4}{1-\nu}\rho c_s\,\Delta a^2\frac{a^2}{2}$$

where n is equal to the total number of separate foundations and x_i denotes the distance from the center line. Introduce the following parameters: $\bar{s} = \omega_c a/c_s$, $\bar{a} = \Delta a/a$, n, $\bar{h} = h/a$, $\bar{m} = m/(\rho a^3)$, I/h^2m, ν, ζ_c, and ζ_g. Express the three equations of motion as a function of these parameters and of ω/ω_c.

(c) The significant relative displacement between the upper and lower ends of the columns leads to large forces in the foundations, in the columns, and in the lower edge beam of the shell. Hence, for increasing seismic loads, these parts of the tower are the first whose design is governed by the loading case of earthquake instead of by the wind. Plot $|u_c|/|u_g|$ at resonance versus $\bar{s}$ ($0.05 < \bar{s} < 10$) for the following values, which correspond to the tower shown in Fig. 4-13: $h = 65.64$ m, $a = 58.54$ m, $\Delta a = 2.5$ m, $n = 36$, $m = 18.550$ Gg, $I = 46{,}500$ Ggm^2, $k_c = 5.05$ GN/m, $\zeta_c = 0.025$, $\nu = 0.4$, $\rho = 2.4$ Mg/m^3, and $\zeta_g = 0.05$. Determine the range of $\bar{s}$ for which the influence of soil–structure interaction can be neglected by plotting the corresponding $|u_o + u_g|/|u_g|$ and $|h\phi|/|u_g|$ versus $\bar{s}$. Plot the properties of an equivalent one-degree-of-freedom system $\tilde{\omega}/\omega_c$, $\tilde{\zeta}$ which are determined from u_c at resonance.

Results:

(a)

$$\begin{bmatrix} \frac{\omega_c^2}{\omega^2}(1+2\zeta_c i)-1 & -1 & -1 \\ -1 & \frac{\omega_h^2}{\omega^2}(1+2\zeta_x i+2\zeta_g i)-1 & -1 \\ -1 & -1 & \left[\frac{\omega_r^2}{\omega^2}(1+2\zeta_\phi i+2\zeta_g i)-1\right]\left(1+\frac{I}{h^2m}\right) \end{bmatrix} \begin{Bmatrix} u_c \\ u_o \\ h\phi \end{Bmatrix} = \begin{Bmatrix} 1 \\ 1 \\ 1 \end{Bmatrix} u_g$$

(b)

$$\begin{bmatrix} \frac{\omega_c^2}{\omega^2}(1+2\zeta_c i)-1 & -1 & -1 \\ -1 & \frac{\omega_c^2}{\omega^2}\frac{8n\bar{a}}{(2-\nu)\bar{s}^2\bar{m}}\left(1+0.58\bar{s}\bar{a}\frac{\omega}{\omega_c}i+2\zeta_g i\right)-1 & -1 \\ -1 & -1 & \left[\frac{\omega_c^2}{\omega^2}\frac{2n\bar{a}}{(1-\nu)\bar{s}^2\bar{m}\bar{h}^2\left(1+\frac{I}{h^2m}\right)}\left(1+\bar{s}\bar{a}\frac{\omega}{\omega_c}i+2\zeta_g i\right)-1\right]\left(1+\frac{I}{h^2m}\right) \end{bmatrix} \begin{Bmatrix} u_c \\ u_o \\ h\phi \end{Bmatrix} = \begin{Bmatrix} 1 \\ 1 \\ 1 \end{Bmatrix} u_g$$

(c) The results are plotted in Fig. P4-7b to e.

5

FUNDAMENTALS OF WAVE PROPAGATION

5.1 ONE-DIMENSIONAL WAVE EQUATION

5.1.1 Significance of Wave Propagation

In the fundamental equation of motion of soil–structure interaction analysis (Eq. 3.9), the dynamic-stiffness matrix $[S_{bb}^g]$ (or $[S_{bb}^f]$) and the free-field motion $\{u_b^f\}$ are the terms associated with the unbounded soil. Both can be determined using the concepts of wave propagation in a continuum. Especially the calculation of $[S_{bb}^g]$ leads to a complex two- or three-dimensional problem involving different types of waves. General procedures to determine $[S_{bb}^g]$ and $\{u_b^f\}$ are described in Chapters 7 and 6, respectively. They are based on the, in general, three-dimensional wave equation, which is derived in Section 5.2. Different types of body waves are identified in the same section. The corresponding stiffness matrices for a layer and for a half-space can then be calculated (Sections 5.3, 5.4, and 5.5). This allows the stiffness matrix of a site consisting of layers resting on a half-space to be established, following the standard assembly process of finite elements. The applications are described in detail in Chapters 6 and 7.

5.1.2 Statement of Problem

Before deriving the wave equation in three dimensions, it is appropriate to investigate the one-dimensional case, using the standard theory of the strength of materials. The vital aspects of wave propagation can be clearly established and their impact on the dynamic-stiffness matrix of the soil and on

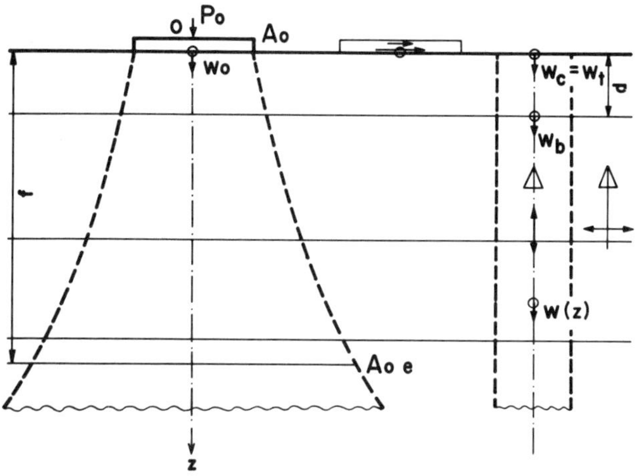

Figure 5-1 Dynamic stiffness and free field modeled as rod with exponentially varying area and as prismatic rod, respectively.

the free-field motion studied. Consider the site consisting of layers resting on a half-space shown in Fig. 5-1. The origin of the z-axis pointing downward is located at the free surface. Material damping is at first disregarded. The vertical free-field motion with the displacement amplitude w_c (the subscript c stands for control motion) is assumed to be specified in the control point located on the free surface. This motion shall be associated with vertically propagating compressional waves (P-waves). To determine the free-field response $w(z)$ at a depth z, a prismatic rod representing a column of soil can be examined. A rigid (massless) base of area A_o also rests on the free surface of the site. The vertical dynamic-stiffness coefficient S_o ($= S^g_{oo}$) of the infinite soil relates the vertical-displacement amplitude w_o to the vertical-force amplitude P_o

$$P_o = S_o w_o \tag{5.1}$$

To take the load distribution arising from P_o approximately into account, the soil can be modeled crudely as a vertical rod with an area $A(z)$ increasing exponentially with depth

$$A(z) = A_o \exp\left(\frac{z}{f}\right) \tag{5.2}$$

The length f represents the depth at which the area equals $A_o e$. Only amplitudes of normal stress σ distributed uniformly over the cross section arise. For a layered site, the rod consists of elements with varying moduli of elasticity E which are selected equal to the constrained moduli of the layer. This representation can be used to model a two- or a three-dimensional problem. The length f is determined in such a way as to achieve the same static-stiffness coefficient in the vertical direction for the (three-dimensional) site and for the rod. For $f \longrightarrow \infty$, the prismatic rod results for any finite value of z. The model used for the analysis of the free-field response is thus contained as a special case in the formulation. It should be realized that the model for the free-field analysis is exact for vertically incident waves, while that for the calculation of the stiffness coefficient is only approximate. The horizontal direction could also be modeled, replacing the area A by the shear area and E by the shear modulus. The rocking behavior could also be approximately represented by selecting two vertical bars (with the area A) spaced apart at a suitable selected distance.

5.1.3 Equation of Motion

The dynamic equation of motion of this rod with variable area is derived for harmonic excitation with frequency ω as follows. The unknowns vary as exp $(i\omega t)$. An infinitesimal element of length dz is shown in Fig. 5-2. Equilibrium is formulated as

$$-\sigma A + (\sigma + \sigma_{,z}\, dz)(A + A_{,z}\, dz) + \omega^2 w\rho(A + \tfrac{1}{2}A_{,z}\, dz)\, dz = 0 \tag{5.3}$$

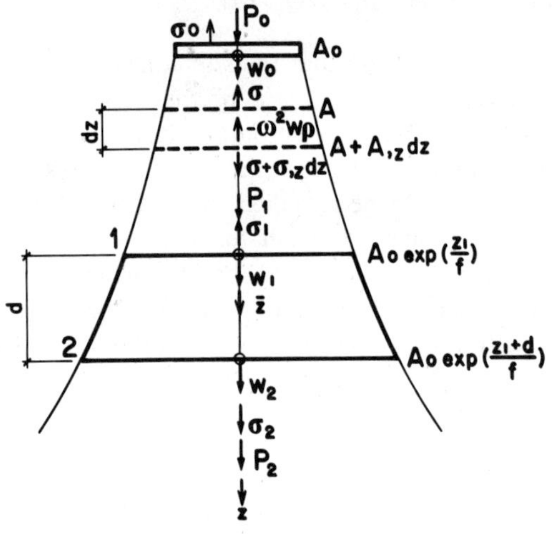

Figure 5-2 Rod with exponentially varying area (infinitesimal element and layer).

A comma denotes the derivative. The harmonic inertial load is calculated on the basis of the average area. The density is denoted by ρ. Equation 5.3 reduces to

$$A\sigma_{,z} + \sigma A_{,z} + \omega^2 w\rho A = 0 \tag{5.4}$$

after neglecting terms in $(dz)^2$. Introducing the stress-displacement relationship of the theory of the strength of materials for a rod

$$\sigma = Ew_{,z} \tag{5.5}$$

in Eq. 5.4 leads to the equation of motion of the rod with variable cross section

$$w_{,zz} + \frac{A_{,z}}{A} w_{,z} + \frac{\omega^2}{c_l^2} w = 0 \tag{5.6}$$

where the velocity of the longitudinal waves in the prismatic rod c_l is defined as

$$c_l^2 = \frac{E}{\rho} \tag{5.7}$$

Specializing for the exponentially varying area (Eq. 5.2) and defining the dimensionless frequency a_o as

$$a_o = \frac{\omega f}{c_l} \tag{5.8}$$

Eq. 5.6 is transformed to

$$w_{,zz} + \frac{1}{f} w_{,z} + \frac{a_o^2}{f^2} w = 0 \tag{5.9}$$

For $f \longrightarrow \infty$, the standard (one-dimensional) wave equation is recovered

$$w_{,zz} + \frac{\omega^2}{c_l^2} w = 0 \tag{5.10}$$

To solve Eq. 5.9, a solution for w of the form $\exp(i\gamma z)$ is considered. As wave propagation is envisaged, i is explicitly included. This results in the quadratic equation

$$\gamma^2 - \frac{i\gamma}{f} - \frac{a_o^2}{f^2} = 0 \tag{5.11}$$

with the solution

$$\gamma_1 = \frac{i}{2f} + \frac{i}{2f}\sqrt{1 - 4a_o^2} \tag{5.12a}$$

$$\gamma_2 = \frac{i}{2f} - \frac{i}{2f}\sqrt{1 - 4a_o^2} \tag{5.12b}$$

The displacement amplitude w is thus equal to

$$w = a \exp(i\gamma_1 z) + b \exp(i\gamma_2 z) \tag{5.13a}$$

or

$$w = a \exp\left(-\frac{1 + \sqrt{1 - 4a_o^2}}{2f} z\right) + b \exp\left(-\frac{1 - \sqrt{1 - 4a_o^2}}{2f} z\right) \tag{5.13b}$$

where a and b represent the constants of integration.

5.1.4 Types of Waves

As material damping is disregarded, c_l and a_o are real. For $1 - 4a_o^2 > 0$ (i.e., $a_o \leq 0.5$) no wave propagates along the z-axis. The motion diminishes exponentially with depth. For higher frequencies, $a_o > 0.5$, this equation is reformulated by changing the sign of the square root as

$$w = a \exp\left(-\frac{z}{2f}\right) \exp\left(-\frac{i\omega z}{c}\right) + b \exp\left(-\frac{z}{2f}\right) \exp\left(+\frac{i\omega z}{c}\right) \tag{5.14}$$

where the phase velocity c equals

$$c = \frac{2a_o c_l}{\sqrt{4a_o^2 - 1}} \tag{5.15}$$

To interpret Eq. 5.14 physically, it is recalled that the harmonic motion is represented as $\exp(+i\omega t)$. An expression of the form $\exp[i\omega(t - z/c_a)]$ represents a wave propagation in the positive z-direction with the apparent velocity c_a. The first term can thus be interpreted as such an outgoing wave with $c_a = c$, whereby its amplitude (specified at $z = 0$ as a) diminishes exponentially with depth. Analogously, the second term with the amplitude b represents a wave propagating with the same velocity c in the negative z-direction. This incoming wave exhibits the same variation of its amplitude with depth. The phase velocity c depends on a_o and thus on the frequency ω. This means that an harmonic wave with frequency ω will only propagate at a specific apparent velocity c_a ($= c$ for zero material damping, governed by Eq. 5.15). Consider, for example, a wave consisting of the motions of two distinct frequencies. As each component will propagate with its own individual apparent velocity, the motions will become increasingly out of phase. The original shape of the wave will become distorted. The motion is thus dispersive. For $a_o = 0.5$, c equals infinity. For increasing ω, c decreases and converges to c_l for $\omega \longrightarrow \infty$. This is shown in Fig. 5-4b (curve $\zeta = 0$ for no material damping), where c_a/c_l is plotted versus a_o. This representation is called a dispersion curve. The frequency at which the motion ceases to propagate is called the cutoff frequency ($a_o = 0.5$). For the special case of the prismatic rod, no cutoff frequency exists. For the whole frequency range, the motion will propagate with a (frequency-independent) $c_a = c_l$. No dispersion and no diminishing of the motion with depth occur for the prismatic rod.

Many other dynamic systems exhibit a cutoff frequency below which no waves propagate. See Problems 5.1, 5.2, 5.3, and 5.4.

5.1.5 Dynamic-Stiffness Matrix of Finite Rod

The dynamic-stiffness matrix of a rod element of length d representing a homogeneous layer of soil is derived next (Fig. 5-2). A local coordinate axis $\bar{z}$ with the origin located at the top interface, at node 1, pointing downward, ($\bar{z} = z - z_1$) is introduced, where the area of the rod equals $A_o \exp(z_1/f)$. The straightforward procedure for calculating the dynamic-stiffness coefficients consists of selecting the integration constants a and b in Eq. 5.13 to satisfy the boundary conditions at the two interfaces, at nodes 1 ($\bar{z} = 0$) and 2 ($\bar{z} = d$). For instance, for the coefficients S_{11}, S_{21} of the first column of the dynamic-stiffness matrix, these are

$$w_1 = 1 \tag{5.16a}$$

$$w_2 = 0 \tag{5.16b}$$

Substituting Eq. 5.13a in Eq. 5.5 leads to the stress amplitudes σ

$$\sigma = iEa\,\gamma_1 \exp(i\gamma_1 z) + iEb\,\gamma_2 \exp(i\gamma_2 z) \tag{5.17}$$

The amplitudes σ_1 and σ_2 at nodes 1 and 2 then follow by setting $\bar{z} = 0$ and $= d$, respectively. The corresponding force amplitudes are equal to

$$P_1 = -\sigma_1 A_0 \exp\left(\frac{z_1}{f}\right) \tag{5.18a}$$

$$P_2 = \sigma_2 A_0 \exp\left(\frac{z_1 + d}{f}\right) \tag{5.18b}$$

As P_1 and σ_1 act in opposite directions, a minus sign is introduced in Eq. 5.18a. The dynamic-stiffness coefficient S_{11} is equal to P_1, S_{21} to P_2. The other coefficients, S_{12} and S_{22}, follow analogously, if one replaces Eq. 5.16 by

$$w_1 = 0 \tag{5.19a}$$

$$w_2 = 1 \tag{5.19b}$$

This procedure can be modified slightly by deriving a transfer matrix as an intermediate step. The displacement and stress amplitudes at node 1 ($\bar{z} = 0$) follow from Eqs. 5.13a and 5.17 as

$$\begin{Bmatrix} w_1 \\ \sigma_1 \end{Bmatrix} = \begin{bmatrix} 1 & 1 \\ iE\gamma_1 & iE\gamma_2 \end{bmatrix} \begin{Bmatrix} a \\ b \end{Bmatrix} \tag{5.20}$$

Expressing the corresponding values w_2 and σ_2 at node 2 as a function of a and b and then eliminating a and b using Eq. 5.20 results in the transfer matrix, which is not symmetric.

$$\begin{Bmatrix} w_2 \\ \sigma_2 \end{Bmatrix} = \frac{1}{\gamma_2 - \gamma_1} \left[\begin{array}{c|c} \gamma_2 \exp(i\gamma_1 d) - \gamma_1 \exp(i\gamma_2 d) & \frac{i}{E}[\exp(i\gamma_1 d) - \exp(i\gamma_2 d)] \\ iE\gamma_1\gamma_2 [\exp(i\gamma_1 d) - \exp(i\gamma_2 d)] & -\gamma_1 \exp(i\gamma_1 d) + \gamma_2 \exp(i\gamma_2 d) \end{array}\right] \begin{Bmatrix} w_1 \\ \sigma_1 \end{Bmatrix} \tag{5.21}$$

It relates the state vector with elements w and σ at node 1 to that at node 2. Performing a partial inversion pivoting on the element in the first row and in the second column leads to

$$\begin{Bmatrix} \sigma_1 \\ \sigma_2 \end{Bmatrix} = \frac{iE}{\exp(i\gamma_1 d) - \exp(i\gamma_2 d)} \times \left[\begin{array}{c|c} \gamma_2 \exp(i\gamma_1 d) - \gamma_1 \exp(i\gamma_2 d) & \gamma_1 - \gamma_2 \\ (\gamma_2 - \gamma_1) \exp[i(\gamma_1 + \gamma_2)d] & \gamma_1 \exp(i\gamma_1 d) - \gamma_2 \exp(i\gamma_2 d) \end{array}\right] \begin{Bmatrix} w_1 \\ w_2 \end{Bmatrix} \tag{5.22}$$

Finally, substituting Eq. 5.18 in Eq. 5.22 results in

$$\begin{Bmatrix} P_1 \\ P_2 \end{Bmatrix} = \frac{EA_o \exp(z_1/f)}{f} \begin{bmatrix} \bar{S}_{11} & \bar{S}_{12} \\ \bar{S}_{21} & \bar{S}_{22} \end{bmatrix} \begin{Bmatrix} w_1 \\ w_2 \end{Bmatrix} \tag{5.23}$$

As will be shown later (Eqs. 5.29 and 5.31), the factor $EA_o \exp(z_1/f)/f$ is equal to the static-stiffness coefficient of the infinite rod at node 1. The matrix $[\bar{S}]$ represents the nondimensionalized dynamic-stiffness matrix of the rod element.

$$\bar{S}_{11} = if\frac{\gamma_1 \exp(i\gamma_2 d) - \gamma_2 \exp(i\gamma_1 d)}{\exp(i\gamma_1 d) - \exp(i\gamma_2 d)} \tag{5.24a}$$

$$\bar{S}_{12} = \bar{S}_{21} = if\frac{\gamma_2 - \gamma_1}{\exp(i\gamma_1 d) - \exp(i\gamma_2 d)} \tag{5.24b}$$

$$\bar{S}_{22} = if \exp\left(\frac{d}{f}\right)\frac{\gamma_1 \exp(i\gamma_1 d) - \gamma_2 \exp(i\gamma_2 d)}{\exp(i\gamma_1 d) - \exp(i\gamma_2 d)} \tag{5.24c}$$

As expected, the dynamic-stiffness matrix is symmetric. A formulation (not using the stiffness matrix) based on the nonsymmetric transfer matrix (Eq. 5.21) leads to larger storage requirements and computational operations. Substituting the definitions of γ_1 and γ_2 (Eq. 5.12) in Eq. 5.24, alternative expressions follow, which are adapted to the two ranges of a_o. For instance, for $\bar{S}_{11}$, with the dimensionless length $\alpha = d/f$:

$$a_o \leq 0.5: \quad \bar{S}_{11} = \tfrac{1}{2}\left[1 + \sqrt{1 - 4a_o^2} \coth\left(\frac{\alpha}{2}\sqrt{1 - 4a_o^2}\right)\right] \tag{5.25a}$$

$$a_o > 0.5: \quad \bar{S}_{11} = \tfrac{1}{2}\left[1 + \sqrt{4a_o^2 - 1} \cot\left(\frac{\alpha}{2}\sqrt{4a_o^2 - 1}\right)\right] \tag{5.25b}$$

For all values of a_o, the stiffness coefficient $\bar{S}_{11}$ (and analogously the others) is real when no material damping is present. This is to be expected, recalling the definition of the dynamic-stiffness matrix of a bounded domain (Eq. 3.1), as the static-stiffness matrix $[K]$ and the mass matrix $[M]$ are real. The applied displacement and the resulting force are thus always in phase (or 180° out-of-phase). This also holds when the thickness of the layer d approaches infinity.

5.1.6 Dynamic-Stiffness Coefficient of Infinite Rod

Besides the dynamic-stiffness matrix of the finite-rod element, that of the infinite rod representing the homogeneous half-space is also needed to model a realistic site. For material damping equal 0, it cannot be derived for the case $a_o > 0.5$ as a limiting case of that of a rod element, letting $d \to \infty$ (and thus also $\alpha \to \infty$). This is apparent from Eq. 5.25b. As the cot function (of a real argument) is periodic with values ranging between $-\infty$ and $+\infty$, any value could result.

To be able to understand this behavior, the solution of the wave equation at infinity is addressed. The infinite rod represents an unbounded domain. Some condition has to be imposed on the solution at infinity. It is not sufficient just to require the solution to go to zero at infinity. This is verified as follows. From Eq. 5.14, it is deduced that both waves die out at $z = \infty$ (i.e., $w \longrightarrow 0$). It is possible to enforce $w = 0$ at $z = z_1$.

$$w = a \exp\left(-\frac{z}{2f}\right)\left[\exp\left(-\frac{i\omega z}{c}\right) - \exp\left(-\frac{2i\omega z_1}{c}\right)\exp\left(\frac{i\omega z}{c}\right)\right] \tag{5.26}$$

This solution is not identical to zero, although it tends to zero at infinity and vanishes at $z = z_1$. This will result, for example, in a nonzero stress σ_1 at $z = z_1$, which is not acceptable. The solution is not unique. The solution of the wave equation specified in Eq. 5.14 has two independent parts. What is needed is a stipulation, called the radiation condition, which will suppress one of them, leading to a unique solution.

When applying a displacement at the top of the infinite rod, no incoming wave exists. Thus, setting $b = 0$ in Eq. 5.20 leads to

$$w_1 = a \tag{5.27a}$$

$$\sigma_1 = iE\gamma_1 a \tag{5.27b}$$

Eliminating a from Eq. 5.27 and substituting the resulting expression for σ_1 in Eq. 5.18a leads to

$$P_1 = -iEA_o \exp\left(\frac{z_1}{f}\right)\gamma_1 w_1 \tag{5.28}$$

or, using the same factor to nondimensionalize as for the layer,

$$P_1 = \frac{EA_o \exp(z_1/f)}{f}\bar{S}_1 w_1 \tag{5.29}$$

The nondimensionalized dynamic-stiffness coefficient $\bar{S}_1$ of the infinite rod equals

$$\bar{S}_1 = -if\gamma_1 \tag{5.30}$$

Finally, substituting Eq. 5.12a leads to

$$\bar{S}_1 = \tfrac{1}{2}(1 + \sqrt{1 - 4a_o^2}) \tag{5.31}$$

For the static case ($a_o = 0$), $\bar{S}_1 = 1$ results. The factor $EA_o \exp(z_1/f)/f$ employed to nondimensionalize the dynamic-stiffness coefficient in Eq. 5.29 is thus the static-stiffness coefficient of the infinite rod. In $\bar{S}_1$ the dynamic behavior is captured. For $a_o \leq \frac{1}{2}$, $\bar{S}_1$ is real, while for $a_o > 0.5$ a complex $\bar{S}_1$ results. To gain physical insight, the generally complex $\bar{S}_1$ is split up into its real and imaginary parts:

$$\bar{S}_1 = k_1 + ia_o c_1 \tag{5.32}$$

The letter k_1 is thus equivalent to the spring coefficient. The first term represents the force which is in phase with the displacement. The force $ia_o c_1 w_1$ arising from the second term $= (f/c_l)c_1 i\omega w_1 = (f/c_l)c_1 \dot{w}_1$ can be interpreted as a damping force, with the damping coefficient being proportional to c_1. (For the sake of conciseness, c_1 is henceforth called the damping coefficient.) This damping force is 90° out of phase. The dynamic-stiffness coefficient of the infinite rod can thus be visualized as a generalized spring, consisting of a spring and a dashpot, with coefficients which are a function of the frequency. For the two ranges of a_o, k_1, and c_1 are determined as (Eqs. 5.12a and 5.30)

$$a_o \leq 0.5: \quad k_1 = \tfrac{1}{2}(1 + \sqrt{1 - 4a_o^2}) \tag{5.33a}$$

$$c_1 = 0 \tag{5.33b}$$

$$a_o > 0.5: \quad k_1 = \tfrac{1}{2} \tag{5.34a}$$

$$c_1 = \sqrt{1 - \frac{1}{4a_o^2}} \tag{5.34b}$$

The spring coefficient k_1 and the damping coefficient c_1 are plotted as a function of frequency a_o in Fig. 5-3. A logarithmic scale is selected on the abscissa. Below the cutoff frequency $a_o = 0.5$, no damping exists. This general trend is typical and also applies to other infinite domains. For increasing frequency, k_1 diminishes and c_1 increases.

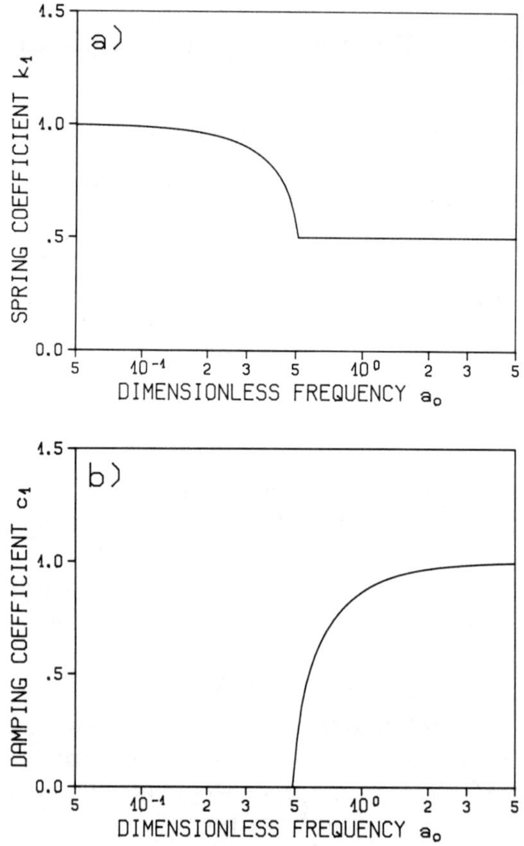

Figure 5-3 Dynamic-stiffness coefficient, without damping.

For a prismatic rod,

$$\frac{P_1}{A_o} = \rho c_l \dot{w}_1 \tag{5.35}$$

results. A dashpot with a damping coefficient (per unit area) ρc_l represents the dynamic stiffness in this case.

5.1.7 Rate of Energy Transmission

The damping coefficient c_1 is a measure of the radiation of energy of the outgoing waves. The rate of energy transmission N through the interface at node 1 is defined as the product of the real part of the force $P_1 \exp(i\omega t)$ and of the real part of the velocity $\dot{w}_1 \exp(i\omega t)$, averaged over a period $2\pi/\omega$:

$$N = \frac{\omega}{2\pi}\int_0^{2\pi/\omega} \text{Re}\,[P_1 \exp(i\omega t)]\, \text{Re}\,[\dot{w}_1 \exp(i\omega t)]\, dt \tag{5.36}$$

The two factors equal

$$\text{Re}\,[P_1 \exp(i\omega t)] = \text{Re}(P_1)\cos\omega t - \text{Im}\,(P_1)\sin\omega t \tag{5.37a}$$

$$\text{Re}\,[\dot{w}_1 \exp(i\omega t)] = \text{Re}\,(\dot{w}_1)\cos\omega t - \text{Im}\,(\dot{w}_1)\sin\omega t \tag{5.37b}$$

Substituting Eqs. 5.37 in Eq. 5.36 and integrating over time leads to

$$N = \tfrac{1}{2}\,[\text{Re}\,(P_1)\,\text{Re}\,(\dot{w}_1) + \text{Im}\,(P_1)\,\text{Im}\,(\dot{w}_1)] \tag{5.38}$$

The scalar N is thus proportional to the scalar product of the two vectors of the amplitudes P_1 and $\dot{w}_1$. With

$$\dot{w}_1 = i\omega w_1 = -\omega\,\text{Im}\,(w_1) + i\omega\,\text{Re}\,(w_1) \tag{5.39}$$

$\text{Re}\,(\dot{w}_1)$ and $\text{Im}\,(\dot{w}_1)$ are identified. Analogously, using Eqs. 5.29 and 5.32, $\text{Re}\,(P_1)$ and $\text{Im}\,(P_1)$ follow from

$$P_1 = \frac{EA_o \exp(z_1/f)}{f}(k_1 + ia_oc_1)[\text{Re}\,(w_1) + i\,\text{Im}\,(w_1)] \tag{5.40}$$

Substituting in Eq. 5.38 results in

$$N = \frac{\omega EA_o \exp(z_1/f)}{2f}\, a_oc_1\,|w_1|^2 \tag{5.41}$$

where $|w_1|^2 = \text{Re}^2\,(w_1) + \text{Im}^2\,(w_1)$ is the square of the (absolute) amplitude of w_1. The spring coefficient k_1 does not appear in the formula for N. The rate of energy transmission is thus proportional to the damping coefficient c_1. For the frequency range below the cutoff frequency $a_o = 0.5$, no energy is radiated. The mechanism with the damping coefficient c_1 is often referred to as radiation damping. It also occurs when no material damping is present.

5.1.8 Material Damping

To demonstrate the changes in the salient features by introducing material damping, the rod with exponentially varying cross section can again be used. All formulas apply if the real modulus of elasticity E is replaced by the complex value E^*:

$$E^* = E(1 + 2\zeta i) \tag{5.42}$$

where ζ is the damping ratio. This concept follows from the correspondence principle discussed in Section 2.4 (Eq. 2.14). It should be noted that E^* enters directly, for example, in the factor used to nondimensionalize the stiffness matrix (Eq. 5.23) and also indirectly, through the longitudinal-wave velocity c_l^* (Eq. 5.7), in the nondimensional frequency a_o^* (Eq. 5.8). The complex values equal

$$c_l^* = c_l\sqrt{1 + 2\zeta i} \tag{5.43}$$

$$a_o^* = \frac{a_o}{\sqrt{1 + 2\zeta i}} \tag{5.44}$$

In the equation of the displacement amplitude w (Eq. 5.13b), the expression $\sqrt{1 - 4a_o^{*2}}$ arises. Using Eq. 5.44 and for small ζ, $1 - 4a_o^{*2}$ is approximated as

$$1 - 4a_o^{*2} \sim 1 - 4a_o^2 + 8a_o^2\zeta i \tag{5.45}$$

While Re $(1 - 4a_o^{*2}) \gtrless 0$, depending on whether $a_o \lessgtr 0.5$, the term Im$(1 - 4a_o^{*2}) \sim 8a_o^2\zeta$ is always positive. The complex number $1 - 4a_o^{*2}$ thus lies either in the first or second quadrant of the complex plane. The square root will always lie in the first quadrant (i.e., Re $\sqrt{1 - 4a_o^{*2}} > 0$, Im $\sqrt{1 - 4a_o^{*2}} > 0$). One part of the exponential function in the first term of Eq. 5.13b (associated with the outgoing wave) is reformulated as

$$\exp\left(-\frac{\sqrt{1 - 4a_o^{*2}}}{2f}z\right) = \exp\left(-\frac{\text{Re}\sqrt{1 - 4a_o^{*2}}}{2f}z\right)\exp\left(-\frac{i\,\text{Im}\sqrt{1 - 4a_o^{*2}}}{2f}z\right) \tag{5.46}$$

The second factor on the right-hand side indicates that a wave propagates for all frequencies (in the positive z-direction). Formulating this factor as $\exp(-i\omega z/c_a)$, the (real) apparent velocity c_a follows as

$$c_a = \frac{2a_o c_l}{\text{Im}\sqrt{1 - 4a_o^{*2}}} \tag{5.47}$$

To eliminate f, use is made of Eq. 5.8. The ratio c_a/c_l depends on the dimensionless frequency a_o and on the damping ratio ζ. This dimensionless apparent velocity is shown for $\zeta = 0.05$ and 0.20 in Fig. 5-4. Below $a_o = 0.5$, c_a/c_l is very large (Fig. 5-4a). Above, the ratio hardly depends on ζ. No cutoff frequency, as in the undamped case, thus exists. The first factor in Eq. 5.46 is associated with an (additional) attenuation of the motion. For a distance z equal to the wavelength $\lambda = 2\pi c_a/\omega$, the (real) additional decay factor per wavelength δ_λ caused by material damping results from the first factor, also using Eq. 5.47, as

$$\delta_\lambda = \exp\left(-2\pi\frac{\text{Re}\sqrt{1 - 4a_o^{*2}}}{\text{Im}\sqrt{1 - 4a_o^{*2}}}\right) \tag{5.48}$$

Analogously, the second term in Eq. 5.13b represents an incoming wave propagating with the same apparent velocity c_a attenuating with the same decay factor

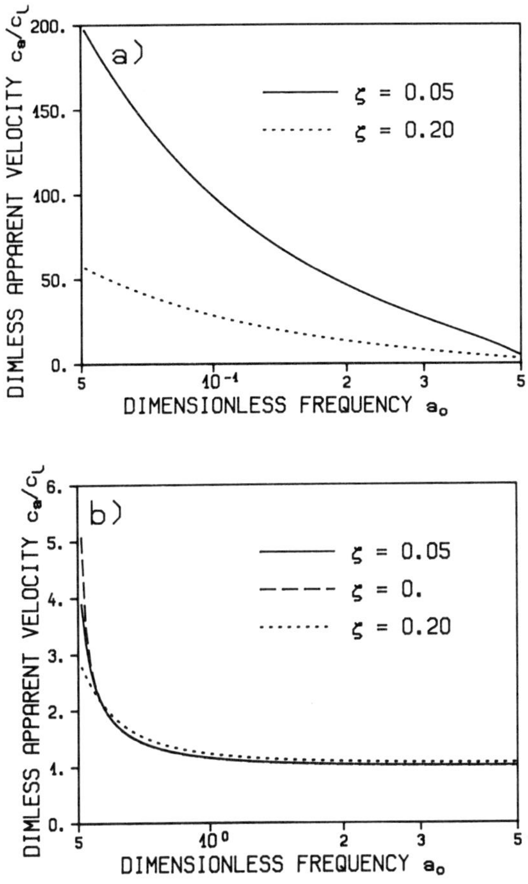

Figure 5-4 Dispersion. (a) $a_o < 0.5$; (b) $a_o > 0.5$.

δ_λ. The change in sign reflects the fact that the wave travels and thus attenuates in the negative z-direction.

For the damped case, the phase velocity c defined in Eq. 5.15 is complex:

$$c = \frac{2a_o c_l}{\sqrt{4a_o^{*2} - 1}} \tag{5.49}$$

It is algebraically straightforward to express c_a and δ_λ as functions of Re c and Im c.

$$c_a = \frac{|c|^2}{\text{Re}\,(c)} \tag{5.50}$$

$$\delta_\lambda = \exp\left[-2\pi \frac{\text{Im}\,(c)}{\text{Re}\,(c)}\right] \tag{5.51}$$

For a wave propagating in a prismatic rod, $c = c_l^*$. As is easily derived from Eq. 5.43, $\text{Im}\,(c_l^*)/\text{Re}\,(c_l^*) = \zeta$ (for small ζ). Thus the ratio Im (c)/Re (c) appearing in Eq. 5.51 can be interpreted for the rod with exponentially varying area as the effective damping ratio. This ratio is plotted in Fig. 5-5 for $\zeta = 0.05$. Very large values are obtained for $a_o < 0.5$, resulting in strong attenuation for this range of frequencies. It follows that below the cutoff frequency $a_o = 0.5$ which exists for the undamped case, the motion in the damped rod is only of academic interest.

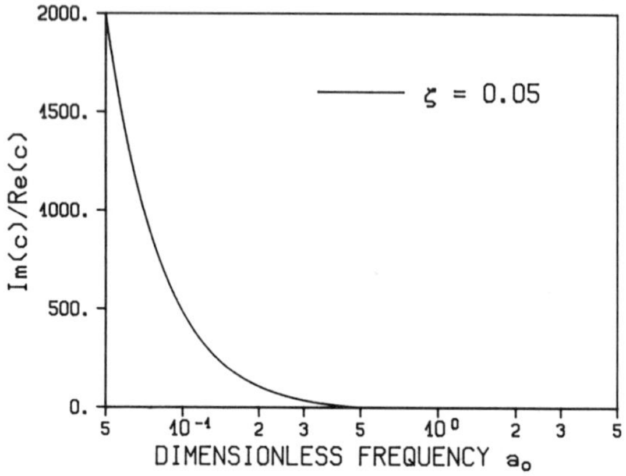

Figure 5-5 Attenuation.

The dynamic-stiffness coefficient of the infinite rod with material damping follows from Eqs. 5.29 and 5.31. Using the same (damping-independent) factor to nondimensionalize the coefficient (Eq. 5.29), $\bar{S}_1$ follows as

$$\bar{S}_1 = \tfrac{1}{2}(1 + 2\zeta i)(1 + \sqrt{1 - 4a_o^{*2}}) \tag{5.52}$$

It is important to note that the dynamic-stiffness coefficient of the undamped case cannot just be multiplied by the factor $1 + 2\zeta i$ to determine that of the damped case. As in the undamped case, $\bar{S}_1$ can be split up into its real and imaginary parts as defined in Eq. 5.32. It should be observed that a_o and not a_o^* appears in this formula. This results in the (real) spring and damping coefficients k_1 and c_1. The formulas apply for the whole range of frequencies.

$$k_1 = \tfrac{1}{2}(1 + \mathrm{Re}\sqrt{1 - 4a_o^{*2}} - 2\zeta\,\mathrm{Im}\sqrt{1 - 4a_o^{*2}}) \tag{5.53a}$$

$$c_1 = \frac{1}{2a_o}(2\zeta + 2\zeta\,\mathrm{Re}\sqrt{1 - 4a_o^{*2}} + \mathrm{Im}\sqrt{1 - 4a_o^{*2}}) \tag{5.53b}$$

The coefficients k_1 and c_1 depend on a_o and ζ. A damping force exists for all frequencies. Introducing damping increases c_1 as expected (especially around $a_o = 0.5$ and below) and decreases k_1 in the higher-frequency range. Even negative values of k_1 can arise for large a_o and ζ. In Fig. 5-6, the (dimensionless) dynamic-stiffness coefficient of the infinite rod is plotted as a function of a_o for a damping ratio $\zeta = 0.05$. The result of the extreme case $\zeta = 0.20$ is also shown. For comparison, the solution for $\zeta = 0$ is indicated as well.

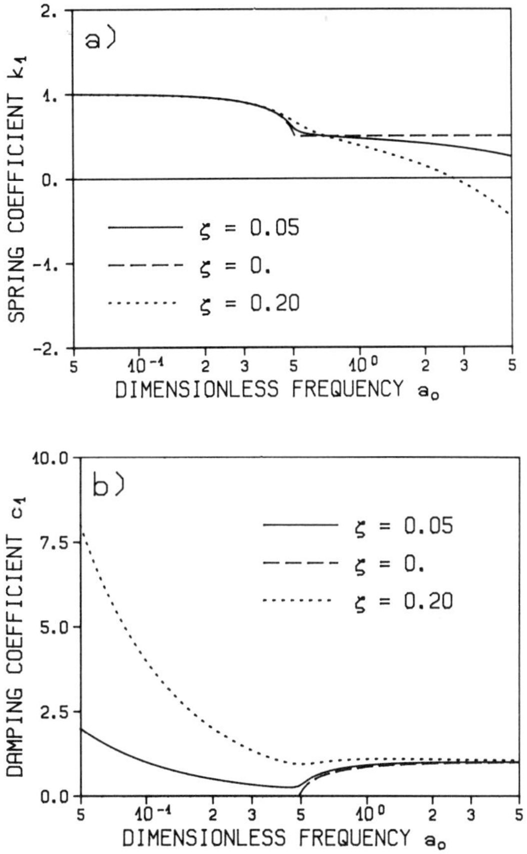

Figure 5-6 Dynamic-stiffness coefficient, with damping.

The rate of the energy transmission N can still be calculated by applying Eq. 5.41. The appropriate damping coefficient c_1 is to be used. Energy is radiated for all frequencies.

Applying the correspondence principle, the dynamic-stiffness matrix of a damped-rod element follows directly from Eqs. 5.23 and 5.24. Using the same factor to nondimensionalize as in the undamped case, for example, $\bar{S}_{11}$ is specified after substituting the definitions of γ_1 and γ_2 (Eq. 5.12) as

$$\bar{S}_{11} = \frac{1+2\zeta i}{2}\left[1 - \sqrt{1-4a_o^{*2}}\,\frac{\exp\left(-\frac{\alpha}{2}\sqrt{1-4a_o^{*2}}\right)+\exp\left(\frac{\alpha}{2}\sqrt{1-4a_o^{*2}}\right)}{\exp\left(-\frac{\alpha}{2}\sqrt{1-4a_o^{*2}}\right)-\exp\left(\frac{\alpha}{2}\sqrt{1-4a_o^{*2}}\right)}\right] \tag{5.54}$$

where $\alpha = d/f$.

5.1.9 Convergence of Dynamic-Stiffness Coefficient of Finite Rod to That of Infinite Rod

When damping is present, the coefficient $\bar{S}_{11}$ of the finite-rod element converges for $d \longrightarrow \infty$ (and thus $\alpha \longrightarrow \infty$) to the dynamic-stiffness matrix $\bar{S}_1$ of the infinite rod. As Re $\sqrt{1 - 4a_o^{*2}} > 0$, as discussed above,

$$\exp\left(-\frac{\alpha}{2}\sqrt{1 - 4a_o^{*2}}\right) = \exp\left(-\frac{\alpha}{2}\text{Re}\sqrt{1 - 4a_o^{*2}}\right)\exp\left(-i\frac{\alpha}{2}\text{Im}\sqrt{1 - 4a_o^{*2}}\right) \tag{5.55}$$

converges to zero for the limiting case of $\alpha \longrightarrow \infty$. The other term in the fraction involving the exponential function equals -1, leading to (Eq. 5.54).

$$\bar{S}_{\underset{\alpha \to \infty}{11}} = \frac{1 + 2\zeta i}{2}(1 + \sqrt{1 - 4a_o^{*2}}) \tag{5.56}$$

This expression is equal to $\bar{S}_1$, as is apparent from Eq. 5.52. This means that in the limit ($\alpha \longrightarrow \infty$), the finite rod with node 2 fixed ($w_2 = 0$) leads to the same force amplitude P_1 when excited by a displacement amplitude w_1 as the infinite rod under the same excitation. The wave propagating in the positive z-direction is reflected at node 2 and propagates back in the negative z-direction, attenuating exponentially with the distance traveled (first factor in Eq. 5.46). For a sufficiently large length d of the rod element the decay of this latter wave is so pronounced that from a practical point of view, no stress is generated in node 1 by this wave. Only the other wave, which is the outgoing wave, leads to stresses in node 1. This is obviously the same wave pattern as that one enforced when deriving $\bar{S}_1$ directly.

The length d of the finite rod has to be selected as quite large and depends, not surprisingly, on the frequency investigated and on the chosen damping ratio. Analogously as for $\bar{S}_1$ in Eq. 5.32, $\bar{S}_{11}$ can be split up into its real and imaginary parts k_{11} and $a_o c_{11}$. The ratio of the spring coefficient of the finite rod k_{11} and that of the infinite rod k_1 is plotted versus the dimensionless length α in Fig. 5-7a. The same representation for the damping coefficients c_{11}/c_1 is shown in Fig. 5-7b. The damping ratio ζ equals 0.05. Convergence is reached when the ratio equals 1. As expected, convergence is obtained for a much smaller α, when $a_o = 0.4$ (thus below the cutoff frequency $a_o = 0.5$ of the undamped case) than for $a_o = 0.6$, where, even for $\zeta = 0$, waves occur. It is interesting to note that k_{11}/k_1 exhibits larger oscillations than c_{11}/c_1 does. The same study of convergence is presented for $a_o = 1$ and $= 5$ in Fig. 5-8. As expected, the oscillation increases for these higher frequencies. For k_{11}/k_1, even the wrong sign can result for an α that is too small. For certain values of α, selecting a larger length results in a ratio that deviates more from 1. This means that increasing the dimensions of the finite-element mesh can also lead to a larger error. Increasing ζ to the value 0.20 improves the properties of convergence, as is visible for $a_o = 1$ and $= 5$ in Fig. 5-9. However, convergence is achieved to the value of the dynamic-

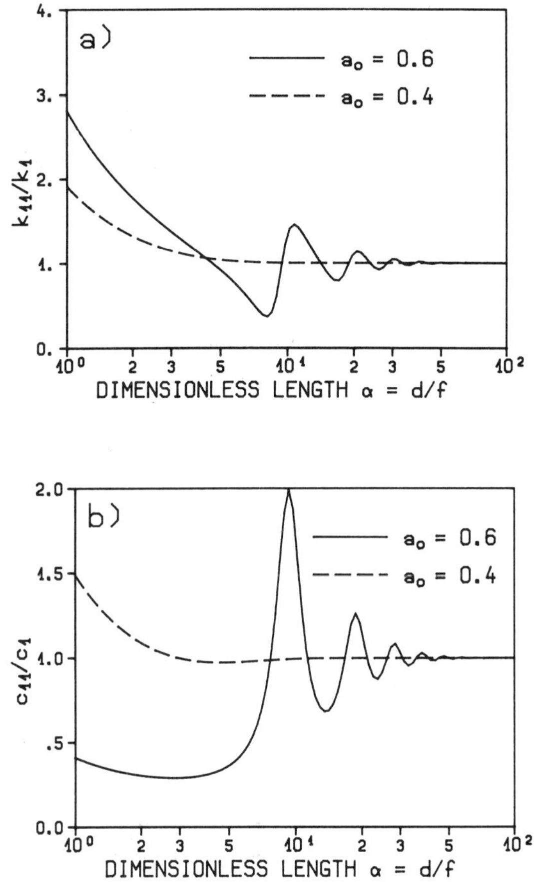

Figure 5-7 Convergence of dynamic-stiffness coefficient of finite rod, low-frequency range, $\zeta = 0.05$.

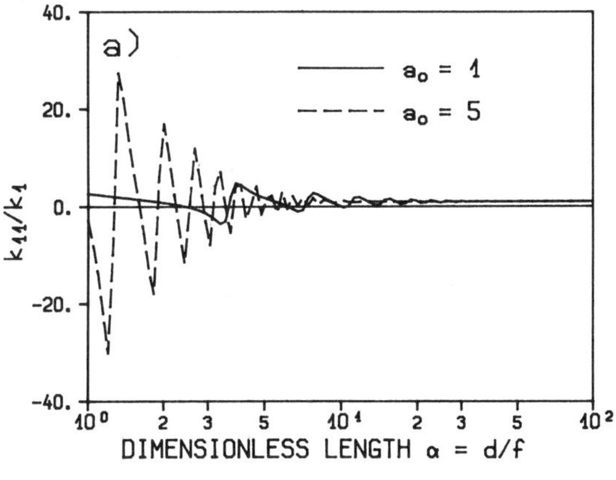

Figure 5-8 Convergence of dynamic-stiffness coefficient of finite rod, high-frequency range, $\zeta = 0.05$.

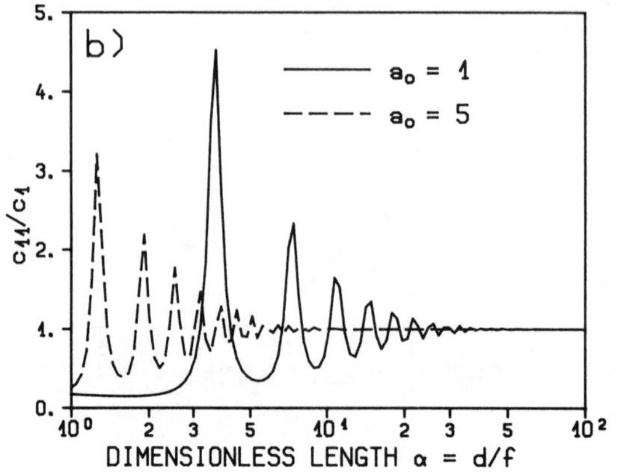

Figure 5-8 (Continued)

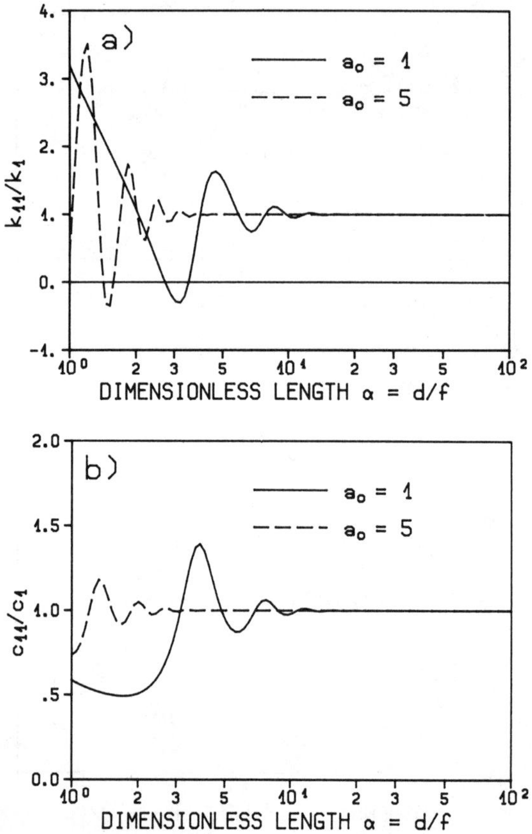

Figure 5-9 Convergence of dynamic-stiffness coefficient of finite rod, high-frequency range, $\zeta = 0.20$.

stiffness coefficient of the infinite rod corresponding to the selected damping coefficient. This value depends quite strongly on the damping coefficient ζ (Eq. 5.53), as is apparent from Table 5-1.

TABLE 5-1 Dynamic-Stiffness Coefficient of the Infinite Rod

Dimensionless Frequency, a_o	Damping Ratio, ζ: 0		0.05		0.20	
	k_1	c_1	k_1	c_1	k_1	c_1
0.40	0.80000	0.00000	0.80114	0.26613	0.81217	1.04458
0.60	0.50000	0.55277	0.52083	0.64343	0.57111	0.98962
1.00	0.50000	0.86603	0.47120	0.91795	0.38837	1.09580
5.00	0.50000	0.99499	0.25408	1.00625	−0.46612	1.05436

5.1.10 Free-Field Response

Returning to the task of determining the free-field motion $w(z)$ for a specified control motion w_c at the free surface, the dynamic-stiffness matrix of the rod element for $f \longrightarrow \infty$ is developed. Employing Eqs. 5.23 and 5.24 for this special case leads to

$$\begin{Bmatrix} P_1 \\ P_2 \end{Bmatrix} = \frac{E^* A_o}{\sin(\omega d/c_l^*)} \frac{\omega}{c_l^*} \begin{bmatrix} \cos \frac{\omega}{c_l^*} d & -1 \\ -1 & \cos \frac{\omega}{c_l^*} d \end{bmatrix} \begin{Bmatrix} w_1 \\ w_2 \end{Bmatrix} \tag{5.57}$$

Assuming a homogeneous site, node 1 can be selected at the free surface in the control point and node 2 at the depth d (Fig. 5-1). (By assembling the stiffness matrices for all rod elements located between the free surface and the depth z and by eliminating all intermediate degrees of freedom, Eq. 5.57 is easily generalized. See Section 6.2 for details.) At the free surface w_1 equals the control motion w_c, which is denoted as w_t, the displacement amplitude at the top of the site. The equation $P_1 = 0$ also applies. The displacement amplitude w_2, which equals w_b (the letter b stands for base), follows from the first equation of Eq. 5.57 as

$$w_b = \cos \frac{\omega}{c_l^*} d \, w_t \tag{5.58}$$

The subscripts t and b are introduced to achieve consistency with the nomenclature used in Chapter 6. For damping equal to zero, w_b is zero for $\omega_j d/c_l = (2j - 1)\pi/2, j = 1, 2, \ldots$, that is, for

$$\omega_j = \frac{2j - 1}{2} \pi \frac{c_l}{d} \tag{5.59}$$

in other words, for the natural frequencies of the soil column of depth d fixed at its base. The amplification within the site from the top of the base can be defined as the ratio of the absolute values of the displacement amplitudes $|w_b|/|w_t|$. This amplification is plotted in Fig. 5-10 versus (another) dimensionless frequency $\omega d/c_l$. The damping ratio $\zeta = 0.05$ is selected. From a practical point of view, the dips still occur at the natural frequencies of the soil column, but the corresponding value is no longer zero, because of the influence of damping. It is apparent that the frequency content of a transient motion is strongly affected by varying the depth.

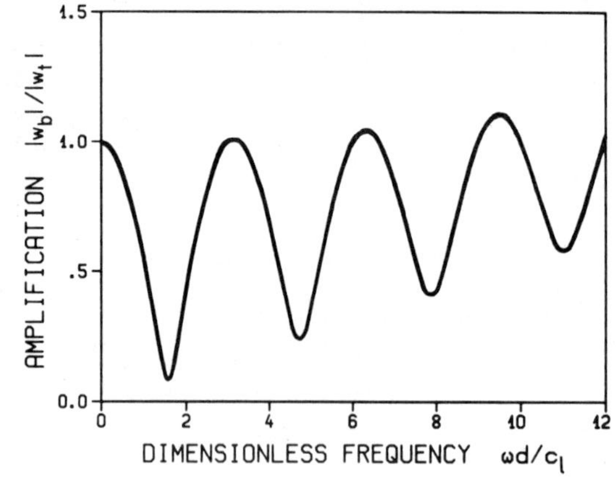

Figure 5-10 Amplification within site, $\zeta = 0.05$.

5.1.11 Dynamic-Stiffness Coefficient of Site

The other task, the calculation of the dynamic-stiffness coefficient S_o of the infinite soil, has already been discussed. For a homogeneous site, S_o is equal to the dynamic-stiffness coefficient of the infinite rod S_1. For a layered site, the dynamic-stiffness matrices of the rod elements and of the infinite rod (representing the half-space on which the layers rest) are assembled. Eliminating all degrees of freedom with the exceptions of that at point O results in S_o. This is discussed in detail in Section 7.2.

Finally, it should be stressed that an actual layered site should not be modeled with this one-dimensional model consisting of a rod with exponentially varying cross section. It is introduced only to discuss the salient features of wave propagation which will also appear in the two- and three-dimensional models but which are much more complicated to treat mathematically. In Problems 5.5 and 5.6, a conical shear beam is examined which can be used to calculate approximately the dynamic-stiffness coefficients in the horizontal and twisting directions. A similar model also exists for the rocking direction.

5.2 THREE-DIMENSIONAL WAVE EQUATION IN CARTESIAN COORDINATES

In the following, the well-known fundamental equations of elastodynamics are summarized. This allows the nomenclature to be defined. Only those aspects are discussed that are used directly in Chapters 6 and 7. In this section an isotropic homogeneous elastic medium is assumed at first. Hysteretic damping is introduced at a later stage.

5.2.1 Equation of Motion in Volumetric Strain and in Rotation Strains

For harmonic excitation with frequency ω the three-dimensional dynamic equilibrium equations in Cartesian coordinates x, y, z are equal to

$$\sigma_{x,x} + \tau_{xy,y} + \tau_{xz,z} = -\rho\omega^2 u \tag{5.60a}$$

$$\tau_{yx,x} + \sigma_{y,y} + \tau_{yz,z} = -\rho\omega^2 v \tag{5.60b}$$

$$\tau_{zx,x} + \tau_{zy,y} + \sigma_{z,z} = -\rho\omega^2 w \tag{5.60c}$$

No body forces are assumed to act. The normal- and shear-stress amplitudes are denoted as σ and τ, respectively. As usual, the first subscript denotes the direction of the stress component, the second the direction of the normal of the infinitesimal area that the stress component acts on. The displacement vector has the component amplitudes u, v, and w. The letter ρ represents the mass density. A comma denotes a partial derivative. All amplitudes are a function of x, y, and z.

The strain-displacement equations are formulated as

$$\epsilon_x = u_{,x} \tag{5.61a}$$

$$\epsilon_y = v_{,y} \tag{5.61b}$$

$$\epsilon_z = w_{,z} \tag{5.61c}$$

$$\gamma_{xy} = u_{,y} + v_{,x} \tag{5.61d}$$

$$\gamma_{xz} = u_{,z} + w_{,x} \tag{5.61e}$$

$$\gamma_{yz} = v_{,z} + w_{,y} \tag{5.61f}$$

The component amplitudes of the normal and shear strains are denoted by ϵ and γ, respectively.

Finally, Hooke's law, the constitutive equation, is specified as

$$\epsilon_x = \frac{1}{E}(\sigma_x - \nu\sigma_y - \nu\sigma_z) \tag{5.62a}$$

$$\epsilon_y = \frac{1}{E}(-\nu\sigma_x + \sigma_y - \nu\sigma_z) \tag{5.62b}$$

$$\epsilon_z = \frac{1}{E}(-\nu\sigma_x - \nu\sigma_y + \sigma_z) \tag{5.62c}$$

$$\gamma_{xy} = \frac{\tau_{xy}}{G} \tag{5.62d}$$

$$\gamma_{xz} = \frac{\tau_{xz}}{G} \tag{5.62e}$$

$$\gamma_{yz} = \frac{\tau_{yz}}{G} \tag{5.62f}$$

The shear modulus G can be expressed as a function of the modulus of elasticity E and Poisson's ratio ν:

$$G = \frac{E}{2(1+\nu)} \tag{5.63}$$

The surface-traction vector with component amplitudes t_x, t_y, and t_z expressed in the global-coordinate system acting on an infinitesimal element with a unit normal vector with components n_x, n_y, and n_z follows as

$$t_x = n_x\sigma_x + n_y\tau_{xy} + n_z\tau_{xz} \tag{5.64a}$$

$$t_y = n_x\tau_{yx} + n_y\sigma_y + n_z\tau_{yz} \tag{5.64b}$$

$$t_z = n_x\tau_{zx} + n_y\tau_{zy} + n_z\sigma_z \tag{5.64c}$$

The 15 components of the (symmetric) stress and strain tensors and of the displacement vector are related by the 15 equations of Eqs. 5.60, 5.61, and 5.62, which can be solved by taking the boundary conditions into account. The latter will enforce prescribed displacements and surface tractions (Eq. 5.64). Eliminating the strains from Eqs. 5.61 and 5.62 and substituting in Eq. 5.60 results in the three equations of motion expressed in the displacements and their derivatives (up to the second). All boundary conditions are also easily expressed as a function of the displacements and their derivatives (substituting in Eq. 5.64). The three displacements are coupled in the equations of motion (see Problem 5.7). They can be uncoupled by eliminating two displacements. The resulting equation is, however, of the fourth order. To avoid this, and at the same time to be able to identify the different types of waves, new variables are introduced: the volumetric strain with amplitude e and the rotation-strain vector $\{\Omega\}$ with the amplitude components Ω_x, Ω_y, and Ω_z. These are defined as

$$e = u,_x + v,_y + w,_z \tag{5.65}$$

and as

$$\Omega_x = \tfrac{1}{2}(w,_y - v,_z) \tag{5.66a}$$

$$\Omega_y = \tfrac{1}{2}(u,_z - w,_x) \tag{5.66b}$$

$$\Omega_z = \tfrac{1}{2}(v,_x - u,_y) \tag{5.66c}$$

Since

$$\Omega_{x,x} + \Omega_{y,y} + \Omega_{z,z} = 0 \tag{5.67}$$

four equations (Eqs. 5.60 and 5.67) exist for the four unknowns (e, Ω_x, Ω_y, Ω_z).

Using the new variables, the three equations of motion (Eq. 5.60) are rewritten as

$$(\lambda + 2G)e_{,x} + 2G(\Omega_{y,z} - \Omega_{z,y}) = -\rho\omega^2 u \tag{5.68a}$$

$$(\lambda + 2G)e_{,y} + 2G(\Omega_{z,x} - \Omega_{x,z}) = -\rho\omega^2 v \tag{5.68b}$$

$$(\lambda + 2G)e_{,z} + 2G(\Omega_{x,y} - \Omega_{y,x}) = -\rho\omega^2 w \tag{5.68c}$$

where the Lamé constant λ is expressed as

$$\lambda = \frac{\nu E}{(1 + \nu)(1 - 2\nu)} \tag{5.69}$$

Eliminating the rotations by differentiating Eq. 5.68a with respect to x, Eq. 5.68b with respect to y, and Eq. 5.68c with respect to z and adding the three relations results in

$$(\lambda + 2G)(e_{,xx} + e_{,yy} + e_{,zz}) = -\rho\omega^2 e \tag{5.70}$$

which is rewritten as

$$\nabla^2 e = -\frac{\omega^2}{c_p^2} e \tag{5.71}$$

The Laplace operator of a scalar a is denoted as $\nabla^2 a$ $(= a_{,xx} + a_{,yy} + a_{,zz})$. The variable c_p which will later be identified as the dilatational wave velocity is specified as

$$c_p^2 = \frac{\lambda + 2G}{\rho} \tag{5.72}$$

Eliminating the volumetric strain by differentiating Eq. 5.68c with respect to y, Eq. 5.68b with respect to z, subtracting the two expressions, and noting that the derivative of Eq. 5.67 with respect to x also vanishes leads to

$$G(\Omega_{x,xx} + \Omega_{x,yy} + \Omega_{x,zz}) = -\rho\omega^2\Omega_x \tag{5.73a}$$

Analogously,

$$G(\Omega_{y,xx} + \Omega_{y,yy} + \Omega_{y,zz}) = -\rho\omega^2\Omega_y \tag{5.73b}$$

$$G(\Omega_{z,xx} + \Omega_{z,yy} + \Omega_{z,zz}) = -\rho\omega^2\Omega_z \tag{5.73c}$$

follow. Introducing c_s, which will be interpreted below as the shear-wave velocity, defined as

$$c_s^2 = \frac{G}{\rho} \tag{5.74}$$

Eq. 5.73 is rewritten as

$$\nabla^2\{\Omega\} = -\frac{\omega^2}{c_s^2}\{\Omega\} \tag{5.75}$$

The equations of motion (for harmonic excitation) are specified in Eq. 5.71 with the amplitude of the volumetric strain e as the unknown and in Eq. 5.75 with the rotation-strain-vector amplitude $\{\Omega\}$. The components of the latter have to satisfy Eq. 5.67. These wave equations are linear partial differential equations of the second order.

5.2.2 P-Wave

The solution of Eq. 5.71 with one variable is addressed first. The result is displayed in Fig. 5-11a. By substituting the trial function

$$e = -\frac{i\omega}{c_p} A_P \exp\left[\frac{i\omega}{c_p}(-l_x x - l_y y - l_z z)\right] \tag{5.76}$$

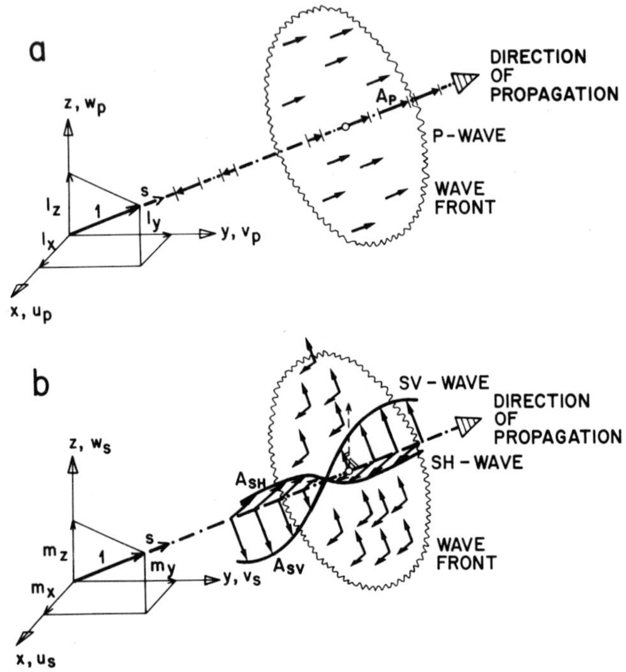

Figure 5-11 Displacements associated with body waves. (a) P-wave; (b) S-wave.

it is verified that the wave equation is satisfied, provided that the equation

$$l_x^2 + l_y^2 + l_z^2 = 1 \tag{5.77}$$

holds. The three scalars l_x, l_y, and l_z may be considered as the direction cosines of a straight line. To interpret the exponential function in Eq. 5.76, it is recalled that the harmonic motion is represented as $\exp(+i\omega t)$. An expression of the form $\exp[i\omega(t - s/c_p)]$ represents a wave propagating in the positive s-direction with the velocity c_p. Comparison with Eq. 5.76 leads to

$$s = l_x x + l_y y + l_z z \tag{5.78}$$

This scalar product shows that the coordinate s is measured along the straight line. It also follows that for a given time $t = t_o$, the amplitude of the volumetric strain is constant if s = constant. This is the equation of a plane (Eq. 5.78) normal to the direction of propagation.

The corresponding amplitudes of the displacements equal

$$u_p = l_x A_P \exp\left[\frac{i\omega}{c_p}(-l_x x - l_y y - l_z z)\right] \tag{5.79a}$$

$$v_p = l_y A_P \exp\left[\frac{i\omega}{c_p}(-l_x x - l_y y - l_z z)\right] \tag{5.79b}$$

$$w_p = l_z A_P \exp\left[\frac{i\omega}{c_p}(-l_x x - l_y y - l_z z)\right] \tag{5.79c}$$

Equation 5.65 is satisfied, as is verified by substituting Eq. 5.76 and making use of Eq. 5.77. It follows from Eq. 5.79 that A_P is the amplitude of a wave whose displacement vector coincides with the direction of propagation. This represents the definition of the dilatational or P-wave. The subscript p has been introduced to denote the corresponding displacements in Eq. 5.79. Summarizing, the particle motion of a P-wave with an amplitude A_P takes place along the direction of propagation (determined by the direction cosines l_x, l_y, l_z) and is constant over a plane perpendicular to it. The velocity of propagation c_p is constant and depends on material properties only.

5.2.3 S-Wave

Turning to the other wave equations (Eq. 5.75), a solution is analogously given to Eq. 5.76 by

$$\{\Omega\} = -\frac{i\omega}{2c_s}\{C\} \exp\left[\frac{i\omega}{c_s}(-m_x x - m_y y - m_z z)\right] \tag{5.80}$$

with

$$m_x^2 + m_y^2 + m_z^2 = 1 \tag{5.81}$$

$$m_x C_x + m_y C_y + m_z C_z = 0 \tag{5.82}$$

The direction of propagation is specified by the direction cosines m_x, m_y, and m_z (Fig. 5-11b). The velocity equals c_s. Equation 5.82 follows from Eq. 5.67. As the scalar product vanishes, the vector $\{C\}$ and thus $\{\Omega\}$ are perpendicular to the direction of propagation.

The corresponding displacement amplitudes equal

$$u_s = (m_z C_y - m_y C_z) \exp\left[\frac{i\omega}{c_s}(-m_x x - m_y y - m_z z)\right] \tag{5.83a}$$

$$v_s = (m_x C_z - m_z C_x) \exp\left[\frac{i\omega}{c_s}(-m_x x - m_y y - m_z z)\right] \tag{5.83b}$$

$$w_s = (m_y C_x - m_x C_y) \exp\left[\frac{i\omega}{c_s}(-m_x x - m_y y - m_z z)\right] \tag{5.83c}$$

Equation 5.66 is satisfied. From Eq. 5.83 it follows that the displacements are proportional to the components of the vector product of $\{C\}$ and the direction of propagation. This means that the particle motion of this wave lies in a plane perpendicular to the direction of propagation. This is the definition of the distortional or S-wave. The subscript s denotes the corresponding displacements in Eq. 5.83.

The trial functions selected for e and $\{\Omega\}$ (Eqs. 5.76 and 5.80) do not represent the most general solution of the wave equations. They correspond to plane waves.

In general, the displacement vector of the S-wave can be further decomposed into a horizontal component with an amplitude A_{SH} and into a component with an amplitude A_{SV} lying in the plane which contains the global vertical z-axis and the direction of propagation (Fig. 5-11b).

$$A_{SH} = \frac{C_z}{\sqrt{m_x^2 + m_y^2}} \tag{5.84a}$$

$$A_{SV} = \frac{m_x C_y - m_y C_x}{\sqrt{m_x^2 + m_y^2}} \tag{5.84b}$$

A_{SH} and A_{SV} are the amplitudes of the SH- and SV-waves, respectively. The displacements in terms of these amplitudes can be reformulated as

$$u_s = \frac{m_x m_z A_{SV} - m_y A_{SH}}{\sqrt{m_x^2 + m_y^2}} \exp\left[\frac{i\omega}{c_s}(-m_x x - m_y y - m_z z)\right] \tag{5.85a}$$

$$v_s = \frac{m_y m_z A_{SV} + m_x A_{SH}}{\sqrt{m_x^2 + m_y^2}} \exp\left[\frac{i\omega}{c_s}(-m_x x - m_y y - m_z z)\right] \tag{5.85b}$$

$$w_s = -\sqrt{m_x^2 + m_y^2}\, A_{SV} \exp\left[\frac{i\omega}{c_s}(-m_x x - m_y y - m_z z)\right] \tag{5.85c}$$

These expressions are easily verified using simple geometric considerations.

For a vertically propagating S-wave (parallel to the z-axis, $m_x = m_y = 0$) the definition of the SH- and SV-waves breaks down. Assuming quite arbitrarily that the SV-wave exhibits a motion in the x-z plane leads to

$$u_s = A_{SV} \exp\left(-\frac{i\omega}{c_s} z\right) \tag{5.86a}$$

$$v_s = A_{SH} \exp\left(-\frac{i\omega}{c_s} z\right) \tag{5.86b}$$

$$w_s = 0 \tag{5.86c}$$

Summarizing, the particle motion of an S-wave occurs in a plane perpendicular to the direction of propagation (determined by the direction cosines m_x, m_y, and m_z) and is constant over this plane. The horizontal component with an amplitude A_{SH} and the component with an amplitude A_{SV} (lying in a plane determined by the z-axis and the direction of propagation) propagate with the constant material-dependent velocity c_s.

In an infinite medium, these inclined body waves occur. As the three components l_x, l_y, l_z or m_x, m_y, m_z are all equal to or less than 1, they can be interpreted as the direction cosines of the direction of propagation of the P-wave or the S-wave, respectively. However, the formulation still remains valid for values larger than 1 and for nonreal values as long as Eq. 5.77 or Eq. 5.81 is satisfied. As will be discussed in depth later, this case corresponds to generalized surface waves.

5.2.4 Material Damping

The effect of material damping, which can be different for P- and S-waves, is examined next. Applying the correspondence principle (Section 2.4), material damping is taken into account by introducing complex material properties in Eqs. 5.72 and 5.74.

$$\lambda^* + 2G^* = (\lambda + 2G)(1 + 2\zeta_p i) \tag{5.87a}$$

$$G^* = G(1 + 2\zeta_s i) \tag{5.87b}$$

An asterisk always denotes a complex value. The ratios of the linear hysteretic damping for P- and S-waves are denoted as ζ_p and ζ_s, respectively. For $\zeta_p \neq \zeta_s$, Poisson's ratio ν will also be complex. This is deduced from Eqs. 5.63 and 5.69. Complex wave velocities result from Eqs. 5.72 and 5.74.

$$c_p^* = c_p\sqrt{1 + 2\zeta_p i} \tag{5.88a}$$

$$c_s^* = c_s\sqrt{1 + 2\zeta_s i} \tag{5.88b}$$

If $\zeta_p = \zeta_s$ is assumed, ζ is written without a subscript.

5.2.5 Total Motion

It is often reasonable to assume that the directions of propagation of the P- and S-waves lie in the same vertical plane (e.g., the x-z plane). Substituting $l_y = m_y = 0$ in Eqs. 5.79 and 5.85, adding the displacements caused by P- and S-waves, and taking damping into consideration leads to

$$u = l_x A_P \exp\left[i\omega\left(-\frac{l_x x}{c_p^*} - \frac{l_z z}{c_p^*}\right)\right] + m_z A_{SV} \exp\left[i\omega\left(-\frac{m_x x}{c_s^*} - \frac{m_z z}{c_s^*}\right)\right] \tag{5.89a}$$

$$v = A_{SH} \exp\left[i\omega\left(-\frac{m_x x}{c_s^*} - \frac{m_z z}{c_s^*}\right)\right] \tag{5.89b}$$

$$w = l_z A_P \exp\left[i\omega\left(-\frac{l_x x}{c_p^*} - \frac{l_z z}{c_p^*}\right)\right] - m_x A_{SV} \exp\left[i\omega\left(-\frac{m_x x}{c_s^*} - \frac{m_z z}{c_s^*}\right)\right] \tag{5.89c}$$

The in-plane displacements with the amplitudes u and w depend only on the P- and SV-waves. The out-of-plane displacement with the amplitude v is caused by the SH-wave and is thus independent of u and w. For this orientation of the directions of propagation, Eqs. 5.77 and 5.81 result in

$$l_x^2 + l_z^2 = 1 \tag{5.90}$$

$$m_x^2 + m_z^2 = 1 \tag{5.91}$$

To be able to analyze the layered half-space used as the model for the soil, boundary conditions (at the surface) and continuity requirements between adjacent layers and between the bottom layer and the half-space have to be formulated. Excitations, either as prescribed motions or as specified external loads, have to be processed. This can be achieved, as in conventional structural analysis, by assembling the contributions of the individual structural elements to form the total system. It is thus necessary to calculate the dynamic-stiffness matrices and the consistent-load vectors of a layer and of a half-space, which form the structural elements in this case. This is performed in the remainder of this chapter.

5.3 DYNAMIC-STIFFNESS MATRIX FOR OUT-OF-PLANE MOTION

5.3.1 Types of Waves

The horizontal layer of constant material properties shown in Fig. 5-12 represents the basic element for analyzing a layered site. The half-space can, as will be discussed, be regarded for certain derivations as the limiting case of a layer with the depth approaching infinity. The origin of the local coordinate system with the z-axis pointing downward is located at the top of the layer of depth d. For SH-waves, the out-of-plane displacement with the amplitude v is specified in Eq. 5.89b. The form of the equation compels the boundary conditions at the top and bottom of the layer to vary also as $\exp[-i\omega m_x x/c_s^*]$. As a total of two boundary conditions has to be satisfied [displacement $v(z, x)$ and/or stress $\tau_{yz}(z, x)$], a second wave with the same variation in x is introduced. For a given value of m_x (which can specify the angle of incidence), m_z can be selected as $\mp\sqrt{1 - m_x^2}$ (Eq. 5.91). Equation 5.89b is then reformulated as

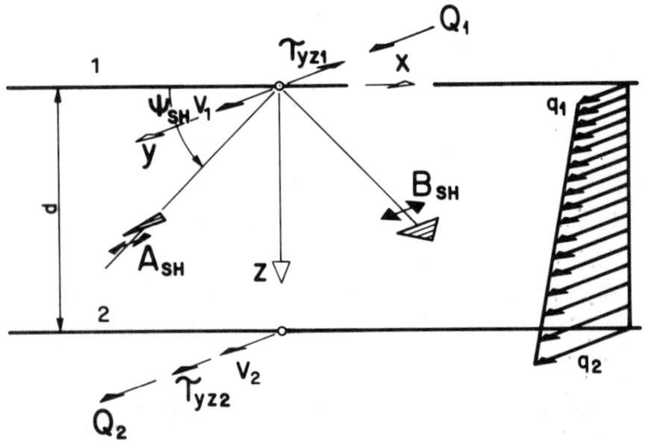

Figure 5-12 Nomenclature of layer for out-of-plane motion.

$$v(z, x) = \left[A_{SH} \exp\left(i\omega\frac{\sqrt{1 - m_x^2}}{c_s^*}z\right) + B_{SH} \exp\left(-i\omega\frac{\sqrt{1 - m_x^2}}{c_s^*}z\right)\right] \exp\left(-i\omega\frac{m_x}{c_s^*}x\right) \tag{5.92}$$

where A_{SH} and B_{SH} are the amplitudes of the waves traveling in the negative and positive z-directions, respectively (Fig. 5-12). In determining the direction of propagation it should be remembered that the harmonic motion is represented as $\exp(+i\omega t)$. The value m_x equals $\cos\psi_{SH}$, whereby ψ_{SH} is the angle of incidence measured from the horizontal. This interpretation of m_x holds only for a real value that is smaller than or equal to 1. Introducing the notation

$$c = \frac{c_s^*}{m_x} \tag{5.93}$$

$$k = \frac{\omega}{c} \tag{5.94}$$

$$t = \sqrt{\frac{1}{m_x^2} - 1} \tag{5.95a}$$

in Eq. 5.92 leads to

$$v(z, x) = [A_{SH} \exp(iktz) + B_{SH} \exp(-iktz)] \exp(-ikx) \tag{5.96}$$

The values c and k are the phase velocity and the wave number, respectively. It follows from Eq. 5.93 that for a layer without material damping, c is equal to the apparent velocity c_a discussed briefly in Section 4.2.1 (Eq. 4.1a and Fig. 4-2). For vertically incident waves ($\psi_{SH} = 90°$), $c = \infty$. The scalar t equals $\tan\psi_{SH}$. For a value of $m_x \leq 1$, t is a real positive value. The same also follows if t is defined slightly differently as

$$t = -i\sqrt{1 - \frac{1}{m_x^2}} \tag{5.95b}$$

As will be shown later, when surface waves are examined (Section 6.4), this definition also results in the correct sign, when $m_x > 1$. In all algebraic transformations, $\sqrt{-1} = +i$ can be used. Admittedly, at this stage of the derivation, Eq. 5.95b appears to be unnecessarily complicated.

5.3.2 Transfer- and Dynamic-Stiffness Matrices of Layer and of Half-Space

For the sake of conciseness, the term describing the variation with z in Eq. 5.96 is denoted as v or $v(z)$. Equation 5.96 can thus be rewritten as

$$v(z, x) = v(z) \exp(-ikx) \tag{5.97}$$

$v(z)$ can be interpreted as the amplitude of the wave traveling in the x-direction:

$$v(z) = A_{SH} \exp(iktz) + B_{SH} \exp(-iktz) \tag{5.98}$$

The amplitude of the shear stress $\tau_{yz}(z)$ calculated from Eqs. 5.62f and 5.61f with $w = 0$ using Eq. 5.87b as

$$\tau_{yz}(z) = G^* v_{,z} \tag{5.99}$$

equals

$$\tau_{yz}(z) = iktG^*[A_{\text{SH}} \exp(iktz) - B_{\text{SH}} \exp(-iktz)] \tag{5.100}$$

The other shear stress with amplitude τ_{xy} is not needed to calculate the stiffness matrix, as this stress does not act on the interface z = constant. The displacement and stress amplitudes at the top of the layer (subscript 1) follow from Eqs. 5.98 and 5.100 as

$$\begin{Bmatrix} v_1 \\ \tau_{yz1} \end{Bmatrix} = \begin{bmatrix} 1 & 1 \\ iktG^* & -iktG^* \end{bmatrix} \begin{Bmatrix} A_{\text{SH}} \\ B_{\text{SH}} \end{Bmatrix} \tag{5.101}$$

Expressing the corresponding values v_2 and τ_{yz2} at the bottom of the layer (subscript 2) as a function of A_{SH} and B_{SH} and then eliminating A_{SH} and B_{SH} using Eq. 5.101 results in the transfer matrix relating the state vector at the top of the layer to that at the bottom.

$$\begin{Bmatrix} v_2 \\ \tau_{yz2} \end{Bmatrix} = \begin{bmatrix} \cos ktd & (ktG^*)^{-1} \sin ktd \\ -ktG^* \sin ktd & \cos ktd \end{bmatrix} \begin{Bmatrix} v_1 \\ \tau_{yz1} \end{Bmatrix} \tag{5.102}$$

The vector composed of v and τ_{yz} describes the state. The transfer matrix is not symmetric. It must be realized that when assembling the stiffness matrix, the applied loads are defined in the global-coordinate system. The local system used to define the stresses is opposite to it on the negative face of an element. Introducing the external load amplitudes $Q_1 = -\tau_{yz1}$ and $Q_2 = \tau_{yz2}$ in Eq. 5.102 and performing a partial inversion (pivoting on the second element of the first row) leads to the (symmetric) dynamic-stiffness matrix of the layer $[S^L_{\text{SH}}]$.

$$\begin{Bmatrix} Q_1 \\ Q_2 \end{Bmatrix} = \frac{ktG^*}{\sin ktd} \begin{bmatrix} \cos ktd & -1 \\ -1 & \cos ktd \end{bmatrix} \begin{Bmatrix} v_1 \\ v_2 \end{Bmatrix} \tag{5.103}$$

The superscript L stands for the (soil) layer. In the dynamic-stiffness matrix, the out-of-plane motion is denoted by the subscript SH.

Applying a load at the free surface of a half-space, only an outgoing wave with amplitude B_{SH} is developed. The radiation condition, stating that no energy can propagate from infinity toward the free surface, excludes the incoming (incident) wave with the amplitude A_{SH}. The subscript o is used to denote the free surface of the half-space. Setting $A_{\text{SH}} = 0$ in Eq. 5.101, eliminating B_{SH} and with $Q_o = -\tau_{yz1}$,

$$Q_o = iktG^* v_o \tag{5.104}$$

results. For an undamped system, the dynamic-stiffness coefficient of a half-space S^R_{SH} is imaginary in contrast to the real $[S^L_{\text{SH}}]$ (for a real value of t). As the half-space is mostly used to represent the (bed)rock, a superscript R is introduced. The matrix $[S^L_{\text{SH}}]$ can be interpreted as a system of springs with frequency-dependent coefficients, for which the applied load is in phase with the displacement. In contrast, S^R_{SH} represents a damper with a coefficient tG/c, indicating that for the half-space, energy is radiated toward infinity. This is verified substituting Eq. 5.94 and with $\dot{v}_o = i\omega v_o$.

For a damped system, the dynamic-stiffness coefficient S^R_{SH} (Eq. 5.104) also follows from the first element of $[S^L_{SH}]$ (Eq. 5.103) for the limiting case $d \longrightarrow \infty$. This can be shown as follows: As discussed in connection with Eq. 5.95, t is a real positive value. Expanding c^*_s (Eq. 5.88b) into a Taylor series, it is verified that the imaginary term of c (Eq. 5.93) is positive and that of k (Eq. 5.94) negative. In the complex value of $kt = \text{Re}\,(kt) + i\,\text{Im}\,(kt)$, $\text{Im}\,(kt)$ is thus negative. The factor appearing in the first element of $[S^L_{SH}]$ is expanded using straightforward algebra as

$$\frac{\cos\,[(\text{Re}\,(kt) + i\,\text{Im}\,(kt))d]}{\sin\,[(\text{Re}\,(kt) + i\,\text{Im}\,(kt))d]}$$
$$= \frac{\sin\,[\text{Re}\,(kt)d]\cos\,[\text{Re}\,(kt)d](1 - \tanh^2\,[\text{Im}\,(kt)d])}{\sin^2\,[\text{Re}\,(kt)d] + \cos^2\,[\text{Re}\,(kt)d]\tanh^2\,[\text{Im}\,(kt)d]} \tag{5.105}$$
$$- i\frac{\tanh\,[\text{Im}\,(kt)d]}{\sin^2\,[\text{Re}\,(kt)d] + \cos^2\,[\text{Re}\,(kt)d]\tanh^2\,[\text{Im}\,(kt)d]}$$

As the $\tanh\,[\text{Im}\,(kt)d]$ (with a negative argument) converges for $d \longrightarrow \infty$ to the value -1, the real and imaginary parts converge to 0 and $+i$, respectively. The first element of $[S^L_{SH}]$ thus converges to $iktG^*$, which is equal to S^R_{SH}. It is worth mentioning that the off-diagonal term of $[S^L_{SH}]$ converges to zero.

It should be stressed that the amplitudes of the displacements v and of the loads Q (or stresses τ_{yz}) vary as $\exp\,(-ikx)$ in the horizontal direction. An expansion in the wave-number domain is thus performed, which is actually a Fourier transform. For a given ω, the stiffness matrix is a function of the phase velocity c (or of the wave number k), of the material properties (G, ρ, ζ_s), and of the depth d of the layer.

5.3.3 Special Cases

The following special cases are of interest, in which indefinite expressions arise in Eqs. 5.103 and 5.104. They occur in the applications in Chapters 6 and 7.

1. $\omega > 0$, $k = 0$

   This corresponds to vertically incident waves ($c = \infty$). The value kt converges to ω/c^*_s.

$$[S^L_{SH}] = \frac{G^*}{\sin\,(\omega d/c^*_s)}\frac{\omega}{c^*_s}\begin{bmatrix} \cos\dfrac{\omega d}{c^*_s} & -1 \\ -1 & \cos\dfrac{\omega d}{c^*_s} \end{bmatrix} \tag{5.106a}$$

$$S^R_{SH} = iG^*\frac{\omega}{c^*_s} \tag{5.106b}$$

   Equation 5.106a corresponds to Eq. 5.57.

2. $\omega = 0, k \neq 0$

$$[S^L_{SH}] = \frac{kG^*}{\sinh kd}\begin{bmatrix} \cosh kd & -1 \\ -1 & \cosh kd \end{bmatrix} \tag{5.107a}$$

$$S^R_{SH} = kG^* \tag{5.107b}$$

3. $\omega = 0, k = 0$

$$[S^L_{SH}] = \frac{G^*}{d}\begin{bmatrix} 1 & -1 \\ -1 & 1 \end{bmatrix} \tag{5.108a}$$

$$S^R_{SH} = 0 \tag{5.108b}$$

4. $\omega \neq 0, k \longrightarrow +\infty$
   The value kt converges to $-ik$, as in case 2, resulting in the same result as in the static case (Eq. 5.107). For the layer, the solution tends to that of the half-space.
5. $\omega \longrightarrow \infty, k \neq 0$
   The value kt converges to ω/c_s^*, as in case 1. For high-frequency excitation, the system behaves the same as for vertically incident waves (Eq. 5.106).

5.3.4 Loaded Layer

Distributed loads acting across the thickness of the layer are calculated as follows: The layer on which the distributed load acts is fixed at the two interfaces. The corresponding reaction forces (external loads) are calculated to achieve this condition, whereby the analysis can be restricted to the loaded layer. They are then applied with the opposite sign to the total system. To this global response, that of the fixed layer has to be added to calculate the total one. The analysis of the fixed layer can be performed as described below.

With the same variation with x [$= \exp(-ikx)$] as for the displacement v, the amplitude $q(z)$ for a linear variation with z equals (Fig. 5-12)

$$q(z) = q_1 + (q_2 - q_1)\frac{z}{d} \tag{5.109}$$

The dynamic-equilibrium equation (Eq. 5.60b) for the out-of-plane motion with a distributed load equals

$$\tau_{yx,x} + \tau_{yz,z} = -\rho\omega^2 v - q \tag{5.110}$$

The out-of-plane motion with the amplitude $v(z, x)$ (Eq. 5.97) leads to shear stresses with amplitudes $\tau_{yz}(z, x)$ (which have already been addressed, Eq. 5.99) and with amplitudes $\tau_{xy}(z, x)$, which are equal to (Eqs. 5.61d and 5.62d)

$$\tau_{xy}(z, x) = G^* v_{,x}(z, x) \tag{5.111}$$

Substituting Eq. 5.97 in Eq. 5.111 results in

$$\tau_{xy}(z, x) = -ikG^* v(z) \exp(-ikx) \tag{5.112a}$$

or

$$\tau_{xy}(z) = -ikG^* v \tag{5.112b}$$

Substituting Eqs. 5.112a and 5.99 in Eq. 5.110 and differentiating and introducing Eq. 5.109 leads to

$$-k^2G^*v + G^*v_{,zz} = -\rho\omega^2 v - \left(1 - \frac{z}{d}\right)q_1 - \frac{z}{d}q_2 \tag{5.113}$$

By inspection, a particular solution equals

$$v^p(z) = \frac{1}{k^2G^* - \rho\omega^2}\left[q_1 - (q_1 - q_2)\frac{z}{d}\right] \tag{5.114a}$$

or

$$v^p(z) = \frac{-1}{G^*k^2t^2}\left[q_1 - (q_1 - q_2)\frac{z}{d}\right] \tag{5.114b}$$

The superscript p denotes the particular solution. At the top and bottom of the layer (at the interfaces), the corresponding displacement amplitudes are

$$v_1^p = \frac{-q_1}{k^2G^*t^2} \tag{5.115a}$$

$$v_2^p = \frac{-q_2}{k^2G^*t^2} \tag{5.115b}$$

Substituting Eq. 5.114b in Eqs. 5.99 and 5.112b leads to

$$\tau_{yz}^p(z) = \frac{q_1 - q_2}{k^2t^2d} \tag{5.116a}$$

$$\tau_{xy}^p(z) = +\frac{i}{kt^2}\left[q_1 - (q_1 - q_2)\frac{z}{d}\right] \tag{5.116b}$$

The amplitudes of the external loads (reactions) at the top ($Q_1^p = -\tau_{yz1}^p$) and bottom of the layer ($Q_2^p = \tau_{yz2}^p$) follow as

$$Q_1^p = \frac{-q_1 + q_2}{k^2t^2d} \tag{5.117a}$$

$$Q_2^p = \frac{q_1 - q_2}{k^2t^2d} \tag{5.117b}$$

To fix the two interfaces, the homogeneous solution (superscript h) corresponding to the negative values of v_1^p and v_2^p (Eq. 5.115) has to be superimposed on the particular one. The external loads (reactions) Q_1^h and Q_2^h follow from Eq. 5.103 using $[S_{SH}^L]$ with $v_1^h = -v_1^p$ and $v_2^h = -v_2^p$. The total external loads Q_1 and Q_2 to be applied to the total system are equal to

$$Q_1 = -Q_1^p - Q_1^h \tag{5.118a}$$

$$Q_2 = -Q_2^p - Q_2^h \tag{5.118b}$$

To determine the local response, the homogeneous solution described by Eq. 5.98 (corresponding to v_1^h and v_2^h) has to be added to that of the particular one (Eq. 5.114). For the sake of conciseness, the equations, whose derivation is straightforward, are not specified explicitly. The case of a linearly distributed load acting on a line inclined with respect to the horizontal is discussed in Section 7.5.4.

5.3.5 Rate of Energy Transmission

Finally, the rate of horizontal energy transmission N for a layer at $x = 0$ is calculated. The $\tau_{yx}(z)$ does work with $v(z)$ on such a vertical plane. The rate of energy transmission N through $x = 0$ is defined as the product of the real parts of the force $\tau_{yx}(z)\,dz \exp(i\omega t)$ and of the velocity $\dot{v}(z) \exp(i\omega t)$, integrated over the depth d of the layer and averaged over a period $2\pi/\omega$

$$N = -\frac{\omega}{2\pi}\int_0^d \int_0^{2\pi/\omega} \mathrm{Re}\,[\tau_{yx}(z) \exp(i\omega t)]\,\mathrm{Re}\,[\dot{v}(z) \exp(i\omega t)]\,dt\,dz \tag{5.119}$$

Recalling that the displacement equals the product of the displacement amplitude (Eq. 5.97) and $\exp(i\omega t)$, it follows that

$$\dot{v}(z) = i\omega v(z) \tag{5.120}$$

The two factors appearing in Eq. 5.119 equal

$$\mathrm{Re}\,[\tau_{yx}(z) \exp(i\omega t)] = \mathrm{Re}\,[\tau_{yx}(z)] \cos \omega t - \mathrm{Im}\,[\tau_{yx}(z)] \sin \omega t \tag{5.121a}$$

$$\mathrm{Re}\,[\dot{v}(z) \exp(i\omega t)] = \mathrm{Re}\,[\dot{v}(z)] \cos \omega t - \mathrm{Im}\,[\dot{v}(z)] \sin \omega t \tag{5.121b}$$

Substituting Eq. 5.121 in Eq. 5.119 and averaging over one period $2\pi/\omega$ results in

$$N = -\tfrac{1}{2}\int_0^d [\mathrm{Re}\,(\dot{v}(z))\,\mathrm{Re}\,(\tau_{yx}(z)) + \mathrm{Im}\,(\dot{v}(z))\,\mathrm{Im}\,(\tau_{yx}(z))]\,dz \tag{5.122}$$

The integrand of this equation can be interpreted as the scalar product of the two vectors $\dot{v}(z)$ and $\tau_{yx}(z)$.

Substituting the definitions of $\dot{v}(z)$ given in Eq. 5.120 and of $\tau_{xy}(z)$ specified in Eq. 5.112b, in Eq. 5.122 leads to

$$N = \frac{\omega}{2}\mathrm{Re}\,(kG^*) \int_0^d |v(z)|^2\,dz \tag{5.123}$$

5.4 DYNAMIC-STIFFNESS MATRIX FOR IN-PLANE MOTION

5.4.1 Types of Waves

The nomenclature for the in-plane motion of the layer of constant material properties is shown in Fig. 5-13. The origin of the local coordinate system with the z-axis pointing downward is located at the top of the layer of depth d. For P- and SV-waves, the in-plane motion with amplitudes u and w are specified in Eqs. 5.89a and 5.89c. The form of these equations compels the boundary conditions at the top and bottom of the layer to vary as $\exp[-i\omega l_x x/c_p^*]$ and as $\exp[-i\omega m_x x/c_s^*]$. To achieve the same variation with x, which allows the analysis to be in effect, concentrated on the variation with z,

$$\frac{l_x}{c_p^*} = \frac{m_x}{c_s^*} \tag{5.124}$$

has to be enforced. As a total of four boundary conditions has to be satisfied (displacements with amplitudes u and w, stresses with amplitudes σ_z and τ_{xz}), a second P- and SV-wave with the same variation in x is introduced in Eqs. 5.89a and 5.89c. For a given value of l_x, l_z can be selected as $\mp\sqrt{1 - l_x^2}$ (Eq. 5.90). The value l_x equals $\cos \psi_P$, whereby ψ_P is the angle of incidence of the P-wave measured from the horizontal. This interpretation of l_x holds only for a real value which is smaller than or equal to 1. The top sign before the radical corresponds to a wave (with the amplitude A_P) traveling in the negative z-direction, as the harmonic motion is represented as $\exp(+i\omega t)$. The bottom sign corresponds to a wave (with the amplitude B_P) propagating in the positive z-direction. Analogously, m_z can be chosen as $\mp\sqrt{1 - m_x^2}$ with $m_x = \cos \psi_{SV}$, whereby ψ_{SV} is the angle of incidence of the SV-wave. The amplitudes A_{SV} and B_{SV} are associated with the SV-waves propagating in the negative and positive z-directions, respectively. This procedure is analogous to the derivation of Eq. 5.92 for the SH-waves. Equations 5.89a and 5.89c are then reformulated as

$$
\begin{aligned}
u(z, x) = {} & l_x\left[A_P \exp\left(i\omega\frac{\sqrt{1 - l_x^2}}{c_p^*}z\right) + B_P \exp\left(-i\omega\frac{\sqrt{1 - l_x^2}}{c_p^*}z\right)\right] \\
& \times \exp\left(-i\omega\frac{l_x}{c_p^*}x\right) - \sqrt{1 - m_x^2}\left[A_{SV} \exp\left(i\omega\frac{\sqrt{1 - m_x^2}}{c_s^*}z\right)\right. \\
& \left. - B_{SV} \exp\left(-i\omega\frac{\sqrt{1 - m_x^2}}{c_s^*}z\right)\right] \exp\left(-i\omega\frac{m_x}{c_s^*}x\right)
\end{aligned}
\tag{5.125a}
$$

$$
\begin{aligned}
w(z, x) = {} & -\sqrt{1 - l_x^2}\left[A_P \exp\left(i\omega\frac{\sqrt{1 - l_x^2}}{c_p^*}z\right) - B_P \exp\left(-i\omega\frac{\sqrt{1 - l_x^2}}{c_p^*}z\right)\right] \\
& \times \exp\left(-i\omega\frac{l_x}{c_p^*}x\right) - m_x\left[A_{SV} \exp\left(i\omega\frac{\sqrt{1 - m_x^2}}{c_s^*}z\right)\right. \\
& \left. + B_{SV} \exp\left(-i\omega\frac{\sqrt{1 - m_x^2}}{c_s^*}z\right)\right] \exp\left(-i\omega\frac{m_x}{c_s^*}x\right)
\end{aligned}
\tag{5.125b}
$$

It is convenient to introduce the following notation:

$$c = \frac{c_p^*}{l_x} = \frac{c_s^*}{m_x} \tag{5.126}$$

$$k = \frac{\omega}{c} \tag{5.127}$$

$$s = -i\sqrt{1 - \frac{1}{l_x^2}} \tag{5.128a}$$

$$t = -i\sqrt{1 - \frac{1}{m_x^2}} \tag{5.128b}$$

where c and k are the phase velocity and the wave number, respectively. It follows from Eq. 5.126 that the phase velocity c (and the wave number) are the same for the P- and SV-waves. For a layer without material damping, c is equal to the apparent velocity c_a (Eq. 4.1). This relationship is illustrated in Fig. 4-2. The values s and t are equal to tan ψ_P and tan ψ_{SV}, respectively. This somewhat strange definition of s and t is discussed in connection with Eq. 5.95b.

For the sake of conciseness, the terms describing the variation with z in Eq. 5.125 are denoted as u or $u(z)$ and w or $w(z)$, respectively. This equation can thus be rewritten as

$$u(z, x) = u(z) \exp(-ikx) \tag{5.129a}$$

$$w(z, x) = w(z) \exp(-ikx) \tag{5.129b}$$

where

$$u(z) = l_x[A_P \exp(iksz) + B_P \exp(-iksz)] - m_x t[A_{SV} \exp(iktz) - B_{SV} \exp(-iktz)] \tag{5.130a}$$

$$w(z) = -l_x s[A_P \exp(iksz) - B_P \exp(-iksz)] - m_x[A_{SV} \exp(iktz) + B_{SV} \exp(-iktz)] \tag{5.130b}$$

Here $u(z)$ and $w(z)$ can be interpreted as the amplitudes of the wave traveling in the x-direction.

5.4.2 Transfer- and Dynamic-Stiffness Matrices of Layer and of Half-Space

Using Eqs. 5.61a, 5.61c, 5.62a, 5.62c, 5.61e, and 5.62e, the amplitudes of the normal stress $\sigma_z(z)$ and the shear stress $\tau_{xz}(z)$ are expressed as

$$\sigma_z(z) = \lambda^*(u_{,x} + w_{,z}) + 2G^* w_{,z} \tag{5.131a}$$

$$\tau_{xz}(z) = G^*(u_{,z} + w_{,x}) \tag{5.131b}$$

Substituting Eqs. 5.129 and 5.130 in Eq. 5.131 and performing the differentiations with respect to x and z leads to

$$\sigma_z(z) = +ikl_x(1 - t^2)G^*[A_P \exp(iksz) + B_P \exp(-iksz)] - i2km_x tG^*[A_{SV} \exp(iktz) - B_{SV} \exp(-iktz)] \tag{5.132a}$$

$$\tau_{xz}(z) = i2kl_x sG^*[A_P \exp(iksz) - B_P \exp(-iksz)] + ikm_x(1 - t^2)G^*[A_{SV} \exp(iktz) + B_{SV} \exp(-iktz)] \tag{5.132b}$$

The other normal stress with the amplitude σ_x is not needed to calculate the stiffness matrix, as this stress does not act on the interface plane z = constant.

The displacement and stress amplitudes at the top $u_1, w_1, \tau_{xz1}, \sigma_{z1}$, and at the bottom of the layer $u_2, w_2, \tau_{xz2}, \sigma_{z2}$ follow from Eqs. 5.130 and 5.132 by introducing $z = 0$ and $= d$, respectively, as a function of the amplitudes A_P, A_{SV} of the incident waves and B_P, B_{SV} of the reflected waves. Eliminating these amplitudes results in the transfer matrix shown as Eq. 5.133 in Table 5-2. The state vector is composed of u, w, τ_{xz}, and σ_z. Introducing the external load amplitudes $P_1 = -\tau_{xz1}$, $R_1 = -\sigma_{z1}$, $P_2 = \tau_{xz2}$, and $R_2 = \sigma_{z2}$ defined in the global-coordinate system in Eq. 5.133 of Table 5-2 and performing a partial inversion leads to the dynamic-stiffness matrix of a layer $[S^L_{P\text{-}SV}]$, shown as Eq. 5.134 in Table 5-3. To achieve symmetry, R_1, R_2, w_1, and w_2 are multiplied by i.

TABLE 5-2 Transfer Matrix of Layer, Eq. 5.133

$$
\begin{Bmatrix} u_2 \\ w_2 \\ \tau_{xz2} \\ \sigma_{z2} \end{Bmatrix} = \frac{1}{1+t^2}
\left[\begin{array}{c:c:c:c}
\begin{matrix} 2\cos ksd \\ +(t^2-1)\cos ktd \end{matrix} & \begin{matrix} +i\dfrac{1-t^2}{s}\sin ksd \\ +\, i2t\sin ktd \end{matrix} & \begin{matrix} +\dfrac{1}{ksG^*}\sin ksd \\ +\dfrac{t}{kG^*}\sin ktd \end{matrix} & \begin{matrix} \dfrac{i}{kG^*}\cos ksd \\ -\dfrac{i}{kG^*}\cos ktd \end{matrix} \\ \hdashline
\begin{matrix} -\, i2s\sin ksd \\ -\, i\dfrac{1-t^2}{t}\sin ktd \end{matrix} & \begin{matrix} (t^2-1)\cos ksd \\ +\, 2\cos ktd \end{matrix} & \begin{matrix} \dfrac{i}{kG^*}\cos ksd \\ -\dfrac{i}{kG^*}\cos ktd \end{matrix} & \begin{matrix} +\dfrac{s}{kG^*}\sin ksd \\ +\dfrac{1}{ktG^*}\sin ktd \end{matrix} \\ \hdashline
\begin{matrix} -\, 4kG^*s\sin ksd \\ -kG^*\dfrac{(1-t^2)^2}{t}\sin ktd \end{matrix} & \begin{matrix} i2kG^*(1-t^2)\cos ksd \\ -\, i2kG^*(1-t^2)\cos ktd \end{matrix} & \begin{matrix} 2\cos ksd \\ +\,(t^2-1)\cos ktd \end{matrix} & \begin{matrix} -\, i2s\sin ksd \\ -\, i\dfrac{1-t^2}{t}\sin ktd \end{matrix} \\ \hdashline
\begin{matrix} i2kG^*(1-t^2)\cos ksd \\ -i2kG^*(1-t^2)\cos ktd \end{matrix} & \begin{matrix} -\, kG^*\dfrac{(1-t^2)^2}{s}\sin ksd \\ -\, 4kG^*t\sin ktd \end{matrix} & \begin{matrix} +i\dfrac{1-t^2}{s}\sin ksd \\ +\, i2t\sin ktd \end{matrix} & \begin{matrix} (t^2-1)\cos ksd \\ +\, 2\cos ktd \end{matrix}
\end{array}\right]
\begin{Bmatrix} u_1 \\ w_1 \\ \tau_{xz1} \\ \sigma_{z1} \end{Bmatrix}
$$

TABLE 5-3 Dynamic-Stiffness Matrix of Layer, Eq. 5.134

$$
\begin{Bmatrix} P_1 \\ iR_1 \\ P_2 \\ iR_2 \end{Bmatrix} = \frac{(1+t^2)kG^*}{D}
\begin{bmatrix}
\begin{array}{l} \frac{1}{t}\cos ksd \sin ktd \\ +s \sin ksd \cos ktd \end{array} &
\begin{array}{l} \frac{3-t^2}{1+t^2}(1-\cos ksd \cos ktd) \\ +\frac{1+2s^2t^2-t^2}{st(1+t^2)}\sin ksd \sin ktd \end{array} &
\begin{array}{l} -s \sin ksd \\ -\frac{1}{t}\sin ktd \end{array} &
\begin{array}{l} \cos ksd \\ -\cos ktd \end{array} \\
\begin{array}{l} \frac{3-t^2}{1+t^2}(1-\cos ksd \cos ktd) \\ +\frac{1+2s^2t^2-t^2}{st(1+t^2)}\sin ksd \sin ktd \end{array} &
\begin{array}{l} \frac{1}{s}\sin ksd \cos ktd \\ +t \cos ksd \sin ktd \end{array} &
\begin{array}{l} -\cos ksd \\ +\cos ktd \end{array} &
\begin{array}{l} -\frac{1}{s}\sin ksd \\ -t \sin ktd \end{array} \\
\begin{array}{l} -s \sin ksd \\ -\frac{1}{t}\sin ktd \end{array} &
\begin{array}{l} -\cos ksd \\ +\cos ktd \end{array} &
\begin{array}{l} \frac{1}{t}\cos ksd \sin ktd \\ +s \sin ksd \cos ktd \end{array} &
\begin{array}{l} \frac{t^2-3}{1+t^2}(1-\cos ksd \cos ktd) \\ +\frac{t^2-2s^2t^2-1}{st(1+t^2)}\sin ksd \sin ktd \end{array} \\
\begin{array}{l} \cos ksd \\ -\cos ktd \end{array} &
\begin{array}{l} -\frac{1}{s}\sin ksd \\ -t \sin ktd \end{array} &
\begin{array}{l} \frac{t^2-3}{1+t^2}(1-\cos ksd \cos ktd) \\ +\frac{t^2-2s^2t^2-1}{st(1+t^2)}\sin ksd \sin ktd \end{array} &
\begin{array}{l} \frac{1}{s}\sin ksd \cos ktd \\ +t \cos ksd \sin ktd \end{array}
\end{bmatrix}
\begin{Bmatrix} u_1 \\ iw_1 \\ u_2 \\ iw_2 \end{Bmatrix}
$$

where

$$D = 2(1 - \cos ksd \cos ktd) + \left(st + \frac{1}{st}\right) \sin ksd \sin ktd$$

As discussed for the out-of-plane motion in Section 5.3, the dynamic-stiffness matrix of the half-space $[S^R_{\text{P-SV}}]$ is derived by suppressing the incoming (incident) waves. Equations 5.130 and 5.132 are formulated for $z = 0$. The condition $A_P = A_{SV} = 0$ is then introduced and finally B_P and B_{SV} are eliminated. This leads to (with $P_o = -\tau_{xz1}$, $R_o = -\sigma_{z1}$)

$$\begin{Bmatrix} P_o \\ iR_o \end{Bmatrix} = kG^* \begin{bmatrix} \dfrac{is(1+t^2)}{1+st} & 2 - \dfrac{1+t^2}{1+st} \\ 2 - \dfrac{1+t^2}{1+st} & \dfrac{it(1+t^2)}{1+st} \end{bmatrix} \begin{Bmatrix} u_o \\ iw_o \end{Bmatrix} \tag{5.135}$$

The subscript o is used to denote the free surface of the half-space, which is denoted with a superscript R (for bedrock).

Again, for a medium without material damping, $[S^R_{\text{P-SV}}]$ is imaginary (for real values of s and t), which corresponds to a damper, in contrast to $[S^L_{\text{P-SV}}]$, which is real and can be interpreted as a spring. For the half-space, energy is radiated toward infinity.

For a damped system, the submatrix of order 2×2 on the diagonal of the dynamic-stiffness matrix of the layer converges to that of the half-space for the limiting case $d \longrightarrow \infty$. The off-diagonal submatrix of $[S^L_{\text{P-SV}}]$ converges to zero. This is demonstrated analogously as for the case of out-of-plane motion (see Eq. 5.105). In a damped layer, the imaginary parts of ks and of kt are both negative. The derivation is not presented in detail.

For a given ω, the stiffness matrix is a function of the phase velocity c (or of the wave number k), of the material properties (λ, G, ρ, ζ_p, ζ_s) and of the depth d of the layer.

5.4.3 Special Cases

Indefinite expressions arise in Eqs. 5.134 and 5.135 in special cases. The following equations then apply; the amplitudes of the displacement and of the external load in the vertical direction are still multiplied by i.

1. $\omega > 0, k = 0$

   This corresponds to vertically incident waves ($c = \infty$). The values ks and kt converge to ω/c_p^* and ω/c_s^*, respectively.

$$[S^L_{\text{P-SV}}] = G^* \frac{\omega}{c_s^*} \begin{bmatrix} \cot \dfrac{\omega d}{c_s^*} & 0 & -\dfrac{1}{\sin(\omega d/c_s^*)} & 0 \\ 0 & \dfrac{c_p^*}{c_s^*} \cot \dfrac{\omega d}{c_p^*} & 0 & -\dfrac{c_p^*}{c_s^*} \dfrac{1}{\sin(\omega d/c_p^*)} \\ -\dfrac{1}{\sin(\omega d/c_s^*)} & 0 & \cot \dfrac{\omega d}{c_s^*} & 0 \\ 0 & -\dfrac{c_p^*}{c_s^*} \dfrac{1}{\sin(\omega d/c_p^*)} & 0 & \dfrac{c_p^*}{c_s^*} \cot \dfrac{\omega d}{c_p^*} \end{bmatrix} \tag{5.136a}$$

TABLE 5-4 Dynamic-Stiffness Matrix of Layer for $\omega = 0$, $k \neq 0$, Eq. 5.137a

$$[S^L_{\text{P-SV}}] = \frac{2kG^*}{D}\left[\begin{array}{c|c|c|c}
\left(1+\frac{c_s^{*2}}{c_p^{*2}}\right)\sinh kd\cosh kd & -\left(1+\frac{c_s^{*2}}{c_p^{*2}}\right)\sinh^2 kd & \left(1-\frac{c_s^{*2}}{c_p^{*2}}\right)kd\cosh kd & kd\left(1-\frac{c_s^{*2}}{c_p^{*2}}\right)\sinh kd \\
-\left(1-\frac{c_s^{*2}}{c_p^{*2}}\right)kd & +D & -\left(1+\frac{c_s^{*2}}{c_p^{*2}}\right)\sinh kd & \\
\hdashline
-\left(1+\frac{c_s^{*2}}{c_p^{*2}}\right)\sinh^2 kd & \left(1+\frac{c_s^{*2}}{c_p^{*2}}\right)\sinh kd\cosh kd & -kd\left(1-\frac{c_s^{*2}}{c_p^{*2}}\right)\sinh kd & -\left(1-\frac{c_s^{*2}}{c_p^{*2}}\right)kd\cosh kd \\
+D & +kd\left(1-\frac{c_s^{*2}}{c_p^{*2}}\right) & & -\left(1+\frac{c_s^{*2}}{c_p^{*2}}\right)\sinh kd \\
\hdashline
\left(1-\frac{c_s^{*2}}{c_p^{*2}}\right)kd\cosh kd & -kd\left(1-\frac{c_s^{*2}}{c_p^{*2}}\right)\sinh kd & \left(1+\frac{c_s^{*2}}{c_p^{*2}}\right)\sinh kd\cosh kd & \left(1+\frac{c_s^{*2}}{c_p^{*2}}\right)\sinh^2 kd \\
-\left(1+\frac{c_s^{*2}}{c_p^{*2}}\right)\sinh kd & & -\left(1-\frac{c_s^{*2}}{c_p^{*2}}\right)kd & -D \\
\hdashline
kd\left(1-\frac{c_s^{*2}}{c_p^{*2}}\right)\sinh kd & -\left(1-\frac{c_s^{*2}}{c_p^{*2}}\right)kd\cosh kd & \left(1+\frac{c_s^{*2}}{c_p^{*2}}\right)\sinh^2 kd & \left(1+\frac{c_s^{*2}}{c_p^{*2}}\right)\sinh kd\cosh kd \\
 & -\left(1+\frac{c_s^{*2}}{c_p^{*2}}\right)\sinh kd & -D & +kd\left(1-\frac{c_s^{*2}}{c_p^{*2}}\right)
\end{array}\right]$$

where

$$D = \left(1+\frac{c_s^{*2}}{c_p^{*2}}\right)^2\sinh^2 kd - k^2d^2\left(1-\frac{c_s^{*2}}{c_p^{*2}}\right)^2$$

$$[S^R_{\text{P-SV}}] = iG^*\frac{\omega}{c_s^*}\begin{bmatrix} 1 & 0 \\ 0 & \dfrac{c_p^*}{c_s^*} \end{bmatrix} \tag{5.136b}$$

The horizontal and vertical directions are uncoupled.

2. $\omega = 0, k \neq 0$
   $[S^L_{\text{P-SV}}]$ is specified as Eq. 5.137a in Table 5-4.

$$[S^R_{\text{P-SV}}] = 2kG^*\begin{bmatrix} \dfrac{1}{1 + c_s^{*2}/c_p^{*2}} & \dfrac{1}{1 + c_p^{*2}/c_s^{*2}} \\ \dfrac{1}{1 + c_p^{*2}/c_s^{*2}} & \dfrac{1}{1 + c_s^{*2}/c_p^{*2}} \end{bmatrix} \tag{5.137b}$$

3. $\omega = 0, k = 0$

$$[S^L_{\text{P-SV}}] = \frac{G^*}{d}\begin{bmatrix} 1 & 0 & -1 & 0 \\ 0 & \dfrac{c_p^{*2}}{c_s^{*2}} & 0 & -\dfrac{c_p^{*2}}{c_s^{*2}} \\ -1 & 0 & 1 & 0 \\ 0 & -\dfrac{c_p^{*2}}{c_s^{*2}} & 0 & \dfrac{c_p^{*2}}{c_s^{*2}} \end{bmatrix} \tag{5.138a}$$

$$[S^R_{\text{P-SV}}] = \begin{bmatrix} 0 & 0 \\ 0 & 0 \end{bmatrix} \tag{5.138b}$$

4. $\omega \neq 0, k \longrightarrow +\infty$
   As in the case of the SH-wave, the system acts statically.
5. $\omega \longrightarrow \infty, k \neq 0$
   Once again, for this high-frequency excitation, the system behaves as for vertically incident waves.

5.4.4 Loaded Layer

Distributed loads acting across the thickness of the layer are calculated analogously as for the out-of-plane motion. For a linear variation with z, the amplitudes $p(z)$ and $r(z)$ in the x- and z-directions (Fig. 5-13) equal

$$p(z) = p_1 + (p_2 - p_1)\frac{z}{d} \tag{5.139a}$$

$$r(z) = r_1 + (r_2 - r_1)\frac{z}{d} \tag{5.139b}$$

The variation in the x-direction is the same as for the other variables [exp $(-ikx)$]. The dynamic-equilibrium equations (Eqs. 5.60a and 5.60c) for the in-plane motion with a distributed load equal

$$\sigma_{x,x} + \tau_{xz,z} = -\rho\omega^2 u - p \tag{5.140a}$$

$$\tau_{zx,x} + \sigma_{z,z} = -\rho\omega^2 w - r \tag{5.140b}$$

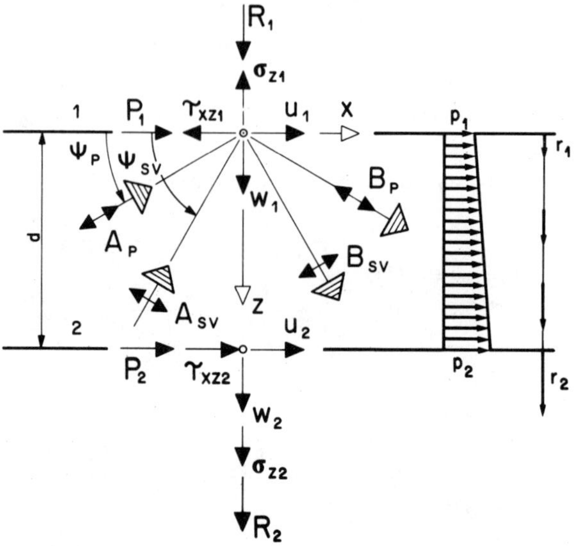

Figure 5-13 Nomenclature of layer for in-plane motion.

Using Eqs. 5.61a, 5.61c, 5.62a, and 5.62c, the amplitude of the (other) normal stress follows.

$$\sigma_x(z) = \lambda^*(u_{,x} + w_{,z}) + 2G^* u_{,x} \tag{5.141}$$

Substituting the stress-displacement relations (Eqs. 5.131 and 5.141), introducing Eq. 5.139 and performing the differentiations with respect to x leads to

$$\begin{aligned} &-(\lambda^* + 2G^*)k^2 u + G^* u_{,zz} - ik(\lambda^* + G^*)w_{,z} \\ &= -\rho\omega^2 u - \left(1 - \frac{z}{d}\right)p_1 - \frac{z}{d}p_2 \end{aligned} \tag{5.142a}$$

$$\begin{aligned} &-ik(\lambda^* + G^*)u_{,z} - k^2 G^* w + (\lambda^* + 2G^*)w_{,zz} \\ &= -\rho\omega^2 w - \left(1 - \frac{z}{d}\right)r_1 - \frac{z}{d}r_2 \end{aligned} \tag{5.142b}$$

A linear variation with z represents a particular solution (superscript p)

$$u^p(z) = a_1 + a_2\frac{z}{d} \tag{5.143a}$$

$$w^p(z) = c_1 + c_2\frac{z}{d} \tag{5.143b}$$

Substituting Eq. 5.143 in Eq. 5.142 and identifying the constant and linear terms leads to

$$a_1 = -\frac{1}{k^2 G^* s^2}\frac{c_s^{*2}}{c_p^{*2}}p_1 + \frac{i}{k^3 G^* d s^2 t^2}\left(1 - \frac{c_s^{*2}}{c_p^{*2}}\right)(r_1 - r_2) \tag{5.144a}$$

$$a_2 = \frac{1}{k^2 G^* s^2} \frac{c_s^{*2}}{c_p^{*2}}(p_1 - p_2) \tag{5.144b}$$

$$c_1 = \frac{i}{k^3 G^* d s^2 t^2}\left(1 - \frac{c_s^{*2}}{c_p^{*2}}\right)(p_1 - p_2) - \frac{1}{k^2 G^* t^2} r_1 \tag{5.144c}$$

$$c_2 = \frac{1}{k^2 G^* t^2}(r_1 - r_2) \tag{5.144d}$$

The corresponding displacements at the top and bottom of the layer follow from Eq. 5.143, setting $z = 0$ and $= d$, respectively. For the sake of consistency, the amplitudes of the loads and of the displacements in the z-direction are multiplied by i.

$$\begin{Bmatrix} u_1^p \\ iw_1^p \\ u_2^p \\ iw_2^p \end{Bmatrix} = \frac{1}{k^2 G^*}$$

$$\times \begin{bmatrix} -\frac{1}{s^2}\frac{c_s^{*2}}{c_p^{*2}} & \frac{1}{kds^2t^2}\left(1 - \frac{c_s^{*2}}{c_p^{*2}}\right) & 0 & -\frac{1}{kds^2t^2}\left(1 - \frac{c_s^{*2}}{c_p^{*2}}\right) \\ -\frac{1}{kds^2t^2}\left(1 - \frac{c_s^{*2}}{c_p^{*2}}\right) & -\frac{1}{t^2} & \frac{1}{kds^2t^2}\left(1 - \frac{c_s^{*2}}{c_p^{*2}}\right) & 0 \\ 0 & \frac{1}{kds^2t^2}\left(1 - \frac{c_s^{*2}}{c_p^{*2}}\right) & -\frac{1}{s^2}\frac{c_s^{*2}}{c_p^{*2}} & -\frac{1}{kds^2t^2}\left(1 - \frac{c_s^{*2}}{c_p^{*2}}\right) \\ -\frac{1}{kds^2t^2}\left(1 - \frac{c_s^{*2}}{c_p^{*2}}\right) & 0 & \frac{1}{kds^2t^2}\left(1 - \frac{c_s^{*2}}{c_p^{*2}}\right) & -\frac{1}{t^2} \end{bmatrix}$$

$$\times \begin{Bmatrix} p_1 \\ ir_1 \\ p_2 \\ ir_2 \end{Bmatrix} \tag{5.145}$$

Substituting Eq. 5.143 in the stress-displacement relations (Eqs. 5.131 and 5.141), $\tau_{xz}^p(z)$, $\sigma_z^p(z)$, and $\sigma_x^p(z)$ follow. The amplitudes of the external loads (reactions) at the top ($P_1^p = -\tau_{xz1}^p$, $R_1^p = -\sigma_{z1}^p$) and bottom of the layer ($P_2^p = \tau_{xz2}^p$, $R_2^p = \sigma_{z2}^p$) are equal to

$$\begin{Bmatrix} P_1^p \\ iR_1^p \\ P_2^p \\ iR_2^p \end{Bmatrix} = \frac{G^*}{d} \begin{bmatrix} 1 & kd & -1 & 0 \\ -kd\left(\frac{c_p^{*2}}{c_s^{*2}} - 2\right) & \frac{c_p^{*2}}{c_s^{*2}} & 0 & -\frac{c_p^{*2}}{c_s^{*2}} \\ -1 & 0 & 1 & -kd \\ 0 & -\frac{c_p^{*2}}{c_s^{*2}} & kd\left(\frac{c_p^{*2}}{c_s^{*2}} - 2\right) & \frac{c_p^{*2}}{c_s^{*2}} \end{bmatrix} \begin{Bmatrix} u_1^p \\ iw_1^p \\ u_2^p \\ iw_2^p \end{Bmatrix} \tag{5.146}$$

Substituting Eq. 5.145 in this equation, the external reactions are expressed as a function of the applied loads. From here on, the procedure is analogous to that described for the out-of-plane motion and will not be repeated. The case of a linearly distributed load acting on a line inclined with respect to the horizontal is sketched in Section 7.5.4.

5.4.5 Rate of Energy Transmission

The rate of horizontal transmission N for a layer at $x = 0$ is calculated next. On such a vertical plane, $\sigma_x(z)$ and $\tau_{zx}(z)$ do work with u and w, respectively. The contribution of the normal stress to N is equal to the product of the real parts of the force $\sigma_x(z)\,dz\exp(i\omega t)$ and of the velocity $\dot{u}(z)\exp(i\omega t)$, integrated over the depth d of the layer and averaged over a period $2\pi/\omega$.

$$N = -\frac{\omega}{2\pi}\int_0^d \int_0^{2\pi/\omega} (\mathrm{Re}\,[\sigma_x(z)\exp(i\omega t)]\,\mathrm{Re}\,[\dot{u}(z)\exp(i\omega t)] + \mathrm{Re}\,[\tau_{zx}(z)\exp(i\omega t)]\,\mathrm{Re}\,[\dot{w}(z)\exp(i\omega t)])\,dt\,dz \tag{5.147}$$

As the displacement is equal to the amplitude times $\exp(i\omega t)$, the velocity amplitudes are equal to

$$\dot{u}(z) = i\omega u(z) \tag{5.148a}$$

$$\dot{w}(z) = i\omega w(z) \tag{5.148b}$$

As an example for determining the factors in Eq. 5.147,

$$\mathrm{Re}\,[\dot{u}(z)\exp(i\omega t)] = \mathrm{Re}\,[\dot{u}(z)]\cos\omega t - \mathrm{Im}\,[\dot{u}(z)]\sin\omega t \tag{5.149}$$

follows. Substituting this type of expression in Eq. 5.147 and averaging over one period $2\pi/\omega$ results in

$$N = -\tfrac{1}{2}\int_0^d (\mathrm{Re}\,[\sigma_x(z)]\,\mathrm{Re}\,[\dot{u}(z)] + \mathrm{Im}\,[\sigma_x(z)]\,\mathrm{Im}\,[\dot{u}(z)] + \mathrm{Re}\,[\tau_{zx}(z)]\,\mathrm{Re}\,[\dot{w}(z)] + \mathrm{Im}\,[\tau_{zx}(z)]\,\mathrm{Im}\,[\dot{w}(z)])\,dz \tag{5.150}$$

The integrand of this equation can be interpreted as the sum of the scalar products of the vectors $\sigma_x(z)$ and $\dot{u}(z)$ and of the vectors $\tau_{zx}(z)$ and $\dot{w}(z)$.

5.5 THREE-DIMENSIONAL WAVE EQUATION IN CYLINDRICAL COORDINATES

For the calculation of the dynamic-stiffness matrix of a three-dimensional foundation, cylindrical coordinates are introduced (Section 7.2.3). To be able to demonstrate that the same dynamic-stiffness matrices arise in cylindrical coordinates as for plane waves in Cartesian coordinates, the derivation of the wave equation in cylindrical coordinates is summarized in the following. As in Section 5.2, the main objective is to define the nomenclature used in these well-known relations.

5.5.1 Equation of Motion in Volumetric Strain and in Rotation Strains

In the cylindrical coordinate system r, θ, z shown in Fig. 5-14, the dynamic-equilibrium equations for harmonic excitation are formulated as

$$\sigma_{r,r} + \frac{1}{r}\tau_{r\theta,\theta} + \tau_{rz,z} + \frac{\sigma_r - \sigma_\theta}{r} = -\rho\omega^2 u \tag{5.151a}$$

$$\tau_{\theta r,r} + \frac{1}{r}\sigma_{\theta,\theta} + \tau_{\theta z,z} + \frac{2}{r}\tau_{\theta r} = -\rho\omega^2 v \tag{5.151b}$$

$$\tau_{zr,r} + \frac{1}{r}\tau_{z\theta,\theta} + \sigma_{zz,z} + \frac{\tau_{zr}}{r} = -\rho\omega^2 w \tag{5.151c}$$

The amplitudes of the displacements u, v, and w are defined in the radial, circumferential, and vertical directions and the stress amplitudes illustrated in Fig. 5-14 are functions of r, θ, and z. These arguments are omitted from the equations for the sake of conciseness.

The strain-displacement equations are equal to

$$\epsilon_r = u_{,r} \tag{5.152a}$$

$$\epsilon_9 = \frac{u}{r} + \frac{1}{r}v_{,\theta} \tag{5.152b}$$

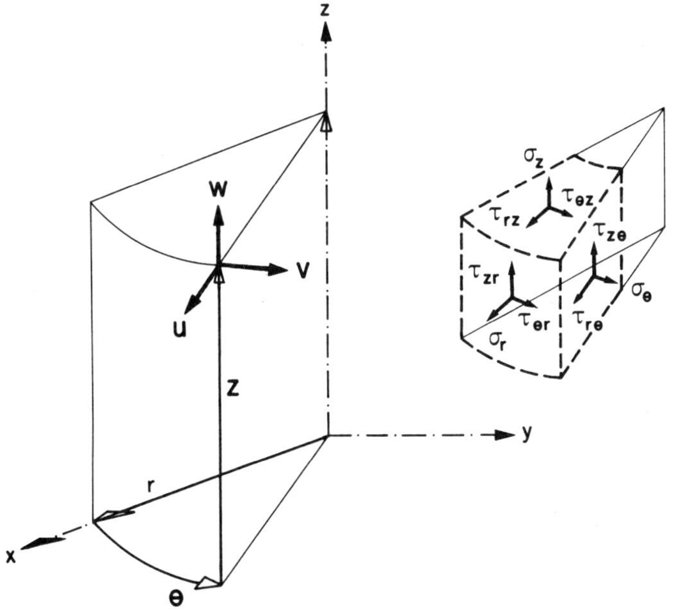

Figure 5-14 Cylindrical coordinate system with displacements and stresses.

$$\epsilon_z = w,_z \tag{5.152c}$$

$$\gamma_{r\theta} = \frac{1}{r}u,_\theta + v,_r - \frac{v}{r} \tag{5.152d}$$

$$\gamma_{rz} = u,_z + w,_r \tag{5.152e}$$

$$\gamma_{\theta z} = v,_z + \frac{1}{r}w,_\theta \tag{5.152f}$$

Hooke's law specified in Eq. 5.62 still applies, changing x and y to r and θ.

As for Cartesian coordinates, the volumetric strain with amplitude e and the rotation-strain vector with the amplitude components Ω_r, Ω_θ, and Ω_z are introduced. In cylindrical coordinates, they are defined as

$$e = u,_r + \frac{u}{r} + \frac{1}{r}v,_\theta + w,_z \tag{5.153}$$

$$\Omega_r = \tfrac{1}{2}\left(-v,_z + \frac{1}{r}w,_\theta\right) \tag{5.154a}$$

$$\Omega_\theta = \tfrac{1}{2}(u,_z - w,_r) \tag{5.154b}$$

$$\Omega_z = \tfrac{1}{2}\left(-\frac{1}{r}u,_\theta + v,_r + \frac{v}{r}\right) \tag{5.154c}$$

The components of the rotation-strain vector satisfy the following condition:

$$\Omega_{r,r} + \frac{\Omega_r}{r} + \frac{1}{r}\Omega_{\theta,\theta} + \Omega_{z,z} = 0 \tag{5.155}$$

The three equations of motion (Eq. 5.151) are expressed in the new variables as

$$(\lambda^* + 2G^*)e,_r + 2G^*\left(\Omega_{\theta,z} - \frac{1}{r}\Omega_{z,\theta}\right) = -\rho\omega^2 u \tag{5.156a}$$

$$\frac{\lambda^* + 2G^*}{r}e,_\theta + 2G^*(-\Omega_{r,z} + \Omega_{z,r}) = -\rho\omega^2 v \tag{5.156b}$$

$$(\lambda^* + 2G^*)e,_z + 2G^*\left(\frac{1}{r}\Omega_{r,\theta} - \Omega_{\theta,r} - \frac{\Omega_\theta}{r}\right) = -\rho\omega^2 w \tag{5.156c}$$

To uncouple these equations of motion, the three potentials with amplitudes φ, ψ, and χ are introduced. The components of the displacements are expressed in terms of the derivatives of these potentials as

$$u = \varphi,_r + \psi,_{rz} + \frac{1}{r}\chi,_\theta \tag{5.157a}$$

$$v = \frac{1}{r}\varphi,_\theta + \frac{1}{r}\psi,_{\theta z} - \chi,_r \tag{5.157b}$$

$$w = \varphi,_z - \psi,_{rr} - \frac{1}{r}\psi,_r - \frac{1}{r^2}\psi,_{\theta\theta} \tag{5.157c}$$

Substituting Eq. 5.157 in Eqs. 5.153 and 5.154 results in

$$e = \nabla^2 \varphi \tag{5.158}$$

$$\Omega_r = \tfrac{1}{2}\left(-\frac{1}{r}\nabla^2 \psi_{,\theta} + \chi_{,rz}\right) \tag{5.159a}$$

$$\Omega_\theta = \tfrac{1}{2}\left(\nabla^2 \psi_{,r} + \frac{1}{r}\chi_{,\theta z}\right) \tag{5.159b}$$

$$\Omega_z = \tfrac{1}{2}(-\nabla^2 \chi + \chi_{,zz}) \tag{5.159c}$$

whereby the Laplace operator of any scalar a is defined as

$$\nabla^2 a = a_{,rr} + \frac{1}{r}a_{,r} + \frac{1}{r^2}a_{,\theta\theta} + a_{,zz} \tag{5.160}$$

Substituting Eqs. 5.158 and 5.159 in Eq. 5.156 leads to the three equations of motion expressed in the potentials. They are satisfied if the following three equations apply:

$$\nabla^2 \varphi = -\frac{\omega^2}{c_p^{*2}}\varphi \tag{5.161a}$$

$$\nabla^2 \psi = -\frac{\omega^2}{c_s^{*2}}\psi \tag{5.161b}$$

$$\nabla^2 \chi = -\frac{\omega^2}{c_s^{*2}}\chi \tag{5.161c}$$

The dilatational wave velocity c_p^* and the shear-wave velocity c_s^* are defined in Eqs. 5.88a and 5.72 and 5.88b and 5.74, respectively. Equations 5.161 represent the uncoupled wave equations.

All amplitudes introduced up to now are a function of the three coordinates [e.g., $u(r, \theta, z)$]. In the following, a bar is used for conciseness to denote an amplitude which is a function of r and z [$\bar{u} = u(r, z)$]. An amplitude that depends on one variable only, mostly on z, is referenced without indicating the argument [$u = u(z)$].

5.5.2 Solution Using Fourier Series Circumferentially and Bessel Function Radially

It is convenient to express the variation of the amplitudes in the circumferential direction θ in terms of a Fourier series. For the displacements, this leads to

$$u(r, \theta, z) = \sum_n \bar{u}_n^s \cos n\theta + \sum_n \bar{u}_n^a \sin n\theta \tag{5.162a}$$

$$v(r, \theta, z) = -\sum_n \bar{v}_n^s \sin n\theta + \sum_n \bar{v}_n^a \cos n\theta \tag{5.162b}$$

$$w(r, \theta, z) = \sum_n \bar{w}_n^s \cos n\theta + \sum_n \bar{w}_n^a \sin n\theta \tag{5.162c}$$

The summation is performed over the integer n ($n = 0, 1, 2, \ldots$). The superscripts s and a denote the symmetric and antimetric terms, respectively. For simplicity, only the symmetric terms are kept. This allows the superscript s to be dropped. The derived equations hold also for the antimetric case without any modifications. In addition, the subscript n is omitted. It should be remembered that all amplitudes with a bar depend on the Fourier term investigated.

Analogous series are formulated for all other amplitudes. Differentiations with respect to θ are performed. For example, substituting Eq. 5.162 in Eq. 5.153, it follows that

$$e = \sum_n \bar{e} \cos n\theta \tag{5.163a}$$

where

$$\bar{e} = \bar{u}_{,r} + \frac{\bar{u}}{r} - \frac{n}{r}\bar{v} + \bar{w}_{,z} \tag{5.163b}$$

All equations specified in this section can also be written using the amplitudes denoted with a bar. For instance, Eq. 5.156a is formulated as

$$(\lambda^* + 2G^*)\bar{e}_{,r} + 2G^*\left(\bar{\Omega}_{\theta,z} + \frac{n}{r}\bar{\Omega}_z\right) = -\rho\omega^2\bar{u} \tag{5.164}$$

In particular, the potentials are expressed as

$$\varphi(r, \theta, z) = \sum_n \bar{\varphi} \cos n\theta \tag{5.165a}$$

$$\psi(r, \theta, z) = \sum_n \bar{\psi} \cos n\theta \tag{5.165b}$$

$$\chi(r, \theta, z) = -\sum_n \bar{\chi} \sin n\theta \tag{5.165c}$$

The uncoupled wave equations (Eqs. 5.161) are equal to

$$\nabla^2\bar{\varphi} = -\frac{\omega^2}{c_p^{*2}}\bar{\varphi} \tag{5.166a}$$

$$\nabla^2\bar{\psi} = -\frac{\omega^2}{c_s^{*2}}\bar{\psi} \tag{5.166b}$$

$$\nabla^2\bar{\chi} = -\frac{\omega^2}{c_s^{*2}}\bar{\chi} \tag{5.166c}$$

where (Eq. 5.160)

$$\nabla^2\bar{a} = \bar{a}_{,rr} + \frac{1}{r}\bar{a}_{,r} - \frac{n^2}{r^2}\bar{a} + \bar{a}_{,zz} \tag{5.167}$$

The wave equations are solved by assuming a product for the function, which separates the variables r and z. Substituting

$$\bar{\varphi}(r, z) = \varphi_r(r)\varphi_z(z) \tag{5.168}$$

in Eq. 5.166a results in

$$\varphi_{r,rr}\varphi_z + \frac{1}{r}\varphi_{r,r}\varphi_z - \frac{n^2}{r^2}\varphi_r\varphi_z + \varphi_r\varphi_{z,zz} = -\frac{\omega^2}{c_p^{*2}}\varphi_r\varphi_z \tag{5.169a}$$

which, after rearranging, is equal to

$$\frac{\varphi_{r,rr}}{\varphi_r} + \frac{1}{r}\frac{\varphi_{r,r}}{\varphi_r} - \frac{n^2}{r^2} = -\frac{\varphi_{z,zz}}{\varphi_z} - \frac{\omega^2}{c_p^{*2}} = -k^2 \tag{5.169b}$$

Since the variables appearing on the respective sides are independent of each other, the common separation value, denoted as $-k^2$, is constant. The right-hand side of this equation,

$$\varphi_{z,zz} + \left(\frac{\omega^2}{c_p^{*2}} - k^2\right)\varphi_z = 0 \tag{5.170}$$

has as solution

$$\varphi_z = a_1 \exp\left(i\sqrt{\frac{\omega^2}{c_p^{*2}} - k^2}\, z\right) + b_1 \exp\left(-i\sqrt{\frac{\omega^2}{c_p^{*2}} - k^2}\, z\right) \tag{5.171}$$

The values a_1 and b_1 are the constants of integration. The left-hand side of Eq. 5.169d is rewritten as

$$r^2\varphi_{r,rr} + r\varphi_{r,r} + (k^2r^2 - n^2)\varphi_r = 0 \tag{5.172}$$

This variable-coefficient differential equation is the so-called Bessel equation of order n with the parameter k. Its solution equals

$$\varphi_r = c_1 J_n(kr) + d_1 Y_n(kr) \tag{5.173}$$

$J_n(kr)$ and $Y_n(kr)$ denote the Bessel functions of order n of the first and second kinds, respectively. As the latter is unbounded in the neighborhood of the origin ($kr = 0$), the integration constant d_1 is set equal to zero. This applies for the tasks solved in this text. (If problems were addressed which did not include the origin, as in the case of a cavity, this condition would not be enforced.)

When solving Eq. 5.172, it is assumed that k^2 is a positive value. If the case $k^2 < 0$ is investigated, so-called modified Bessel functions of order n of the first kind $I_n(kr)$ and of the second kind $K_n(kr)$ arise. As I_n and K_n are unbounded for $kr \longrightarrow \infty$ and for $kr \longrightarrow 0$, respectively, this solution can be disregarded in the context examined in this text (see problem 7.15).

Substituting Eqs. 5.173 and 5.171 in Eq. 5.168 leads to

$$\bar{\varphi} = J_n(kr)\left[A_1 \exp\left(i\sqrt{\frac{\omega^2}{c_p^{*2}} - k^2}\, z\right) + B_1 \exp\left(-i\sqrt{\frac{\omega^2}{c_p^{*2}} - k^2}\, z\right)\right] \tag{5.174a}$$

Analogously, the solutions of the other two wave equations (Eqs. 5.166b and 5.166c) are as follows:

$$\bar{\psi} = J_n(kr)\left[A_2 \exp\left(i\sqrt{\frac{\omega^2}{c_s^{*2}} - k^2}\, z\right) + B_2 \exp\left(-i\sqrt{\frac{\omega^2}{c_s^{*2}} - k^2}\, z\right)\right] \tag{5.174b}$$

$$\bar{\chi} = J_n(kr)\left[A_3 \exp\left(i\sqrt{\frac{\omega^2}{c_s^{*2}} - k^2}\, z\right) + B_3 \exp\left(-i\sqrt{\frac{\omega^2}{c_s^{*2}} - k^2}\, z\right)\right] \tag{5.174c}$$

The resulting integration constants are denoted as $A_1, B_1, \ldots, B_3$.

Formulating Eq. 5.157 for the amplitudes denoted with a bar, and substituting Eq. 5.174, the displacement amplitudes $\bar{u}$, $\bar{v}$, and $\bar{w}$ are expressed as a function of r and z as follows:

$$\bar{u} = J_n(kr)_{,r}\left[A_1 \exp\left(i\sqrt{\frac{\omega^2}{c_p^{*2}} - k^2}\,z\right) + B_1 \exp\left(-i\sqrt{\frac{\omega^2}{c_p^{*2}} - k^2}\,z\right)\right]$$
$$+ J_n(kr)_{,r} i\sqrt{\frac{\omega^2}{c_s^{*2}} - k^2}\left[A_2 \exp\left(i\sqrt{\frac{\omega^2}{c_s^{*2}} - k^2}\,z\right) - B_2 \exp\left(-i\sqrt{\frac{\omega^2}{c_s^{*2}} - k^2}\,z\right)\right]$$
$$-\frac{n}{r}J_n(kr)\left[A_3 \exp\left(i\sqrt{\frac{\omega^2}{c_s^{*2}} - k^2}\,z\right) + B_3 \exp\left(-i\sqrt{\frac{\omega^2}{c_s^{*2}} - k^2}\,z\right)\right] \tag{5.175a}$$

$$\bar{v} = \frac{n}{r}J_n(kr)\left[A_1 \exp\left(i\sqrt{\frac{\omega^2}{c_p^{*2}} - k^2}\,z\right) + B_1 \exp\left(-i\sqrt{\frac{\omega^2}{c_p^{*2}} - k^2}\,z\right)\right]$$
$$+ \frac{n}{r}J_n(kr) i\sqrt{\frac{\omega^2}{c_s^{*2}} - k^2}\left[A_2 \exp\left(i\sqrt{\frac{\omega^2}{c_s^{*2}} - k^2}\,z\right) - B_2 \exp\left(-i\sqrt{\frac{\omega^2}{c_s^{*2}} - k^2}\,z\right)\right]$$
$$- J_n(kr)_{,r}\left[A_3 \exp\left(i\sqrt{\frac{\omega^2}{c_s^{*2}} - k^2}\,z\right) + B_3 \exp\left(-i\sqrt{\frac{\omega^2}{c_s^{*2}} - k^2}\,z\right)\right] \tag{5.175b}$$

$$\bar{w} = J_n(kr) i\sqrt{\frac{\omega^2}{c_p^{*2}} - k^2}\left[A_1 \exp\left(i\sqrt{\frac{\omega^2}{c_p^{*2}} - k^2}\,z\right) - B_1 \exp\left(-i\sqrt{\frac{\omega^2}{c_p^{*2}} - k^2}\,z\right)\right]$$
$$+ J_n(kr)k^2\left[A_2 \exp\left(i\sqrt{\frac{\omega^2}{c_s^{*2}} - k^2}\,z\right) + B_2 \exp\left(-i\sqrt{\frac{\omega^2}{c_s^{*2}} - k^2}\,z\right)\right] \tag{5.175c}$$

The following nomenclature is introduced, which is defined for the case of the plane waves in Sections 5.3 and 5.4.

$$c = \frac{\omega}{k} \tag{5.176}$$

$$l_x = \frac{c_p^*}{c} = \frac{c_p^* k}{\omega} \tag{5.177a}$$

$$m_x = \frac{c_s^*}{c} = \frac{c_s^* k}{\omega} \tag{5.177b}$$

$$s = -i\sqrt{1 - \frac{1}{l_x^2}} = -i\sqrt{1 - \frac{\omega^2}{c_p^{*2}k^2}} \tag{5.178a}$$

$$t = -i\sqrt{1 - \frac{1}{m_x^2}} = -i\sqrt{1 - \frac{\omega^2}{c_s^{*2}k^2}} \tag{5.178b}$$

In addition, the integration constants are expressed as functions of other values which can, as will be shown later, be identified as the amplitudes of the P-, SV-, and SH-waves.

$$A_1 = \frac{l_x}{k} A_P \tag{5.179a}$$

$$B_1 = \frac{l_x}{k} B_P \tag{5.179b}$$

$$A_2 = i\frac{m_x}{k^2} A_{SV} \tag{5.179c}$$

$$B_2 = i\frac{m_x}{k^2} B_{SV} \tag{5.179d}$$

$$A_3 = -\frac{1}{k} A_{SH} \tag{5.179e}$$

$$A_4 = -\frac{1}{k} B_{SH} \tag{5.179f}$$

Using Eqs. 5.177, 5.178, and 5.179, Eq. 5.175 is reformulated as

$$\bar{u} = \frac{1}{k} J_n(kr)_{,r} u(z) + \frac{n}{kr} J_n(kr) v(z) \tag{5.180a}$$

$$\bar{v} = \frac{n}{kr} J_n(kr) u(z) + \frac{1}{k} J_n(kr)_{,r} v(z) \tag{5.180b}$$

$$\bar{w} = -J_n(kr)[iw(z)] \tag{5.180c}$$

with

$$\begin{aligned} u(z) = {} & l_x[A_P \exp(iksz) + B_P \exp(-iksz)] \\ & - m_x t[A_{SV} \exp(iktz) - B_{SV} \exp(-iktz)] \end{aligned} \tag{5.181a}$$

$$v(z) = A_{SH} \exp(iktz) + B_{SH} \exp(-iktz) \tag{5.181b}$$

$$\begin{aligned} w(z) = {} & -l_x s[A_P \exp(iksz) - B_P \exp(-iksz)] \\ & - m_x[A_{SV} \exp(iktz) + B_{SV} \exp(-iktz)] \end{aligned} \tag{5.181c}$$

It is important to point out that the amplitudes $v(z)$ and $u(z)$, $w(z)$ are identical to those arising for plane waves in out-of-plane (Eq. 5.98) and in-plane motions (Eq. 5.130), respectively. While for plane waves, the variation in the horizontal x-direction is described by $\exp(-ikx)$, Bessel functions with the argument kr arise for that one in the radial direction, applicable to three-dimensional waves formulated in cylindrical coordinates. The variation in the circumferential direction is specified as a Fourier series.

Using straightforward algebra, it is easy to show that the following equation holds for the stresses, analogous to that applicable to the displacements (Eq. 5.180):

$$\bar{\tau}_{rz} = \frac{1}{k} J_n(kr)_{,r} \tau_{xz}(z) + \frac{n}{kr} J_n(kr) \tau_{yz}(z) \tag{5.182a}$$

$$\bar{\tau}_{\theta z} = \frac{n}{kr} J_n(kr) \tau_{xz}(z) + \frac{1}{k} J_n(kr)_{,r} \tau_{yz}(z) \tag{5.182b}$$

$$\bar{\sigma}_z = -J_n(kr)[i\sigma_z(z)] \tag{5.182c}$$

The amplitudes $\tau_{yz}(z)$ and $\sigma_z(z)$, $\tau_{xz}(z)$ are specified in Eq. 5.100 and Eq. 5.132, respectively.

5.5.3 Dynamic-Stiffness Matrix

It also follows that the same dynamic-stiffness matrices as in the plane-wave cases apply.

$$\begin{Bmatrix} Q_1 \\ Q_2 \end{Bmatrix} = \begin{Bmatrix} -\tau_{\theta z1} \\ \tau_{\theta z2} \end{Bmatrix} = \begin{bmatrix} S^L_{\text{SH}} \end{bmatrix} \begin{Bmatrix} v_1 \\ v_2 \end{Bmatrix} \tag{5.183a}$$

with $[S^L_{\text{SH}}]$ specified for the out-of-plane motion in Eq. 5.103, and

$$\begin{Bmatrix} P_1 \\ R_1 \\ P_2 \\ R_2 \end{Bmatrix} = \begin{Bmatrix} -\tau_{rz1} \\ -\sigma_{z1} \\ \tau_{rz2} \\ \sigma_{z2} \end{Bmatrix} = \begin{bmatrix} S^L_{\text{P-SV}} \end{bmatrix} \begin{Bmatrix} u_1 \\ w_1 \\ u_2 \\ w_2 \end{Bmatrix} \tag{5.183b}$$

with $[S^L_{\text{P-SV}}]$ specified for the in-plane motion in Eq. 5.134. The subscripts 1 and 2 denote the interfaces $z = 0$ and $z = d$, respectively. In cylindrical coordinates, the displacement and stress amplitudes in the z-direction are not multiplied by i. The dynamic-stiffness matrices are independent of the Fourier index n. For the half-space, analogous expressions apply.

Distributed loads acting across the thickness of the layer can also be calculated using the same concepts. The equations are not derived in this text.

SUMMARY

1. As an introductory example, the wave propagation in a rod with exponentially increasing area is investigated. The motion diminishes along the axis of the rod. For an undamped system, a cutoff frequency exists below which the motion ceases to propagate. Above the cutoff frequency, the waves propagate with a frequency-dependent apparent velocity (dispersion). The dynamic-stiffness matrix of an undamped finite rod (in which waves propagate in both directions) is real for the whole range of frequency. To achieve a unique solution for the infinite rod, it is not sufficient that the solution tends to zero at infinity. The dynamic-stiffness coefficient of the undamped infinite rod (in which only outgoing waves exist) is real for an excitation below the cutoff frequency (spring with frequency-dependent coefficient) and complex above (spring and damper with frequency-dependent coefficients). In general, for increasing frequency, the spring and damping coefficients diminish and increase, respectively. The damping coefficient is a measure of the radiation of the energy of the outgoing waves. In a rod with material damping, the motion exhibits an additional attenuation and a cutoff frequency no longer exists (the motion, however, decays strongly for excitations below the cutoff

frequency of the undamped case). The coefficient of the material damping leads directly to a complex elastic-material modulus and affects indirectly the dimensionless frequency. The dynamic-stiffness matrix of the damped finite rod is complex and converges for the limiting case of an infinite length to that of the damped infinite rod. The convergence is, in general, quite slow and exhibits oscillations. Introducing material damping increases the damping coefficient, especially for small frequencies, and decreases the spring coefficient in the higher-frequency range.

2. In a layered half-space, two types of (inclined) body waves exist: P- and S-waves. For a P-wave, involving a volumetric strain only, the particle motion coincides with the direction along which the wave propagates with the (material-dependent) dilatational-wave velocity and is constant over a plane perpendicular to it. For an S-wave, with a distortional strain only, the particle motion takes place in a plane perpendicular to the direction of propagation and is constant over this plane. This motion, which propagates with the (material-dependent) shear-wave velocity, can be decomposed into a horizontal SH-wave and a vertical SV-wave. Assuming the directions of propagation of the P- and S-waves to lie in a vertical plane, the SH-wave will result in a (horizontal) out-of-plane displacement, which is independent of the in-plane displacements caused by the P- and SV-waves. The phase velocity (and wave number) is common to the P- and SV-waves. Introducing material damping results in complex wave velocities.
3. Separately, for the out-of-plane and in-plane motions, the (symmetric) dynamic-stiffness matrices of a layer and of the half-space can be established. For a given frequency, they depend on the wave number (or the phase velocity), the material properties, and the depth of the layer. The stiffness matrices of an undamped layer and of an undamped half-space are real and imaginary, respectively, the latter indicating that energy is radiated toward infinity. For a damped system, the diagonal terms of the dynamic-stiffness matrix of a layer converge to those of the half-space for the limiting case of infinite depth.
4. In the horizontal direction, all displacements and stresses for plane waves vary exponentially, whereby an expansion in the wave-number domain (Fourier transform) is performed. In cylindrical coordinates, the displacements and stresses are decomposed circumferentially in a Fourier series and radially in a Bessel function. In the vertical direction, the variation is the same for Cartesian coordinates (plane waves) and for cylindrical coordinates, resulting in the same dynamic-stiffness matrix.
5. Distributed loads acting across the thickness of the layer are straightforwardly processed.
6. The rate of horizontal energy transmission is equal to the sum of the scalar products of the stress components and of the corresponding velocity components acting on the plane.

PROBLEMS

5.1. It is a key feature of soil–structure interaction that for certain unbounded dynamic systems, a cutoff frequency exists below which (for an undamped system) no waves propagate and thus no radiation of energy is possible. Consider the taut string of area A and mass density ρ under the specified normal force N (tension) and which rests on an elastic foundation with a spring coefficient k shown in Fig. P5-1a. The vertical displacement with an amplitude w does not affect N. Discuss the different types of waves that can occur as a function of the excitation frequency ω, introducing the phase velocity c. Identify a dimensionless frequency denoted as a_o. For the semi-infinite string, determine the dynamic-stiffness coefficient S_o [i.e., the vertical-force amplitude P_o (applied at one end) which will result in a unit displacement amplitude w_o in the same point]. Decompose the nondimensional $\bar{S}_o$ into a spring with a coefficient k_o and a damper with a coefficient c_o.

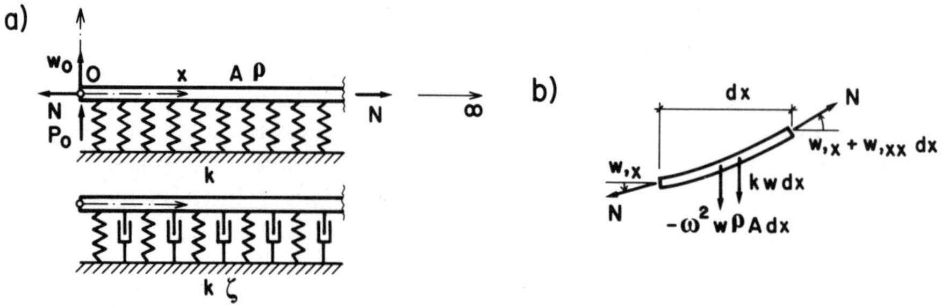

Figure P5-1 String resting on elastic foundation. (a) Semi-infinite string on elastic and viscoelastic foundation; (b) equilibrium of infinitesimal element.

Solution:

Formulating the equilibrium equation in the vertical direction (Fig. P5-1b) leads to the equation of motion

$$w_{,xx} - \frac{k}{N}w + \frac{\omega^2}{c_l^2}w = 0$$

with the velocity c_l defined as

$$c_l^2 = \frac{N}{\rho A}$$

Selecting a solution for w as $\exp(i\gamma x)$ results in

$$w = a \exp\left(-i\sqrt{\frac{k}{N}}\sqrt{a_o^2 - 1}\,x\right) + b \exp\left(+i\sqrt{\frac{k}{N}}\sqrt{a_o^2 - 1}\,x\right)$$

With the dimensionless frequency a_o.

$$a_o = \frac{\omega\sqrt{N}}{c_l\sqrt{k}}$$

Introducing the phase velocity c

$$c = \frac{c_l}{\sqrt{1 - 1/a_o^2}}$$

leads to

$$w = a \exp\left(-i\omega\frac{x}{c}\right) + b \exp\left(+i\omega\frac{x}{c}\right)$$

For an a_o that results in a real value of c,

$$a_o > 1 \quad \longrightarrow \quad \omega > c_l\sqrt{\frac{k}{N}}$$

waves propagate with a velocity c; the terms with a and b travel in the positive and negative x-directions, respectively. The corresponding motion is dispersive $[c(\omega)]$. For $a_o < 1$ the motion diminishes exponentially with x, but does not propagate. For a_o equal to the cutoff frequency $= 1$, $w = a + b$ (i.e., there is no spatial variation of the motion).

When applying a load at 0 on the semi-infinite string, no incoming wave (propagating in the negative x-direction) exists. Thus setting $b = 0$ and enforcing w_o at $x = 0$ leads to

$$w = w_o \exp\left(-i\sqrt{\frac{k}{N}}\sqrt{a_o^2 - 1}\,x\right)$$

The amplitude P_o is equal to $-Nw_{,x}$ evaluated at $x = 0$

$$P_o = -i\sqrt{kN}\sqrt{a_o^2 - 1}\,w_o$$

The static-stiffness coefficient $\sqrt{kN}$ is used to nondimensionalize S_o, leading to

$$\bar{S}_o = i\sqrt{a_o^2 - 1}$$

The coefficient $\bar{S}_o$ can be split up into its real and imaginary parts as

$$\bar{S}_o = k_o + ia_o c_o$$

where

$$\text{for } a_o < 1: \quad k_o = \sqrt{1 - a_o^2}$$

$$c_o = 0$$

$$\text{for } a_o > 1: \quad k_o = 0$$

$$c_o = \sqrt{1 - \frac{1}{a_o^2}}$$

These spring and damping coefficients are plotted versus the dimensionless frequency a_o in Fig. P5-2 ($\zeta = 0$).

5.2. Assuming that the taut string of Fig. P5-1 rests on a viscoelastic foundation with a spring coefficient k and a hysteretic damping ratio ζ, repeat the discussion of Problem 5.1. In addition, calculate the apparent velocity c_a and the decay factor per wavelength δ_λ. Select $\zeta = 0.05$ for the plots.

Solution:

In all equations, k is replaced by $k(1 + 2\zeta i)$. The phase velocity is complex

$$c = \frac{c_l}{\sqrt{1 - [(1 + 2\zeta i)/a_o^2]}} = \frac{c_l}{\operatorname{Re}\sqrt{} + i\operatorname{Im}\sqrt{}}$$

While the real part of the argument of the square root in the denominator can be positive or negative, the imaginary part is always negative, resulting in a

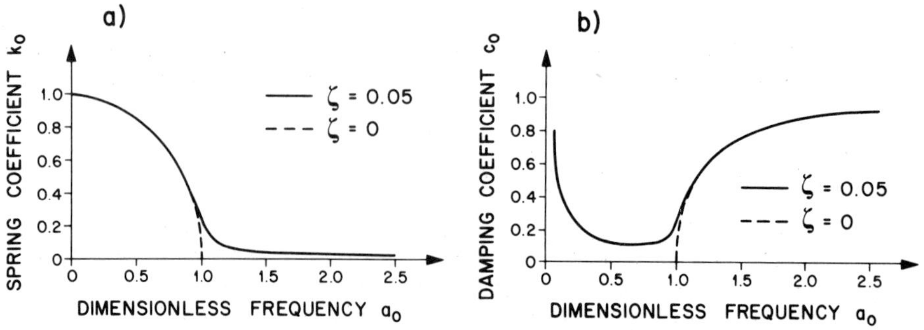

Figure P5-2 Dynamic-stiffness coefficient of semi-infinite string on elastic and viscoelastic foundation.

negative value of Re $\sqrt{\;}$ and positive one of Im $\sqrt{\;}$. This leads for the w-term with the integration constant b to

$$w = b \exp\left(-\frac{\omega \operatorname{Im} \sqrt{\;}}{c_l} x\right) \exp\left(\frac{i\omega \operatorname{Re} \sqrt{\;}}{c_l} x\right)$$

The second term indicates that the wave propagates (for all ω) in the positive x-direction with

$$c_a = -\frac{c_l}{\operatorname{Re} \sqrt{\;}}$$

and the first term describes the attenuation caused by the material damping of the foundation with

$$\delta_\lambda = \exp\left(-2\pi \frac{\operatorname{Im} \sqrt{\;}}{\operatorname{Re} \sqrt{\;}}\right)$$

No cutoff frequency exists for the damped case, although for $a_o < 1$ δ_λ is very large.

The dynamic-stiffness coefficient S_o is equal to

$$S_o = -i\sqrt{kN}\sqrt{1 + 2\zeta i}\sqrt{\frac{a_o^2}{1 + 2\zeta i} - 1}$$

Nondimensionalizing with the same frequency-independent factor $\sqrt{kN}$ and then decomposing into k_o and c_o leads to

$$k_o = \operatorname{Re} \sqrt{1 - a_o^2 + 2\zeta i}$$

$$c_o = \operatorname{Re} \sqrt{1 - \frac{1 + 2\zeta i}{a_o^2}}$$

In Fig. P5-2, the spring coefficient k_o and the damping coefficient c_o are plotted as a function of a_o for the damped and undamped cases.

5.3. Prove that the dynamic behavior of the shear beam having shear area $\bar{A}$, shear modulus G, area A, and density ρ and resting on an elastic foundation with spring constant k (Fig. P5-3a) is analogous to the string supported on an elastic foundation (Fig. P5-1a). By transforming the corresponding variables, specify the equations for the dynamic-stiffness coefficient S_o of the semi-infinite beam.

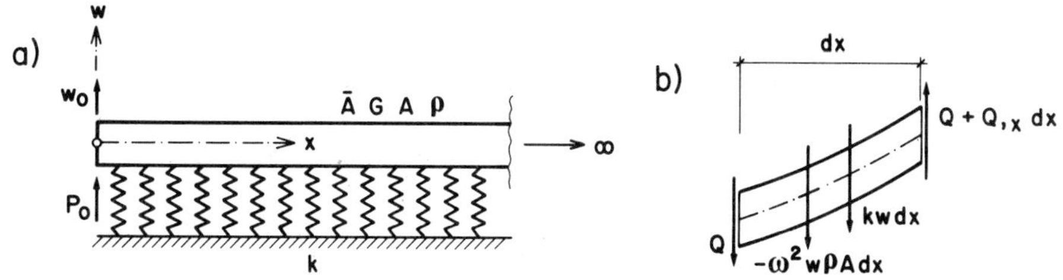

Figure P5-3 Shear beam resting on elastic foundation. (a) Semi-infinite shear beam on elastic foundation; (b) equilibrium of infinitesimal element.

Solution:

Formulating the equilibrium equation in the vertical direction and substituting for the amplitude of the transverse shear force,

$$Q = G\bar{A}w_{,x}$$

leads to

$$w_{,xx} - \frac{k}{G\bar{A}}w + \frac{\omega^2}{c_s^2}w = 0$$

with the somewhat generalized shear-wave velocity c_s defined as

$$c_s^2 = \frac{G\bar{A}}{\rho A}$$

Comparing this equation of motion with that of the string leads to

$$N = G\bar{A}$$

$$c_l^2 = c_s^2$$

The cutoff frequency is equal to

$$c_l\sqrt{\frac{k}{N}} = c_s\sqrt{\frac{k}{G\bar{A}}} = \sqrt{\frac{k}{\rho A}}$$

that is, the natural frequency of the rigid beam resting on the springs of the foundation.

Introducing the dimensionless frequency a_o,

$$a_o = \omega\sqrt{\frac{\rho A}{k}}$$

the dynamic-stiffness coefficient S_o follows:

$$S_o = \sqrt{G\bar{A}k}\,(k_o + ia_oc_o)$$

where

$$\text{for } a_o < 1: \quad k_o = \sqrt{1 - a_o^2}$$

$$c_o = 0$$

$$\text{for } a_o > 1: \quad k_o = 0$$

$$c_o = \sqrt{1 - \frac{1}{a_o^2}}$$

applies.

5.4. As another example of an unbounded dynamic system which exhibits dispersion and a cutoff frequency, examine the semi-infinite beam with moment of inertia I, modulus of elasticity E, area A, and density ρ and which rests on an elastic foundation with spring constant k (Fig. P5-4a). Disregard the work of the transverse shear and the effect of rotational inertia. Calculate the moment amplitude M_o which has to be applied at the one end of the semi-infinite system to cause a rotation amplitude $\phi_o = 1$ with $w_o = 0$ (i.e., one element of the dynamic-stiffness matrix). Introduce dimensionless variables where appropriate. Discuss the characteristics of the waves for the domain below the cutoff frequency.

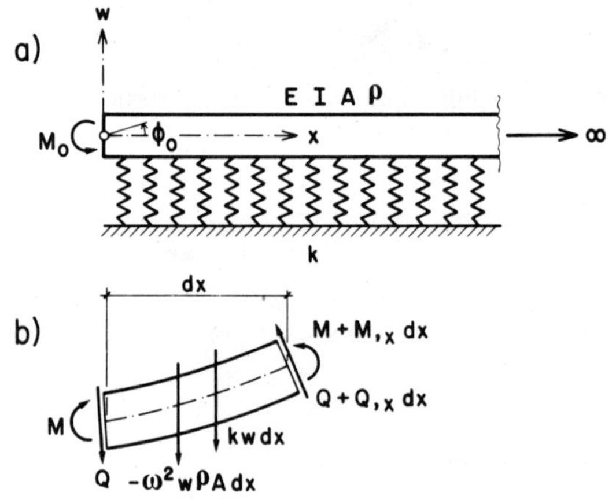

Figure P5-4 Beam resting on elastic foundation. (a) Semi-infinite beam on elastic foundation; (b) equilibrium of infinitesimal element.

Solution:

Formulating the two equilibrium equations (Fig. P5-4b) and using

$$M = EIw_{,xx}$$

the equation of motion in the amplitude of the vertical displacement w results

$$w_{,xxxx} + \frac{k}{EI}w - \omega^2\frac{\rho A}{EI}w = 0$$

Choosing a solution for w as exp $(i\gamma x)$ leads to

$$\gamma^4 = \omega^2\frac{\rho A}{EI} - \frac{k}{EI}$$

or introducing the dimensionless frequency a_o,

$$a_o^2 = \frac{\omega^2 A\rho}{E}$$

this results in

$$\gamma_{1,2} = \mp\sqrt[4]{\frac{a_o^2}{I} - \frac{k}{EI}}$$

$$\gamma_{3,4} = \pm i\sqrt[4]{\frac{a_o^2}{I} - \frac{k}{EI}}$$

The solution for w is equal to

$$w = a \exp\left(-i\sqrt[4]{\frac{a_o^2}{I} - \frac{k}{EI}}x\right) + b \exp\left(+i\sqrt[4]{\frac{a_o^2}{I} - \frac{k}{EI}}x\right)$$
$$+ c \exp\left(-\sqrt[4]{\frac{a_o^2}{I} - \frac{k}{EI}}x\right) + d \exp\left(+\sqrt[4]{\frac{a_o^2}{I} - \frac{k}{EI}}x\right)$$

For $a_o > \sqrt{k/E}$, the first and second terms represent waves propagating in the positive and negative x-directions, respectively. The apparent velocity c_a is specified as

$$c_a = \frac{1}{\sqrt[4]{A\rho/\omega^2 EI - k/\omega^4 EI}}$$

The third and fourth terms decay and increase, repectively, exponentially with increasing x. To calculate the dynamic-stiffness matrix of the semi-infinite beam, only the first and third terms with the integration constants a and c are kept.

For $a_o < \sqrt{k/E}$ a very interesting behavior is observed. Introducing the positive constant f,

$$f = \frac{k}{EI} - \frac{a_o^2}{I}$$

leads to

$$\gamma_{1,2} = \mp(1 - i)\frac{\sqrt[4]{f}}{\sqrt{2}}$$

$$\gamma_{3,4} = \pm\,(1 + i)\frac{\sqrt[4]{f}}{\sqrt{2}}$$

The corresponding solution for w is equal to

$$w = a \exp\left(-\frac{\sqrt[4]{f}}{\sqrt{2}}x\right) \exp\left(-i\frac{\sqrt[4]{f}}{\sqrt{2}}x\right) + b \exp\left(+\frac{\sqrt[4]{f}}{\sqrt{2}}x\right) \exp\left(+i\frac{\sqrt[4]{f}}{\sqrt{2}}x\right)$$
$$+ c \exp\left(-\frac{\sqrt[4]{f}}{\sqrt{2}}x\right) \exp\left(+i\frac{\sqrt[4]{f}}{\sqrt{2}}x\right) + d \exp\left(+\frac{\sqrt[4]{f}}{\sqrt{2}}x\right) \exp\left(-i\frac{\sqrt[4]{f}}{\sqrt{2}}x\right)$$

The first term represents a wave propagating in the positive x-direction with the apparent velocity c_a:

$$c_a = \frac{\sqrt{2}}{\sqrt[4]{k/\omega^4 EI - A\rho/\omega^2 EI}}$$

Its amplitude decays exponentially with increasing x. The second and fourth terms correspond to waves with amplitudes that grow exponentially. The third term represents a wave propagating in the negative x-direction with an amplitude that decays exponentially with increasing x. Thus to calculate the dynamic-stiffness matrix of the semi-infinite beam, the first and third terms with the integration constants a and c are used (so-called standing wave).

The one element of the dynamic-stiffness matrix is calculated by enforcing the boundary conditions $w(x = 0) = w_o = 0$ and $w_{,x}(x = 0) = \phi_o = 1$, which determines the integration constants a and c. The corresponding dynamic-stiffness coefficient S_o which is equal to the moment with the amplitude M_o in the global coordinate system is specified as

$$S_o = -EIw_{,xx}(x = 0)$$

The static value K_o is equal to

$$K_o = \sqrt{2}\, EI \sqrt[4]{\frac{k}{EI}}$$

Nondimensionalizing S_o and decomposing into the real and imaginary parts as

$$S_o = K_o(k_o + i a_o c_o)$$

the following equations apply:

$$a_o < \sqrt{\frac{k}{E}}: \quad k_o = \sqrt[4]{1 - \frac{a_o^2 E}{k}}$$

$$c_o = 0$$

$$a_o > \sqrt{\frac{k}{E}}: \quad k_o = \frac{1}{\sqrt{2}} \sqrt[4]{\frac{a_o^2 E}{k} - 1}$$

$$c_o = \frac{1}{\sqrt{2}} \sqrt[4]{\frac{E}{a_o^2 k} - \frac{1}{a_o^4}}$$

For the range below the cutoff frequency, a standing wave is present. This also occurs when rotary inertia terms (and the influence of the shear deformations) are taken into consideration.

5.5. The rod with an exponentially varying area is used in Section 5.1 to illustrate certain features of the dynamic-stiffness coefficient of an unbounded domain. It should, however, not actually be used to model the soil, in contrast to the conical shear beam, which represents an acceptable approximation for calculating the horizontal and twisting dynamic-stiffness coefficients of a rigid circular base (disk) on the surface of a half-space.

The shear beam with circular cross section is shown with the nomenclature corresponding to the horizontally excited system in Fig. P5-5a. The horizontal planes of the beam are assumed to displace only in the horizontal direction. Determine the dynamic-stiffness coefficient S_o as a function of the radius a at the surface, the angle α, the shear modulus G, Poisson's ratio ν, and the dimensionless frequency $a_o = \omega a / c_s$, with the shear-wave velocity $c_s = \sqrt{G}/\sqrt{\rho}$ (ρ = density). Calculate α by enforcing the static-stiffness coefficient $8Ga/(2-\nu)$. Split the nondimensionalized dynamic-stiffness coefficient into the real and imaginary parts k_o and $a_o c_o$. Discuss the influence of a hysteretic-damping ratio ζ. Select $\zeta = 0.05$ for the plots.

Solution:

The origin $z = 0$ is chosen at a distance $a \cot \alpha$ above the disk. In analogy to Eq. 5.6, the equation of motion of the shear beam with the shear area A $(= \pi \tan^2 \alpha\, z^2)$ is equal to

$$u_{,zz} + \frac{A_{,z}}{A} u_{,z} + \frac{\omega^2}{c_s^2} u = 0$$

$$u_{,zz} + \frac{2}{z} u_{,z} + \frac{\omega^2}{c_s^2} u = 0$$

or

$$(zu)_{,zz} + \frac{\omega^2}{c_s^2}(zu) = 0$$

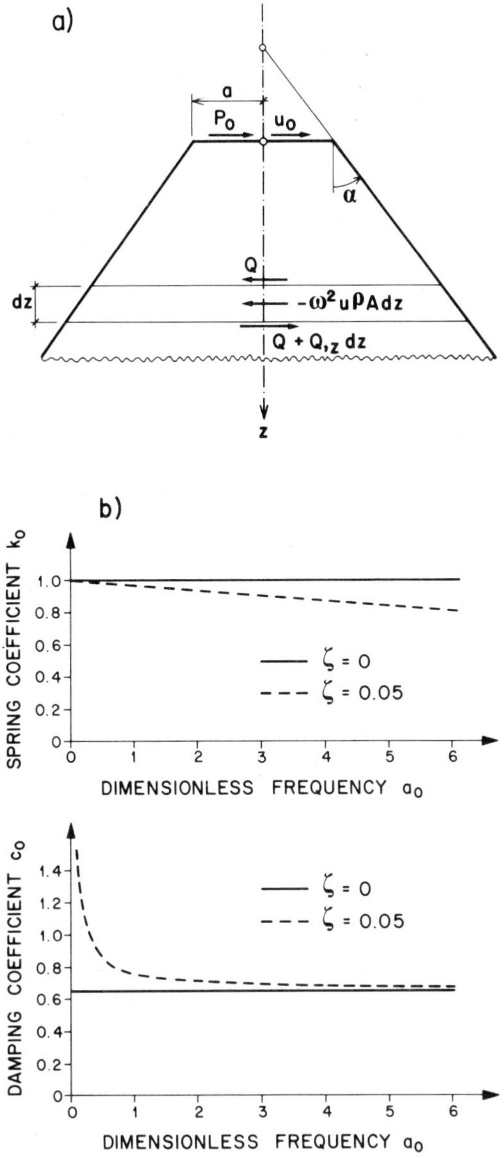

Figure P5-5 Conical shear beam used for horizontal dynamic-stiffness coefficient. (a) Semi-infinite shear beam; (b) dynamic-stiffness coefficient.

The solution is equal to

$$u = \frac{a}{z} \exp\left(-\frac{i\omega}{c_s} z\right)$$

whereby the term with $\exp(+i\omega z/c_s)$ is deleted to suppress the incoming wave. Enforcing the boundary condition $u(z = a \cot \alpha) = u_o$ and calculating the load amplitude P_o which is equal to the negative value of that of the transverse shear force at the same location,

$$P_o = -G\pi a^2 u_{,z}(z = a \cot \alpha)$$

results in

$$P_o = \frac{\pi G a}{\cot \alpha}(1 + ia_o \cot \alpha)u_o$$

Setting $\pi Ga/\cot \alpha$ equal to the static-stiffness coefficient leads to

$$\tan \alpha = \frac{8}{(2 - \nu)\pi}$$

The nondimensionalized dynamic-stiffness coefficient $\bar{S}_o$ is decomposed into

$$\bar{S}_o = k_o + ia_o c_o$$

where the spring and damping coefficients are equal to

$$k_o = 1$$

$$c_o = \frac{2 - \nu}{8}\pi$$

They are both frequency independent. The agreement with the exact values shown in Fig. 7-19 (for $\nu = 0.33$) is astonishingly good.

For a damped half-space, the dynamic-stiffness coefficient, normalized with the same static value, is equal to

$$\bar{S}_o = (1 + 2\zeta i)\left(1 + ia_o^* \frac{2 - \nu}{8}\pi\right)$$

with

$$a_o^* = \frac{a_o}{\sqrt{1 + 2\zeta i}} \sim a_o(1 - \zeta i)$$

This results in

$$k_o = 1 - a_o \frac{2 - \nu}{8}\pi\zeta$$

$$c_o = \frac{2 - \nu}{8}\pi + \frac{2\zeta}{a_o}$$

The dynamic-stiffness coefficients are plotted in Fig. P5-5b. The important differences when compared to the undamped case are well captured, as can be verified using the exact values presented in Fig. 7-20 ($\zeta = 0.05$).

It is worth mentioning that the equation of motion can also be solved by using so-called half-order Bessel functions.

5.6. The conical shear beam shown in Fig. P5-6a can be used to calculate approximately the dynamic-stiffness coefficient for twisting of a rigid circular base (disk) on the surface of a half-space. The horizontal planes are assumed to rotate around the vertical axis. The variation of the shear stress is proportional to the distance from the axis. Determine the dynamic-stiffness coefficient S_o as a function of the radius a at the surface, the angle α, the shear modulus G, and the dimensionless frequency $a_o = \omega a/c_s$. Calculate α by enforcing the static-stiffness

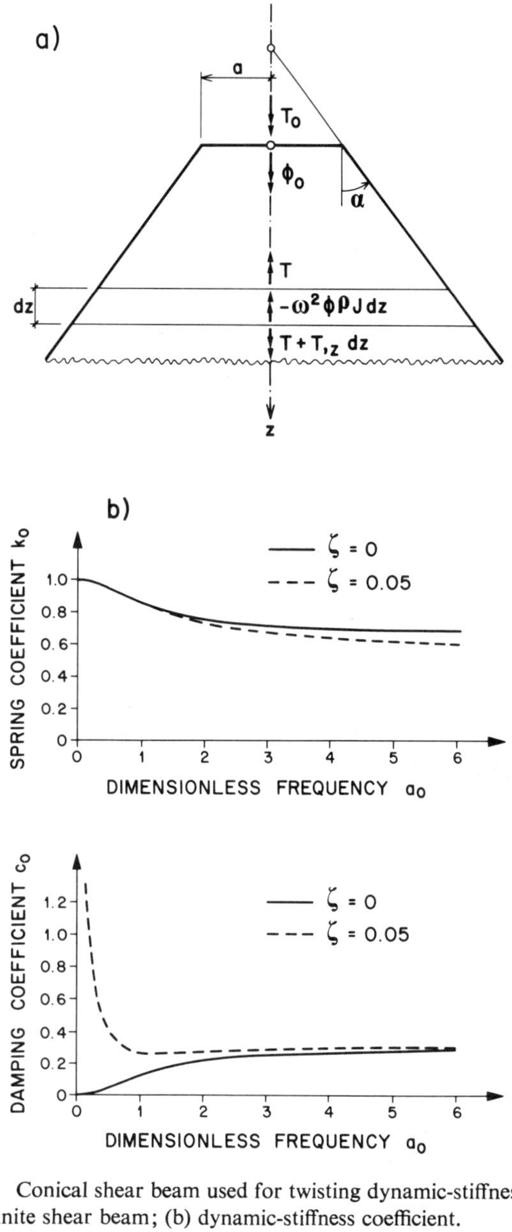

Figure P5-6 Conical shear beam used for twisting dynamic-stiffness coefficient. (a) Semi-infinite shear beam; (b) dynamic-stiffness coefficient.

coefficient of the half-space $16Ga^3/3$. Split the nondimensionalized dynamic-stiffness coefficient into the real and imaginary parts k_o and $a_o c_o$. Plot k_o and c_o versus a_0 for the undamped system and then introduce material damping with a ratio $\zeta = 0.05$.

Solution:

The origin $z = 0$ is selected at a distance $a \cot \alpha$ above the disk. In analogy to Eq. 5.6, the equation of motion of the torsional-shear beam with the polar moment of inertia $J\ (= \pi \tan^4 \alpha\ z^4/2)$ is formulated as

$$\phi,_{zz} + \frac{J,_z}{J}\phi,_z + \frac{\omega^2}{c_s^2}\phi = 0$$

$$\phi,_{zz} + \frac{4}{z}\phi,_z + \frac{\omega^2}{c_s^2}\phi = 0$$

or introducing a dimensionless coordinate ξ in the vertical direction,

$$\xi = \frac{z\omega}{c_s}$$

$$\phi,_{\xi\xi} + \frac{4}{\xi}\phi,_\xi + \phi = 0$$

The general solution equals

$$\phi = \frac{1}{\xi^2}\left[a\left(-\frac{1}{i\xi} - 1\right)\exp(-i\xi) + b\left(\frac{1}{i\xi} - 1\right)\exp(+i\xi)\right]$$

Deleting the incoming wave ($b = 0$) and changing the integration constant to d leads to

$$\phi = d\frac{1 + i\xi}{\xi^3}\exp(-i\xi)$$

Enforcing the boundary condition at the disk $\phi\ (\xi = a \cot \alpha\ \omega/c_s) = \phi_o$ results in

$$\phi = \phi_o \frac{a^3 \cot^3 \alpha\omega^3}{c_s^3(1 + ia \cot \alpha\omega/c_s)} \frac{1 + i\xi}{\xi^3} \exp\left[-i\left(\xi - a \cot \alpha \frac{\omega}{c_s}\right)\right]$$

Determining the amplitude of the twisting moment T_o which is equal to the negative value of that of the torsional moment of the beam at the same location,

$$T_o = -\frac{G\pi a^4}{2}\phi,_z(z = a \cot \alpha) = -\frac{G\pi a^3}{2} a_o \phi,_\xi\left(\xi = a \cot \alpha \frac{\omega}{c_s}\right)$$

leads to

$$T_o = \frac{3\pi G a^3}{2 \cot \alpha}\left[1 - f(a_o) + ia_o\frac{9\pi}{32}f(a_o)\right]\phi_o$$

The function $f(a_o)$ is specified as

$$f(a_o) = \frac{1}{3}\frac{[(9\pi/32)a_o]^2}{1 + [(9\pi/32)a_o]^2}$$

Setting $3\pi G a^3/(2 \cot \alpha)$ equal to the static-stiffness coefficient results in

$$\tan \alpha = \frac{32}{9\pi}$$

The nondimensionalized dynamic-stiffness coefficient $\bar{S}_o$ is decomposed into

$$\bar{S}_o = k_o + ia_o c_o$$

where (for the undamped case) the spring and damping coefficients are equal to

$$k_o = 1 - f(a_o)$$

$$c_o = \frac{9\pi}{32} f(a_o)$$

They are plotted as a function of a_o in Fig. P5-6b. When comparing these curves with the exact values shown in Fig. 7-19, it is observed that the results are similar. The agreement is, however, not as good as for the horizontal dynamic-stiffness coefficient (see Problem 5.5).

When introducing material damping, G is replaced by $G(1 + 2\zeta i)$. Nondimensionalizing $\bar{S}_o$ with the same static value, k_o and $a_o c_o$ follow as the real and imaginary parts.

$$k_o = 1 - f(a_o)\left(1 + \frac{9\pi}{32} a_o \zeta\right)$$

$$c_o = \frac{9\pi}{32} f(a_o)\left[1 + 12\left(\frac{32}{9\pi a_o}\right)^3 \zeta + 4 \frac{32}{9\pi a_o} \zeta\right]$$

The two coefficients are plotted in Fig. P5-6b for $\zeta = 0.05$.

5.7. Derive the coupled equations of motion for harmonic excitation expressed in the displacement amplitudes

$$G\nabla^2 u + (\lambda + G)(u_{,xx} + v_{,yx} + w_{,zx}) = -\rho\omega^2 u$$
$$G\nabla^2 v + (\lambda + G)(u_{,xy} + v_{,yy} + w_{,zy}) = -\rho\omega^2 v$$
$$G\nabla^2 w + (\lambda + G)(u_{,xz} + v_{,yz} + w_{,xz}) = -\rho\omega^2 w$$

5.8. Verify that the off-diagonal term of the dynamic-stiffness matrix of a damped layer $[S^L_{\text{SH}}]$ (Eq. 5.103) converges to zero for the thickness d approaching infinity.

Solution:

The off-diagonal term is inversely proportional to sin ktd. Expanding this expression leads to

$$\sin\,[\text{Re}\,(ktd) + i\,\text{Im}\,(ktd)]$$
$$= \sin\,\text{Re}\,(ktd)\cos i\,\text{Im}\,(ktd) + \sin i\,\text{Im}\,(ktd)\cos\,\text{Re}\,(ktd)$$
$$= \sin\,\text{Re}\,(ktd)\cosh\,\text{Im}\,(ktd) + i\cos\,\text{Re}\,(ktd)\sinh\,\text{Im}\,(ktd)$$

As both sinh Im (ktd) and cosh Im (ktd) converge to $+\infty$ for $d \longrightarrow \infty$, the off-diagonal term will vanish in the limit.

5.9. In the vertical direction, the variation of the out-of-plane motion $v(z)$ is specified by an exponential function. If the wavelength is large compared to the depth of the layer, a linearization is possible. The transcendental functions are replaced by algebraic ones. By expanding the trigonometric functions sin ktd and cos ktd in a Taylor series, derive the discrete dynamic-stiffness matrix of a layer $[S^L_{\text{SH}}]$.

Solution:

Substituting

$$\sin ktd = ktd - \frac{k^3 t^3 d^3}{6}$$

$$\cos ktd = 1 - \frac{k^2 t^2 d^2}{2}$$

in Eq. 5.103 and using Eqs. 5.95b, 5.93, and 5.94 leads to

$$[S_{\mathrm{SH}}^{L}] = \frac{G}{d}\begin{bmatrix} 1 & -1 \\ -1 & 1 \end{bmatrix} + \frac{k^2 dG}{6}\begin{bmatrix} 2 & 1 \\ 1 & 2 \end{bmatrix} - \omega^2 \frac{\rho d}{6}\begin{bmatrix} 2 & 1 \\ 1 & 2 \end{bmatrix}$$

The first and last terms are associated with the standard static-stiffness and the consistent mass matrices, respectively, for a linear expansion in the vertical direction of the displacement.

5.10. Assuming the wavelength to be large compared to the depth of the layer, verify that the dynamic-stiffness matrix of the layer $[S_{\text{P-SV}}^{L}]$ is equal to

$$[S_{\text{P-SV}}^{L}] = \frac{1}{d}\begin{bmatrix} G & & -G & \\ & \lambda + 2G & & -(\lambda + 2G) \\ -G & & G & \\ & -(\lambda + 2G) & & \lambda + 2G \end{bmatrix}$$

$$- \frac{k}{2}\begin{bmatrix} & \lambda - G & & -\lambda - G \\ \lambda - G & & \lambda + G & \\ & \lambda + G & & -\lambda + G \\ -\lambda - G & & -\lambda + G & \end{bmatrix}$$

$$+ \frac{k^2 d}{6}\begin{bmatrix} 2(\lambda + 2G) & & \lambda + 2G & \\ & 2G & & G \\ \lambda + 2G & & 2(\lambda + 2G) & \\ & G & & 2G \end{bmatrix} - \omega^2 \frac{\rho d}{6}\begin{bmatrix} 2 & & 1 & \\ & 2 & & 1 \\ 1 & & 2 & \\ & 1 & & 2 \end{bmatrix}$$

See also Problem 5.9.

6

FREE-FIELD RESPONSE OF SITE

6.1 DEFINITION OF TASK

6.1.1 Three Aspects When Determining Seismic Environment

The first part of any soil–structure interaction analysis where the loads are not applied directly to the structure is the calculation of the free-field response of the site, that is, the spatial and temporal variation of the motion before excavating the soil and superimposing the structure. Theoretically, this task can be achieved using a model which also includes the source. However, for seismic excitation, the many uncertainties in the source mechanism and in the geological parameters along the transmission path, and the sheer size of this model, dictate a much simpler approach. Starting from the control motion assumed to act at a specific location in a single control point, the seismic environment is calculated. This task cannot be solved without making far-reaching assumptions as to the wave pattern. The same control motion acting in the same control point can arise from many different types of waves, for example, from vertically incident or inclined body waves or surface waves. The choice will, in general, affect the variation of the motion with depth and in the horizontal direction. In connection with Fig. 4-2, it is discussed quantitatively that inclined body waves result in the motion propagating horizontally, in contrast to vertical incidence of the waves. This results in a rotational component in the kinematic-interaction analysis of a surface structure (Fig. 4-3), modifying significantly the inertial loads of the inertial-interaction analysis.

The first aspect, the selection of the control motion, is discussed in Section

3.3.3. The second, the choice of the control point, is examined immediately below. The third aspect (i.e., identifying the wave pattern) is probably the least understood. The direction of propagation and the orientation of the planes of incidence will depend on the location of the site relative to the source of the earthquake. However, as the engineering seismologist cannot give advice to the analyst in sufficient detail, extreme cases are calculated assuming, for example, the control motion to arise from only one wave pattern with limiting parameters. The free-field motions (or even the final results of the total soil–structure interaction analyses) are then compared. Conservative choices of the wave patterns have to be made, which can turn out to be different for the various results used for design. It is therefore important to examine in detail the various wave patterns which possibly can occur for a specific site, and to gain insight into the expected free-field response. This is achieved by extensive parametric studies in this chapter. Of course, these three aspects are interrelated.

6.1.2 Location of Control Point

Three choices exist for the location of the control point associated with the site, which are illustrated in Fig. 6-1. The site consists of layers of soil resting on bedrock. The control point should be selected on the ground surface of the free field (point *A*) or at an assumed rock outcrop (point *B*), that is, on the level of the rock, but assuming there is no soil on top, or, less frequently, at the ground surface of another layered system which can have different soil properties (point *C*). The latter choice allows the earthquake records of a seismic station with known soil properties (which will, in general, differ from those of the site) to be used as the control motion. In all cases, the control point is located at the

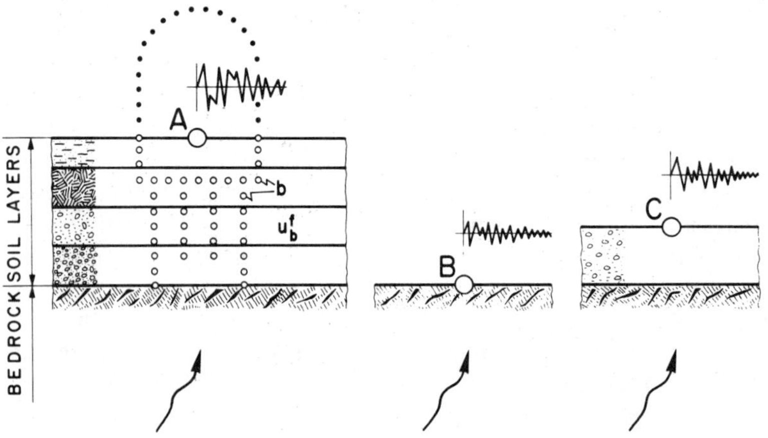

Figure 6-1 Selection of control point for seismic input (after Ref. [5]).

surface, where the strong-motion earthquakes are normally recorded. Under no circumstances should the control point be selected at some (arbitrary) depth below the surface, for example, within the site on the level where subsequently the basemat of the embedded structure will be placed. At such a location within the site, the frequency content of the motion will depend strongly on the depth. This is illustrated in Fig. 5-10, where significant dips (at the natural frequencies of the soil column built in at the depth) appear in the amplification from the top to the depth, which leads to peaks in that in the other direction (from the depth to the top). Assuming a broad-banded spectrum at the depth will thus result in unrealistically high values at the surface for the same frequencies. The whole motion will be distorted. The same occurs if a historic earthquake recorded at the surface or (as an exception) within a site with different material properties is used. The reflections of the waves at the free surface cause this effect. Many other examples, also for more general wave patterns, are found in this chapter.

Starting from the control motion with an assumed wave pattern in the selected control point, the spatial and temporal variation of the free-field response of the site has to be calculated on the line which will subsequently form the structure–soil interface. The latter is shown in Fig. 6-1 for an embedded structure founded on piles. This free-field motion with the amplitudes $\{u_b^f\}$ appears in the load vector of the basic equation of motion (Eq. 3.9). The subscript b denotes the nodes on the structure–soil interface. For the selection of the control points B and C, it is assumed that the source (the incident waves) in the bedrock is the same as for the site. The properties of the bedrock are also identical.

The equations of motion for calculating the spatial variation of the free-field response for the three possible locations of the control points are derived in Section 6.2. The mathematical formulation for inclined body and surface waves is developed. The vertical variation is characterized by various amplifications, the horizontal by dispersion and attenuation. The mathematical formulation is then applied to the half-space in Section 6.3 and to the single layer resting on a half-space in Section 6.4. As the calculation of the free-field response of a site represents the most important factor in soil–structure interaction, it is appropriate to discuss this aspect in depth. To identify the key features, a vast parametric study is performed for the out-of-plane and in-plane motions in Sections 6.5 and 6.6, respectively. Harmonic excitations for sites consisting of a layer with homogeneous properties and with increasing stiffness resting on bedrock are investigated. An actual soft site and a rock site are also examined in Sections 6.7 and 6.8. These detailed analyses allow well-founded conclusions to be stated in the summary at the end of this chapter. The reader who does not place great value on understanding the physical phenomenon and gaining experience can simply browse through Sections 6.5 to 6.8. The mathematical formulation is fully developed and sufficiently applied in the earlier sections.

6.2 AMPLIFICATION, DISPERSION, AND ATTENUATION

For all three choices of the control point, the displacement and the surface traction, which is equal to zero, are prescribed simultaneously along the boundary (free surface). Along the remaining boundary, no conditions are imposed. This is not the classical boundary-value problem.

The equations governing the out-of-plane and in-plane motions are uncoupled (Sections 5.3 and 5.4). This allows the computational procedure to determine the spatial variation of the free-field motion to be developed using the out-of-plane motion only. Only those aspects that are not analogous or cannot straightforwardly be transferred will be discussed for the in-plane motion.

Taking the Fourier transform of the control motion in the time domain leads to frequency-dependent amplitudes. As will become apparent in this section, making certain assumptions on the wave pattern determines the wave number k. For example, for vertically incident waves, $k = 0$. This is analogous to a spatial Fourier transform into the wave-number domain of the motion. The spatial variation of the motion (in the k-domain) can be calculated assembling the stiffness matrices of the individual layers and of the half-space for plane waves. Use is made of reference subsystems of the soil as in Section 3.1.

6.2.1 Dynamic-Stiffness Matrix of Site

The site with a stiffness varying with depth shown in Fig. 6-2 is discretized with $n - 1$ layers of constant material properties resting on the half-space with index n. The nodes coinciding with the interfaces of the layers are numbered from 1 at the free surface to n at the top of the half-space. The dynamic-stiffness matrices for the out-of-plane motion of the layer i $[S^L_{SH}]_i$ and of the half-space

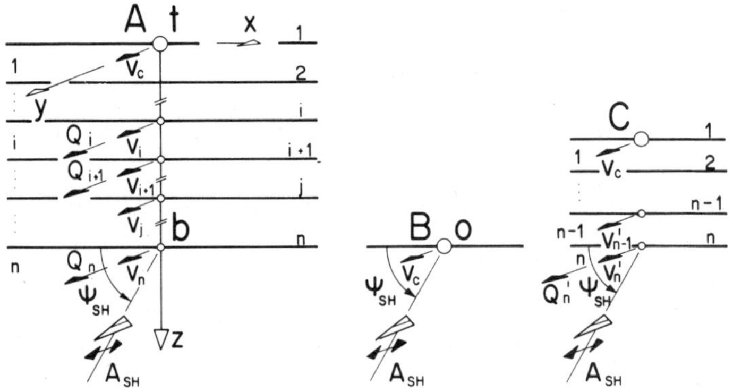

Figure 6-2 Layered site and reference soil system with nomenclature for out-of-plane motion (after Ref. [5]).

S^R_{SH} are specified in Eqs. 5.103 and 5.104, respectively. With the nomenclature illustrated in Fig. 6-2, the force–displacement relationships are formulated as

$$\begin{Bmatrix} Q_i \\ Q_{i+1} \end{Bmatrix} = \begin{bmatrix} S^L_{SH} \end{bmatrix}_i \begin{Bmatrix} v_i \\ v_{i+1} \end{Bmatrix} \tag{6.1a}$$

$$Q_n = S^R_{SH} v_n \tag{6.1b}$$

It is important to realize that $[S^L_{SH}]_i$ is the rigorous dynamic-stiffness matrix of a homogeneous layer of finite depth d for all excitation frequencies. In general, the discretization will consist of only very few layers which have different physical properties. Moreover, the rock is directly represented as a half-space and not modeled as a series of layers.

As discussed in Section 5.3.1, the variation of the displacement amplitude $v(z, x)$ in the x-direction is determined by k (Eq. 5.97) and is thus constant with depth for the layer. The boundary conditions at the interface of two layers compel the value of k to be constant for the total system. For a given ω, the phase velocity c also has to be constant (Eq. 5.94). This has to be taken into consideration when determining $[S^L_{SH}]_i$ and S^R_{SH}.

For the in-plane motion, $[S^L_{P\text{-}SV}]_i$ relates the displacement amplitude u and w at nodes i and $i + 1$ to the load amplitudes P and R at the same nodes (Eq. 5.134). The matrix $[S^R_{P\text{-}SV}]_i$ is specified in Eq. 5.135. Again, k and c have to be constant for all layers and for the half-space.

Assembling the matrices of the individual elements, the discretized dynamic-equilibrium equation of the site results in

$$[S_{SH}]\{v\} = \{Q\} \tag{6.2}$$

The dynamic-stiffness matrix of the total system is denoted by $[S_{SH}]$; $\{v\}$ is the vector of the displacement amplitudes with elements v_1 to v_n and $\{Q\}$ the vector of the external load amplitudes. As in the assembling process, $[S^L_{SH}]_i$ of two adjacent layers are partly overlapped, the symmetric $[S_{SH}]$ is tridiagonal. Also, $[S_{P\text{-}SV}]$ is strongly banded.

6.2.2 Site Amplification for Body Waves

The site-amplification problem for body waves is discussed first, for which incident waves in the half-space propagating from the source to the site exist. For an assumed wave pattern consisting of body waves only, the control point is selected at the ground surface of the site (point A in Fig. 6-1), at an assumed rock outcrop (point B), or at the ground surface of another layered system (point C). The angle of incidence in the half-space ψ_{SH} determines the phase velocity c (Eq. 5.93, with $m_x = \cos \psi_{SH}$). The wave number k, which enters in the determination of the dynamic-stiffness matrices, is thus also specified. As m_x in the half-space is always smaller than or equal to 1, c will lie between c^*_s of the half-space ($\psi_{SH} = 0$) and infinity ($\psi_{SH} = 90°$) for body waves. This also applies when calculating the site response using points B or C, as it is assumed that the incident waves in the bedrock and its material properties are the same as those present at the site.

The location of the control point influences only $\{Q\}$ in Eq. 6.2. For a prescribed control motion v_c in point A, all exterior-load amplitudes $\{Q\}$ are equal to zero. Setting $v_1 = v_c$ and using the standard Gaussian elimination, the free-field-displacement amplitude v_j in node j results. The amplitude v_j is in this case independent of the properties of the system below node j. This of course also holds for the stresses and strains at node j.

For a prescribed control motion v_c in point B, $\{Q\}$ can be calculated analogously as the load vector in the basic equation of soil–structure interaction, Eq. 3.9. For the site, the layers and the bedrock represent the two substructures, the first corresponding to the structure and the second to the soil in the soil–structure system. This determines the left-hand side of the equations of motion. The bedrock is regarded as the reference system of the soil. The load vector will thus consist of the product of the dynamic-stiffness matrix of the bedrock S^R_{SH} and the (free-field)-motion vector v_c. The nth element of $\{Q\}$, Q_n, can be formulated as

$$Q_n = S^R_{\text{SH}} v_c \tag{6.3}$$

all other elements being zero. The displacement amplitude v_j, which again follows from Eq. 6.2, is a function of the properties of the total system. See Problem 6.1 for an application.

For a prescribed control motion v_c in point C, the layers of the actual site and the bedrock still represent the two substructures. The bedrock and the layers (with point C) are regarded as the reference soil system. The free-field motion of this system is calculated first. As the control motion is prescribed on the ground surface, the procedure discussed in connection with the control point A, but applied to the reference soil system, is used. As the layers of this reference system will, in general, be different from those of the site, a dash is used to denote variables associated with the reference system. The only nonzero element Q_n is calculated as

$$Q_n = S^R_{\text{SH}} v'_n + Q'_n \tag{6.4a}$$

The amplitude v'_n is the free-field motion of the reference-soil system in node n. The load amplitude Q'_n is calculated from Eq. 6.1a as the product of the lower row of $[S^L_{\text{SH}}]'_{n-1}$ and the vector with the two elements $v_1 = v'_{n-1}$ and $v_2 = v'_n$. The first and second terms on the right-hand side of Eq. 6.4a represent the load amplitudes of the elements below and above node n, respectively, calculated as the products of the corresponding stiffness matrices and the free-field-motion vector of the reference soil system. Denoting the dynamic-stiffness matrix of the reference soil system in node n as S'^f_n, Q_n can also be written formally as

$$Q_n = S'^f_n v'^f_n \tag{6.4b}$$

with $v'^f_n = v'_n$. The agreement with the load vector of the basic equation of motion (Eq. 3.9) becomes visible. The right-hand side is equal to the product of the dynamic-stiffness coefficient of the free field $S_n'^f$ (discretized in the node n at which the soil layers of the actual site are introduced) and of the free-field motion $v_n'^f$ in the same node. Both factors are determined in the reference soil system. For the control point C, the dynamic equation of motion (Eq. 6.2 with Eq. 6.4b) is illustrated schematically in Fig. 6-3. The analogy to the physical interpretation of the basic equation of motion shown in Fig. 3-4 is obvious. The choice of the control point at the ground surface of another site represents an exception. This case is thus not addressed further in this text. Selecting the control motion to act in an outcrop of rock (control point B) follows as a special case.

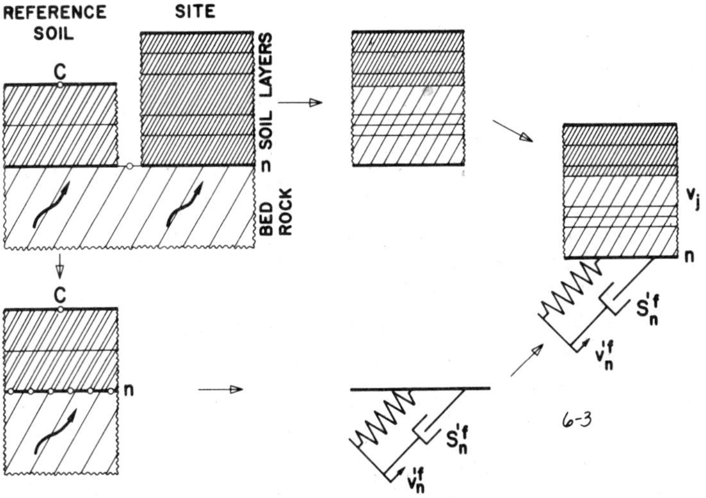

Figure 6-3 Physical interpretation of equation of motion of site with control point on ground surface of reference soil.

For the in-plane motion, selecting, for example, the angle of incidence in the half-space ψ_P of the P-waves (Fig. 6-4) determines not only ψ_{SV} but also the wave number k ($l_x = \cos \psi_P$, Eqs. 5.126 and 5.127). The procedure for calculating the free-field response is analogous.

6.2.3 Surface Waves

For a wave pattern consisting of surface waves, for which no source exists, no energy is propagated from infinity toward the site. As discussed in connection with the derivation of the dynamic-stiffness matrix of the half-space (Eq. 5.104), this condition leads to excluding the incident waves. If A_{SH} were set equal to zero, a vanishing control motion in point B (Fig. 6-1) would result. The control point can thus not be selected at the outcrop of rock. For an assumed wave pattern consisting of surface waves, only the control point A at the ground surface of the site is meaningful. Suppressing the incident waves leads to $Q_n = Q_1 = 0$. Setting the amplitude-load vector $\{Q\} = \{0\}$ in the discretized dynamic-equilibrium equations of the site (Eq. 6.2), an eigenvalue problem arises. The surface-wave motion is thus equal to the natural modes of wave propagation of the site. For a given frequency ω, the only "free" parameter is the phase velocity c; only distinct values of c associated with the various so-called modes exist. For these values of $c(\omega)$, the determinant of the dynamic-stiffness matrix $[S_{SH}]$ vanishes, leading to nontrivial solutions for the displacement amplitudes $\{v\}$. These are then scaled to form the control motion. Details are described in subsequent sections.

To discuss the possible range of c for surface waves, damping is neglected. For a surface wave, the motion in the half-space has to decay for increasing depth. Examining this motion $B_{SH} \exp(-iktz)$ specified in Eq. 5.98 (with $A_{SH} = 0$) leads to the condition that the value t has to be negative and imaginary. This results in $m_x > 1$, based on Eq. 5.95b. Finally, with Eq. 5.93, $c < c_s$ of the half-space follows. It can be shown that the lower bound of the phase velocity equals the shear-wave velocity of the top layer.

As the "exact" expressions for the layer-stiffness matrices are used, the eigenvalue problem is transcendental and has to be solved iteratively by search techniques. For no damping, and in the phase-velocity range of interest, the determinant is real. The various so-called dispersion curves $c(\omega)$ are determined as follows. For each curve, the cutoff frequency is calculated first by setting $c = c_s$ of the half-space and varying ω. By increasing ω and decreasing c simultaneously, the dispersion curve then follows, up to the ω of interest. For nonzero damping, the determinant and the phase velocity are complex. For a specified (real) ω, the real and imaginary parts of c are iterated until the real and imaginary parts of the determinant vanish. It is computationally efficient to vary c and ω simultaneously based on a steepest-descent procedure.

For the in-plane motion, analogous methods are applicable to calculate the surface waves. The range of the phase velocity c is somewhat different and will be discussed in Section 6.6.4.

Valuable physical insight can be gained by examining body and surface waves systematically in undamped and damped sites. It is possible to determine in which cases the variables are real, imaginary, or complex. Without loss of generality, this investigation can be performed for a half-space (Section 6.3) and for a single layer on a half-space (Section 6.4).

6.2.4 Amplifications

To describe the variation of the motion with depth, various amplifications can be defined. Any two locations within the site or at a fictitious outcrop of rock could be selected. For the out-of-plane motion, results will be shown for three ratios of the absolute values of the displacement amplitudes. [Besides the amplitudes of the waves A_{SH} and B_{SH}, the displacement and stress amplitudes $v(z)$, $\tau_{xy}(z)$, and $\tau_{yz}(z)$ are complex values.] First, within the site (Fig. 6-2) from the top ($v_t = v_1$) to the base ($v_b = v_n$): $|v_b|/|v_t|$; second, from outcropping bedrock to the top of the site: $|v_t|/|v_o|$; and third (as a redundant quantity), from outcropping bedrock to the base of the site: $|v_b|/|v_o|$. For the in-plane motion, results will be displayed for two ratios of the absolute values of the displacement amplitudes. First, within the site (Fig. 6-4) from the top (u_t, w_t) to the base (u_b, w_b): $|u_b|/|u_t|$, $|w_b|/|w_t|$; and second from outcropping bedrock (u_o, w_o) to the top of the site: $|u_t|/|u_o|$, $|w_t|/|w_o|$. In addition, results are plotted for the ratios of the absolute values of the displacement amplitudes at the top or at the base of the site and the absolute values of the amplitudes of the incident waves (A_P, A_{SV}) in the half-space: $|u_t|/|A_{SV}|$, $|w_t|/|A_P|$, $|u_b|/|A_{SV}|$, and $|w_b|/|A_P|$.

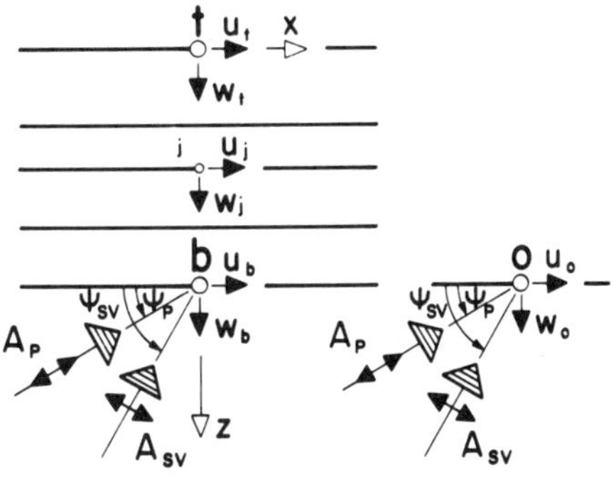

Figure 6-4 Layered site and reference soil system with nomenclature for in-plane motion.

6.2.5 Apparent Velocity and Decay Factors

The variation in the horizontal direction of the displacements and stresses depends only on the wave number k, as can be seen from Eq. 5.97 and 5.129. As is visible from Eqs. 5.93 and 5.94, k and the phase velocity c are in general complex. This factor can be split up as follows:

$$\begin{aligned}\exp(-ikx) &= \exp\left[-\frac{i\omega x}{\operatorname{Re}(c) + i\operatorname{Im}(c)}\right] \\ &= \exp\left[-\frac{i\omega\operatorname{Re}(c)}{|c|^2}x\right]\exp\left[-\frac{\omega\operatorname{Im}(c)}{|c|^2}x\right]\end{aligned} \tag{6.5}$$

The first part describes the propagation with a (real) apparent velocity c_a in the positive x-direction ($\exp i\omega[t - x/c_a]$)

$$c_a = \frac{|c^2|}{\operatorname{Re}(c)} \tag{6.6}$$

the second the attenuation of the motion. For a distance x equal to the wavelength $\lambda = 2\pi c_a/\omega$, the (real) decay factor per wavelength δ_λ results.

$$\delta_\lambda = \frac{v(x + 2\pi c_a/\omega)}{v(x)} = \frac{u(x + 2\pi c_a/\omega)}{u(x)} = \frac{w(x + 2\pi c_a/\omega)}{w(x)}$$
$$= \exp\left[-2\pi \frac{\text{Im}(c)}{\text{Re}(c)}\right] \tag{6.7a}$$

For a distance d, the decay factor equals

$$\frac{v(x+d)}{v(x)} = \frac{u(x+d)}{u(x)} = \frac{w(x+d)}{w(x)} = \exp\left[-\frac{\omega \, \text{Im}(c)}{|c|^2} d\right] \tag{6.7b}$$

For a real value of c, $c_a = c$ and no attenuation occurs. In formulating Eq. 6.7, the same phase velocity c is assumed for the out-of-plane and in-plane motions. The equations for c_a and δ_λ are identical to those for the rod with increasing area (Eqs. 5.50 and 5.51).

The apparent velocity c_a also determines (at a specific location) the spatial variation of the free-field motion. For a structure with a rigid base, it is meaningful to describe the seismic input acting on the structure using the so-called scattered motion $\{u_0^g\}$, which is equal to that of kinematic interaction $\{u_0^k\}$ (Eq. 3.29b). This determines the contribution of the seismic excitation to the load vector of the basic equation of motion formulated in total displacements (Eq. 3.20) or that of inertial interaction (Eq. 3.32). This scattered motion $\{u_0^g\}$ is calculated as a function of that of the free-field $\{u_b^f\}$ using Eq. 3.19. For a surface structure with a rigid basemat, the dynamic-stiffness matrix of the soil and its inverse appear in this equation. The influence of the stiffness of the soil will thus be very small. The wave effects can be characterized by the ratio of a representative length of the basemat and of the apparent wavelength $2\pi c_a/\omega$ (Fig. 4-2). For a circular basemat with radius a, $\omega a/c_a$ can be used to characterize the wave effects. As already discussed qualitatively in connection with Fig. 4-3, the out-of-plane motion will result essentially in a translational component with amplitude v_0^g and in a torsional component with amplitude γ_0^g (see Problem 4.3). The in-plane motion, propagating in the horizontal x-direction, leads to two translational components with amplitudes u_0^g and w_0^g and to the rocking component with an amplitude β_0^g (see Problems 4.2 and 4.4). The dependency of these amplitudes on $\omega a/c_a$ is discussed in Section 7.4.3 (Figs. 7-30 and 7-31).

6.3 HALF-SPACE

The dynamic behavior of the half-space represents an essential element in the site-response analysis. In addition to being the simplest site, the motion on the free surface of the half-space (multiplied by the dynamic-stiffness matrix) determines the load vector of the equations of motion of the total site if the control point is selected at the rock outcrop (Eq. 6.3 and analogously for the in-plane motion). As the half-space in this text is referred to primarily in the latter case, the superscript R is introduced when appropriate. Complete agreement is thus reached with the notation applied in the vast parametric study in Sections 6.5

and 6.6. The subscript o (for outcropping) denotes the motions at the free surface of the half-space.

6.3.1 Incident SH-Waves

The out-of-plane motion is simple to analyze. The displacement amplitude at the free surface v_o of the half-space can be expressed as a function of the incident-wave amplitude A^R_{SH}. Substituting $\tau_{yz1} = 0$ in Eq. 5.101 results in

$$B^R_{SH} = A^R_{SH} \tag{6.8}$$

$$v_o = 2A^R_{SH} \tag{6.9}$$

The out-of-plane displacement is thus independent of the angle of incidence ψ^R_{SH} and is, also for a damped system, in phase with the amplitudes of the incident and reflected waves, the two amplitudes being the same. The phase velocity c is specified by the angle of incidence ψ^R_{SH} (Eq. 5.93 with $m^R_x = \cos\psi^R_{SH}$). It is always larger than the (complex) shear wave velocity c^{*R}_s. The same relationships also apply at the free surface of a layer.

No surface waves exist for the out-of-plane motion. Setting the dynamic-stiffness coefficient of the half-space S^R_{SH} (Eq. 5.104) equal to zero results in $c = c^{*R}_s$. This is not a new wave pattern, but represents an inclined SH-wave with $\psi^R_{SH} = 0$.

6.3.2 Incident P-Waves

The in-plane motion is examined next. Introducing the boundary conditions at the free surface $[\tau_{xz}(z = 0) = 0, \sigma_z(z = 0) = 0]$ in Eq. 5.132, the amplitudes of the reflected waves B^R_P and B^R_{SV} can be expressed as a function of the incident waves A^R_P and A^R_{SV}.

$$\begin{Bmatrix} B^R_P \\ B^R_{SV} \end{Bmatrix} = \frac{1}{4s^Rt^R + [1-(t^R)^2]^2} \begin{bmatrix} 4s^Rt^R - [1-(t^R)^2]^2 & \dfrac{c^{*R}_s}{c^{*R}_p}4t^R[1-(t^R)^2] \\ -\dfrac{c^{*R}_p}{c^{*R}_s}4s^R[1-(t^R)^2] & 4s^Rt^R - [1-(t^R)^2]^2 \end{bmatrix} \begin{Bmatrix} A^R_P \\ A^R_{SV} \end{Bmatrix} \tag{6.10}$$

While for an unbounded domain, the P- and SV-waves can propagate independently, it can be seen from Eq. 6.10 that introducing a free surface e.g., an incident P-wave leads to a reflected P- and SV-wave. This phenomenon is referred to as "mode conversion." Substituting Eq. 6.10 in Eq. 5.130 formulated for $z = 0$ results in

$$\begin{Bmatrix} u_o \\ w_o \end{Bmatrix} \frac{l^R_x[1+(t^R)^2]}{4s^Rt^R + [1-(t^R)^2]^2} \begin{bmatrix} 4s^Rt^R & \dfrac{c^{*R}_s}{c^{*R}_p}2t^R[1-(t^R)^2] \\ 2s^R[1-(t^R)^2] & -\dfrac{c^{*R}_s}{c^{*R}_p}4s^Rt^R \end{bmatrix} \begin{Bmatrix} A^R_P \\ A^R_{SV} \end{Bmatrix} \tag{6.11}$$

This represents the motion at the free surface as a function of the amplitudes of the incident waves. It applies to a half-space and to a layer. In particular, u_o and w_o denote the outcropping motion of the bedrock (Fig. 6-4).

To study the characteristics of the motion at the free surface caused separately by an incident P- and an incident SV-wave, $A^R_{SV} = 0$ and $A^R_P = 0$, respectively, are introduced into Eq. 6.11, leading to

$$\frac{w_o}{u_o} = \frac{1-(t^R)^2}{2t^R} \tag{6.12a}$$

for the incident P-wave, and to

$$\frac{w_o}{u_o} = -\frac{2s^R}{1-(t^R)^2} \tag{6.12b}$$

for the incident SV-wave.

The real phase velocity c is specified by the angle of incidence ψ^R_P for the incident P-wave (Eq. 5.126 with $l^R_x = \cos\psi^R_P$), or by ψ^R_{SV} for the incident SV-wave ($m^R_x = \cos\psi^R_{SV}$). The direction cosine of the angle of incidence of the reflected wave of the other type also follows from Eq. 5.126. For instance, for the incident P-wave with prescribed l^R_x, $m^R_x = \cos\psi^R_{SV}$ of the reflected SV-wave is determined. For P- and SV-waves, c is always larger than c^{*R}_p and c^{*R}_s, respectively. Besides depending on ψ^R_P and ψ^R_{SV}, the coefficient matrix on the right-hand side of Eq. 6.11 is a function of Poisson's ratio ν^R and of the damping ratios ζ^R_p and ζ^R_s.

At first, the motion caused by incident P-waves at the free surface of a half-space ($A^R_{SV} = 0$) is discussed. In Fig. 6-5, the ratio of the vertical-displacement amplitude and of the amplitude of the incident P-wave, $|w_o|/|A^R_P|$ and the ratio $|u_o|/|A^R_P|$ are plotted versus the angle of incidence ψ^R_P for no damping (Eq. 6.11). Poisson's ratio ν^R is varied. As expected, $|w_o| = 2|A^R_P|$ results for vertical incidence. Decreasing ψ^R_P reduces $|w_o|$. Over a significant range of ψ^R_P, $|u_o|$ bears comparison with $|w_o|$ for smaller values of ν^R. The horizontal amplitude $|u_o|$ depends strongly on ν^R. As the ratio w_o/u_o in Eq. 6.12a is a negative real value

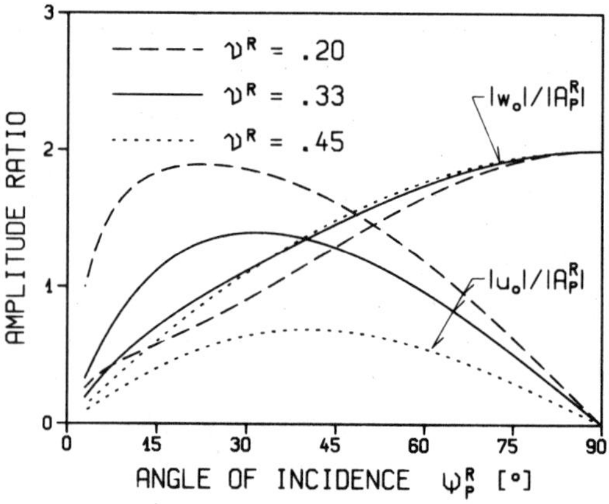

Figure 6-5 Free-surface motion versus angle of incidence, incident P-wave, $\zeta^R = 0$.

for all ψ_P^R and ν^R, w_o and u_o are 180° out of phase. This is also verified by projecting the particle motion of the P-wave onto the global axes. The component on the x-axis has the opposite sign of that on the z-axis (Fig. 6-4). The motion resulting from the incident P-wave is symbolically displayed in Fig. 6-6 for $\psi_P^R = 45°$ and, for the sake of comparison, for vertical incidence. The ratios shown in Fig. 6-5 do not change for a damped half-space if $\zeta_s^R = \zeta_p^R$. For this case, the angle of incidence of the reflected SV-wave ψ_{SV}^R is also real (m_x^R in Eq. 5.126), as c_s^*/c_p^* is real (Eq. 5.88). For $\zeta_s^R \neq \zeta_p^R$, $|w_o|/|A_P^R|$ and $|u_o|/|A_P^R|$ are modified somewhat and w_o and u_o are no longer 180° out of phase. The angle ψ_{SV}^R becomes complex. For instance, for $\zeta_p^R = 0$, $\zeta_s^R = 0.20$ and for $\psi_P^R = 45°$, $c_p^R/c_s^R = 2$ (corresponding to $\nu^R = 0.33$ for the undamped case), $|u_o|/|A_P^R|$ and $|w_o|/|A_P^R|$ change by a factor of 1.06 and 1.03, respectively, while w_o and u_o are 161° out of phase. For this example the phase velocity c is real. The decay factor is 1 (Eq. 6.7) and no attenuation occurs in the horizontal direction. Of course, for $\zeta_p^R \neq 0$, attenuation occurs.

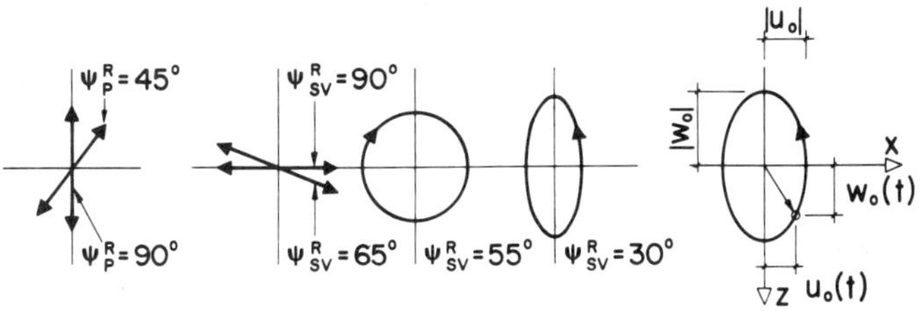

Figure 6-6 Free-surface motion of half-space, $\nu^R = 0.33$, $\zeta^R = 0$.

6.3.3 Incident SV-Waves

An incident P-wave will always lead to reflected P- and SV-waves. However, a reflected P-wave does not exist for a sufficiently shallow incident SV-wave, as $l_x^R = \cos\psi_P^R$ determined from Eq. 5.126 would turn out to be larger than 1. The (limiting) angle of incidence of the SV-wave, for which a reflected P-wave still arises, is called the critical angle ψ_{cr}. The cosine of this critical angle follows from Eq. 5.126 by setting $l_x^R = 1$:

$$\psi_{cr} = \arccos\frac{c_s^{*R}}{c_p^{*R}} = \arctan\sqrt{\frac{1}{1-2\nu^R}} \tag{6.13}$$

For $\psi_{SV}^R < \psi_{cr}$, s of Eq. 5.128a would be a negative imaginary value. It is visible from Eq. 5.130 that the motion associated with B_P decays with depth. (Actually, for the case $\zeta_p \neq \zeta_s$, s is complex with a negative imaginary term.)

Next, the motion at the free surface of a half-space caused by an incident SV-wave ($A_P^R = 0$) is examined. The ratios $|u_o|/|A_{SV}^R|$ and $|w_o|/|A_{SV}^R|$ are plotted versus the angle of incidence ψ_{SV}^R for no damping in Fig. 6-7 (Eq. 6.11). The

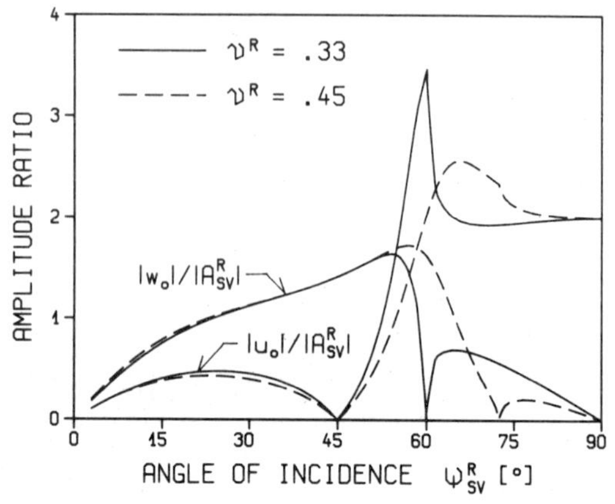

Figure 6-7 Free-surface motion versus angle of incidence, incident SV-wave, $\zeta^R = 0$.

results are presented for two Poisson's ratios ν^R. For vertical incidence $|u_o| = 2|A^R_{\text{SV}}|$. The critical angle ψ_{cr} depends only on Poisson's ratio ν^R (Eq. 6.13) and varies from 45° ($\nu^R = 0$) to 90° ($\nu^R = 0.5$). For $\nu^R = 0.33$, $\psi_{\text{cr}} = 60°$. For $\psi_{\text{cr}} < \psi^R_{\text{SV}} < 90°$, the motion is in phase, as can be verified from Eq. 6.12b. For decreasing ψ^R_{SV} in this range, $|u_o|$ will, in general, increase. The vertical amplitude $w_o = 0$ for $\psi^R_{\text{SV}} = \psi_{\text{cr}}$. For $45° < \psi^R_{\text{SV}} < \psi_{\text{cr}}$, the motion is called prograde; that is, the vertical motion lags behind the horizontal by 90°, the total motion being clockwise. The horizontal amplitude $u_o = 0$ for $\psi^R_{\text{SV}} = 45°$. For $\psi^R_{\text{SV}} < 45°$, a retrograde motion with respect to the direction of propagation results; that is, the horizontal motion lags behind the vertical by 90° for the coordinate system selected, the total motion being counterclockwise. The three different motions of the incident SV-wave occurring for $\psi^R_{\text{SV}} = 65°$, 55°, and 30° are shown in Fig. 6-6. If $\zeta^R_s = \zeta^R_p$, the curves in Fig. 6-7 still apply, based on the same reasoning as explained for P-waves. If $\zeta^R_s \neq \zeta^R_p$, $|u_o|/|A^R_{\text{SV}}|$ and $|w_o|/|A^R_{\text{SV}}|$ are changed slightly and the phase angle between w_o and u_o is modified, similarly as for the case of P-waves. However, close to the critical angle, large differences arise. At $\psi^R_{\text{SV}} = 60°$, the vertical motion bears comparison with the horizontal, which is, however, reduced when compared to the case of uniform damping. These results are not shown.

6.3.4 Rayleigh Waves

For surface waves, no incident waves are present ($A_{\text{P}} = A_{\text{SV}} = 0$). Inverting Eq. 6.11 and enforcing this condition by setting the determinant equal to zero results in

$$[1 - (t^R)^2]^2 = -4s^R t^R \tag{6.14}$$

(Obviously, directly equating the determinant in Eq. 6.11 to zero leads to the same equation.) This so-called Rayleigh wave equation is also derived by setting the determinant of the dynamic-stiffness matrix of the half-space $[S^R_{\text{P-SV}}]$ (Eq. 5.135) equal to zero. The corresponding motion of the surface waves, called Rayleigh waves for the in-plane motion, at the free surface follows from the inverse of Eq. 6.11 using Eq. 6.14 as

$$\frac{w_o}{u_o} = \frac{2s^R}{1-(t^R)^2} \tag{6.15}$$

Substituting Eqs. 5.126 and 5.128 in Eq. 6.14, the phase velocity c (normalized with c_s^{*R}), called the Rayleigh wave velocity, follows as a function of ν^R.

$$\left[2 - \frac{c^2}{(c_s^{*R})^2}\right]^2 - 4\sqrt{1-\frac{c^2}{(c_s^{*R})^2}}\sqrt{1-\frac{c^2}{(c_p^{*R})^2}} = 0 \tag{6.16}$$

As the frequency does not appear in this equation, the phase velocity c is a constant and thus the Rayleigh wave (R-wave) of the half-space is nondispersive.

At first, no damping is assumed. Assuming that $c = \epsilon c_s^R$, where ϵ is a very small number and substituting in Eq. 6.16 leads to $-2 + 2\epsilon^2(c_s^R/c_p^R)^2$, which is negative. Assuming that $c = c_s^R$ results in unity. It thus follows that Eq. 6.16 has at least one real root between $c = 0$ and $c = c_s^R$. Three roots actually exist, but two of them can be eliminated, as the corresponding motions associated with B_{P} and B_{SV} in Eq. 5.130 do not decay with depth. Hence, only one R-mode exists. For $\nu^R = 0$, $c = 0.874c_s^R$; for $\nu^R = 0.33$, $c = 0.933c_s^R$; and for $\nu^R = 0.5$, $c = 0.955c_s^R$ results. The (frequency-independent) ratio of the displacement amplitudes on the free surface w_o/u_o (Eq. 6.15) is a function of ν^R only ($\nu^R = 0 \longrightarrow w_o = i1.272u_o$, $\nu^R = 0.33 \longrightarrow w_o = i1.565u_o$, $\nu^R = 0.5 \longrightarrow w_o = i1.839u_o$). The retrograde particle motion (as a function of time) at the top of the half-space of the R-wave is illustrated together with that of the P- and SV-waves for different angles of incidence in Fig. 6-6. For a damped half-space with $\zeta_p^R = \zeta_s^R$, the phase velocity c becomes complex. The (real) ratio c/c_s^{*R} is the same as c/c_s^R for the undamped case discussed above (Eq. 6.16). By introducing uniform damping, the phase velocity c is affected in the same manner as for an inclined body wave. The apparent velocity c_a and the decay factor δ_λ follow from Eqs. 6.6 and 6.7. The amplitude ratio w_o/u_o also remains unchanged (Eq. 6.15). For $\zeta_p^R \neq \zeta_s^R$, solving Eq. 6.16 leads to different results than for the undamped case. The motion is no longer 90° out of phase. The values c and w_o/u_o are still frequency independent. For instance, for $\zeta_p^R = 0$, $\zeta_s^R = 0.20$, and for $c_p^R/c_s^R = 2$ (corresponding to $\nu^R = 0.33$ for the undamped case), the ratio c/c_s^{*R} hardly changes, but $|w_o|/|u_o| = 2.08$ (compared to 1.565 for the undamped case) with a phase angle $= 76.4°$. The imaginary parts of r and s are negative, leading to a motion which (eventually) decays with depth.

6.3.5 Displacements and Stresses versus Depth

Finally, the variation of the displacements and stresses with depth is studied. For specified displacement amplitudes u_o and w_o (related by Eq. 6.15), Eq. 5.130 is formulated for $z = 0$. This allows B_{P} and B_{SV} to be expressed as a function of u_o and w_o ($A_P = A_{SV} = 0$). The amplitudes of the displacements $u(z)$ and $w(z)$ and of the stresses $\sigma_x(z)$, $\sigma_z(z)$, and $\tau_{xz}(z)$ then follow from Eqs. 5.130 and Eqs. 5.141 and 5.132, respectively. In Fig. 6-8, the horizontal and

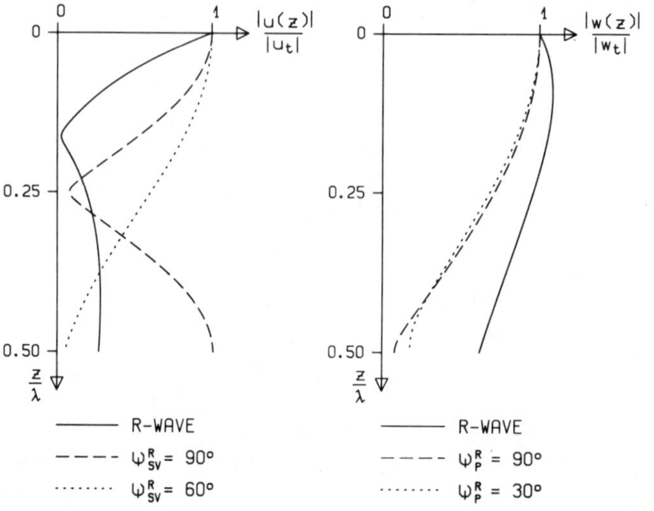

Figure 6-8 Displacement amplitudes versus depth, half-space, $\nu^R = 0.33$, $\zeta^R = 0.05$.

vertical displacement amplitudes of the R-wave, scaled by the value at the free surface, are plotted versus the nondimensionalized depth z/λ with the wavelength $\lambda = c_s 2\pi/\omega$. The subscript o is replaced by t. The horizontal amplitude decays rapidly with depth near the surface. As it actually changes sign at about $z = 0.2\lambda$, the motion below this depth is prograde. The vertical amplitude increases somewhat with depth and then decays rapidly. For comparison, the corresponding horizontal values for the vertically incident and the inclined SV-wave with $\psi^R_{SV} = 60°$ are also plotted on the left-hand side of Fig. 6-8. On the right-hand side, analogously, the vertical displacements of the two P-waves (vertical incidence and $\psi^R_P = 30°$) are also shown. Large discrepancies between the R-wave and the body waves arise. In particular, associating the horizontal motion at the top with the vertically propagating SV-wave, the motion will from a practical point of view be significantly larger at any depth than that arising from the R-wave. In Fig. 6-9 the variation of the amplitude of the normal stress σ_x and of the shear stress τ_{xz} is plotted. Dimensionless stress amplitudes $\bar{\sigma}_x$ and $\bar{\tau}_{xz}$ are defined as $\sigma_x/[(c^R_s)^2\rho^R]$ and $\tau_{xz}/[(c^R_s)^2\rho^R]$, respectively. The variation with depth is shown for $u_o = 0.25\lambda$. The amplitude of the maximum shear stress τ_{max} (calculated from σ_x, σ_z, and τ_{xz}) is also plotted. For comparison, the corresponding results for a vertically incident and for an inclined SV-wave with $\psi^R_{SV} = 60°$ are also indicated. The R-wave leads to a significant τ_{max} close to the free surface.

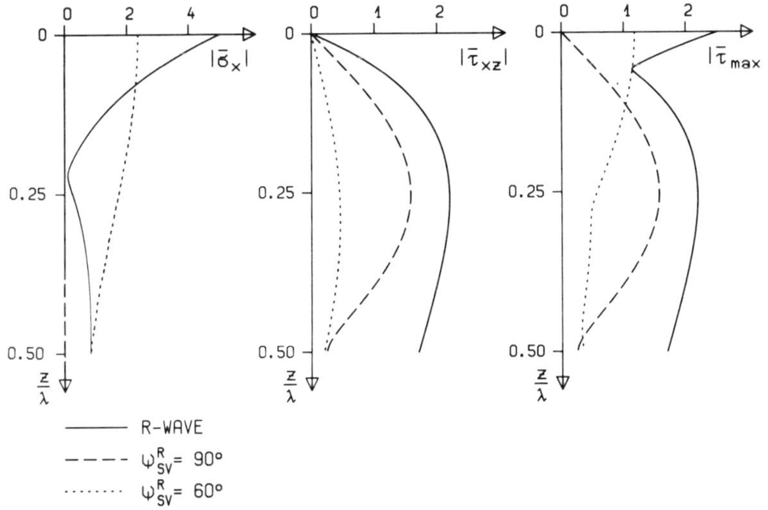

Figure 6-9 Stress amplitudes versus depth, half-space, $\nu^R = 0.33$, $\zeta^R = 0.05$.

6.4 SINGLE LAYER ON HALF-SPACE

By way of illustration the site consisting of a single homogeneous layer resting on a half-space is analyzed. This case forms the basis for the parametric study conducted in Sections 6.5 and 6.6. Thus no examples are presented in this section, with the exception of the dispersion of the surface waves of the layer resting on a rigid half-space. Especially for the out-of-plane case the derivation is so simple that it can be followed easily in all details. The equations allow generally applicable conclusions to be stated.

6.4.1 SH-Waves

Turning first to the out-of-plane motion, the nomenclature of the single soil layer of depth d (superscript L) resting on an infinite bedrock (superscript R) is illustrated in Fig. 6-10. With

$$m_x^R = \cos \psi_{SH}^R \tag{6.17}$$

the phase velocity c of the site follows (Eq. 5.93):

$$c = \frac{c_s^{*R}}{m_x^R} \tag{6.18}$$

The wave number k is specified in Eq. 5.94. The value c is constant for the site. Formulating Eq. 6.18 for the layer leads to

$$m_x^L = \frac{c_s^{*L}}{c} = \frac{c_s^{*L}}{c_s^{*R}} m_x^R \tag{6.19}$$

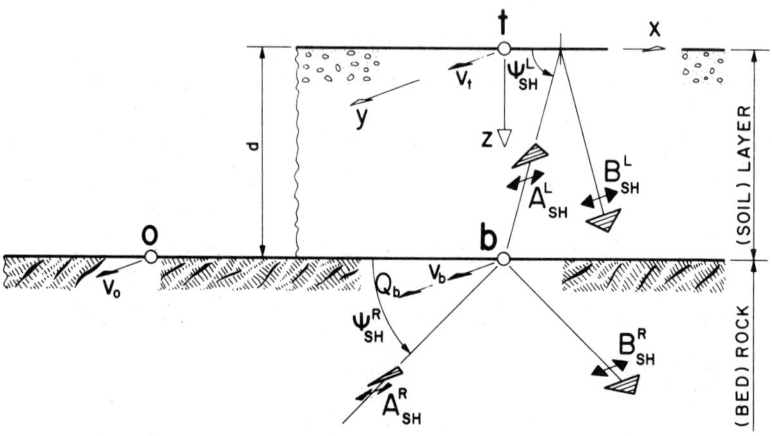

Figure 6-10 Single layer on half-space with nomenclature for out-of-plane motion (after Ref. [5]).

From Eq. 5.95b,

$$t^R = -i\sqrt{1 - \frac{1}{(m_x^R)^2}} \tag{6.20a}$$

$$t^L = -i\sqrt{1 - \frac{1}{(m_x^L)^2}} \tag{6.20b}$$

whereby $t^R = \tan \psi^R_{SH}$.

Assembling the dynamic-stiffness matrices of the layer $[S^L_{SH}]$ (Eq. 5.103) and of the half-space S^R_{SH} (Eq. 5.104), the dynamic-equilibrium equations of the site become (Eq. 6.2)

$$\frac{kt^L G^{*L}}{\sin kt^L d}\begin{bmatrix} \cos kt^L d & -1 \\ -1 & \cos kt^L d + ip \sin kt^L d \end{bmatrix}\begin{Bmatrix} v_t \\ v_b \end{Bmatrix} = \begin{Bmatrix} 0 \\ Q_b \end{Bmatrix} \tag{6.21}$$

where the impedance ratio is defined as

$$p = \frac{t^R G^{*R}}{t^L G^{*L}} \tag{6.22a}$$

which can also be written as

$$p = \frac{\sqrt{1-(m_x^R)^2}\sqrt{G^{*R}\rho^R}}{\sqrt{1-(m_x^L)^2}\sqrt{G^{*L}\rho^L}} \tag{6.22b}$$

From the first row of Eq. 6.21, the ratio of v_b and v_t follows:

$$\frac{v_b}{v_t} = \cos kt^L d \tag{6.23}$$

For a given k, v_b/v_t depends only on the properties of the layer.

For no damping, this ratio is zero for $kt^L d = (2j - 1)\pi/2$, $j = 1, 2, \ldots,$ that is, for

$$\omega_j = \frac{2j - 1}{2}\pi\frac{c_s^L}{d \sin \psi_{SH}^L} \tag{6.24}$$

For vertically incident waves ($\psi_{SH}^L = 90°$), these are the natural frequencies of the layer fixed at its base (expressed in radians).

For a small damping value ζ^L, $|v_t|/|v_b|$ evaluated at ω_j equals (for inclined body waves)

$$\left(\frac{|v_t|}{|v_b|}\right)_{\max} = \frac{2}{(2j - 1)\pi\zeta^L} \tag{6.25}$$

The subscript "max" has been added to indicate that these amplifications are (approximately) the maximum. To derive this equation, $c_s^L\sqrt{1 + 2\zeta^L i}$ is expanded into the first two terms of a Taylor series. These amplifications decrease with increasing frequency ω_j (j augmenting).

For body waves, the ratio of v_t and the outcropping motion v_o (Fig. 6-10) is calculated from Eq. 6.21, using Eqs. 6.3 and 5.104 to determine

$$Q_b = ikt^R G^{*R} v_o \tag{6.26}$$

as

$$\frac{v_t}{v_o} = \frac{1}{\cos kt^L d + \dfrac{i}{p} \sin kt^L d} \tag{6.27a}$$

Analogously,

$$\frac{v_b}{v_o} = \frac{1}{1 + (i/p) \tan kt^L d} \tag{6.27b}$$

For no damping, the ratio $|v_t|/|v_o|$ of the absolute values of the displacement amplitudes equals

$$\frac{|v_t|}{|v_o|} = \frac{1}{\sqrt{\cos^2 kt^L d + \dfrac{\sin^2 kt^L d}{p^2}}} \tag{6.28}$$

The denominator will never become zero for a finite p (elastic rock). This means that resonance does not occur, even if there is no damping in the soil. The amplitude $|v_t|/|v_o|$ is always smaller than or equal to $|v_t|/|v_b|$ (inverse of Eq. 6.23). This is due to the reflected waves B_{SH}^R, which radiate energy downward through the rock toward infinity (radiation damping). In the definitions of $|v_t|/|v_o|$ and $|v_b|/|v_o|$ (but not $|v_t|/|v_b|$) this effect is captured. The stationary values of Eq. 6.28 are obtained at

$$\sin 2kt^L d = 0 \tag{6.29}$$

leading to the same ω_j as in Eq. 6.24 for the maxima of Eq. 6.28. These corresponding maxima are equal to

$$\left(\frac{|v_t|}{|v_o|}\right)_{\max} = p \tag{6.30}$$

For a small damping value ζ^L, $|v_t|/|v_o|$ evaluated at ω_j equals

$$\left(\frac{|v_t|}{|v_o|}\right)_{\max} = \frac{1}{\dfrac{1}{p} + \dfrac{(2j-1)\pi}{2}\zeta^L} \tag{6.31}$$

The radiation damping is represented by p in this equation. With increasing frequency (j augmenting), the effects of the radiation and material dampings will decrease and increase, respectively. The minima of $|v_b|/|v_o|$ for no damping also occur at ω_j (Eq. 6.24). This ratio is always smaller than 1 (for elastic rock).

6.4.2 Love Waves

Setting the determinant of the coefficient matrix of Eq. 6.21 to zero leads to the so-called frequency equation of the surface waves,

$$\tan kt^L d = ip \tag{6.32}$$

The surface waves of the out-of-plane motion are called Love waves. Omitting damping and restricting the investigation to real values of k (and c), the following properties of Love waves are established. As $c < c_s^R$, $m_x^R > 1$ (Eq. 6.18). To determine the frequencies at which the different modes start, c is set equal to c_s^R. This leads to $m_x^R = 1$, $t^R = 0$ (Eq. 6.20a), $p = 0$ (Eq. 6.22a), and from Eq. 6.32

$$kt^L d = (j-1)\pi, \qquad j = 1, 2, \ldots \tag{6.33}$$

Using Eqs. 5.94, 6.20b, and 6.19, Eq. 6.33 is transformed to

$$\frac{\omega d}{c_s^L} = \frac{(j-1)\pi}{\sqrt{1 - \dfrac{(c_s^L)^2}{(c_s^R)^2}}} \tag{6.34}$$

The motion associated with the ω where a new mode starts corresponds to an inclined SH-wave with $\psi_{\text{SH}}^R = 0$.

Examining the asymptotic behavior of the dispersion curves for $\omega \longrightarrow \infty$, the frequency equation 6.32 can easily be rewritten as

$$\sqrt{1 - \frac{(c_s^L)^2}{c^2}}\, d = \frac{c_s^L}{\omega} \arctan \frac{t^R G^R}{t^L G^L} \tag{6.35}$$

For $\omega \longrightarrow \infty$, $c \longrightarrow c_s^L$ for all modes. This is to be expected, as for large ω the motion is restricted to the layer.

It is interesting to note that for a specific c, the frequency difference $\Delta\omega$ between two consecutive modes equals

$$\Delta\omega = \frac{c_s^L}{d} \frac{\pi}{\sqrt{1 - (c_s^L)^2/c^2}} \tag{6.36}$$

Using this equation, the dispersion curves of all higher modes can directly be calculated starting with that of the first.

It is worth stating that the ratio of v_b and v_t depends on c (and not directly on ω). For a specific c, all modes will lead to the same result. This can easily be seen by substituting Eq. 6.32 in Eq. 6.23:

$$\frac{v_b}{v_t} = \frac{1}{\sqrt{1 - p^2}} \tag{6.37}$$

For the special case of a homogeneous layer resting on a rigid half-space, the frequency equation 6.32 is reduced to

$$kt^L d = \frac{2j - 1}{2}\pi \tag{6.38}$$

To study the dependence of the phase velocity c on the frequency ω (dispersion), for this case the real and imaginary parts of the dimensionless phase velocity c/c_s^L of the first mode are plotted versus the dimensionless frequency $\bar{\omega} = \omega d/c_s^L$ in Fig. 6-11a. For no damping in the layer, a cutoff frequency coinciding with the fundamental frequency of the layer exists, corresponding to $\bar{\omega} = \pi/2$. For $\bar{\omega} > \pi/2$, $c(\omega)$ is real and thus the mode propagates horizontally without an attenuation. The decay factor $\delta_\lambda = 1$ (Eq. 6.7a). For $\bar{\omega} < \pi/2$, $c(\omega)$ is imaginary; thus the motion decays exponentially and does not propagate (Eq. 6.5). The corresponding rate of horizontal energy transmission is zero (Eq. 5.123), as k is imaginary. When studying the seismic-site response, this range of the Love waves is of no interest. Introducing damping, $c(\omega)$ is complex for all ω, leading to a motion that propagates and attenuates horizontally. No cutoff frequency exists. However, the motion attenuates significantly for $\bar{\omega} < \pi/2$. While the

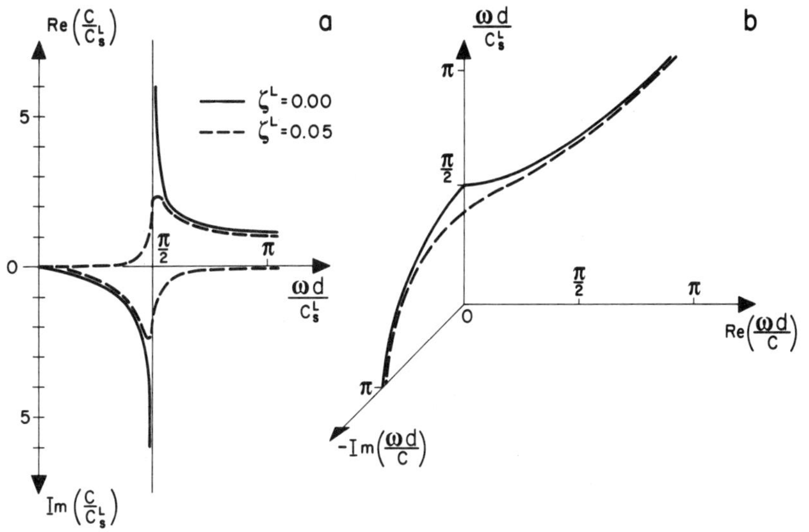

Figure 6-11 First Love mode, homogeneous layer on rigid half-space (Ref. [5]). (a) Dimensionless phase velocity versus dimensionless frequency; (b) frequency spectrum.

decay factor per wavelength δ_λ for large ω equals 0.73 (Eq. 6.7a), it is only 0.001 for $\bar{\omega} = \pi/2$. It should be remembered that a small decay factor represents large attenuation. Near the fundamental frequency of the layer and below, the motion for free-field-response analysis is of academic interest only. In Fig. 6-11b, the same information is presented as a familiar frequency spectrum. For no damping $c = \infty$ at $\bar{\omega} = \pi/2$ (Fig. 6-11a), resulting in $k = 0$ in Fig. 6-11b. Turning to higher modes, the cutoff frequency of the jth mode for no damping coincides with the jth natural frequency of the layer. The properties of the jth mode correspond to those of the first.

6.4.3 Physical Interpretation of Variables

It has been seen that there are some variables that can be real, imaginary, or complex. This is systematically investigated and the physical interpretation of the variables is discussed. The first case shown in Table 6-1 consists of a body wave propagating in an undamped site. The angle of incidence in the half-space ψ_{SH}^R is selected between 0 and 90° and is thus real, resulting in a real direction cosine $m_x^R \leq 1$. The phase velocity c and the wave number k are also real. The wave thus travels horizontally with an apparent velocity $c_a = c$ without any attenuation. The variables ψ_{SH}^L and $m_x^L \leq 1$ are again real. The products kt^R and kt^L, which describe the variation with depth, are both real. The variation with depth is thus harmonic. The sites discussed in Table 6-1 assume that the half-space is stiffer than the layer. For the case of a layer stiffer than the rock and of a wave having a shallow angle of incidence ψ_{SH}^R, $c < c_s^L$ can arise (not shown in Table 6-1). In this case $m_x^L > 1$, and ψ_{SH}^L and t^L are imaginary. As kt^L is imaginary, the amplitude $v(z)$ will no longer be harmonic in the layer (Eq. 5.98), but will vary exponentially with depth.

In the second case, only the layer is damped. The phase velocity c remains real; the motion thus does not attenuate horizontally. The variable m_x^L is complex (Eq. 6.19), resulting also in complex ψ_{SH}^L and t^L. To interpret a complex angle of incidence, only the incident wave in Eq. 5.96 is examined. Substituting $t^L = \text{Re}(t^L) + i\,\text{Im}(t^L)$ in Eq. 5.96 leads to

$$v^L(z, x) = A^L \exp[-k\,\text{Im}(t^L)z] \exp[ik\,\text{Re}(t^L)z] \exp(-ikx) \tag{6.39}$$

The second and third exponential functions are interpreted as a harmonic wave propagating with a real angle of incidence determined by the arctan Re (t^L). As Im (t^L) turns out to to be <0, the first exponential function expresses that the motion increases exponentially with z. As the motion will decay in the direction of propagation because of the damping, the "amplitude" along the wave front will vary, as shown in Fig. 6-12, resulting in no decay in the horizontal direction. In this figure the "amplitude" in two distinct locations of the wave front is presented.

In case 3 of Table 6-1, damping also arises in the half-space. The phase velocity c is complex, leading to an attenuation of the motion in the horizontal

TABLE 6-1 Interpretation of Variables

Case	Type of Wave	Damping								
		Half-space, ζ^R	Layer, ζ^L	ψ_{SH}^R	m_x^R	$\frac{k}{c}$	ψ_{SH}^L	m_x^L	kt^R	kt^L
1	Body wave	No	No	Re	Re, ≤ 1	Re	Re	Re, <1	Re	Re
2		No	Yes	Re	Re, ≤ 1	Re	Com	Com	Re	Com
3		Yes	Yes	Re	Re, ≤ 1	Com	Com	Com	Com	Com
4	Surface wave	No	No	Im	Re, >1	Re	Re	Re, <1	Im	Re
5		Yes	Yes	Com	Com	Com	Com	Com	Com	Com

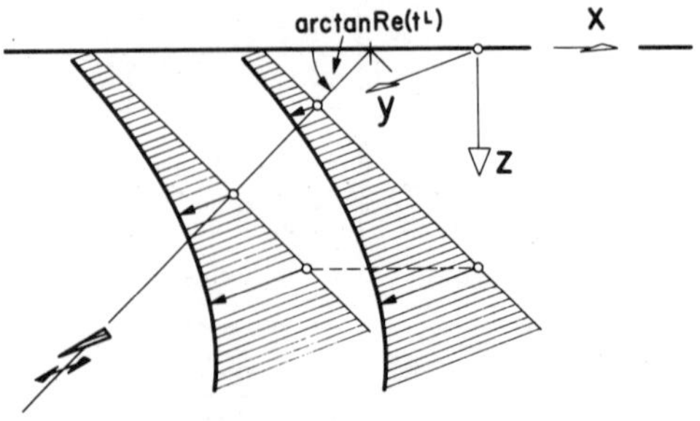

Figure 6-12 Wave pattern for complex angle of incidence (Ref. [5]).

direction. The decay factor for body waves thus depends on the damping of the half-space. For the special case of uniform damping ($\zeta^L = \zeta^R$), not shown in Table 6-1, ψ^L_{SH} and m^L_x are real.

Case 4 deals with Love waves in the undamped site. As $c^L_s < c^R_s$, $m^L_x < 1 < m^R_x$ results, leading to an imaginary ψ^R_{SH} and a real ψ^L_{SH}. As kt^L is real, a harmonic motion with depth takes place in the layer. In the half-space the motion decays exponentially with depth.

In case 5, the site is damped. All variables are complex. The motion attenuates horizontally, and the decay factor for Love waves depends on ζ^L and ζ^R. Again, the motion decays exponentially with depth in the half-space.

6.4.4 P- and SV-Waves

The in-plane motion is analyzed next. The nomenclature is illustrated in Fig. 6-13. For an incident SV-wave with a specified ψ^R_{SV},

$$m^R_x = \cos \psi^R_{\text{SV}} \tag{6.40}$$

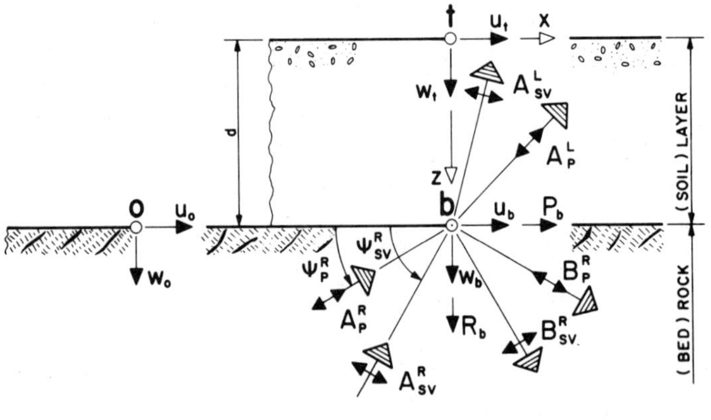

Figure 6-13 Single layer on half-space with nomenclature for in-plane motion (after Ref. [6]).

The phase velocity c of the site follows from Eq. 5.126:

$$c = \frac{c_s^{*R}}{m_x^R} \tag{6.41}$$

For $\zeta^R \neq 0$, c becomes complex and the motion thus attenuates in the horizontal direction. The wave number k is specified in Eq. 5.127.

Formulating Eq. 5.126 for the layer and for the half-space determines ψ_{P}^L, ψ_{SV}^L, and ψ_{P}^R (angle of incidence of the reflected P-wave).

$$c = \frac{c_p^{*L}}{l_x^L} = \frac{c_s^{*L}}{m_x^L} = \frac{c_p^{*R}}{l_x^R} \tag{6.42}$$

For $\psi_{\text{SV}}^R > \psi_{\text{cr}}^R$ and $\zeta^L = \zeta^R$, ψ_{P}^L, ψ_{SV}^L, and ψ_{P}^R are all real. For $\psi_{\text{SV}}^R < \psi_{\text{cr}}^R$, ψ_{P}^R is imaginary, resulting in an associated motion which decays exponentially with depth. For $\zeta^L \neq \zeta^R$, ψ_{P}^L and ψ_{SV}^L are complex. The physical interpretation of a complex ψ^L is discussed in depth in connection with Fig. 6-12. The variables s^R, s^L and t^R, t^L are specified in Eq. 5.128. All terms appearing in the dynamic-stiffness matrices of the layer $[S_{\text{P-SV}}^L]$ (Eq. 5.134) and of the half-space $[S_{\text{P-SV}}^R]$ (Eq. 5.135) are thus determined. For an incident P-wave, analogous formulas apply.

Assembling the dynamic-stiffness matrices $[S_{\text{P-SV}}^R]$ and $[S_{\text{P-SV}}^L]$ leads to the dynamic equations of motion of the site:

$$[S_{\text{P-SV}}]\begin{Bmatrix} u_t \\ iw_t \\ u_b \\ iw_b \end{Bmatrix} = \begin{Bmatrix} 0 \\ 0 \\ P_b \\ iR_b \end{Bmatrix} \tag{6.43}$$

where $[S_{\text{P-SV}}]$ denotes the assembled dynamic-stiffness matrix and P_b and R_b are the component amplitudes of the load vector at the base. The latter are present only if the control point is selected at the outcropping rock. As explained in connection with Eq. 6.3,

$$\begin{Bmatrix} P_b \\ iR_b \end{Bmatrix} = [S_{\text{P-SV}}^R]\begin{Bmatrix} u_o \\ iw_o \end{Bmatrix} \tag{6.44}$$

where $[S_{\text{P-SV}}^R]$ is defined in Eq. 5.135 and u_o and w_o denote the prescribed outcropping motion. The motion throughout the site, including the amplifications, is determined by solving Eq. 6.43.

6.4.5 Rayleigh Waves

For Rayleigh waves (R-waves), the load vector is zero. Setting the determinant of $[S_{\text{P-SV}}]$ in Eq. 6.43 equal to zero leads to the frequency equation. This determines the phase velocities c as a function of ω for the different modes. For an undamped site, and restricting the discussion to real values of c, the following properties of R-waves are established. The first R-mode starts at $\omega = 0$ with $c =$ R-wave velocity of the bedrock (half-space) and converges for $\omega \longrightarrow \infty$ to that of the layer. The higher modes exhibit cutoff frequencies with $c = c_s^R$ and converge, in general, for $\omega \longrightarrow \infty$ to $c = c_s^L$. It follows from Eq. 6.42 that ψ_P^R (and ψ_{SV}^R) are imaginary. The corresponding motion thus decays exponentially with depth. The angles ψ_P^L and ψ_{SV}^L can be real or imaginary, depending on c. For an undamped layer on rigid bedrock, the R-modes all start at the natural frequencies of the layer in the horizontal and vertical directions. Thus below the fundamental horizontal frequency, no R-waves exist. This is analogous to the Love wave. For a damped site, $c(\omega)$ is complex. The motion decays horizontally and the decay factor depends on ζ^L and ζ^R. All angles of incidence are complex. Again, the motion decays exponentially with depth in the bedrock.

6.5 PARAMETRIC STUDY OF OUT-OF-PLANE MOTION

6.5.1 Scope of Investigation

To be able to identify the key parameters governing the free-field response, vast parametric studies are performed. This allows the effects to be evaluated and the conditions to be established for which a specific aspect can dominate. The computational experience presented in this and the next sections should contribute to the engineering judgment of the analyst, which is required if the analyst is to select reasonable parameters to capture the motion of the site even before the structure is built. The results of the free-field analysis should be able to be anticipated to a certain degree, even allowing the calculation to be avoided in some cases. The important aspect of the link to the seismic loading applied to the structure is also addressed.

A site consisting of a single soil layer resting on a half-space representing the bedrock is examined extensively (Fig. 6-10). For a homogeneous layer, the dynamic system is characterized by the following dimensionless parameters:

Ratio of shear-wave velocities: $\bar{c}_s = c_s^R/c_s^L$

Ratio of mass densities: $\bar{\rho} = \rho^R/\rho^L$

Damping ratios: ζ^R, ζ^L

It should be noted that Poisson's ratios do not explicitly arise in the equations for the out-of-plane motion. To describe the response for harmonic motion, the following variables are defined:

Dimensionless frequency: $\bar{\omega} = \omega d/c_s^L$

Dimensionless apparent velocity: $\bar{c}_a = c_a/c_s^L$

In addition to this rather theoretical case of a homogeneous layer, a site consisting of a soil layer having a shear-wave velocity that increases linearly with depth and having a total depth d resting on a homogeneous infinite bedrock is examined. This site is denoted later as "increasing stiffness." The shear velocity of the half-space c_s^R is equal to the shear-wave velocity of the layer at its base. Only the results obtained when the ratio of the shear-wave velocity of the layer at its base to that at its top is equal to 5 are discussed. To be able to calculate $\bar{\omega}$ and $\bar{c}_a$, an average c_s^L for this site with increasing stiffness is calculated, which is defined as follows:

$$c_s^L = \frac{d}{\int_0^d dz/c_s(z)} \tag{6.45}$$

The propagation time of the SH-wave for the (fictitious) homogeneous layer is set equal to that of the layer with increasing stiffness. For our case, this results in $\bar{c}_s = 2.012$. The mass density ρ is assumed to be constant over the whole site. In addition to $\zeta^R = \zeta^L = 0.05$, the damping ratio of the layer is selected to decrease linearly with depth, varying from 0.05 at the top to 0.0 at the base ($\zeta^R = 0$). The layer with increasing stiffness is discretized with six layers of constant material properties.

The homogeneous half-space and the single layer built in at its base represent the two limiting cases. They are contained in the general formulation as special cases ($\bar{c}_s = \bar{\rho} = 1$ and $\bar{c}_s = \infty$).

6.5.2 Vertically Incident SH-Waves

The ratio of the absolute values of the displacement amplitudes at the top of the homogeneous layer and at outcropping of the bedrock $|v_t|/|v_o|$ is plotted in Fig. 6-14a versus the dimensionless frequency $\bar{\omega}$ for vertically incident SH-waves. The parameters are specified in the figure and in its caption. The peaks occur at the $\bar{\omega}$ corresponding to the frequencies of the layer fixed at its base (Eq. 6.24 with $\psi_{SH}^L = 90°$). The increased radiation damping occurring by means of the half-space for decreasing $\bar{c}_s$ reduces the peaks. The peak values are specified in Eq. 6.31, whereby for vertically incident waves, $p = \bar{c}_s\bar{\rho}$ (Eq. 6.22b). The effect of the radiation damping decreases with increasing ω. In Fig. 6-14b, the same ratio is plotted for the site with increasing stiffness. The peaks no longer occur at the natural frequencies of the layer fixed at its base. The fundamental frequency corresponds to an $\bar{\omega} = 2.07$. For comparison, the curves for the homogeneous layer for $\bar{c}_s = 2$ and 5, taken from Fig. 6-14a, are also shown. The amplification for the site with increasing stiffness tends to be more uniform over the frequency range. Even for this most simple wave pattern the more realistic site with increasing stiffness shows a different behavior from that of the homogeneous layer with suitably selected parameters.

The ratio $|v_b|/|v_t|$ is plotted as a function of $\bar{\omega}$ in Fig. 6-15. As discussed in Section 6.2.2, the results are independent of the properties of the rock. The curve for the homogeneous layer is equal to the inverse values of $|v_t|/|v_o|$ for $\bar{c}_s = \infty$ shown in Fig. 6-14a. The minima arise at the natural frequencies of the layer fixed at its base for both sites. The minima for the homogeneous layer are specified as the inverse of Eq. 6.25. For this case, the peaks of $|v_b|/|v_t|$ are > 1,

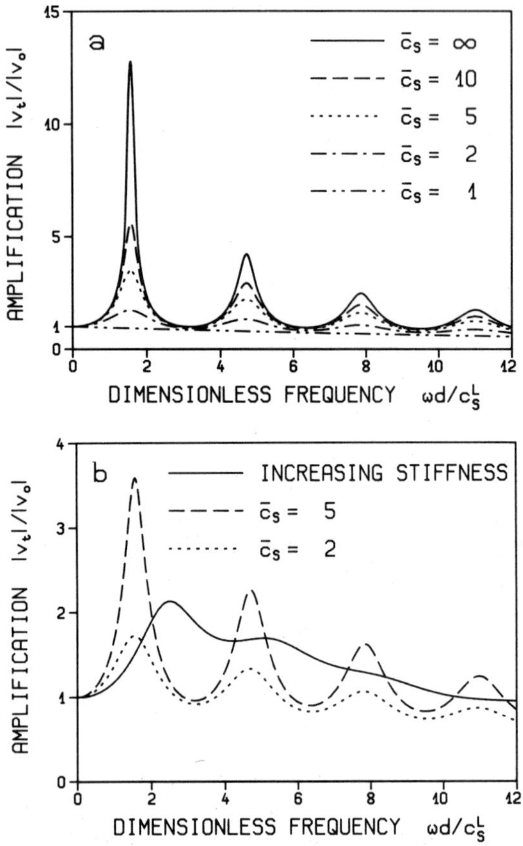

Figure 6-14 Amplification outcropping, vertical incidence, $\bar{\rho} = 1$, $\zeta^R = \zeta^L = 0.05$ (Ref. [5]). (a) Homogeneous layer; (b) increasing stiffness.

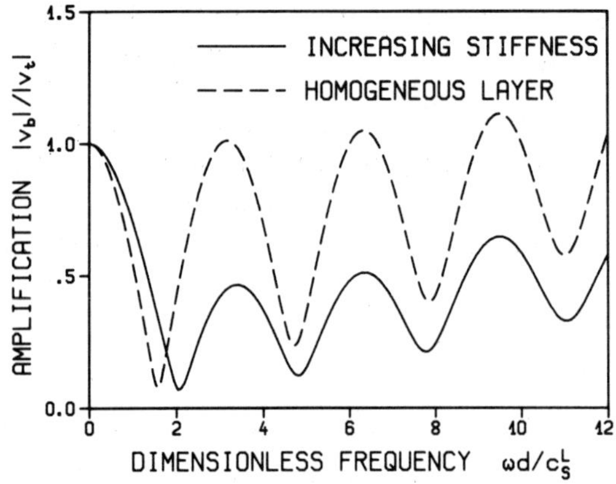

Figure 6-15 Amplification within, vertical incidence, $\zeta^L = 0.05$.

this tendency being more pronounced in the higher-frequency range. For a site with hysteretic damping, the ratio $|v_b|/|v_t|$ will increase without bound for increasing $\bar{\omega}$. This means that the larger the depth d is, the more the amplitudes of the high frequencies increase. Figure 6-15 demonstrates how much the frequency content of the motions at a specific depth is influenced by the reflections of the waves at the free surface. Obviously, as discussed in Section 6.1.2, the control point with a control motion containing all frequencies in the range of interest could not be selected at the base within the site (at point b).

6.5.3 Inclined SH-Waves

Amplification within. In contrast to vertically incident waves, inclined body waves propagate horizontally across the site with an apparent velocity c_a (Eq. 6.6) and attenuate with a decay factor δ_λ (Eq. 6.7a). The (complex) phase velocity c in these two formulas equals $c_s^{*R}/\cos \psi_{SH}^R$. For $\zeta^R = 0$ and for any value of ζ^L, c is real and the motion does not attenuate horizontally. Varying the angle of incidence in the layer ψ_{SH}^L, the amplification within the layer $|v_b|/|v_t|$ is shown in Fig. 6-16. The amplification $|v_b|/|v_t|$ depends only on the properties of the layer and on ψ_{SH}^L. Compared to the case for vertical incidence, the frequencies at which the minima occur are shifted by the factor $1/\sin \psi_{SH}^L$ (Eq. 6.24). The minima themselves are independent of ψ_{SH}^L (Eq. 6.25). The range of physically admissible ψ_{SH}^R is limited, as $\psi_{SH}^R > 0$, resulting in $\arccos(1/\bar{c}_s) \leq \psi_{SH}^L \leq 90°$ (Eqs. 6.17 and 6.19). In practical cases, the direction of propagation of the wave in the layer will be quite steep. This results in the frequency shifts being negligible (with the exception of a site approaching an elastic half-space).

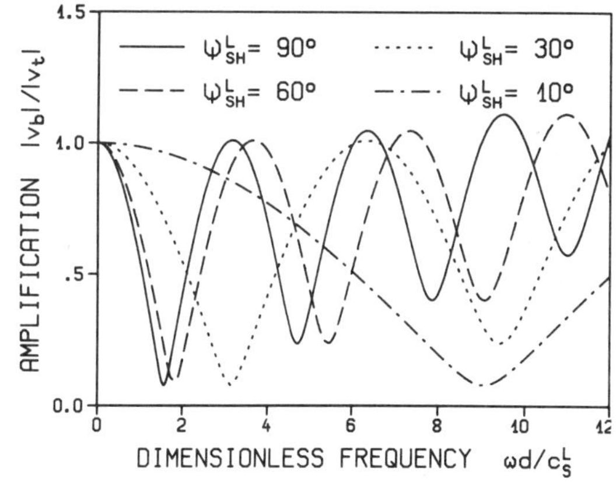

Figure 6-16 Amplification within, homogeneous layer, $\zeta^L = 0.05$ (Ref. [5]).

Amplification outcropping. In contrast, the amplifications defined with respect to the outcropping rock are strongly influenced by ψ_{SH}^R, although the direction of the wave propagation in the layer will also in this case be quite

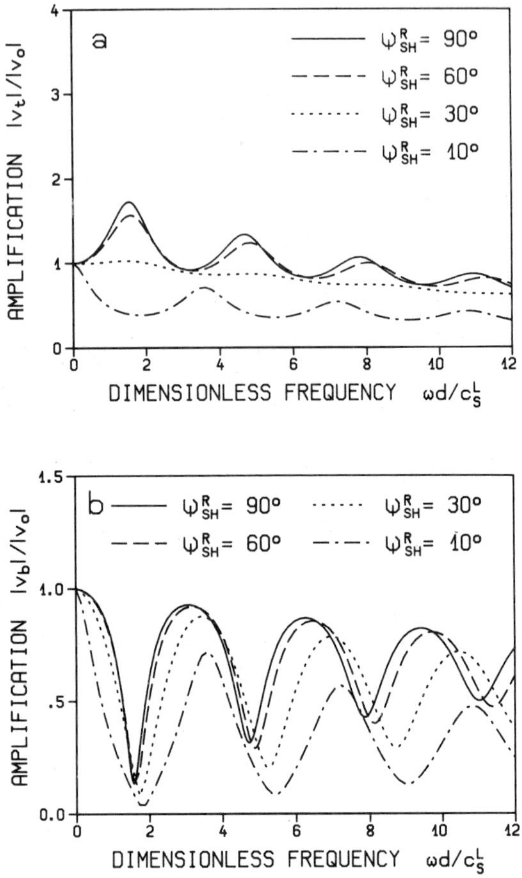

Figure 6-17 Amplification outcropping, homogeneous layer, $\bar{c}_s = 2$, $\bar{\rho} = 1$, $\zeta^R = \zeta^L = 0.05$ (Ref. [5]). (a) To top; (b) to base.

steep. The ratios to the top $|v_t|/|v_o|$ and to the base $|v_b|/|v_o|$ are shown for the homogeneous layer in Figs. 6-17 and 6-18 for $\bar{c}_s = 2$ and $\bar{c}_s = 5$, respectively. The frequencies at which the maxima of $|v_t|/|v_o|$ and the minima of $|v_b|/|v_o|$ occur are again shifted by the same factor, compared to the case of vertical incidence (Eq. 6.24). The maxima of $|v_t|/|v_o|$ are specified in Eq. 6.31. The maxima of $|v_t|/|v_o|$ and the minima of $|v_b|/|v_o|$ are decreased strongly for decreasing ψ^R_{SH}. The corresponding ratios for the outcropping motion are plotted in Fig. 6-19 for the site with increasing stiffness. The same general tendencies as for the site with a homogeneous layer are visible. As $|v_b|/|v_o|$ shows good agreement with the corresponding amplification for the homogeneous layer with $\bar{c}_s = 2$ (Fig. 6-17b), selecting average soil properties seems to be adequate for the motions at the base, but not at the top of the layer (Fig. 6-19a versus

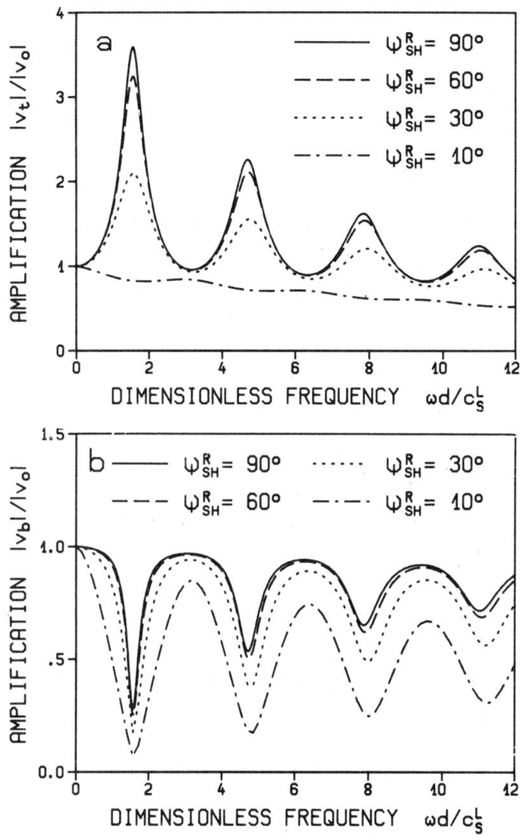

Figure 6-18 Amplification outcropping, homogeneous layer, $\bar{c}_s = 5$, $\bar{\rho} = 1$, $\zeta^R = \zeta^L = 0.05$ (Ref. [5]). (a) To top; (b) to base.

Fig. 6-17a). The variation of the displacement amplitude with depth is studied for inclined body waves in the next subsection.

6.5.4 Love Waves

The surface waves of the homogeneous layer with and without damping resting on a rigid half-space are discussed in Section 6.4.2. The dispersion is illustrated in Fig. 6-11.

Dispersion and attenuation. For a flexible half-space, the dispersion curves of the first three modes are shown in Fig. 6-20 for the indicated parameters. Without damping, the phase velocity c is real and is thus equal to the apparent velocity c_a. No attenuation takes place. The dispersion curves start

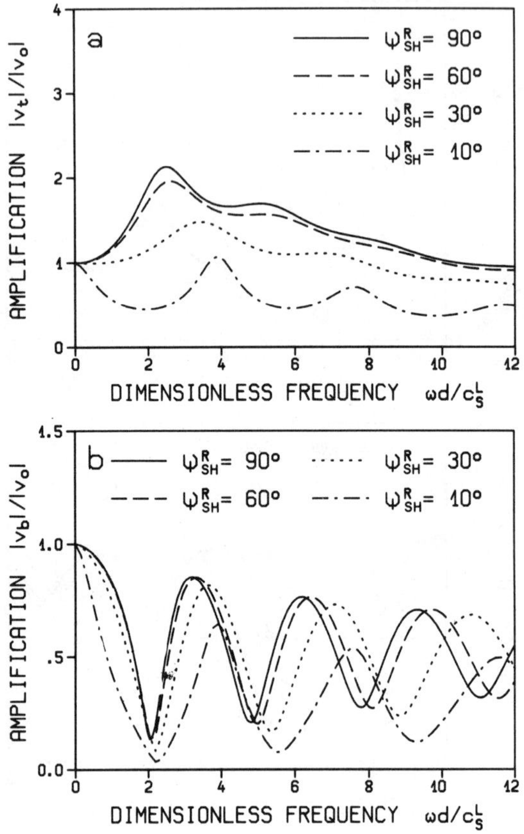

Figure 6-19 Amplification outcropping, increasing stiffness, $\zeta^R = \zeta^L = 0.05$ (Ref. [5]). (a) To top; (b) to base.

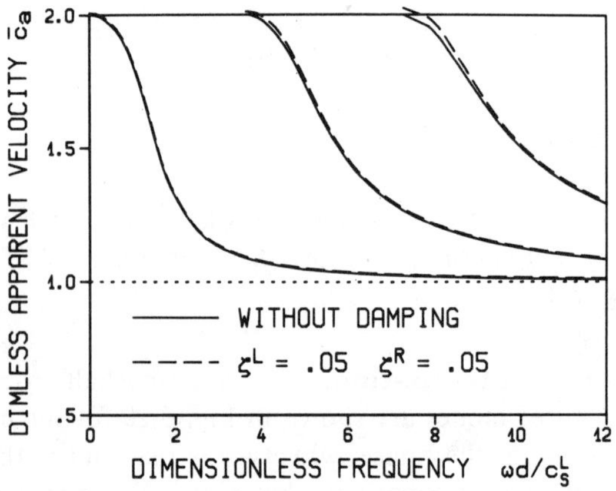

Figure 6-20 Dispersion, homogeneous layer, $\bar{c}_s = 2$, $\bar{\rho} = 1$ (Ref. [5]).

at the ω specified in Eq. 6.34 with the value $c_a = c_s^R$, and all converge to $c_a = c_s^L$ for $\omega \longrightarrow \infty$. In contrast to the layer resting on a rigid half-space, the dispersion curves do not start at the frequencies of the layer fixed at its base ($\bar{\omega} = \pi/2$, $3\pi/2$, $5\pi/2$). In particular, the first mode starts at $\omega = 0$ (with $c_a = c_s^R$). This, however, does not mean that significant radiation damping (effective in the horizontal direction) takes place in the frequency range below the fundamental frequency of the layer, as most of the rate of (horizontal) energy transmission for the first mode for frequencies below $\bar{\omega} = \pi/2$ is present in the half-space. Equation 5.123 for the rate of energy transmission N is evaluated at a specific frequency for the layer and for the half-space. The two rates of energies are then scaled to achieve the sum of 1 ($\bar{N}^R + \bar{N}^L = 1$). This is indicated with a bar. The values are specified at three frequencies in Table 6-2. Above $\bar{\omega} = \pi/2$, the opposite applies, as most of the horizontal energy transmission occurs in the layer. This will also strongly affect the dynamic-stiffness matrix of the soil (Sections 7.3 and 7.4). For the higher modes, the same property applies, as can be seen for the second mode in Table 6-2.

TABLE 6-2 Scaled Rate of Horizontal Energy Transmission, $\bar{c}_s = 2$, $\bar{\rho} = 1$, $\zeta^R = \zeta^L = 0$

Dimensionless Frequency, $\omega d/c_s^L$	First Mode		Second Mode	
	Layer, $\bar{N}^L$	Half-space, $\bar{N}^R$	Layer, $\bar{N}^L$	Half-space, $\bar{N}^R$
$\pi/4$	0.08	0.92		
$3\pi/2$	0.99	0.01	0.37	0.63
$5\pi/2$	1.00	0.00	0.97	0.03

Introducing damping results in a complex c. The apparent velocity c_a hardly changes (Fig. 6-20), but attenuation of the motion in the horizontal direction arises. The direction of energy propagation is no longer horizontal. For instance, for a damped layer on an undamped half-space, energy propagates vertically from the half-space to the layer. This energy transmission in the vertical direction can be verifed as follows. Using Eq. 5.120, Eq. 5.104 with $Q_o = -\tau_o = -\tau_{yz1}$ results in

$$\tau_o = -\frac{kt^R G^{*R}}{\omega}\dot{v}_o \tag{6.46}$$

For an undamped system, t^R for a Love wave is imaginary (Eq. 6.20a, with $m_x^R > 1$), resulting in τ_o and $\dot{v}_o$ being 90° out of phase. The scalar product of these two vectors which defines the rate of energy transmission in the vertical direction (analogously to Eq. 5.122) is thus zero. For a damped system (even with an undamped half-space), however, τ_o and $\dot{v}_o$ will no longer be perpendicular, leading to propagation of energy.

The decay factor per wavelength δ_λ (Eq. 6.7a) depends on the ratio of Im (c) and Re (c). For an inclined SH-wave propagating in a medium with a damping ratio ζ, the phase velocity $c = c_s^*/m_x$. The decay factor per wavelength in the direction of propagation δ_λ equals $\exp[-2\pi \operatorname{Im}(c_s^*)/\operatorname{Re}(c_s^*)]$, where $\operatorname{Im}(c_s^*)/\operatorname{Re}(c_s^*) = \zeta$ (for small ζ). Thus the ratio Im (c)/Re (c) can be interpreted for Love waves as the effective damping ratio. It is plotted for the first two modes versus $\bar{\omega}$ in Fig. 6-21a. For the case of the damped layer and the undamped half-space (solid line), the motion is attenuated, in contrast to body waves. The effective damping ratio starts increasing from zero for both modes. This is to be expected, as the influence of the (undamped) half-space dominates. The effective damping ratio also converges for high frequencies to ζ^L, as the motion is restricted to the layer. However, in between, the values are larger than ζ^L, this trend being more pronounced for the second mode. It is interesting to

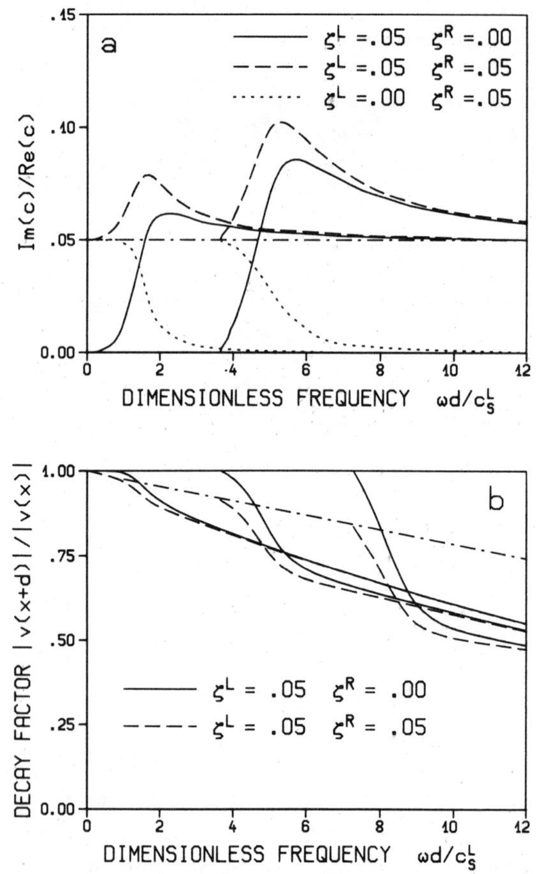

Figure 6-21 Attenuation, homogeneous layer, $\bar{c}_s = 2$, $\bar{\rho} = 1$ (Ref. [5]). (a) Effective damping ratio; (b) decay factor per length d.

note that the peak of the effective damping occurs at a frequency where the inclination of the dispersion curve (Fig. 6-20) is large. For the case of uniform damping (dashed line), the curves start at ζ^R. The other properties are the same. For the hypothetical case of a damped half-space and an undamped layer (dotted line), the limiting values are as expected. In Fig. 6-21b, the decay factor per length d (Eq. 6.7b) is plotted for the first three modes. This displayed information follows from that shown in Figs. 6-20 and 6-21a. For comparison, the line corresponding to an SH-wave with $\psi_{SH}^R = 0$ ($\zeta^R = 0.05$) is also shown as a dashed–dotted line. From Fig. 6-21 it is also visible that the higher modes in the frequency range where they "start" decay less than the first mode.

The shape of the dispersion curves for the different sites is discussed next. In Fig. 6-22, the abscissa represents, as usual, $\bar{\omega}$. To be able to represent the

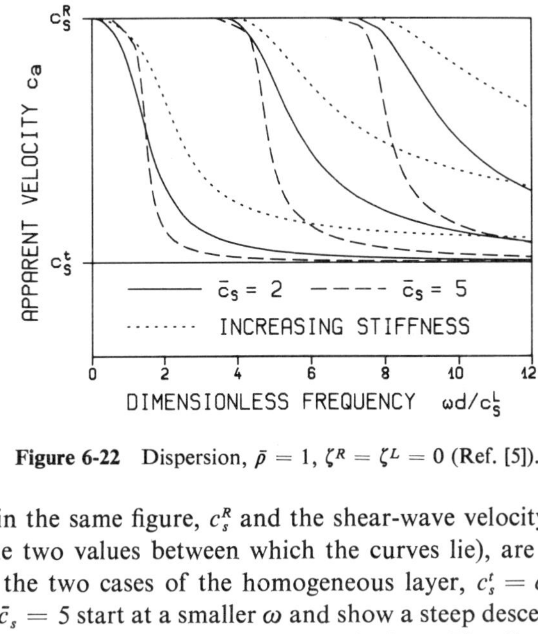

Figure 6-22 Dispersion, $\bar{\rho} = 1$, $\zeta^R = \zeta^L = 0$ (Ref. [5]).

various cases in the same figure, c_s^R and the shear-wave velocity of the layer at the top, c_s^t (the two values between which the curves lie), are specified as the ordinate. For the two cases of the homogeneous layer, $c_s^t = c_s^L$. As expected, the curves for $\bar{c}_s = 5$ start at a smaller ω and show a steep descent in the vicinity of the natural frequencies of the layer fixed at its base. The dispersion curve for increasing stiffness converges to c_s^t more slowly. Including damping would hardly affect the dispersion curves shown in Fig. 6-22. The decay factor per horizontal distance d is shown for the homogeneous layer with $\bar{c}_s = 5$ in Fig. 6-23a. For comparison, the line corresponding to an SH-wave with $\psi_{SH}^R = 0$ ($\zeta^R = 0.05$) is also shown as a dashed–dotted line. The peaks of the effective damping visible in Fig. 6-21a for $\bar{c}_s = 2$ are less pronounced than those corresponding to $\bar{c}_s = 5$, resulting in the dips shown in Fig. 6-23a. In Fig. 6-23b, the decay factor is shown for the first three modes of the site with increasing stiffness. The damping of the layer diminishes from 0.05 at the top linearly with depth to 0.0 at the base. The half-space is not damped. The higher modes

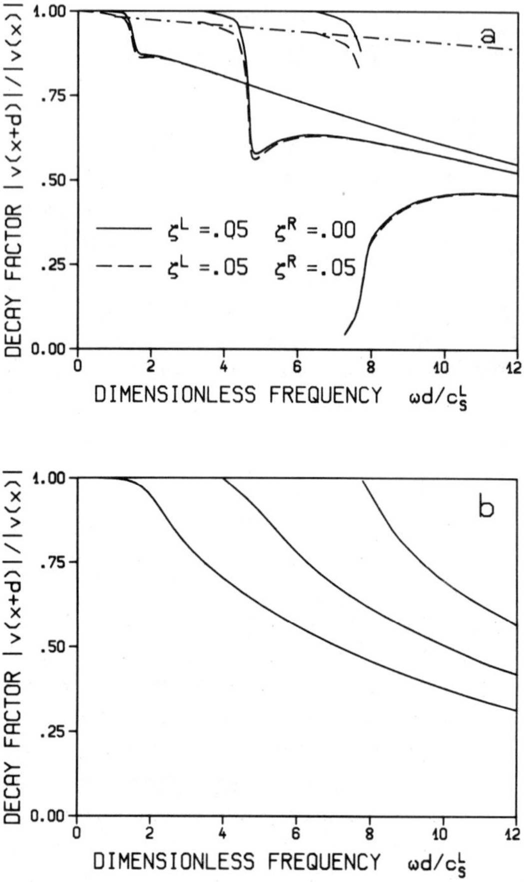

Figure 6-23 Attenuation, $\bar{\rho} = 1$ (Ref. [5]). (a) Homogeneous layer, $\bar{c}_s = 5$; (b) increasing stiffness, $\zeta^R = 0$, $\zeta^L = 0 - 0.05$.

attenuate less, as for a specific frequency, the corresponding motion extends farther down into the region which is less damped.

Displacements and stresses versus depth. To study the motion's variation in the vertical direction, the amplification within the site $|v_b|/|v_t|$ and the amplitude ratio $|v(z)|/|v_t|$ for selected $\bar{\omega}$ are plotted for the homogeneous layer with $\bar{c}_s = 2$ in Fig. 6-24a and b, respectively. For increasing $\bar{\omega}$, the motion of a specific mode is more concentrated in the layer. In addition to the values of the Love wave, the results for the body waves with vertical incidence and with $\psi_{SH}^R = 10°$ ($\psi_{SH}^L = 60.5°$) are shown. The shifts in the frequencies where the minima for the inclined SH-waves occur, compared to those for vertical incidence (Eq. 6.24), are apparent in Fig. 6-24a. For the frequency where a mode "starts,"

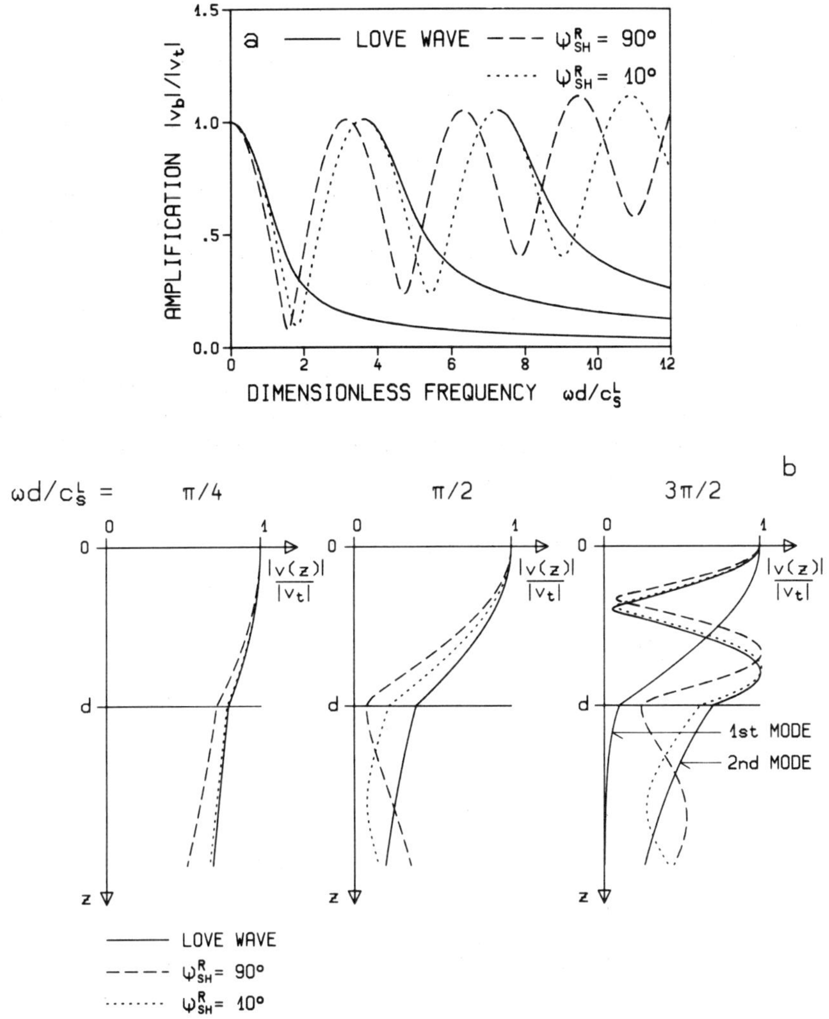

Figure 6-24 Variation with depth, homogeneous layer, $\bar{c}_s = 2$, $\bar{\rho} = 1$, $\zeta^R = \zeta^L = 0.05$ (Ref. [5]). (a) Amplification within; (b) amplitude versus depth.

and somewhat above, the motion of this same mode is very similar to that of the inclined SH-wave (Fig. 6-24). In these frequency ranges, the attenuation is small (Fig. 6-21b). For the remaining ranges of frequency, the discrepancies are large; however, as the Love wave decays strongly, this is of minor practical importance. For instance, in Fig. 6-24b for $\bar{\omega} = 3\pi/2$, the variation with depth of the inclined SH-wave agrees well with that of the second Love mode, but not with that of the first. It is visible from the amplification $|v_b|/|v_t|$ for the case $\bar{c}_s = 5$ (Fig. 6-25) that the two curves of the body waves coincide from a practical

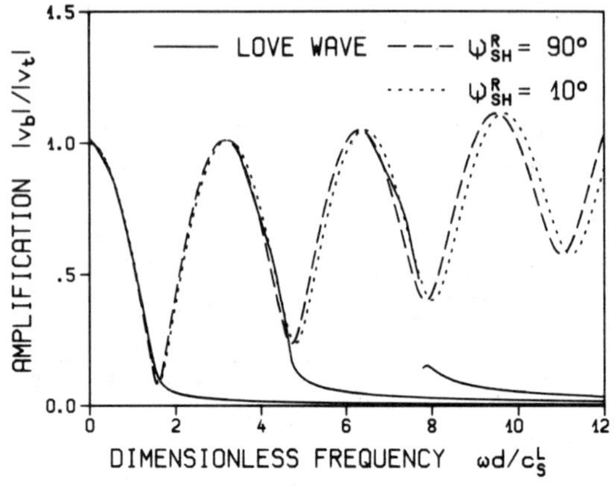

Figure 6-25 Amplification within, homogeneous layer, $\bar{c}_s = 5$, $\bar{\rho} = 1$, $\zeta^R = \zeta^L = 0.05$.

point of view. In addition, there is excellent agreement of $|v_b|/|v_t|$ of the Love mode in the frequency range varying from the frequency where the mode starts up to the next higher natural frequency of the layer with $|v_b|/|v_t|$ of the vertically incident SH-wave.

The corresponding comparisons for the site with increasing stiffness are plotted in Fig. 6-26. Up to the fundamental frequency of the layer fixed at its base ($\bar{\omega} = 2.07$), the three curves agree well. For higher frequencies, the variation with depth is different for the body waves. Where a Love mode starts and just above, the agreement between this mode and the inclined SH-wave is again good.

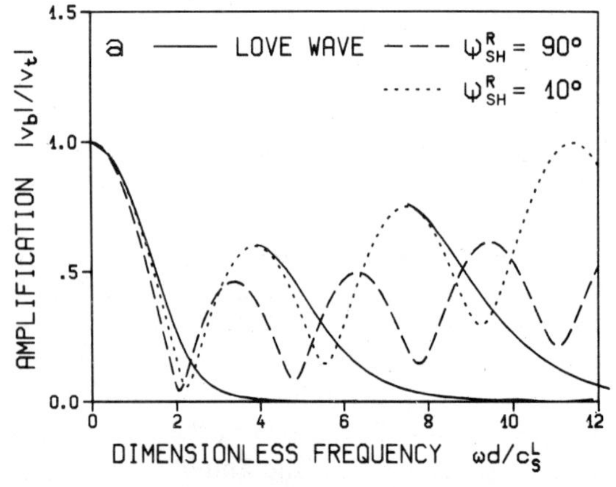

Figure 6-26 Variation with depth, increasing stiffness, $\zeta^R = 0$, $\zeta^L = 0 - 0.05$ (Ref. [5]). (a) Amplification within;

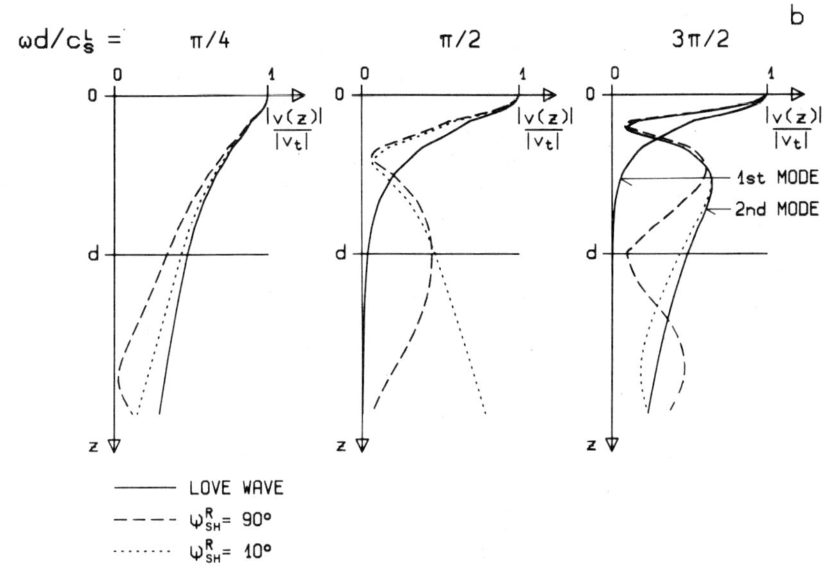

Figure 6-26 (Continued) (b) amplitude versus depth.

The corresponding variation of the shear-stress amplitudes with depth is shown in Fig. 6-27. Dimensionless shear-stress amplitudes $\bar{\tau}$ are defined as $\tau/[(c_s^L)^2\rho^L]$. The results are presented for $|v_t|/d = 1$. The shear-stress amplitude τ_{yz} (Eq. 5.100) is smaller for the Love wave and for the inclined body wave than

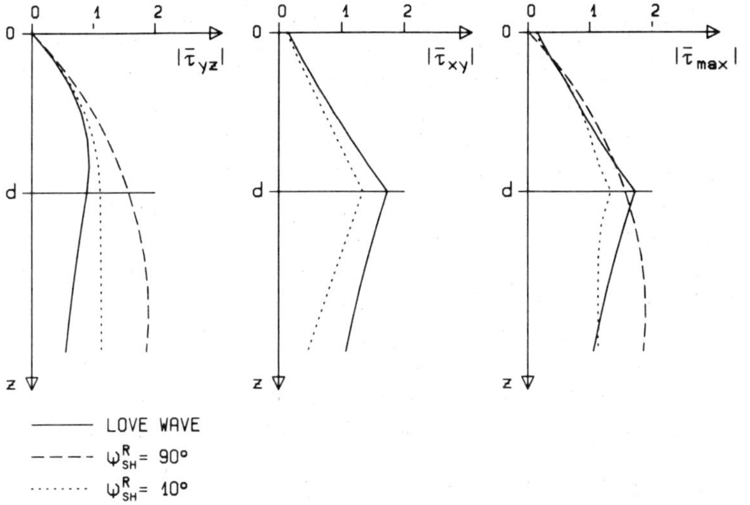

Figure 6-27 Shear-stress amplitudes versus depth, increasing stiffness, $\zeta^R = 0$, $\zeta^L = 0 - 0.05$, $\omega d/c_s^L = \pi/2$ (Ref. [5]).

for the vertically incident wave. Horizontally propagating waves also lead to a shear-stress amplitude τ_{xy} (Eq. 5.112b) which is not zero at the free surface. The amplitude of the maximum shear stress τ_{max} calculated from τ_{yz} and τ_{xy} is also plotted in Fig. 6-27. It is remarkable that all three waves result in a very similar τ_{max} value. This value is in addition only slightly larger than τ_{xy} for the Love wave and for the inclined body wave as τ_{xy} and τ_{yz} are nearly perpendicular.

Scattered motion. Finally, the link of the horizontally propagating free-field motion to the loading applied to a surface structure with a rigid circular basemat of radius a is examined. As discussed in Section 6.2.5, the ratio $\omega a/c_a$ determines the translational and torsional components of the scattered motion resulting from the out-of-plane motion. These are shown in Figs. 7-30 and 7-31. Based on $c_a(\omega)$ presented for the site with increasing stiffness in Fig. 6-22, the curves for the first and second Love modes are constructed. They are presented for $a = 0.5d$ in Fig. 6-28. The corresponding decay factors calculated for a horizontal distance $10a$ ($= 5d$) are also indicated. For comparison, the straight line corresponding to an SH-wave with $\psi^R_{SH} = 0$ is also shown. The latter does not decay ($\zeta^R = 0$). For a specific ω, $\omega a/c_a$ (which characterizes the wave effects) is larger for Love waves than for the body wave. The attenuation is significant but not overwhelming. Considerable wave effects on the seismic input can potentially arise. The influence on the structural response is discussed in depth in Section 9.2.2. The corresponding curves $\omega a/c_a$ versus $\bar{\omega}$ for the site with a homogeneous layer for $\bar{c}_s = 2$ and $= 5$ can be drawn using the information shown in Figs. 6-22, 6-21b, and 6-23a.

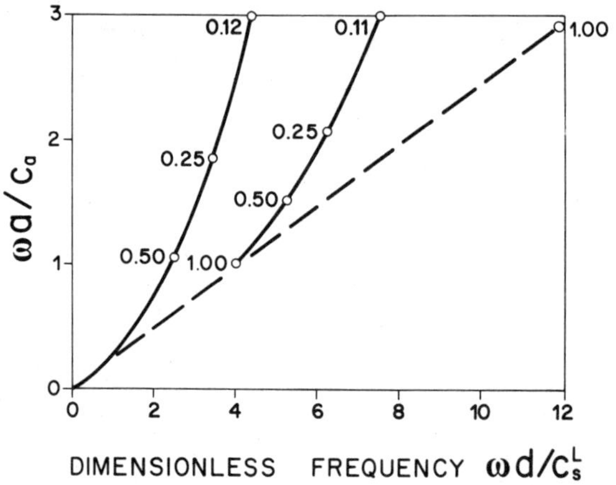

Figure 6-28 Applied seismic-input motion, increasing stiffness, $\zeta^R = 0$, $\zeta^L = 0 - 0.05$ (Ref. [5]).

6.6 PARAMETRIC STUDY OF IN-PLANE MOTION

6.6.1 Scope of Investigation

A site consisting of a single soil layer with depth d resting on a half-space representing the bedrock is analyzed extensively (Fig. 6-13). In Section 6.5 the same dynamic system is examined parametrically for the out-of-plane motion. To be able to demonstrate agreement, and especially differences in the dynamic behavior, a certain duplication cannot be avoided. At the same time this section becomes self-contained. As expected, it is more difficult to establish the key features of the in-plane motion, as the latter consists of two components. Besides their absolute values the phase angle between them is important. For a homogeneous layer, the site is characterized for the in-plane motion by the following dimensionless parameters:

Ratio of shear-wave velocities:	$\bar{c}_s = c_s^R / c_s^L$
Ratio of mass densities:	$\bar{\rho} = \rho^R / \rho^L$
Poisson's ratios:	ν^R, ν^L
Damping ratios:	ζ^R, ζ^L

(Different damping ratios for P- and S-waves are indicated by subscripts p and s.)

To describe the response for harmonic motion, the following variables are defined:

Dimensionless frequency:	$\bar{\omega} = \omega d / c_s^L$
Dimensionless apparent velocity:	$\bar{c}_a = c_a / c_s^L$

Reference is also made to the limiting cases, the half-space ($\bar{c}_s = \bar{\rho} = 1$) and the single layer on a rigid half-space ($\bar{c}_s = \infty$).

In addition to the homogeneous layer on a half-space, a site consisting of a soil layer having a shear-wave velocity that increases linearly with depth and having a total depth d resting on a homogeneous half-space representing the bedrock is investigated. This generally more realistic site is denoted later as "increasing stiffness." The shear-wave velocity of the half-space c_s^R is equal to the shear-wave velocity of the layer at its base. Only the results obtained when the ratio of the shear-wave velocity of the layer at its base to that at its top is equal to 5 are discussed. To be able to calculate $\bar{\omega}$ and $\bar{c}_a$, an average c_s^L for this site with increasing stiffness is calculated.

$$c_s^L = \frac{d}{\int_0^d dz/c_s(z)} \tag{6.47}$$

For our case, this results in $\bar{c}_s = 2.012$. The mass density ρ and Poisson's ratio ν are assumed to be constant over the whole site. In addition to $\zeta^R = \zeta^L =$

0.05, the damping ratio of the layer is selected to decrease linearly with depth, varying from 0.05 at the top to 0.0 at the base ($\zeta^R = 0$). The layer with increasing stiffness is discretized with six layers of constant material properties.

6.6.2 Vertically Incident SV- and P-Waves

For vertical incidence, the amplification for SV-waves is obviously identical to that for SH-waves as discussed in Section 6.5.2. For P-waves, this would also apply if the P-wave velocity c_p instead of c_s were introduced in $\omega d/c_s^L$ and in c_s^R/c_s^L. To be able to discuss combinations of P- and SV-waves, the dimensionless parameters defined in Section 6.6.1 are not modified. As for the selected Poisson's ratio = 0.33, $c_p = 2c_s$, the curves for the P-waves are generated from those of the SH-waves (Figs. 6-14 and 6-15) by multiplying the abscissa by 2. As an example, the ratio of the absolute values of the vertical-displacement amplitudes at the top of the homogeneous layer and at outcropping of the bedrock $|w_t|/|w_o|$ is plotted in Fig. 6-29 versus the dimensionless frequency $\bar{\omega}$ for vertically incident P-waves. The parameters are specified in the figure and in its caption. The peaks occur at the $\bar{\omega}$ corresponding to the natural frequencies of the layer fixed at its base (π, 3π, . . .). The increased radiation damping occurring by means of the half-space for decreasing $\bar{c}_s$ reduces the peaks. The effect of the radiation damping decreases with increasing ω.

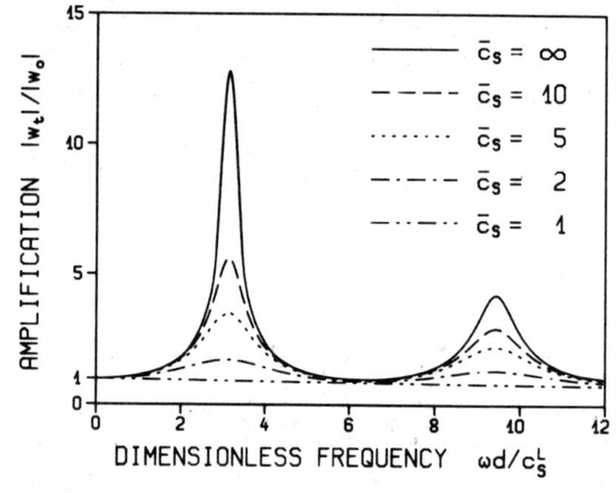

Figure 6-29 Amplification outcropping, P-wave, vertical incidence, $\bar{\rho} = 1$, $\nu^R = \nu^L = 0.33$, $\zeta^R = \zeta^L = 0.05$.

For the site with increasing stiffness (not shown), the peaks of $|w_t|/|w_o|$ no longer occur at the natural frequencies of the layer fixed at its base. The amplification for the site with increasing stiffness tends to be more uniform over the frequency range.

The ratio of the vertical motions within the site $|w_b|/|w_t|$ is independent of the properties of the rock. The curve for the homogeneous layer is equal to the inverse values of $|w_t|/|w_o|$ for $\bar{c}_s = \infty$ shown in Fig. 6-29. The minima arise at the natural frequencies of the layer fixed at its base. This also applies for the layer with increasing stiffness (not shown).

6.6.3 Inclined P- and SV-Waves

In contrast to vertically incident waves, inclined body waves propagate horizontally across the site with an apparent velocity c_a (Eq. 6.6) and attenuate with a decay factor δ_λ (Eq. 6.7a). The (complex) phase velocity c in these two equations equals $c_p^{*R}/\cos \psi_P^R$ for incident P-waves and $c_s^{*R}/\cos \psi_{SV}^R$ for incident SV-waves.

Incident P-waves. At first, the site motion caused by an incident P-wave in the bedrock ($A_{SV}^R = 0$) is discussed. The corresponding outcropping motion u_o, w_o (Fig. 6-4) is discussed in Section 6.3.2 (Fig. 6-5). In this case the dynamic system consists of a half-space.

In contrast to the outcropping motion of the elastic half-space, the motion (caused by an incident P-wave in the half-space, $A_{SV}^R = 0$) at the top of the layer resting on a half-space (Fig. 6-13) depends on the frequency of excitation. The ratios $|w_t|/|A_P^R|$ and $|u_t|/|A_P^R|$ and the phase angle of the ratio w_t/u_t are plotted in Fig. 6-30 for the site with increasing stiffness. For the limiting case of zero dimensionless frequency, the results of the half-space shown in Fig. 6-5 are recovered. The elastic half-space results for $d \longrightarrow 0$, which also leads to $\bar{\omega} \longrightarrow 0$. This arises for all ω. The peaks of $|w_t|$ and $|u_t|$ do not occur at the fundamental frequencies of the layer fixed at its base in the vertical ($\bar{\omega} = 4.14$) and horizontal directions ($\bar{\omega} = 2.07$), respectively. Reducing ψ_P^R leads to a diminution of $|w_t|$ for the whole frequency range. For the frequency range of interest in earthquake

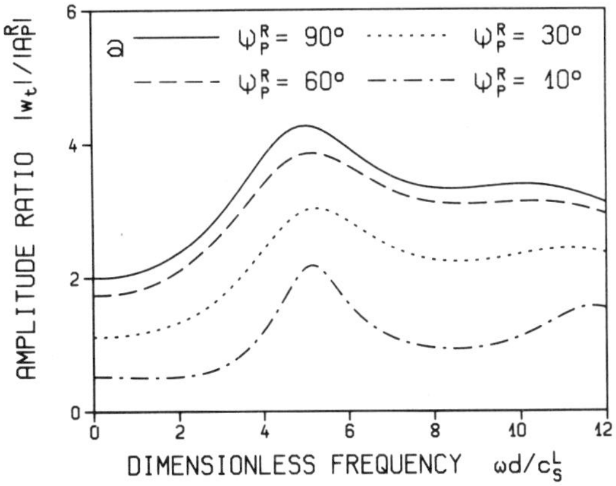

Figure 6-30 Motion at top of layer, incident P-wave, increasing stiffness, $\nu^R = \nu^L = 0.33$, $\zeta^R = \zeta^L = 0.05$ (Ref. [6]). (a) Vertical;

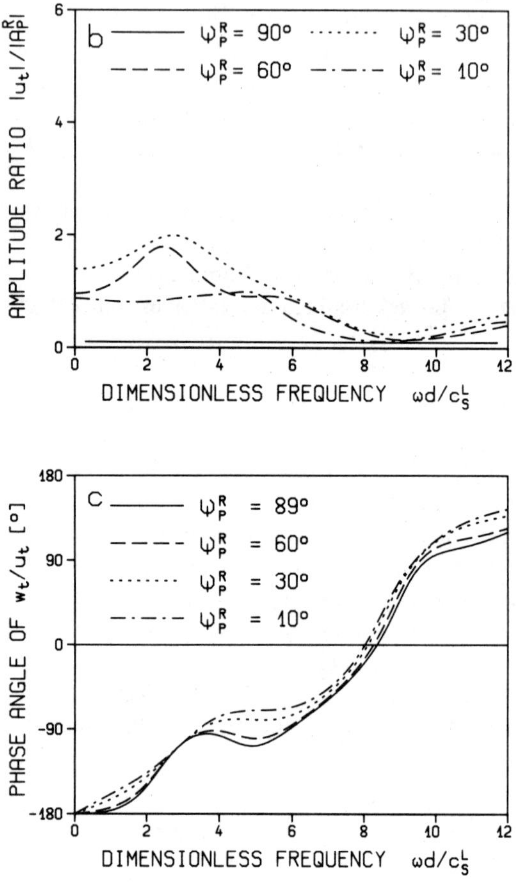

Figure 6-30 (Continued) (b) horizontal; (c) phase angle.

engineering, $|u_t|$ bears comparison with $|w_t|$ for intermediate ψ_P^R. The P-wave will be steep at the top of the layer and will thus not contribute significantly to u_t. But the incident P-wave in the half-space also creates SV-waves in the layer, which lead to the considerable horizontal motion. For the sake of clarity, the zero value for vertical incidence is shown with a small offset. It is apparent from Fig. 6-30c that the phase angle between the vertical and horizontal motions is almost independent of ψ_P^R. The motion in the frequency range of interest ($\bar{\omega} < 2\pi$) is prograde with respect to the direction of propagation (or close to it); that is, the vertical motion lags behind the horizontal by 90° for the coordinate system selected (clockwise). The amplification for the outcropping motion can be calculated on the basis of the information in Figs. 6-5 and 6-30. As an example, $|w_t|/|w_o|$ is plotted in Fig. 6-31. With the exception of the very shallow angle of incidence $\psi_P^R = 10°$, $|w_t|/|w_o|$ hardly depends on ψ_P^R. It is interesting to note that $|w_t|/|w_o|$ is larger for the inclined waves than for vertical incidence for the

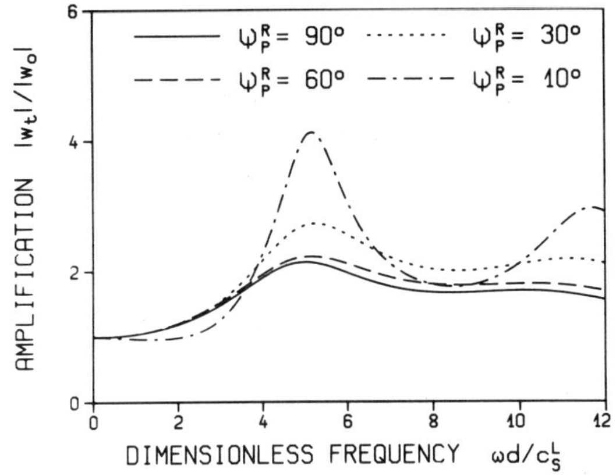

Figure 6-31 Vertical amplification outcropping, incident P-wave, increasing stiffness, $\nu^R = \nu^L = 0.33$, $\zeta^R = \zeta^L = 0.05$ (Ref. [6]).

whole range of frequency. This is in marked contrast to the corresponding amplification for SH-waves (Figs. 6-17, 6-18 and 6-19). The corresponding motion at the base of the layer $|w_b|/|A^R_P|$ and $|u_b|/|A^R_P|$ is shown in Fig. 6-32. Dips of $|w_b|$ occur at frequencies which, for vertical incidence, are equal to the vertical natural frequencies of the layer fixed at its base. A decrease in ψ^R_P shifts these frequencies to a somewhat higher value (Fig. 6-32a). The same applies to those frequencies where the dips of $|u_b|$ arise relative to the natural frequencies in the horizontal direction (Fig. 6-32b). The amplification for the vertical motion within the layer $|w_b|/|w_t|$ follows from the information shown in Figs. 6-30a and 6-32a. As is visible in Fig. 6-33, this ratio is not very sensitive when ψ^R_P is varied.

The tendencies established for the site with increasing stiffness (for an incident P-wave in the half-space) in connection with Figs. 6-30 to 6-33 also

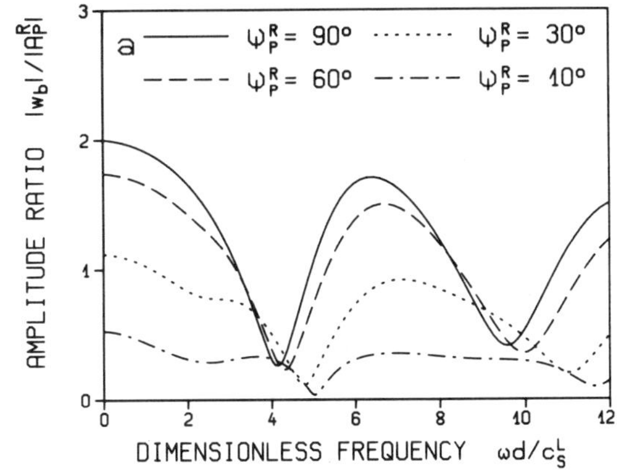

Figure 6-32 Motion at base of layer, incident P-wave, increasing stiffness, $\nu^R = \nu^L = 0.33$, $\zeta^R = \zeta^L = 0.05$ (Ref. [6]). (a) Vertical;

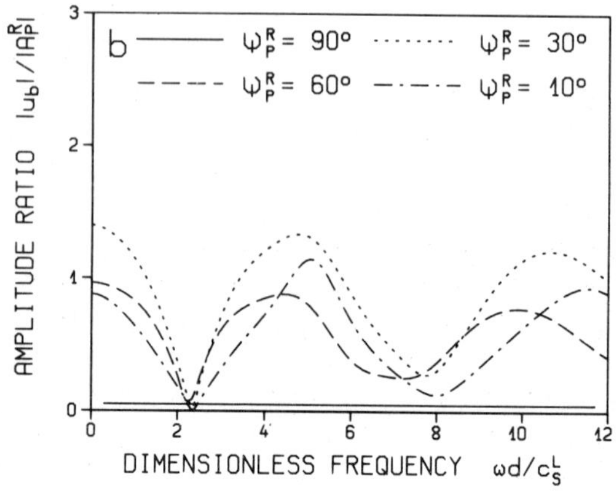

Figure 6-32 (Continued) (b) horizontal.

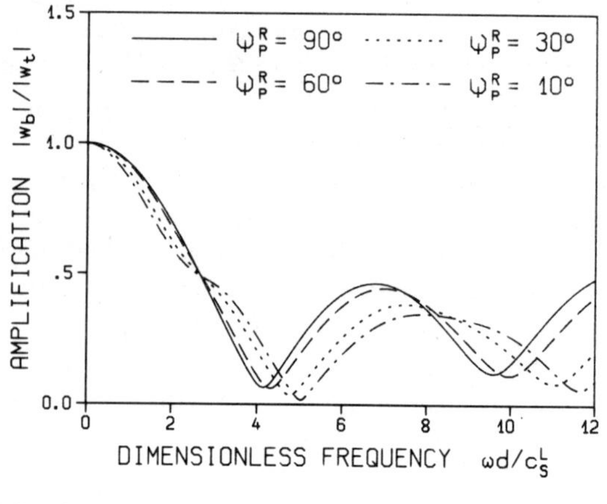

Figure 6-33 Vertical amplification within, incident P-wave, increasing stiffness, $\nu^R = \nu^L = 0.33$, $\zeta^R = \zeta^L = 0.05$ (Ref. [6]).

hold for the site with the homogeneous layer, varying $\bar{c}_s$. As an example, the ratio of the vertical-displacement amplitudes at the top of the layer caused by a P-wave with $\psi_P^R = 30°$ $|w_t^{30}|$ and with $\psi_P^R = 90°$ $|w_t^{90}|$ is shown in Fig. 6-34 for $\bar{c}_s = 2$ and $= 5$. By way of comparison, for the site with increasing stiffness, the same ratio, which can be determined from Fig. 6-30a, is also plotted. It is apparent that the three curves are very similar. In addition, this ratio depends only weakly on $\bar{\omega}$.

Incident SV-waves. Turning to SV-waves, the outcropping motion u_o, w_o at the free surface of the bedrock resulting from an incident SV-wave (Fig. 6-4) is examined in Section 6.3.3 (Fig. 6-7). It is independent of the frequency of excitation.

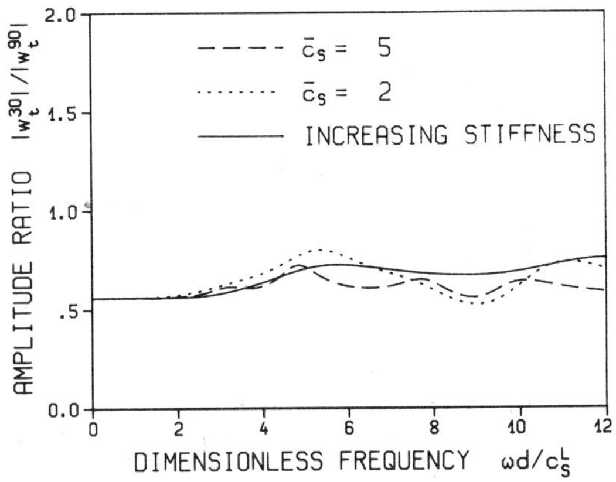

Figure 6-34 Vertical amplitude ratio at top, incident P-wave, $\bar{\rho} = 1$, $\nu^R = \nu^L = 0.33$, $\zeta^R = \zeta^L = 0.05$ (Ref. [6]).

The motion (caused by an incident SV-wave in the half-space, $A_P^R = 0$) at the top of the layer resting on a half-space (Fig. 6-13) is studied next. The ratios $|u_t|/|A_{SV}^R|$ and $|w_t|/|A_{SV}^R|$ and the phase angle of the ratio w_t/u_t are shown in Fig. 6-35 for the site with increasing stiffness. As in the case of P-waves, the results of the half-space shown in Fig. 6-7 are recovered in Fig. 6-35 for $\bar{\omega} = 0$. Again, the peaks of $|u_t|$ and $|w_t|$ do not arise at the fundamental frequencies of the layer in the horizontal and vertical directions, respectively. It is interesting to compare the curves of $|u_t|$ and $|w_t|$ for the critical angle $\psi_{SV}^R = 60°$ and for vertical incidence. While for $\omega = 0$, $\psi_{SV}^R = 60°$ results in higher values of u_t (as discussed above), vertically incident waves lead for increased ω to larger u_t. The vertical amplitude w_t is equal to zero only for $\bar{\omega} = 0$. The nonnegligible

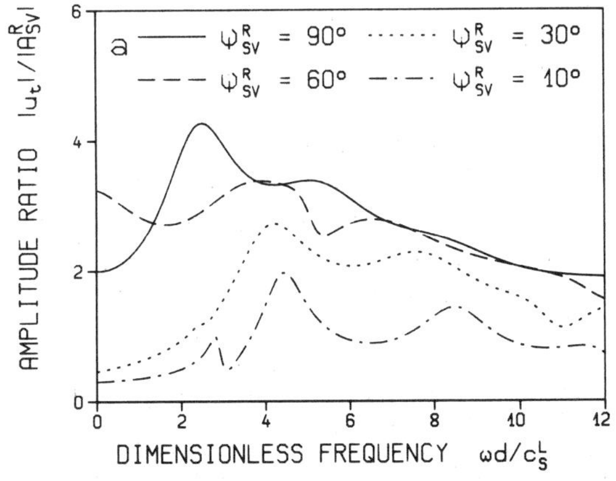

Figure 6-35 Motion at top of layer, incident SV-wave, increasing stiffness, $\nu^R = \nu^L = 0.33$, $\zeta^R = \zeta^L = 0.05$ (after Ref. [6]). (a) Horizontal;

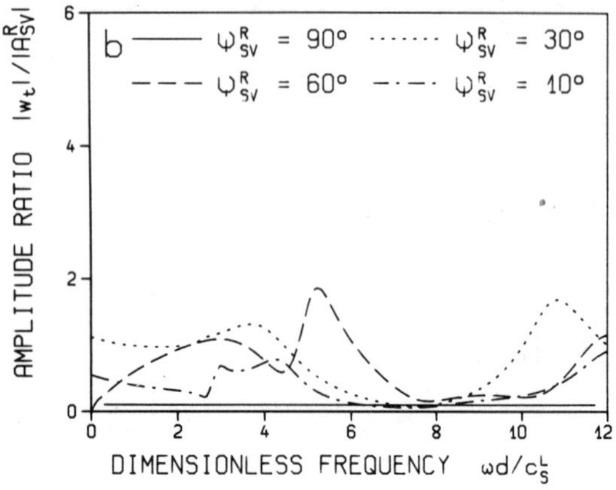

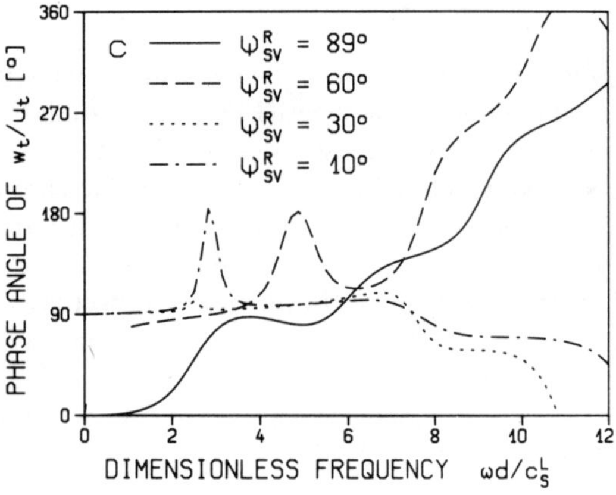

Figure 6-35 (Continued) (b) vertical; (c) phase angle.

$|w_t|$ in Fig. 6-35b are caused mainly by P-waves in the layer. For small ω, the phase angle between w_t and u_t shown in Fig. 6-35c corresponds to that of the outcropping motion (for $\psi^R_{SV} = 60°$, the phase angle is indeterminate for $\bar{\omega} = 0$, as $w_t = 0$). Based on the information contained in Figs. 6-7 and 6-35a, the horizontal amplification for the outcropping motion $|u_t|/|u_o|$ follows (Fig. 6-36). In the light of the fact that for the critical angle $\psi^R_{SV} = 60°$, $|u_o| = 3.46\,|A^R_{SV}|$, it is not surprising that the curve for this angle lies below that for vertical incidence for all frequencies. For the two other shallow angles, the values agree well with those for $\psi^R_{SV} = 90°$ up to approximately the horizontal fundamental frequency of the layer. For larger frequencies $|u_t|/|u_o|$ for $\psi^R_{SV} = 10°$ and $= 30°$ are significantly larger than for vertical incidence. This is not the case for the corresponding values for P-waves (Fig. 6-31) and is contrary to the results of SH-waves (Fig. 6-19). The corresponding motion at the base of the layer $|u_b|/|A^R_{SV}|$ and $|w_b|/|A^R_{SV}|$ is plotted in Fig. 6-37. For $\bar{\omega} = 0$, again the outcropping motion

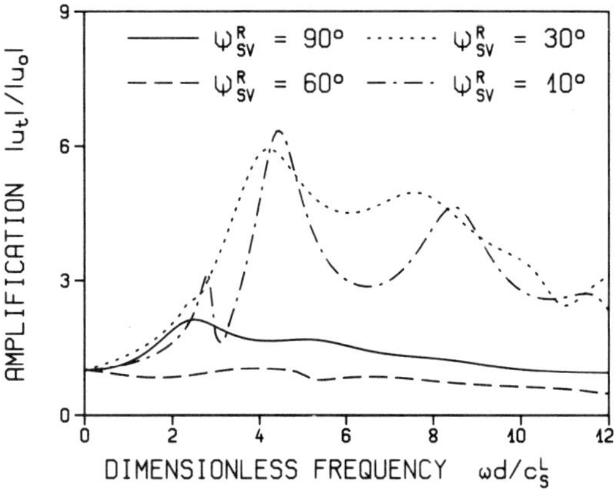

Figure 6-36 Horizontal amplification outcropping, incident SV-wave, increasing stiffness, $\nu^R = \nu^L = 0.33$, $\zeta^R = \zeta^L = 0.05$ (Ref. [6]).

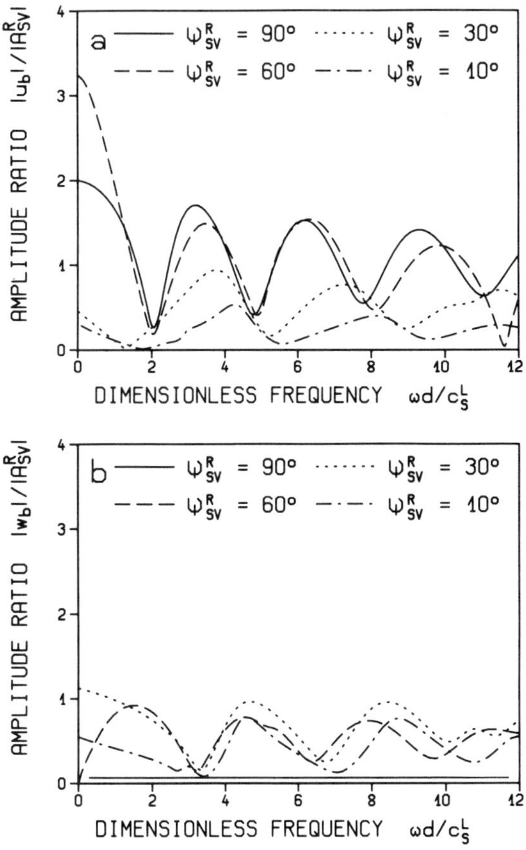

Figure 6-37 Motion at base of layer, incident SV-wave, increasing stiffness, $\nu^R = \nu^L = 0.33$, $\zeta^R = \zeta^L = 0.05$. (a) Horizontal; (b) vertical.

shown in Fig. 6-7 results. For $\psi^R_{SV} \geq 60°$, the dips of $|u_b|$ arise close to the horizontal natural frequencies of the layer (Fig. 6-37a). For $\psi^R_{SV} < 60°$, this is no longer the case. The dips of $|w_b|$ occur below the vertical natural frequencies of the layer for all inclined waves (Fig. 6-37b). The horizontal amplification within the layer $|u_b|/|u_t|$ which follows from Figs. 6-35 and 6-37a is shown in Fig. 6-38. Again, for $\psi^R_{SV} \geq 60°$, which is the critical angle, the curves for an inclined SV-wave in the half-space agree well with those for vertical incidence. In contrast to the corresponding values for P-waves (Fig. 6-33), this no longer holds for shallow angles of incidence. Large discrepancies arise throughout the frequency range, also below the horizontal fundamental frequency of the layer, as is not the case for the corresponding horizontal amplification caused by inclined SH-waves (Fig. 6-26).

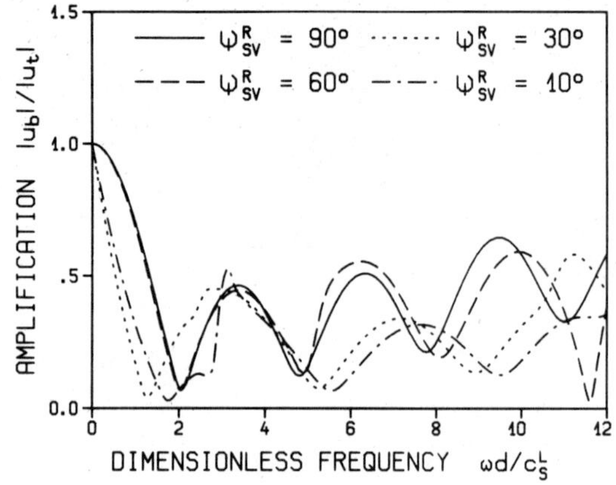

Figure 6-38 Horizontal amplification within, incident SV-wave, increasing stiffness, $\nu^R = \nu^L = 0.33$, $\zeta^R = \zeta^L = 0.05$ (Ref. [6]).

The tentative tendencies established for the site with increasing stiffness in connection with Figs. 6-35 to 6-38 (for an incident SV-wave in the half-space) are also visible for the site with the homogeneous layer, varying $\bar{c}_s$. Compared to the dynamic behavior for incident P-waves, the response for the incident SV-wave is less consistent. The plot of $|u_t^{30}|/|u_t^{90}|$ (corresponding to Fig. 6-34) depends strongly on the frequency and differs for the various sites (not shown). As an example of the homogeneous layer with $\bar{c}_s = 5$, $|u_t|/|A^R_{SV}|$ is plotted in Fig. 6-39. The maximum value of $|u_t|$ occurs approximately at the horizontal fundamental frequency of the layer for all ψ^R_{SV}. In Fig. 6-40, $|u_b|/|u_t|$ is shown for the same parameters. In contrast to the site with increasing stiffness (Fig. 6-38), good agreement for all angles of incidence results up to the horizontal fundamental frequency ($\bar{\omega} = \pi/2$).

Combinations of incident P- and SV-waves. Combinations of incident P- and SV-waves in the half-space (Fig. 6-4), resulting in a common

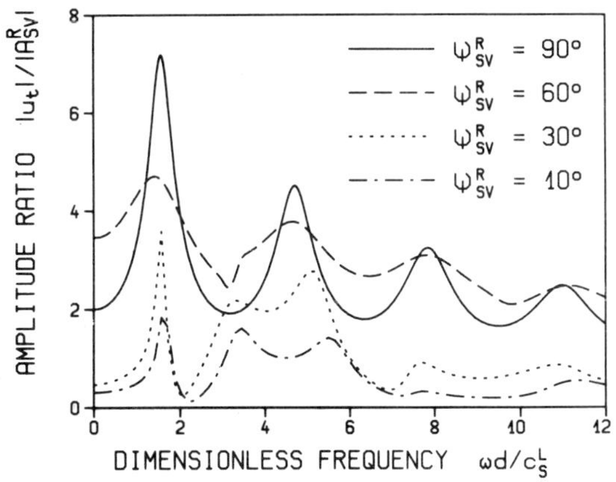

Figure 6-39 Horizontal motion at top of layer, incident SV-wave, homogeneous layer, $\bar{c}_s = 5$, $\bar{\rho} = 1$, $\nu^R = \nu^L = 0.33$, $\zeta^R = \zeta^L = 0.05$.

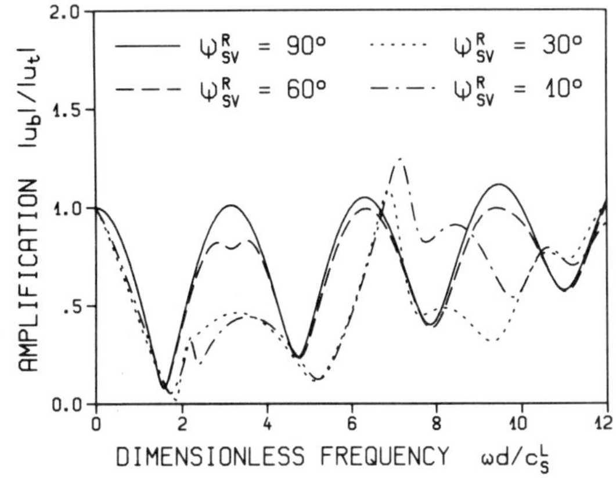

Figure 6-40 Horizontal amplification within, incident SV-wave, homogeneous layer, $\bar{c}_s = 5$, $\bar{\rho} = 1$, $\nu^R = \nu^L = 0.33$, $\zeta^R = \zeta^L = 0.05$.

phase velocity c, are examined next. The complex horizontal and vertical amplitudes u and w of the motion in the control point define the (complex) amplitudes A^R_P and A^R_{SV} of the incident waves for a specified phase velocity (which, of course, has to be larger than c^{*R}_p for an incident P-wave to be able to be present). For the special case of the control point at the outcrop of bedrock, the inverse of Eq. 6.11 determines this relationship. For the control point at the ground surface of the site the procedure is analogous. Not only the absolute values, but also the phase angle, is important. Whereas for vertical incidence u and w are independent, this no longer applies for inclined body waves. The amplification in each direction thus depends on the total motion (i.e., on u and w). This is demonstrated in the following for a combination of incident P- and SV-waves with

$\psi_P^R = 30°$ and $\psi_{SV}^R = 64.3°$, resulting in a common phase velocity. The amplification for the outcropping motion in the horizontal direction is shown in Fig. 6-41a for the homogeneous layer with the indicated parameters for the following cases: first, for zero vertical motion in the control point (solid line); second, for a retrograde motion with equal horizontal and vertical absolute values of the amplitudes (dashed line); and third, with equal horizontal and vertical absolute values of the amplitudes and a phase angle which leads to the maximum and minimum amplification (dotted lines). For the third case, the phase angle is different for the two extremes for each ω. Figure 6-41b shows the corresponding amplification curves in the vertical direction (varying u_o). For the site with increasing stiffness, the corresponding amplifications, based on the same assumption for the incident inclined body waves in the half-space, are

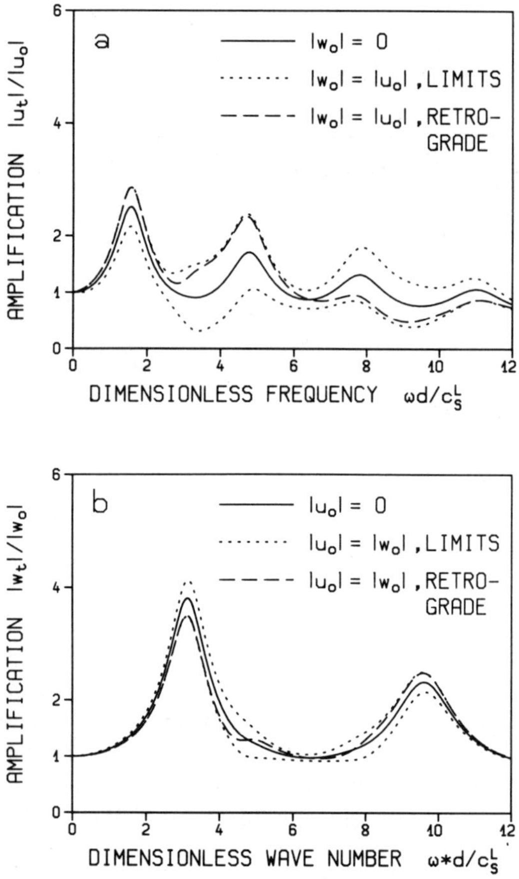

Figure 6-41 Amplification outcropping, incident P- and SV-waves, $\psi_P^R = 30°$, $\psi_{SV}^R = 64.3°$, homogeneous layer, $\bar{c}_s = 5$, $\bar{\rho} = 1$, $\nu^R = \nu^L = 0.33$, $\zeta^R = \zeta^L = 0.05$ (Ref. [6]). (a) Horizontal; (b) vertical.

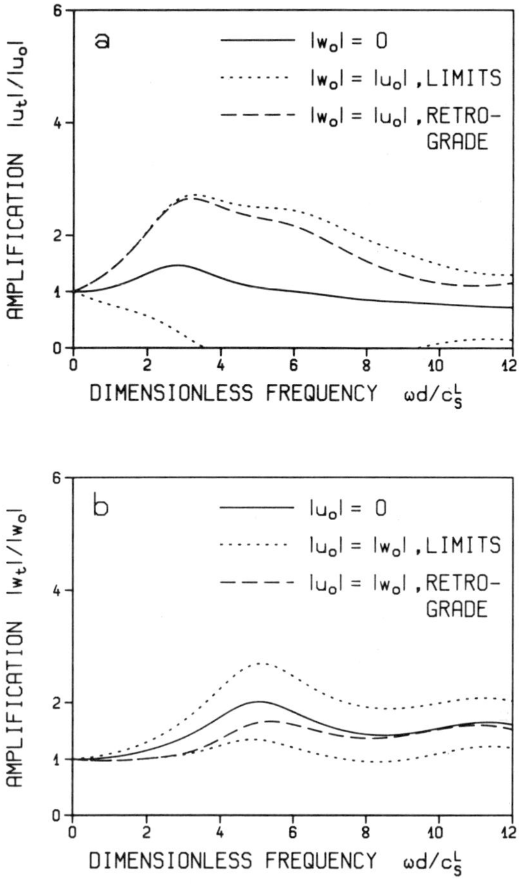

Figure 6-42 Amplification outcropping, incident P- and SV-waves, $\psi_P^R = 30°$, $\psi_{SV}^R = 64.3°$, increasing stiffness, $\nu^R = \nu^L = 0.33$, $\zeta^R = \zeta^L = 0.05$ (Ref. [6]). (a) Horizontal; (b) vertical.

presented in Fig. 6-42. Whereas the influence of the prescribed motion on the vertical amplification for all frequencies and on the horizontal amplification up to the horizontal fundamental frequency of the layer ($\bar{\omega} = \pi/2$) is negligible for the homogeneous layer, this no longer applies for the site with increasing stiffness; especially for $|u_t|/|u_o|$, the influence of w_o is strong and cannot be disregarded. Prescribing the motion at the top of the homogeneous layer, the corresponding amplification for the motion within the site is plotted for the same angles of incidence of the inclined body waves in the half-space in Fig. 6-43. It is apparent that below the fundamental frequencies in the horizontal and vertical directions, $|u_b|/|u_t|$ and $|w_b|/|w_t|$, respectively, are practically independent of the assumed motion. This is confirmed to a somewhat lesser extent by the results of the site with increasing stiffness (not shown). The variation

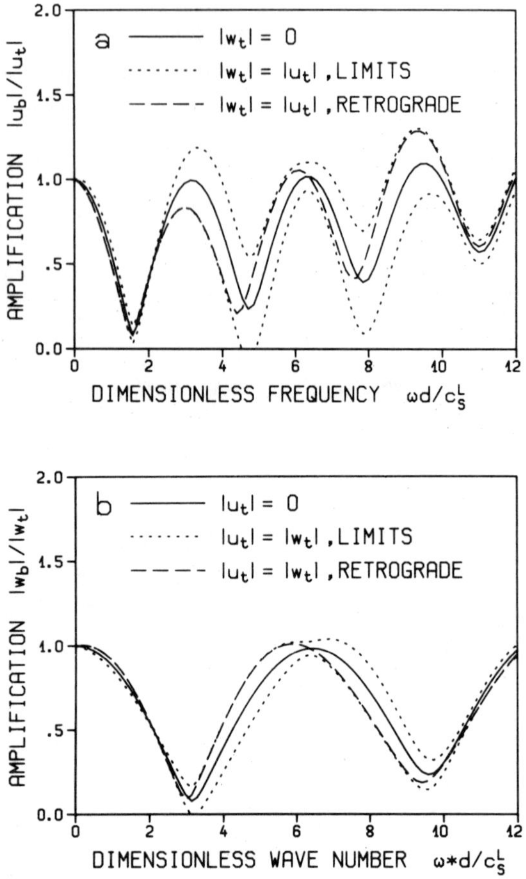

Figure 6-43 Amplification within, incident P- and SV-waves, $\psi_P^R = 30°$, $\psi_{SV}^R = 64.3°$, homogeneous layer, $\bar{c}_s = 5$, $\bar{\rho} = 1$, $\nu^R = \nu^L = 0.33$, $\zeta^R = \zeta^L = 0.05$. (a) Horizontal; (b) vertical.

of the motion with depth for inclined body waves is compared to that of surface waves in the next subsection.

6.6.4 Rayleigh Waves

The properties of the only R-mode in a half-space with a retrograde motion, which propagates with a frequency-independent velocity, are summarized in Section 6.3.4.

Dispersion and attenuation. For a homogeneous layer resting on the half-space, the dispersion curves of the first three modes are shown for the indicated parameters in Fig. 6-44. Without damping, the phase velocity c is real

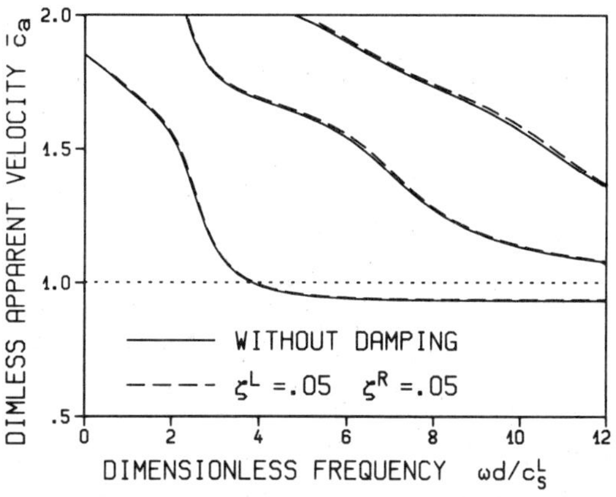

Figure 6-44 Dispersion, homogeneous layer, $\bar{c}_s = 2$, $\bar{\rho} = 1$, $\nu^R = \nu^L = 0.33$ (Ref. [6]).

and is thus equal to the apparent velocity c_a (Eq. 6.6). No attenuation takes place ($\delta_\lambda = 1$, Eq. 6.7a). The first (fundamental) R-mode begins at the R-wave velocity of the rock (half-space) for $\omega = 0$ and converges to that of an elastic half-space with the properties of the layer for $\omega \longrightarrow \infty$. The higher R-modes start at the value $c = c_s^R$, and all converge to c_s^L for $\omega \longrightarrow \infty$. In contrast to the layer resting on a rigid half-space, the dispersion curves do not start at the natural frequencies of the layer fixed at its base. In particular, the first mode starts at $\omega = 0$. This, however, does not mean that significant radiation damping (effective in the horizontal direction) takes place in the frequency range below the horizontal fundamental frequency of the layer, as most of the rate of (horizontal) energy transmission for the first mode for $\bar{\omega} = \pi/2$ is present in the half-space. Equation 5.150 for the rate of energy transmission N is evaluated at a specific frequency for the layer and the half-space. The results are then scaled to achieve $\bar{N}^L + \bar{N}^R = 1$. This is indicated with a bar. The values are specified at three frequencies in Table 6-3. For higher frequencies most of the horizontal energy transmission occurs in the layer. For the higher modes, the same property

TABLE 6-3 Scaled Rate of Horizontal Energy Transmission, $\bar{c}_s = 2$, $\bar{\rho} = 1$, $\zeta^R = \zeta^L = 0$

Dimensionless Frequency, $\omega d/c_s^L$	First Mode		Second Mode	
	Layer, $\bar{N}^L$	Half-space, $\bar{N}^R$	Layer, $\bar{N}^L$	Half-space, $\bar{N}^R$
$\pi/4$	0.09	0.91		
π	0.89	0.11	0.68	0.32
2π	1.00	0.00	0.79	0.21

applies, as can be seen for the second mode in Table 6-3. The same feature is described for Love waves in Table 6-2.

Introducing damping results in a complex c. The apparent velocity c_a hardly changes (Fig. 6-44), but attenuation of the motion in the horizontal direction arises. The direction of energy propagation is no longer horizontal. For instance, for a damped layer on an undamped half-space, energy propagates vertically from the half-space to the layer. This energy transmission in the vertical direction can be verified in a manner analogous to that discussed for Love waves in Section 6.5.4. The decay factor per wavelength δ_λ (Eq. 6.7a) depends on the ratio of Im (c) and Re (c). For instance, for an SV-wave propagating in a medium with a damping ratio ζ, the phase velocity $c = c_s^*/m_x$. The decay factor per wavelength in the direction of propagation equals exp $[-2\pi \text{ Im}(c_s^*)/\text{Re}(c_s^*)]$ where $\text{Im}(c_s^*)/\text{Re}(c_s^*) = \zeta$ (for small ζ). Thus the ratio Im (c)/Re (c) can be interpreted for R-waves as the effective damping ratio. It is plotted for the first mode versus $\bar{\omega}$ in Fig. 6-45a. For the case of the damped layer and the undamped half-space (solid line), the motion is attenuated, in contrast to body waves. The effective damping ratio starts increasing from zero. This is to be expected, as the influence of the (undamped) half-space dominates. The effective damping ratio also converges for high frequencies to ζ^L, as the motion is restricted to the layer. However, in between, the values are larger than ζ^L. For the case of uniform damping (dashed line), the curves start at ζ^R. The other properties are the same. It is interesting to note that the peak of the effective damping occurs at a frequency where the inclination of the dispersion curve (Fig. 6-44) is large. For the hypothetical case of a damped half-space and an undamped layer (dotted line), the limiting values are as expected. Love waves show the same property (Fig. 6-21a). For the higher modes, as shown in Fig. 6-45b, for the second and the third, the effective damping behaves identically as for the first mode in the higher-

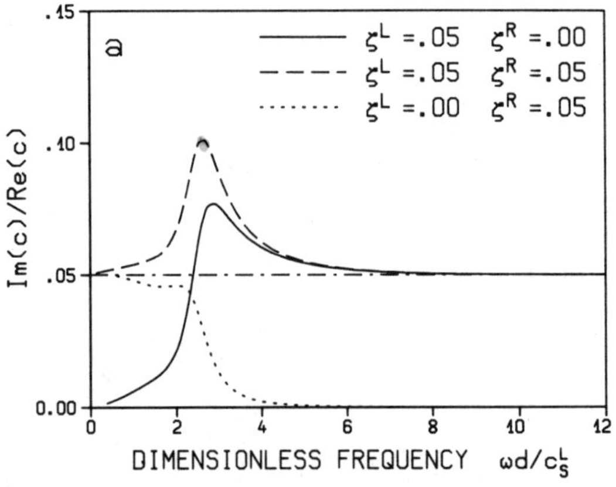

Figure 6-45 Attenuation, homogeneous layer, $\bar{c}_s = 2$, $\bar{\rho} = 1$, $\nu^R = \nu^L = 0.33$ (Ref. [6]). (a) Effective damping ratio, first mode;

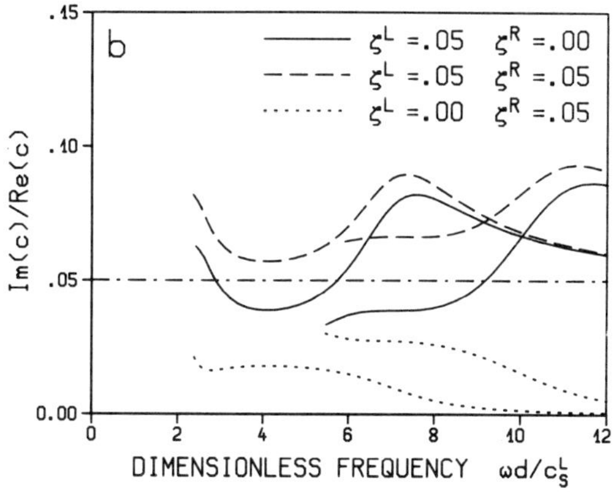

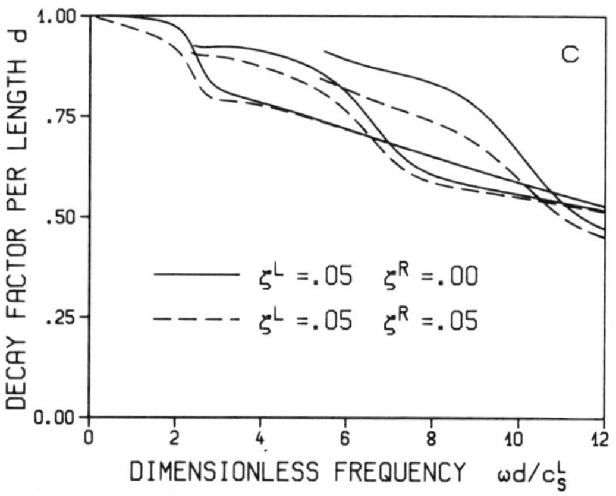

Figure 6-45 (Continued) (b) effective damping ratio, second and third modes; (c) decay factor per length *d*, first three modes.

frequency range. Whereas for an undamped system the cutoff frequencies (above which the motion does not attenuate) of the higher modes can be determined, this no longer applies for a damped system, as attenuation occurs throughout the frequency range. This is discussed for Love waves in connection with Fig. 6-11. In Fig. 6-45b, the curves are drawn only for the frequency range of interest, where the search procedure to determine the roots of the frequency equation described in Section 6.2.3 can easily be applied. In Fig. 6-45c, the decay factor per length *d* (Eq. 6.7b) is plotted for the first three modes. This displayed information follows from that shown in Figs. 6-44 and 6-45a and b. From Fig. 6-45c it is visible that the higher modes in the frequency range where they "start" decay less than the first mode. This is also the case for Love waves (Fig. 6-21b), but the frequency range where this applies is smaller than for R-waves.

The displacement amplitudes at the top of the homogeneous layer are

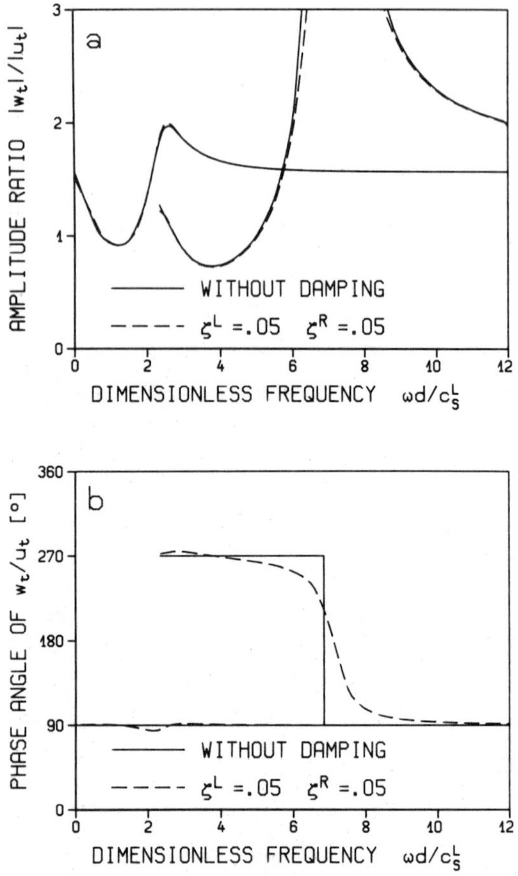

Figure 6-46 Motion at top of layer, homogeneous layer, $\bar{c}_s = 2$, $\bar{\rho} = 1$, $\nu^R = \nu^L = 0.33$ (Ref. [6]). (a) Amplitude ratio; (b) phase angle.

examined next. The amplitude ratio $|w_t|/|u_t|$ and the phase angle of w_t/u_t for the first two modes are plotted in Fig. 6-46. For no damping, w_t and u_t are always $\pm 90°$ out of phase. The first mode is for this case retrograde throughout the frequency range (Fig. 6-46b). The ratio $|w_t|/|u_t| = 1.565$ for $\omega = 0$ (Fig. 6-46a), the same value as for the elastic half-space ($d \longrightarrow 0$, $\bar{\omega} \longrightarrow 0$). The same value results for $\omega \longrightarrow \infty$, as the motion of the first mode is concentrated at the free surface ($\nu^L = \nu^R$). The second mode exhibits prograde and retrograde motions. Damping changes the motion hardly at all.

For the homogeneous layer with $\bar{c}_s = 5$, the first two modes are examined. In contrast to the case $\bar{c}_s = 2$, for no damping, the motion of the first mode is no longer retrograde in the whole range of frequency. A prograde motion arises for $1.67 < \bar{\omega} < 2.80$, a range of practical interest (not shown). The corresponding dispersion curve is shown in Fig. 6-47a. If the damping is increased, the

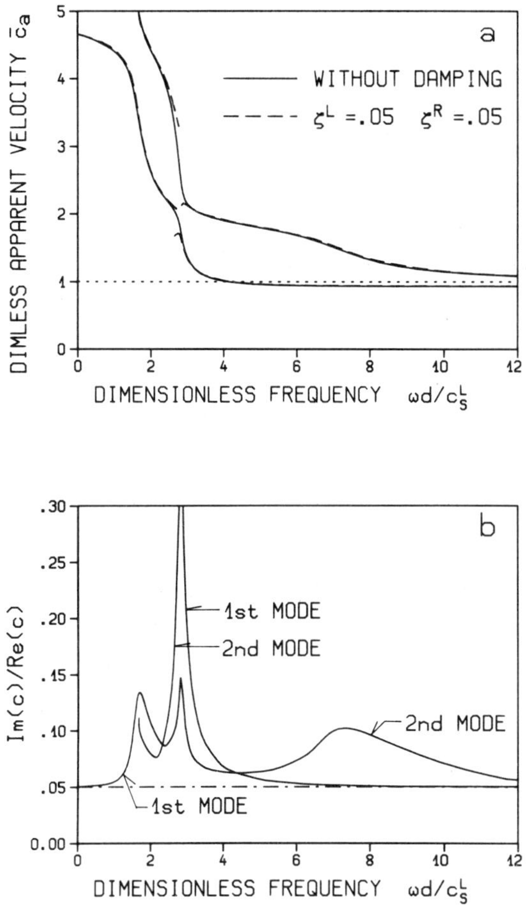

Figure 6-47 Homogeneous layer, $\bar{c}_s = 5$, $\bar{\rho} = 1$, $\nu^R = \nu^L = 0.33$ (Ref. [6]). (a) Dispersion; (b) effective damping.

dispersion curves of the first and second modes become interconnected. While for $\zeta^R = \zeta^L = 0.03$, the dispersion curves are very similar to those for the undamped case (not shown), for $\zeta^R = \zeta^L = 0.05$ the following occurs (Fig. 6-47a): The lower-frequency branch of the first mode is connected to the upper-frequency branch of the second. If $\bar{\omega}$ is decreased, the upper-frequency branch of the first mode becomes highly damped and thus loses its practical importance. This is illustrated in Fig. 6-47b, where the effective damping is plotted. The same happens to the lower-frequency branch of the second mode for increasing $\bar{\omega}$.

For the site with increasing stiffness, the attenuation of the motion for a horizontal distance d is for all higher modes less than that of the first throughout the frequency range (not shown). This is analogous to the attenuation of Love waves (Fig. 6-23b).

Displacements and stresses versus depth. To study the motion's variation in the vertical direction, the amplification within the site $|u_b|/|u_t|$, $|w_b|/|w_t|$ and the amplitude ratios $|u(z)|/|u_t|$, $|w(z)|/|w_t|$ for selected $\bar{\omega}$ are plotted for the homogeneous layer with $\bar{c}_s = 2$ in Figs. 6-48 and 6-49, respectively. In addition to the values of the R-wave, the results for vertically incident and inclined body waves are also shown. For the horizontal and vertical motions, an inclined SV-wave with $\psi_{SV}^R = 60°$ and an inclined P-wave with $\psi_P^R = 30°$, respectively, are selected. Up to the horizontal fundamental frequency of the layer fixed at its base ($\bar{\omega} = \pi/2$), the variation with depth of the horizontal motion of the first R-mode agrees well with that of the vertically incident and inclined SV-waves (Fig. 6-48a, left-hand side of Fig. 6-49a). Discrepancies exist, however, in the vertical motions (Fig. 6-48b, right-hand side of Fig. 6-49a) for the same frequency range. This tendency is even more pronounced for the site

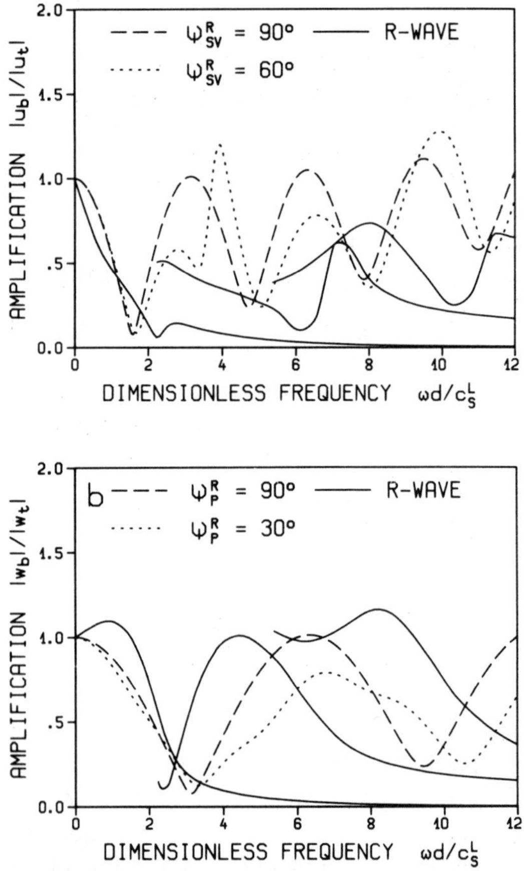

Figure 6-48 Amplification within, homogeneous layer, $\bar{c}_s = 2$, $\bar{\rho} = 1$, $\nu^R = \nu^L = 0.33$, $\zeta^R = \zeta^L = 0.05$ (after Ref. [6]). (a) Horizontal; (b) vertical.

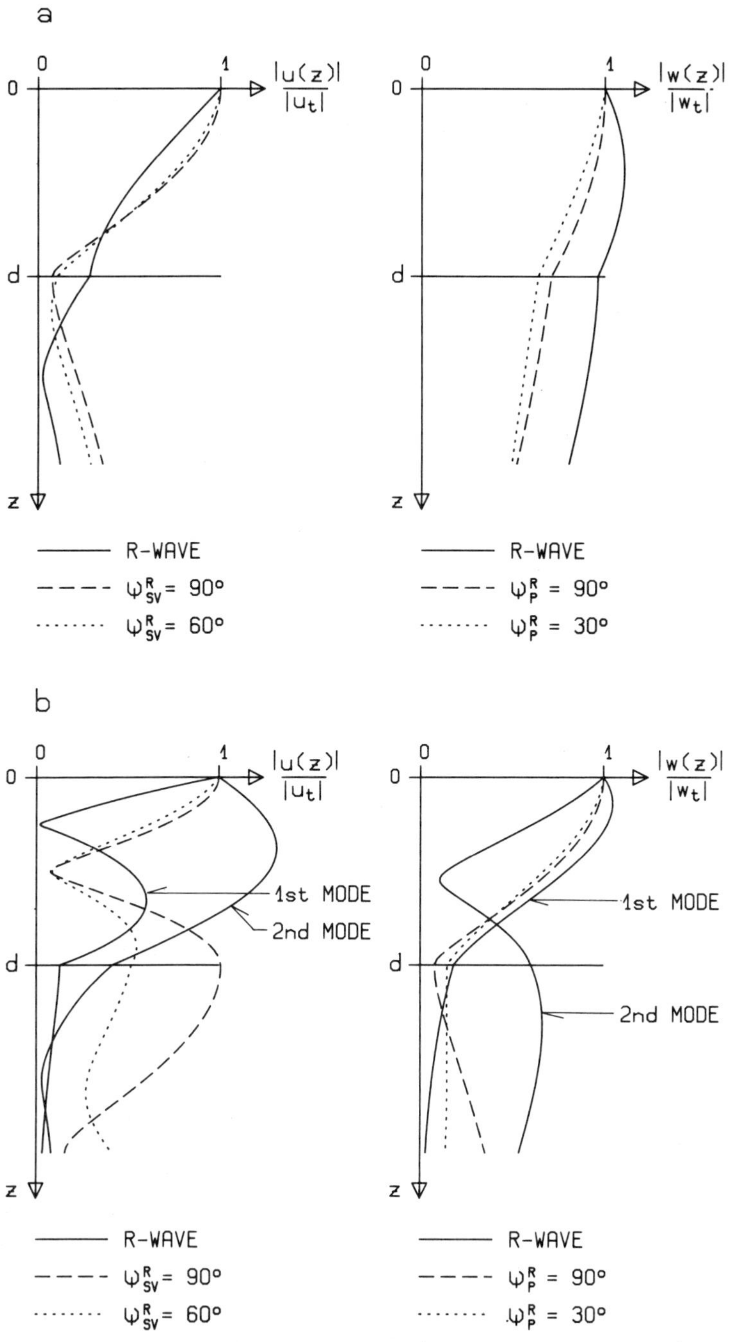

Figure 6-49 Displacement amplitudes versus depth, homogeneous layer, $\bar{c}_s = 2$, $\bar{\rho} = 1$, $\nu^R = \nu^L = 0.33$, $\zeta^R = \zeta^L = 0.05$ (Ref. [6]). (a) $\bar{\omega} = \pi/2$; (b) $\bar{\omega} = \pi$.

with increasing stiffness (not shown), in contrast to the well-known behavior of the half-space. At the vertical fundamental frequency of the layer ($\bar{\omega} = \pi$), the variation with depth of the vertical motion of the first R-mode agrees well with that of the vertically incident and inclined P-waves (right-hand side of Fig. 6-49b). For the other frequency ranges, large differences exist. In particular, the motion of the higher R-mode for the frequency where it "starts" and somewhat above is different from that of the inclined body waves. This is in contrast to the behavior of the motion of the Love wave when compared to that of the inclined SH-wave (Fig. 6-24).

The corresponding variation of the amplitudes of the normal stress σ_x and of the shear stress τ_{xz} with depth is examined next. Dimensionless stress amplitudes $\bar{\sigma}_x$ and $\bar{\tau}_{xz}$ are defined as $\sigma_x/[(c_s^L)^2\rho^L]$ and $\tau_{xz}/[(c_s^L)^2\rho^L]$, respectively. The variation with depth is presented for $|u_t|/d = 1$ in Fig. 6-50 for the homogeneous layer. The shear-stress amplitude τ_{xz} (Eq. 5.131b) is smaller for the R-wave and the inclined SV-wave than for the vertically incident wave. Horizontally propagating waves lead also to a normal-stress amplitude σ_x (Eq. 5.141) which is not zero at the free surface. The amplitude σ_x is significant, especially for the R-wave, and has to be taken into account when designing retaining walls. The amplitude of the maximum shear stress τ_{max} (determined from σ_x, σ_z, and τ_{xz}) is also plotted in Fig. 6-50. R-waves, especially, result in a significant τ_{max} close to the free surface. For larger depths, τ_{max} is only slightly larger than τ_{xz} for the R-wave

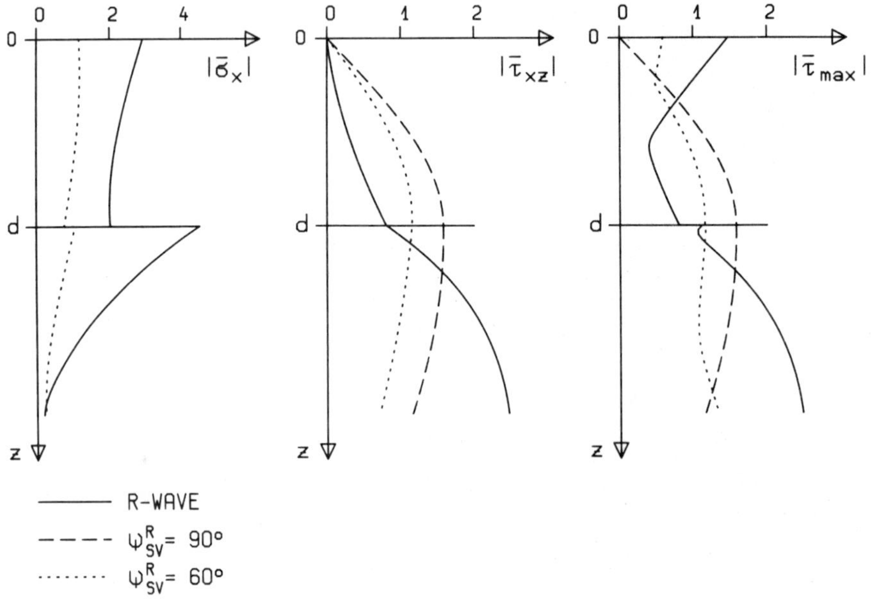

Figure 6-50 Stress amplitudes versus depth, homogeneous layer, $\bar{c}_s = 2$, $\bar{\rho} = 1$, $\nu^R = \nu^L = 0.33$, $\zeta^R = \zeta^L = 0.05$, $\bar{\omega} = \pi/2$ (Ref. [6]).

and inclined SV-wave. Analogous behavior results for the site with increasing stiffness, as is visible from the corresponding plots in Fig. 6-51, with the exception that τ_{max} of the R-wave close to the free surface does not dominate to the same extent. For a half-space, however, τ_{max} close to the free surface (Fig. 6-9) dominates even more than for the homogeneous layer (Fig. 6-50).

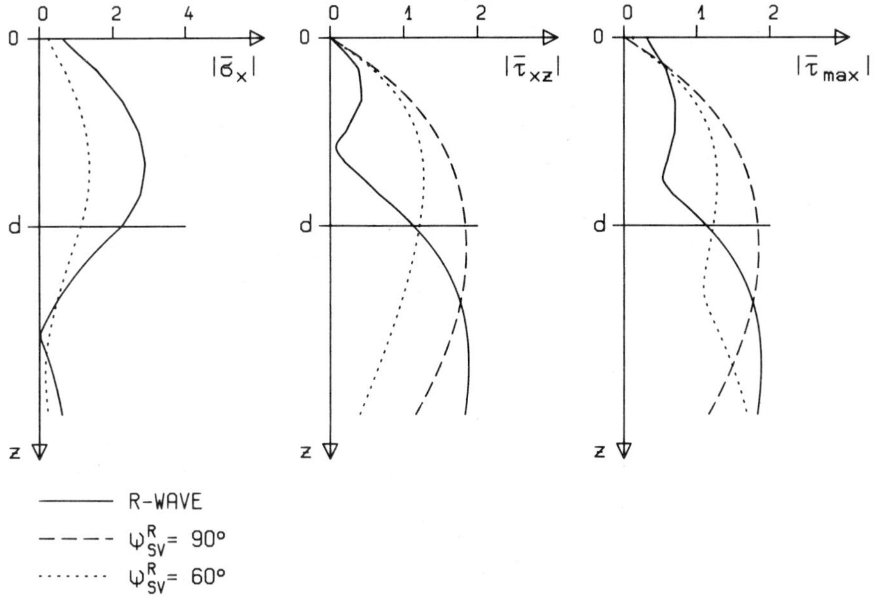

Figure 6-51 Stress amplitudes versus depth, increasing stiffness, $\nu^R = \nu^L = 0.33$, $\zeta^R = 0$, $\zeta^L = 0 - 0.05$, $\bar{\omega} = 2.0$ (Ref. [6]).

Scattered motion. Finally, the link of the horizontally propagating free-field motion to the seismic loading applied to a surface structure is examined. The radius of the rigid basemat is denoted as a. As discussed in Section 6.2.5, the ratio $\omega a/c_a$ determines the two translational and the one rocking components of the scattered motion, resulting from the in-plane motion. These are shown in Figs. 7-30 and 7-31. Based on the dispersion curves $c_a(\omega)$, the curves for the first three R-modes are constructed (for the site with increasing stiffness) and are shown in Fig. 6-52. The radius $a = 0.5d$ is selected. The corresponding decay factors calculated for a horizontal distance $10a$ ($= 5d$) are also indicated. For comparison, the dashed straight line corresponding to an incident SV-wave with $\psi^R_{SV} = 10°$ and the dotted straight line of the incident P-wave with $\psi^R_P = 10°$ are also shown. As the half-space is not damped, the two body waves do not decay. For a specific ω, $\omega a/c_a$ (which characterizes the wave effects) is larger for R-waves than for the body waves. The attenuation is significant but not overwhelming. Considerable wave effects on the seismic input can potentially arise.

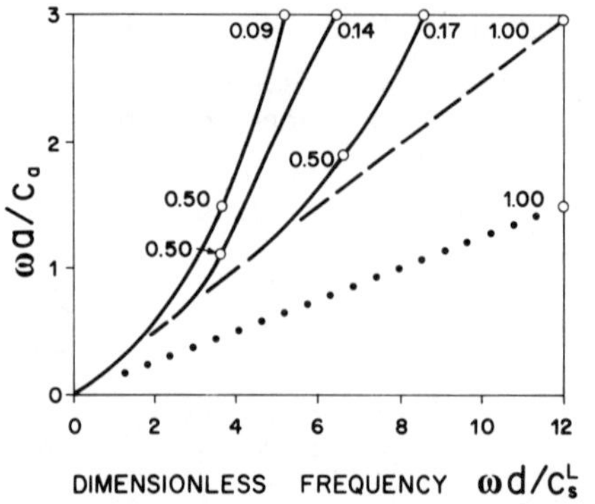

Figure 6-52 Applied seismic-input motion, increasing stiffness, $\nu^R = \nu^L = 0.33$, $\zeta^R = 0$, $\zeta^L = 0 - 0.05$ (after Ref. [6]).

The influence on the structural response is discussed in depth in Section 9.2, where it is shown that the difference of the phase angles of the horizontal and vertical motions is of paramount importance. The corresponding curves $\omega a/c_a$ versus $\bar{\omega}$ for the site with a homogeneous layer for $\bar{c}_s = 2$ and $= 5$ can be drawn using the information shown in Figs. 6-44, 6-45c, and 6-47a and b, respectively.

6.7 SOFT SITE

To illustrate the practical application of the material we have been discussing, the response of an actual site of a nuclear power plant is discussed. The response caused by inclined body waves and by surface waves is compared to that arising from vertically incident waves. Harmonic excitation and artificial earthquake time histories are considered.

6.7.1 Description of Site and of Control Motion

The (strain-compatible) properties of this soft site, consisting of layers of sand and gravel with increasing stiffness and decreasing damping with depth resting on poor rock, are specified in Table 6-4. The origin of the vertical z-axis is selected at the free surface. The shear velocity c_s varies from 200 to 1500 m/s, the damping ratio ζ from 0.07 to 0.02. The first three natural frequencies in the horizontal direction of the total soil layer of 50 m depth fixed at its base equal 3.0, 7.0, and 10.8 Hz, and the fundamental in the vertical direction equals 7.3 Hz. The basemat of the reactor building is located at a depth of 10.0 m. The fundamental frequency of the two top layers fixed at $z = 10.0$ m equals 6.0 Hz. These frequencies are indicated on the abscissas in the following plots, where appropriate.

The two horizontal and the one vertical control motions consist of artificial 30-s-acceleration time histories, the response spectra of which follow U.S. NRC Regulatory Guide 1.60, normalized to 0.30g and 0.20g, respectively

TABLE 6-4 Free-Field Properties of Soft Site

Depth, z (m)	Shear Modulus, G (MN/m^2)	Mass Density, ρ (Mg/m^3)	Shear Wave Velocity, c_s (m/s)	Poisson's Ratio, ν	Damping Ratio, ζ
0					
	80	2.0	200	0.40	0.07
5					
	125	2.0	250	0.40	0.06
10					
	245	2.0	350	0.40	0.05
20					
	550	2.2	500	0.40	0.05
30					
	1408	2.2	800	0.40	0.05
40					
	2400	2.4	1000	0.35	0.04
50					
	5625	2.5	1500	0.30	0.02
∞					

(Section 3.3.3, Fig. 3-17). The response spectra are shown for a damping ratio of 5% in Fig. 6-54a and Fig. 6-58 for the out-of-plane and in-plane motions, respectively. The control point in which this motion is assumed to act is located either at a (fictitious) rock outcrop at a depth of 50 m or at the free surface. For the in-plane motion, these two control points are shown in Fig. 6-53. For this actual site, no dimensionless parameters are introduced.

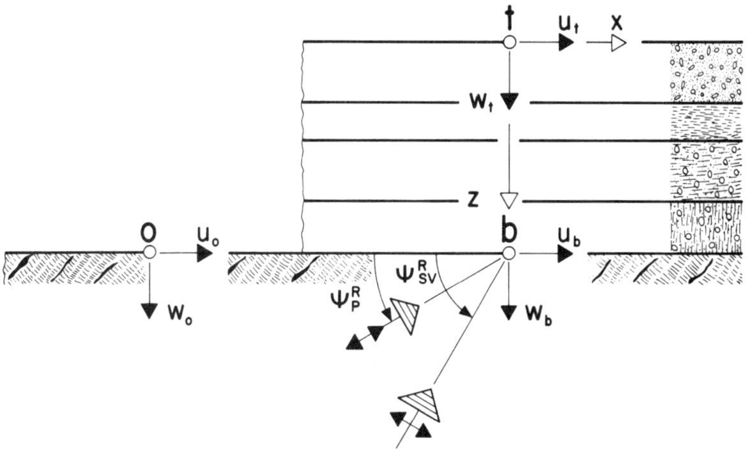

Figure 6-53 Layered site with nomenclature for in-plane motion.

6.7.2 Vertically Incident and Inclined SH-Waves

At first, body waves are examined. To study the seismic environment resulting for the case of out-of-plane motion, the control point is selected at the top of the rock ($z = 50.0$ m). At first, the control motion is assumed to act at the outcrop of this rock. For vertically incident SH-waves, the response spectra of the resulting motions at selected levels within the site are shown in Fig. 6-54a. As shown in the parametric study in Section 6.5 for harmonic motion for the site with increasing stiffness, the frequencies at which the dips of the curve at the base ($z = 50.0$ m) occur coincide with the natural frequencies of the site (Fig. 6-19b). The peaks of the curve at the top of the layer ($z = 0$) are shifted

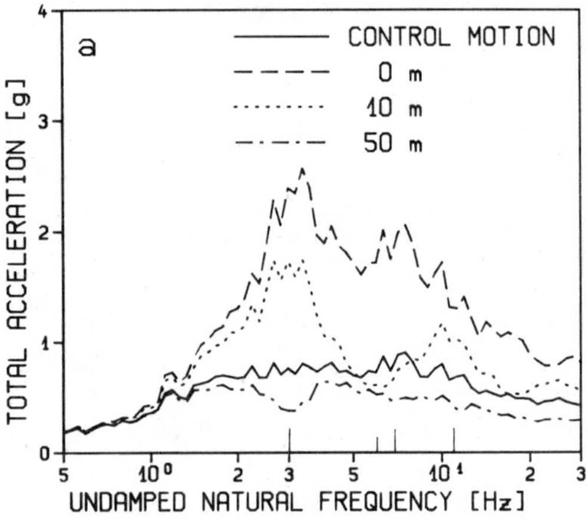

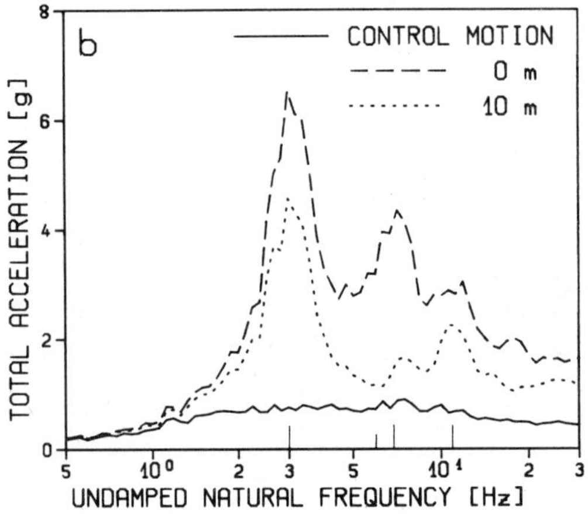

Figure 6-54 Response spectra (5% damping), soft site, out-of-plane motion, vertical incidence, control point at top rock (Ref. [5]). (a) Control motion outcropping; (b) control motion within the site.

somewhat to a higher frequency (Fig. 6-19a). The influence of the free surface on the frequency content of the motion at $z = 10.0$ m, resulting in a dip of the curve at the corresponding frequency, is clearly visible. Specifying the same (broad-banded) control motion to act as within the site at the same depth of 50 m leads to completely unrealistic values, as is visible in Fig. 6-54b. This again illustrates the fact that a control point cannot be chosen within the site.

Selecting the outcropping motion to arise from inclined SH-waves with an angle of incidence in the rock $\psi^R_{SH} = 10°$ (Fig. 6-55) results in values at the free surface ($z = 0$) which are drastically smaller than those arising from the same motion for vertical incidence. This confirms the trend established for harmonic motion (Fig. 6-19a).

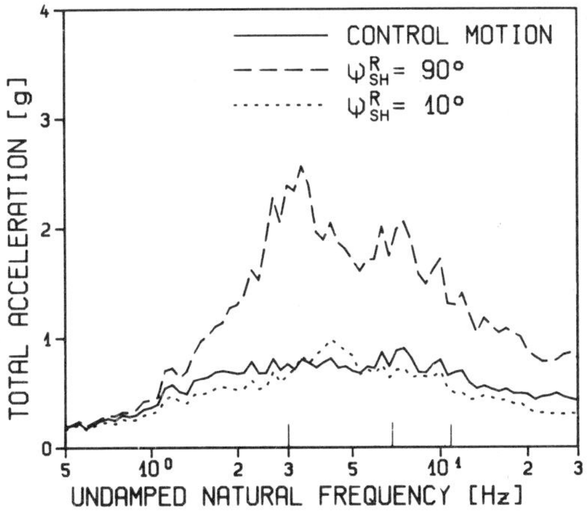

Figure 6-55 Response spectra (5% damping), soft site, out-of-plane motion, free surface, control motion at rock outcrop (Ref. [5]).

6.7.3 Vertically Incident and Inclined P- and SV-Waves

Turning to the case of in-plane motion, the control motion is again assumed to act at the outcrop of the rock ($z = 50.0$ m). The control motion with the horizontal and vertical components u_o and w_o is associated with different wave patterns (Fig. 6-53). The motion given by u_t and w_t at the free surface can be determined. For harmonic excitation of circular frequency ω, the amplifications from outcropping bedrock to the top of the site in the horizontal and vertical directions, $|u_t|/|u_o|$ and $|w_t|/|w_o|$, are studied first. For instance, $|u|$ denotes the absolute value of the corresponding horizontal displacement amplitude. Assuming u_o and w_o to arise from vertically incident SV- and P-waves, respectively, the amplifications $|u_t|/|u_o|$ and $|w_t|/|w_o|$ are plotted as solid lines in Fig. 6-56a and b, respectively. The ratios selecting inclined body waves with the angle of incidence ψ of the SV- and P-waves in the rock (superscript R) are also shown. The observations stated in the parametric study of Section 6.6

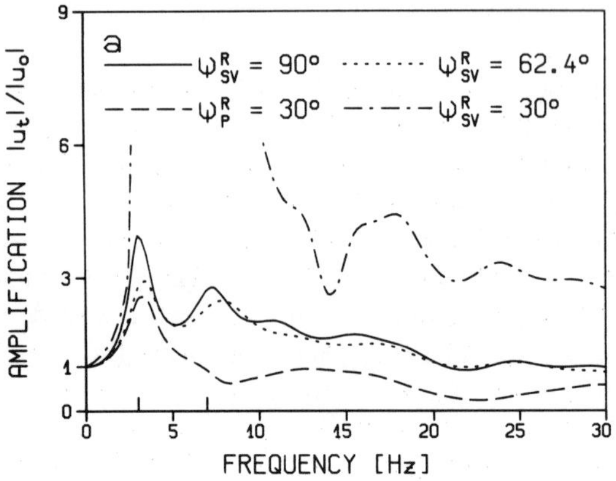

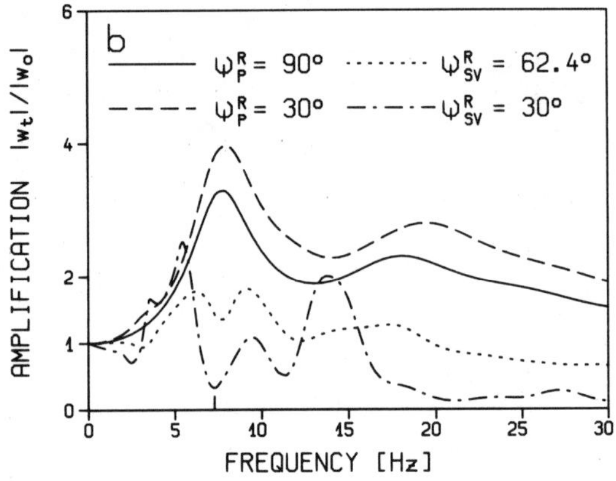

Figure 6-56 Amplification outcropping, soft site, in-plane motion (Ref. [7]). (a) Horizontal; (b) vertical.

(Figs. 6-31 and 6-36) are confirmed for this actual site. Compared with vertical incidence, large differences arise, especially for the shallow SV-wave ($\psi^R_{SV} = 30°$). For this case, however, $|w_o|/|u_o| \sim 2.5$, which of course is quite unrealistic and can thus be disregarded. Matching the prescribed control motion with a combination of P- and SV-waves leads to amplifications which depend on both components of the control motion. This is demonstrated in Fig. 6-57 for incident P- and SV-waves with $\psi^R_P = 30°$ and $\psi^R_{SV} = 62.4°$, resulting in a common apparent velocity $c_a = 3241$ m/s. For the horizontal amplification (Fig. 6-57a) the following three cases are discussed: first, for zero vertical motion in the control point (solid line); second, for a retrograde motion with equal horizontal and vertical absolute values of the amplitudes (dashed line); and third, with equal horizontal and vertical absolute values of the amplitudes and with a phase

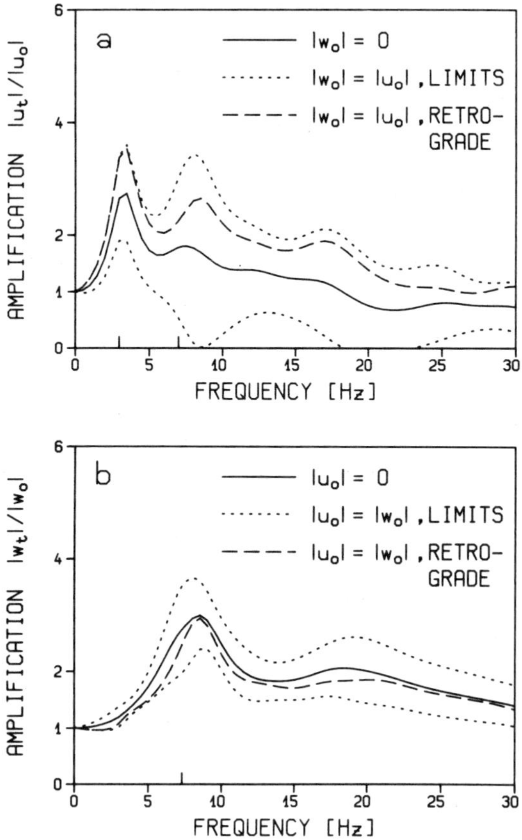

Figure 6-57 Amplification outcropping, incident P- and SV-waves, $\psi_P^R = 30°$, $\psi_{SV}^R = 62.4°$, soft site (Ref. [7]). (a) Horizontal; (b) vertical.

angle which leads to the maximum and minimum amplification (dotted lines). For the third case, the phase angle is different for the two extremes for each frequency. This procedure is discussed in depth in connection with Fig. 6-41a. Figure 6-57b shows the corresponding amplification curves in the vertical direction (varying u_o). In particular, the influence of the vertical motion on the horizontal amplification cannot be neglected. Turning to the transient seismic excitations, the resulting response spectra for the horizontal and vertical motions at the free surface are shown for different wave patterns in Fig. 6-58a and b, respectively. As expected, peaks arise for the vertically incident waves at the natural frequencies of the layer. First, the outcropping motion is assumed to arise from incident SV-waves with $\psi_{SV}^R = 62.4°$. The horizontal time history is matched. This determines the vertical time history (peak acceleration $= 0.14g$), which will be different from the vertical control motion. This wave pattern results

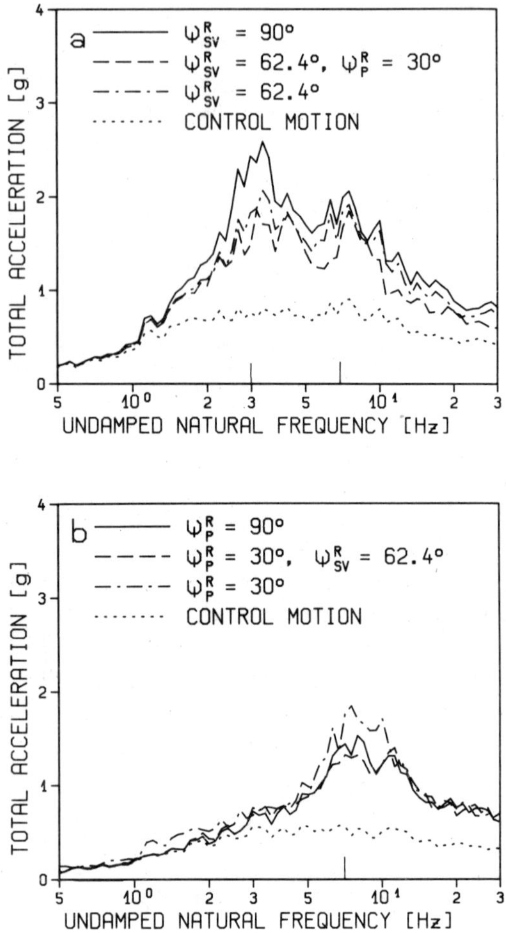

Figure 6-58 Response spectra (5% damping), soft site, in-plane motion, free surface, control motion at rock outcrop (Ref. [7]). (a) Horizontal; (b) vertical.

at the free surface in a horizontal response, which is smaller than that for vertical incidence (Fig. 6-58a). This confirms the trend established for harmonic motion (Fig. 6-56a). Second, if one interprets the vertical control motion as being generated by incident P-waves with $\psi^R_P = 30°$ (which leads to the same apparent velocity), the corresponding horizontal time history ($0.29g$) does not match the horizontal control motion. As expected from the results of the harmonic response (Fig. 6-56b), the vertical response spectrum at the free surface is larger than that for vertical incidence (Fig. 6-58b). Finally, matching both components of the control motion by assuming a combination of SV-waves with $\psi^R_{SV} = 62.4°$ and of P-waves with $\psi^R_P = 30°$ results in a horizontal response spectrum which is smaller than that for vertical incidence, while the two vertical-response spectra bear comparison with one another. The drastic reduction arising for inclined SH-waves (Fig. 6-55) does not occur for inclined P- and SV-waves.

6.7.4 Love Waves

Next, surface waves are examined. Addressing the out-of-plane motion, the dispersion curves for the first three Love modes are shown in Fig. 6-59. The shear-wave velocities of the top layer and of the rock are indicated as dotted lines. Again, the frequency range of steepest descent contains the natural frequencies of the total layer. The corresponding decay factors at a horizontal distance = 500 m are plotted for harmonic excitation in Fig. 6-60(a). For comparison, the curve for an inclined body wave with $\psi^R_{SH} = 10°$ is also specified as a dashed line. For this soft site, Love waves attenuate significantly.

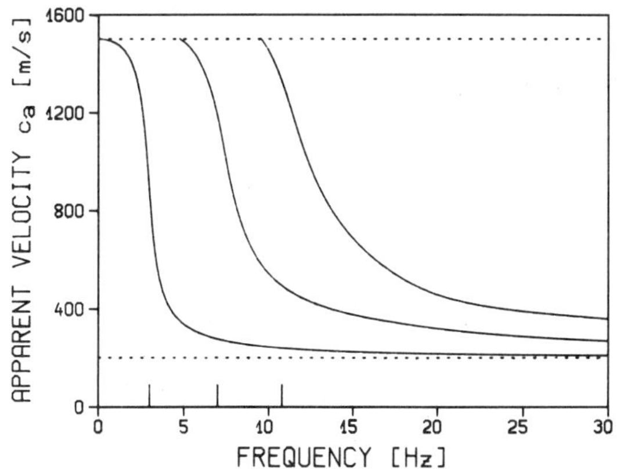

Figure 6-59 Dispersion, soft site, out-of-plane motion (Ref. [5]).

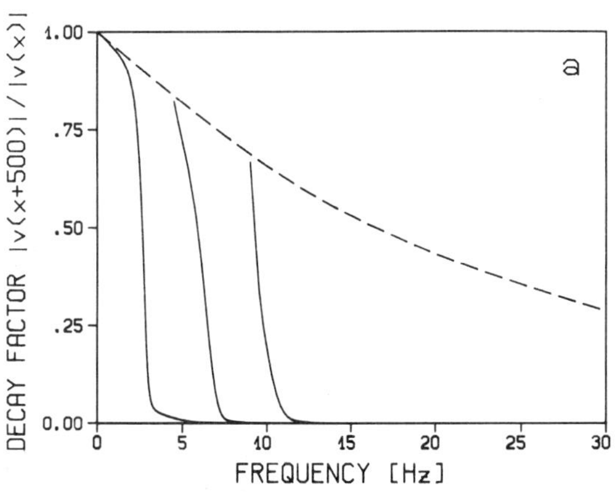

Figure 6-60 Attenuation, soft site, out-of-plane motion (Ref. [5]). (a) Decay factor 500 m distance;

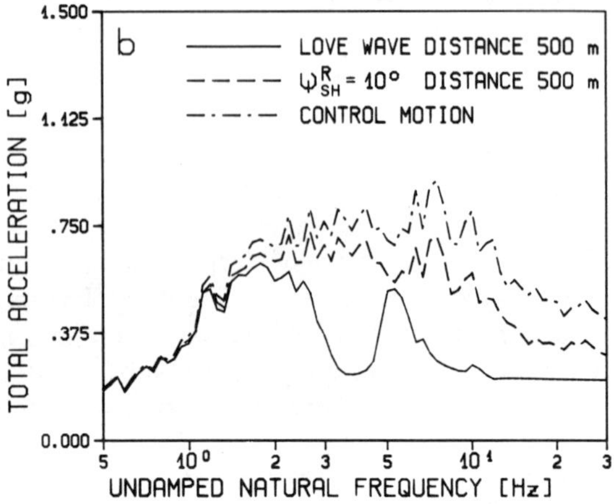

Figure 6-60 (Continued) (b) Response spectra (5% damping).

When interpreting the control motion to consist of Love waves only, the control point is chosen at the free surface. As each Love mode attenuates strongly in its higher-frequency range, the control motion is associated with a certain mode up to the frequency where the next higher mode starts. The first three modes are used (the first from 0 to 5 Hz, the second from 5 to 9.8 Hz, and the third for the remaining frequencies). The response spectra at the control point and at a horizontal distance = 500 m are plotted in Fig. 6-60b. The lower attenuation of the second and, to a lesser extent, of the third mode is clearly visible. For comparison, the same motion at the free surface in the control point is associated with inclined SH-waves with $\psi^R_{SH} = 10°$. As expected, this wave train attenuates less, as is visible from the response spectrum at the same horizontal distance.

To study the variation of the motion with depth, the ratio of the amplitude at the base ($z = 50.0$ m) and at the free surface $|v_b|/|v_t|$ for harmonic response is plotted in Fig. 6-61a. Results are shown for the first three Love modes, for the vertically incident SH-wave and for the inclined SH-wave with $\psi^R_{SH} = 10°$. The same properties as discussed in connection with the site with increasing stiffness used in the parametric study are visible (Fig. 6-26). The corresponding response spectra at the base of the total layer ($z = 50.0$ m) are shown in Fig. 6-61b. The Love wave is associated with the first three modes, as discussed above. The three curves are quite similar; in particular, the dips at the natural frequencies of the total layer are apparent. The discrepancy of the curves for the Love wave and for the body waves for the frequencies ranging from a natural frequency to the frequency where the next higher mode starts, clearly visible for harmonic excitation (Fig. 6-61a), is also present, but is less pronounced for transient loading (Fig. 6-61b).

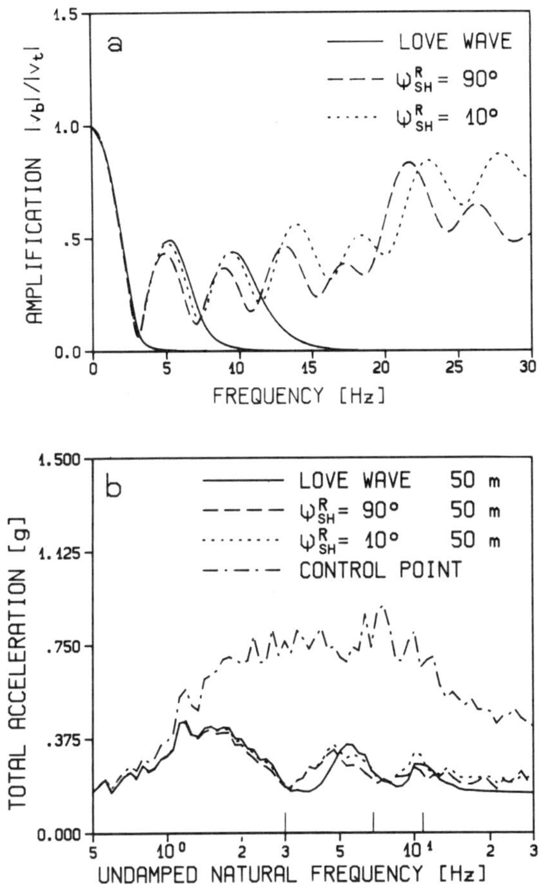

Figure 6-61 Variation with depth, soft site, out-of-plane motion (Ref. [5]). (a) Amplification within; (b) response spectra (5% damping).

6.7.5 Rayleigh Waves

Examining the in-plane motion, the dispersion curves for the first three Rayleigh modes are shown in Fig. 6-62. The shear-wave velocities of the top layer and of the rock are indicated as dotted lines. For comparison, the undamped site has also been analyzed. Material damping hardly affects the apparent velocity $c_a(\omega)$ in the range of interest. The corresponding effective damping ratio Im (c)/Re (c) and the decay factor for 500 m are plotted in Fig. 6-63a and b, respectively. The ratio of the imaginary and real parts of the phase velocity c describes the decay factor per wavelength, as explained in connection with Fig. 6-45. For comparison, the plots for an incident SV-wave with $\psi^R_{SV} = 10°$ and for an incident P-wave with $\psi^R_P = 10°$ are also specified as a dashed and

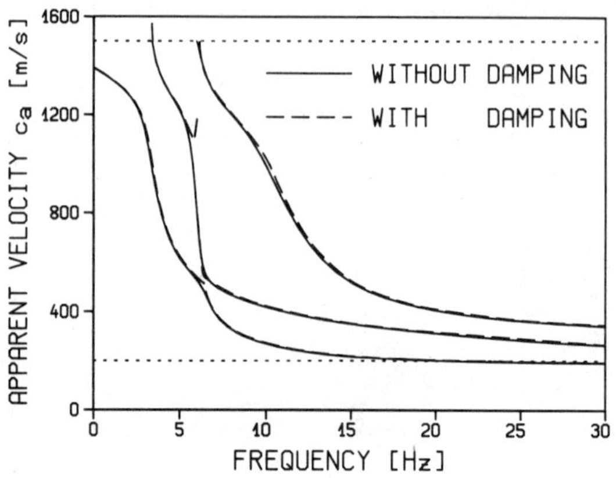

Figure 6-62 Dispersion, soft site, in-plane motion (Ref. [7]).

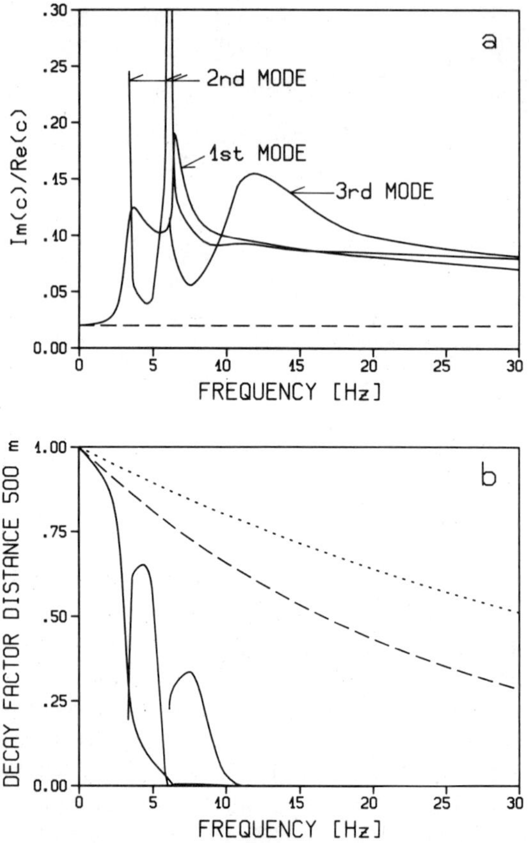

Figure 6-63 Attenuation, soft site, in-plane motion (Ref. [7]). (a) Effective damping ratio; (b) decay factor 500 m distance;

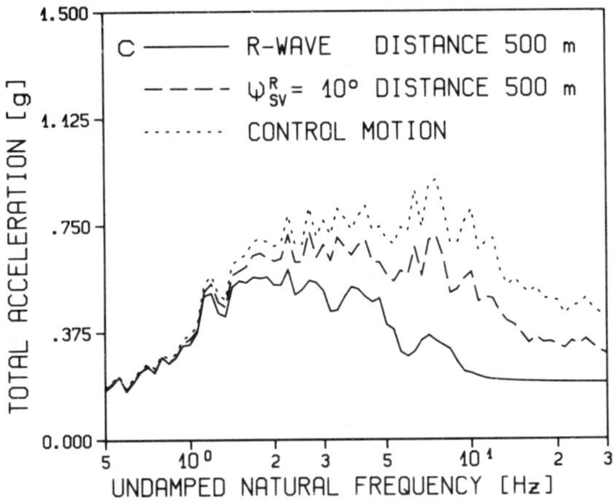

Figure 6-63 (Continued) (c) horizontal response spectra (5% damping).

dotted line, respectively, in Fig. 6-63b. In Fig. 6-63a, the two lines coincide (dashed line). The second mode actually consists of two branches separated by a frequency range with very high effective damping. For this soft site, R-waves attenuate significantly and are virtually nonexistent in the higher-frequency range.

When one interprets the control motion to consist of R-waves only, the control point is chosen at the free surface. As each R-mode attenuates strongly in its higher-frequency range, one component (e.g., the horizontal one of the motion) is associated with that of a certain mode up to the frequency at which the next higher mode starts. This defines the other one, in this example the vertical component, which will, in general, not match the vertical control motion. The first three modes are used (the first from 0 to 3.4 Hz, the second from 3.4 to 6.1 Hz, and third for the remaining frequencies). The horizontal response spectra at the control point and at a horizontal distance = 500 m are plotted in Fig. 6-63c. For comparison, the same horizontal motion at the free surface in the control point is associated with the horizontal motion of the incident SV-waves with $\psi^R_{SV} = 10°$. As expected, this wave train attenuates less, as is visible from the response spectrum at the same horizontal distance.

6.7.6 In-Plane Displacements and Stresses versus Depth

To study the variation of the motion with depth, the amplitude ratios $|u(z)|/|u_t|$ and $|w(z)|/|w_t|$ for selected frequencies are plotted in Fig. 6-64. In addition to the values of the R-waves, the results for vertically incident and inclined body waves are also shown. The amplifications from the free surface to the depth $z = 50$ m in the horizontal and vertical directions, $|u_b|/|u_t|$ and $|w_b|/|w_t|$, are plotted in Fig. 6-65. Similar to the procedure described in connection with Fig. 6-57, a combination of P- and SV-waves with $\psi^R_P = 30°$ and

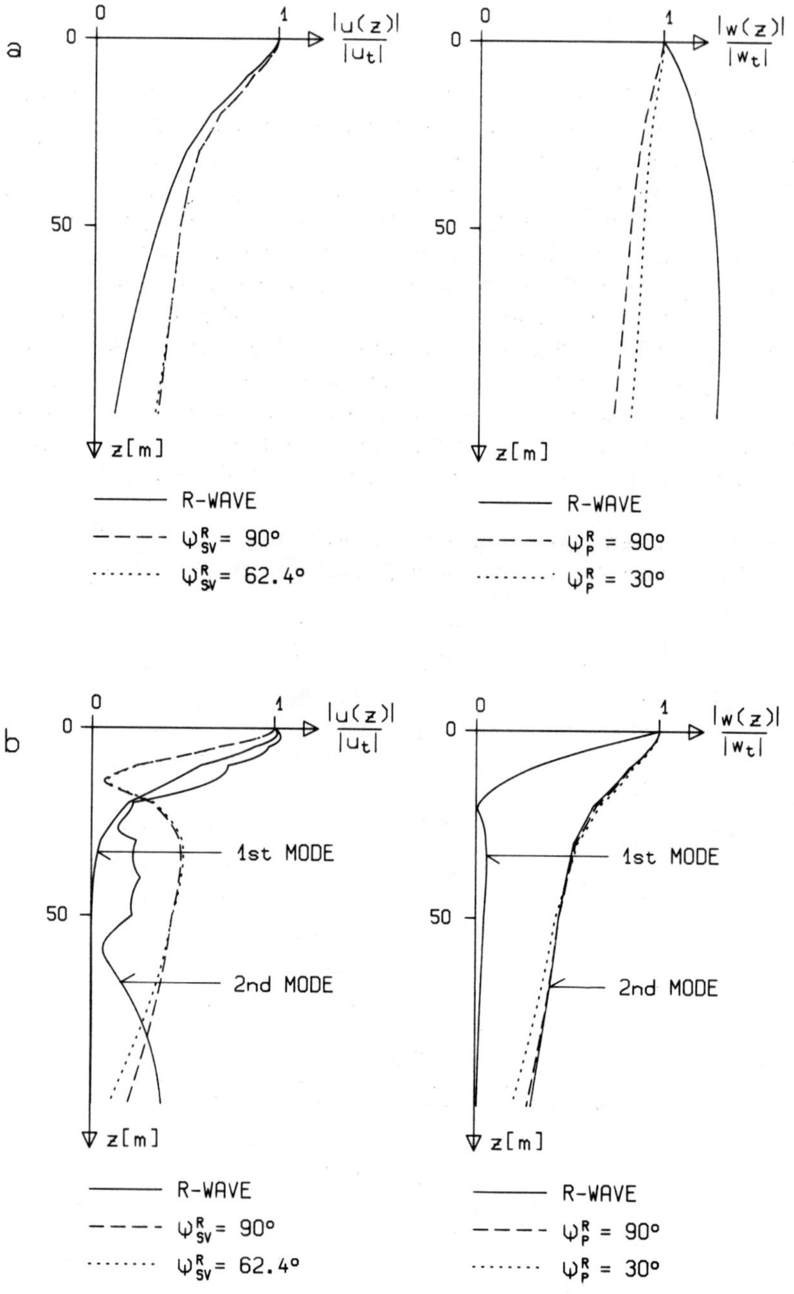

Figure 6-64 Amplitudes versus depth, soft site, in-plane motion (Ref. [7]). (a) Frequency 2 Hz; (b) frequency 5 Hz.

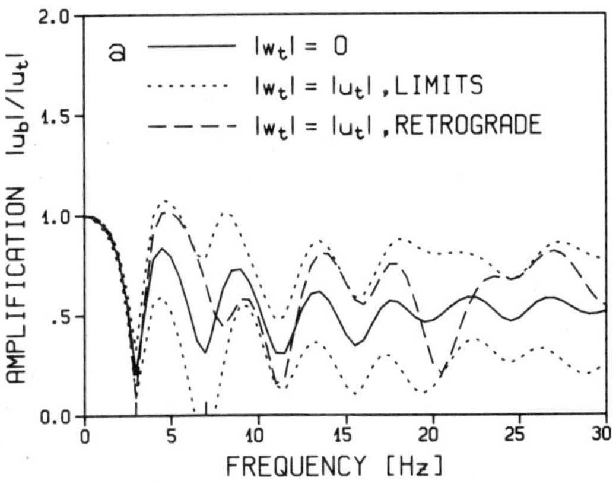

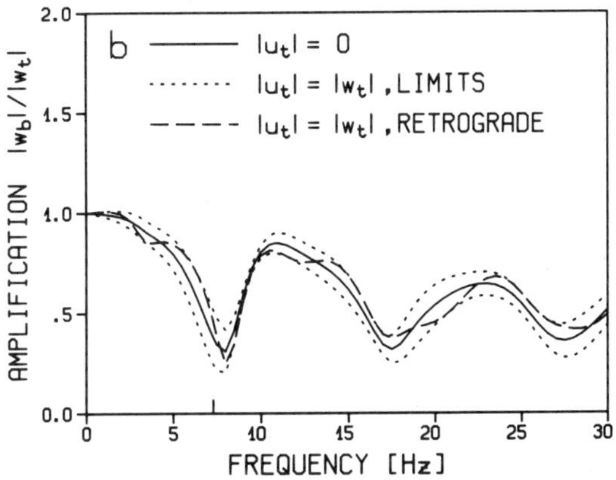

Figure 6-65 Amplification within, incident P- and SV-waves, $\psi_P^R = 30°$, $\psi_{SV}^R = 62.4°$, soft site (Ref. [7]). (a) Horizontal; (b) vertical.

$\psi_{SV}^R = 62.4°$ is selected which matches the prescribed motion at the free surface. The amplification $|w_b|/|w_t|$ is, from a practical point of view, independent of the prescribed wave pattern throughout the frequency range. For $|u_b|/|u_t|$ this holds only for the range up to the fundamental frequency in the horizontal direction. The horizontal- and vertical-response spectra for various assumptions at the base of the total layer ($z = 50$ m) are shown in Fig. 6-66a and b, respectively. The control point is selected at the free surface. At first, the horizontal and vertical control motions are assumed to arise from vertically incident waves. The dips appearing at the natural frequencies of the layer are clearly visible. Second, the control motion is interpreted as being generated by a combination of P- and SV-waves with $\psi_P^R = 30°$ and $\psi_{SV}^R = 62.4°$. The corresponding response spectra in the horizontal and vertical directions agree well with those of the vertically incident waves. For the third interpretation of the control motion,

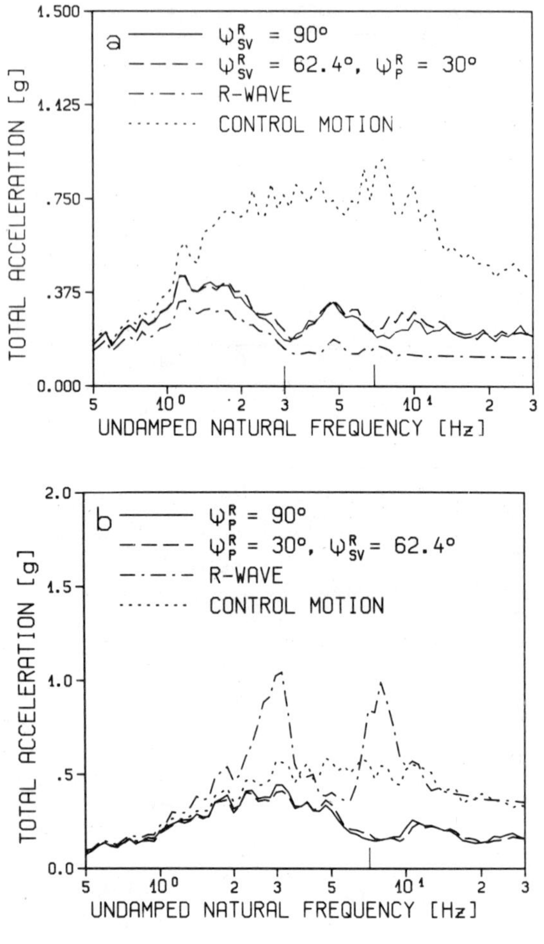

Figure 6-66 Response spectra (5% damping), soft site, in-plane motion, depth 50 m, control motion at free surface (Ref. [7]). (a) Horizontal; (b) vertical.

only one component is matched. If one identifies the horizontal component of the control motion with the first three R-modes, as described above, the horizontal response spectrum at 50 m (Fig. 6-66a) lies below that of vertically incident waves. This can also be seen in Fig. 6-64 for harmonic excitation. Matching analogously the vertical component of the control motion with the first three R-modes results in an unrealistically high horizontal acceleration. This leads to large values of the vertical response spectra at 50 m (Fig. 6-66b). The good agreement for Love waves and vertically incident SH-waves demonstrated in Fig. 6-61b is no longer present.

Examining the variation of the stresses with depth, the amplitudes of the normal stress σ_x and of the shear stress τ_{xz} are plotted for various wave assumptions for an harmonic excitation of 2 Hz in Fig. 6-67. The amplitude of the

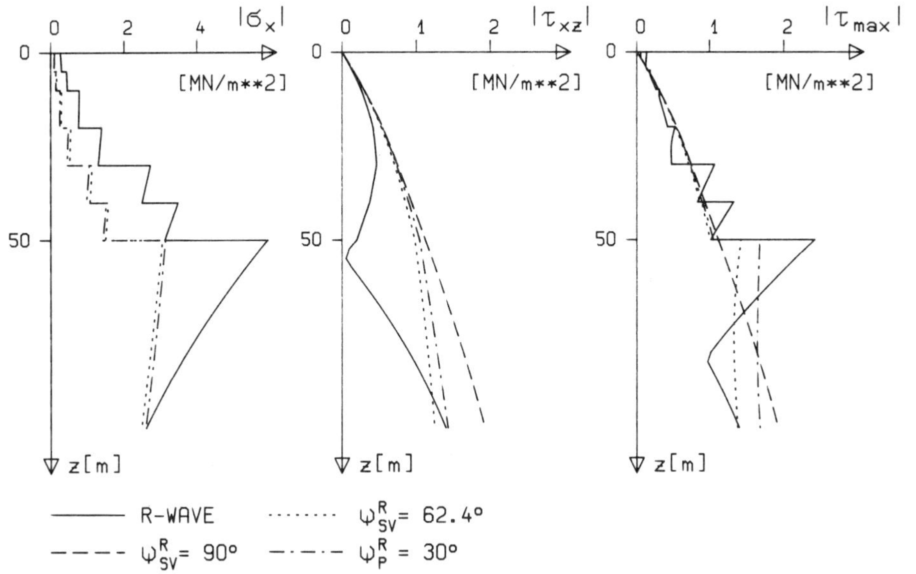

Figure 6-67 Stress amplitudes versus depth, soft site, in-plane motion, frequency 2 Hz (Ref. [7]).

maximum shear stress τ_{max} (determined from σ_x, σ_z, and τ_{xz}) is also drawn. The variation with depth is presented for $|u_t| = 0.1$ m. The discontinuities of σ_x arise from the layers of the site. The trends established in the parametric study of Section 6.6.4 (Figs. 6-50 and 6-51) also apply to this actual site. In Fig. 6-68, the maximum normal stress $\sigma_{x\,max}$ is plotted versus depth for the earthquake time histories for the following wave assumptions indicated in the figure. Identifying the horizontal motion at the free surface with the first three R-modes results in a considerable $\sigma_{x\,max}$. For the combination of SV- and P-waves, both components of the control motion are matched. Negligible $\sigma_{x\,max}$ arises for vertically incident P-waves. For comparison, σ_x from dead weight is also plotted.

6.7.7 Scattered Motion

For a surface structure with a rigid basemat, for the out-of-plane motion the loading applied to the structure arising from horizontally propagating waves consists essentially of a reduced translational and an additional torsional component compared to the loading arising from vertical incidence. For the in-plane motion, the seismic loading is made up of the two reduced translational components and an additional rocking component. For both motions the seismic input motion depends essentially on the ratio $\omega a/c_a$. For a radius $a = 30$ m of the rigid basemat of a surface structure, this ratio is specified for the first three Love modes and the SH-wave with $\psi_{SH}^R = 10°$ in Fig. 6-69, and for the first three R-modes, for an SV-wave with $\psi_{SV}^R = 10°$ (dashed line) and for a P-wave

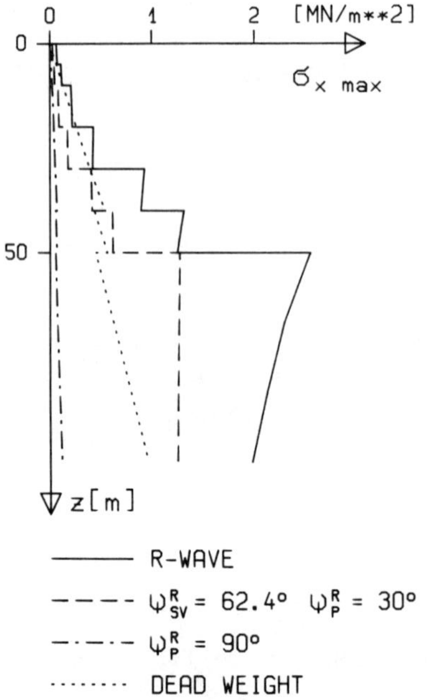

Figure 6-68 Maximum normal stress versus depth, soft site, in-plane motion (Ref. [7]).

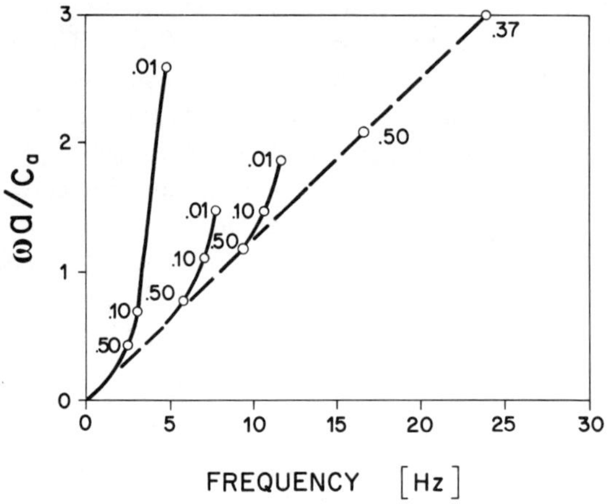

Figure 6-69 Applied seismic-input motion, soft site, out-of-plane motion (Ref. [5]).

with $\psi_P^R = 10°$ (dotted line) in Fig. 6-70. The decay factors for a horizontal distance = 500 m are also given. In the range of frequencies where the c_a (and thus the ratio) differ for the inclined body and surface waves, the latter attenuate strongly.

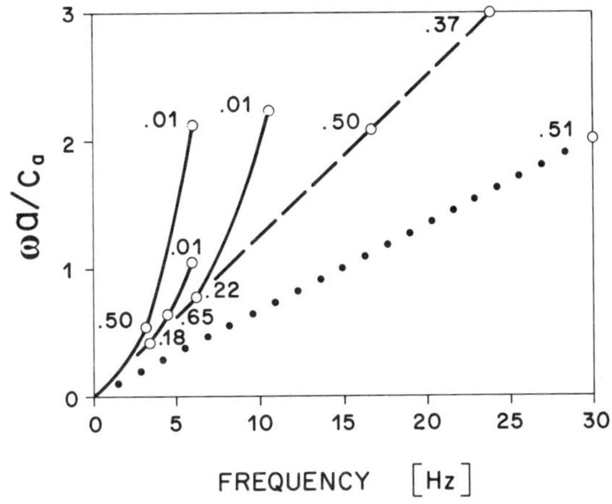

Figure 6-70 Applied seismic-input motion, soft site, in-plane motion (Ref. [7]).

6.8 ROCK SITE

6.8.1 Description of Site and of Control Motion

As the motion of the surface waves attenuates strongly for the soft site discussed in the preceding section, the influence of the traveling wave on the seismic-input motion is significantly reduced. This may, however, not be the case for a rock site which is damped less and which has higher apparent velocities. The site of an actual nuclear power plant is investigated. The dynamic properties of the rock site, consisting of layers on top of a half-space, are specified in Table 6-5. The shear-wave velocity increases from $c_s = 1074$ m/s at the free surface to $c_s = 4385$ m/s below 120 m. The damping ratio $\zeta = 0.02$ is constant. The first two natural frequencies in the horizontal direction of the total layer of 120 m depth fixed at its base equal 6.9 Hz and 15.4 Hz; the fundamental in the vertical direction equals 13.4 Hz. The same control motion as that discussed in the preceding section is used. The response spectra of the horizontal and vertical motions are plotted in Figs. 6-55 and 6-58b, respectively.

6.8.2 Love Waves

The dispersion curves are plotted in Fig. 6-71. The two dotted lines correspond to the shear-wave velocities of the top layer and of the half-space. From the decay factors shown in Fig. 6-72a for a horizontal distance of 500 m, it is apparent that the motion attenuates less than for the soft site (Fig. 6-60a). Besides the results for the first two Love modes, those for an inclined SH-wave with $\psi^R_{SH} = 10°$ in the half-space at $z = 120$ m are also indicated as a dashed

TABLE 6-5 Free-Field Properties of Rock Site

Depth, z (m)	Shear Modulus, G (GN/m²)	Mass Density, ρ (Mg/m³)	Shear Wave Velocity, c_s (m/s)	Poisson's Ratio, ν	Damping Ratio, ζ
0					
	3.0	2.6	1074	0.35	0.02
5					
	3.4	2.6	1144	0.35	0.02
10					
	5.0	2.6	1387	0.40	0.02
20					
	8.0	2.6	1754	0.40	0.02
30					
	12.0	2.6	2148	0.40	0.02
40					
	18.0	2.6	2631	0.35	0.02
50					
	27.0	2.6	3223	0.30	0.02
60					
	31.0	2.6	3453	0.29	0.02
80					
	38.5	2.6	3848	0.27	0.02
100					
	46.0	2.6	4206	0.26	0.02
120					
	50.0	2.6	4385	0.25	0.02
∞					

line. Associating the control motion at the free surface with Love waves, the first mode applies up to 15 Hz (where the second one starts) and the second mode for the remaining. The smaller horizontal attenuation is also visible in

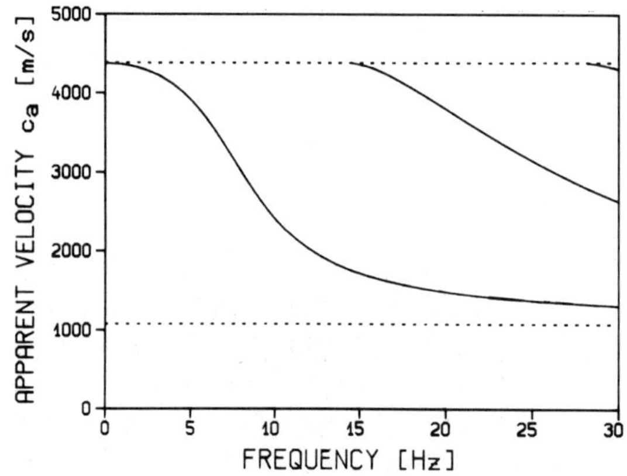

Figure 6-71 Dispersion, rock site, out-of-plane motion (Ref. [5]).

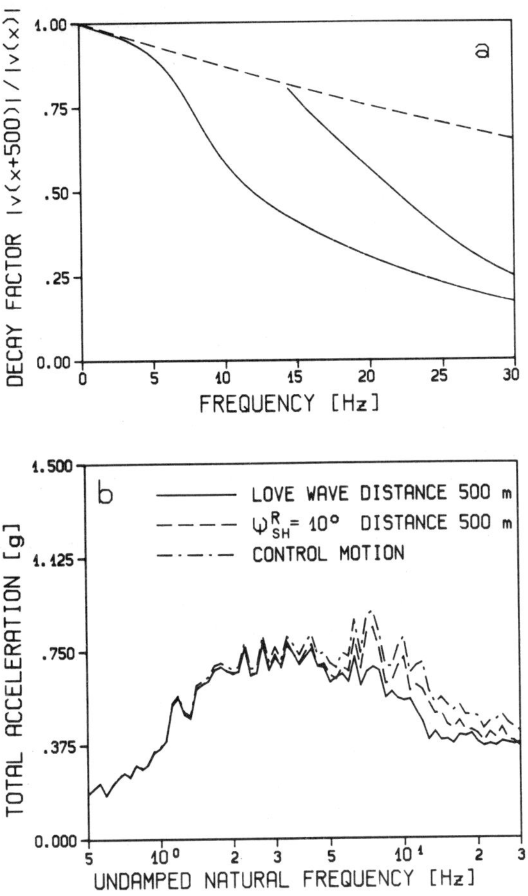

Figure 6-72 Attenuation, rock site, out-of-plane motion (Ref. [5]). (a) Decay factor 500 m distance; (b) response spectra (5% damping).

Fig. 6-72b. The variation with depth (which for a rock site is less important and which is not shown here) confirms the trends established for the soft site (Fig. 6-61).

6.8.3 Rayleigh Waves

The dispersion curves for the first two R-modes are shown in Fig. 6-73. The two dotted lines correspond to the shear-wave velocities of the top layer and of the half-space. Although damping is small, the motion in the higher-frequency range is attenuated considerably, as can be seen from the decay factor at a distance of 500 m from the origin in Fig. 6-74. For comparison, the plots for an incident SV-wave with $\psi^R_{SV} = 10°$ and for an incident P-wave $\psi^R_P = 10°$ are also shown as a dashed and a dotted line, respectively.

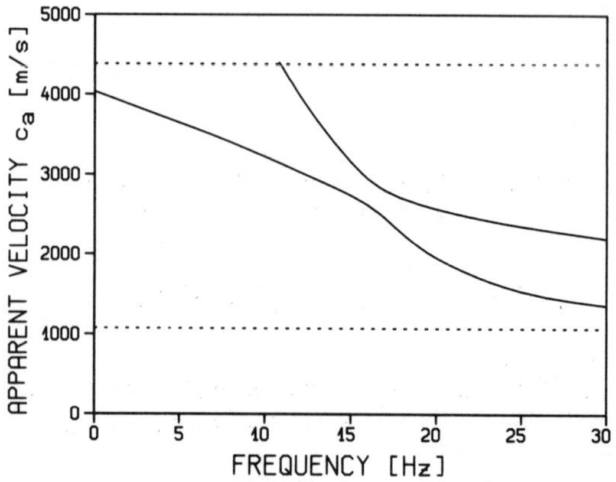

Figure 6-73 Dispersion, rock site, in-plane motion (Ref. [7]).

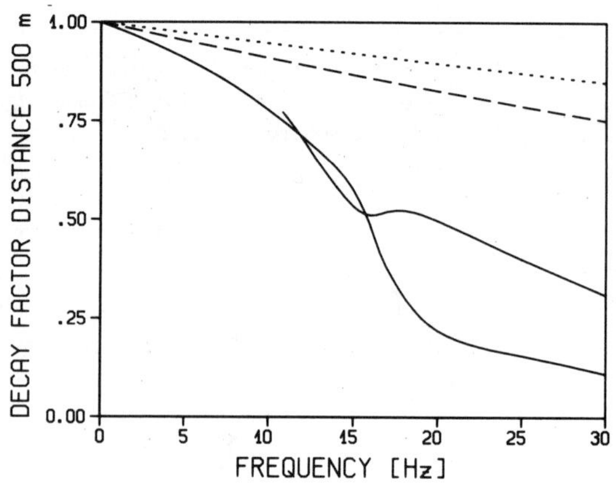

Figure 6-74 Attenuation, rock site, in-plane motion (Ref. [7]).

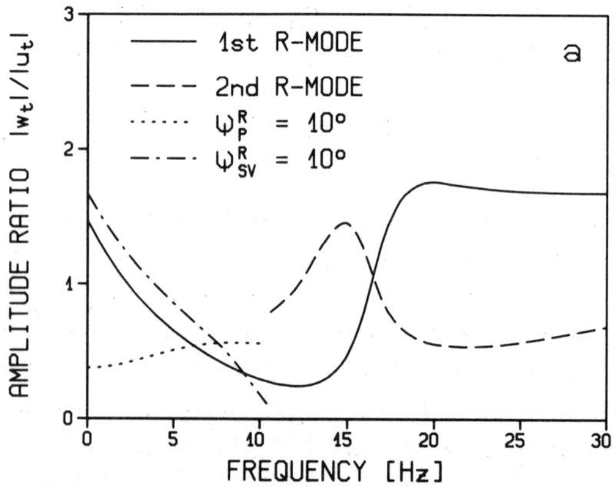

Figure 6-75 Motion at free surface, rock site, in-plane motion (Ref. [7]). (a) Amplitude ratio;

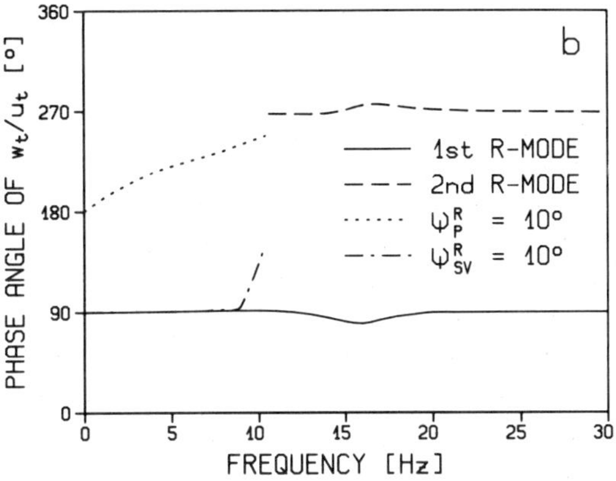

Figure 6-75 (Continued) (b) phase angle.

To study the characteristics of the motion for different wave assumptions, the amplitude ratio $|w_t|/|u_t|$ and the phase angle of w_t/u_t of the R-waves at the free surface are shown in Fig. 6-75a and b. The horizontal and vertical components of the first and second R-modes are almost perpendicular. For all frequencies, the motions of the first and second R-modes are nearly retrograde and prograde, respectively. For comparison, the motions of an incident SV-wave in the half-space with $\psi_{SV}^R = 10°$ and of an incident P-wave in the half-space with $\psi_P^R = 10°$ are also shown for the range up to the frequency at which the second R-mode starts. The motion of the SV-wave in this range is very similar to that of the first R-mode (Fig. 6-75).

6.8.4 Assumed Wave Patterns

The earthquake motion at the free surface is interpreted as a wave train made up, at least partially, of R-waves. Two schemes are investigated. First, the prescribed horizontal component is associated with that of the first R-mode. This defines the vertical component, which will, in general, not match the prescribed vertical-earthquake excitation. The corresponding horizontal seismic-input response spectra at the control point (located on the free surface) and at a distance of 500 m are plotted in Fig. 6-76. As expected from Fig. 6-74, the motion attenuates especially in the higher-frequency range. For comparison, the same horizontal motion of the free surface in the control point is associated with the horizontal motion of the incident SV-wave with $\psi_{SV}^R = 10°$. The response spectrum at the same distance shows that this wave train attenuates less. Love waves exhibit the same property (Fig. 6-72b). Second, when a second wave train is added to the first R-mode, both the prescribed horizontal- and vertical-earthquake excitations can be matched. For frequencies above 10.8 Hz, a second R-mode exists (Fig. 6-73) and can be included in the analysis. For frequencies below 10.8 Hz, only body waves are available in addition to the first R-mode. The obvious choice for this range of frequency consists of body

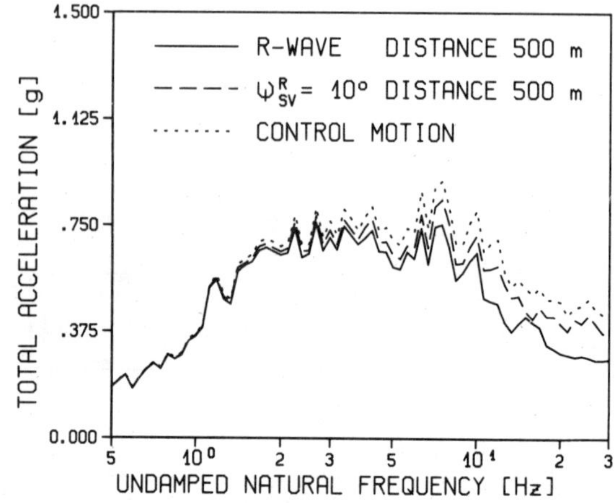

Figure 6-76 Horizontal response spectra (5% damping), rock site, free surface, motion first R-mode (Ref. [7]).

waves with the smallest apparent velocity [i.e., the SV-wave with $\psi^R_{SV} = 10°$ ($c_a = 4454$ m/s)]. As the motion of the SV-wave and the first R-mode are very similar (Fig. 6-75), the equations which have to be solved when simultaneously matching the horizontal and vertical components will be ill-conditioned. This results in large amplitudes of these two waves. The corresponding maximum accelerations of these two wave trains calculated separately equal approximately $2.5g$. This scheme thus has to be abandoned. Instead of the SV-wave, a P-wave with $\psi^R_P = 10°$ ($c_a = 7716$ m/s) is used for frequencies below 10.8 Hz. As is visible from Fig. 6-75, its motion is sufficiently different from that of the first

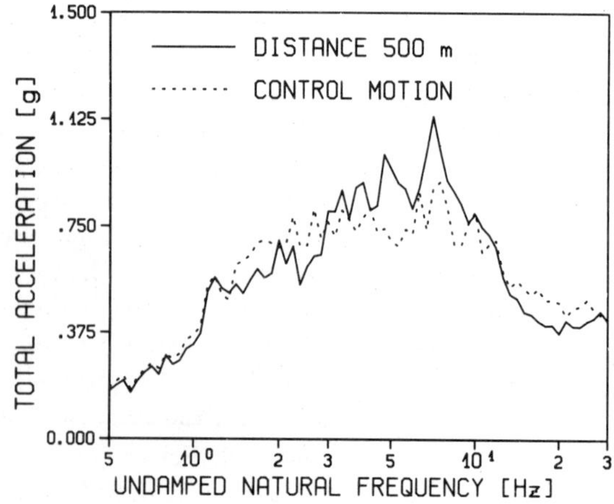

Figure 6-77 Horizontal response spectra (5% damping), rock site, free surface, motion R- and P-waves (Ref. [7]).

R-mode. The resulting motion will, however, be strongly dispersive because of the large differences in c_a. Up to approximately 10 Hz, the response spectrum at a distance of 500 m over- and undershoots that in the control point because of the dispersion of the motion (Fig. 6-77). In the higher-frequency range, the attenuation caused by damping again dominates.

6.8.5 Scattered Motion

Finally, the ratio $\omega a/c_a$, which determines the wave-passage effects, is calculated for this rock site. The radius $a = 30$ m is used for the results in Figs. 6-78 and 6-79. In Fig. 6-78, this ratio is specified for the first two Love modes.

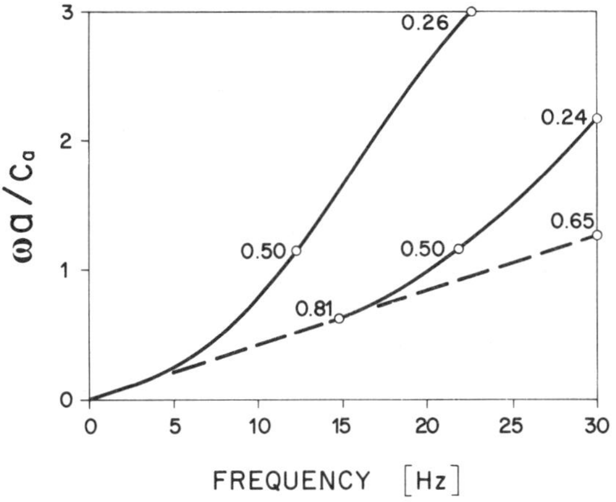

Figure 6-78 Applied seismic-input motion, rock site, out-of-plane motion (Ref. [5]).

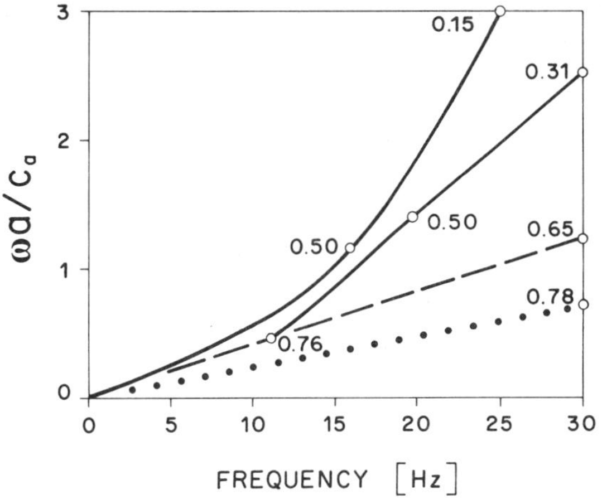

Figure 6-79 Applied seismic-input motion, rock site, in-plane motion (Ref. [7]).

For comparison, the results for an inclined SH-wave with $\psi^R_{SH} = 10°$ are also shown (dashed line). In Fig. 6-79, the ratio $\omega a/c_a$ for the first two R-modes is plotted. An incident SV-wave with $\psi^R_{SV} = 10°$ (dashed line) and an incident P-wave with $\psi^R_P = 10°$ (dotted line) are also included. The decay factors at a distance of 500 m are also indicated. The incident body waves lead to a smaller ratio $\omega a/c_a$ than for the soft site (Figs. 6-69 and 6-70), as c_a is significantly larger. For a specific frequency, the difference in the ratios for the Love-mode and the inclined SH-wave are more important, as the decay factor is smaller. The same applies to the difference in the ratios for the R-mode and the inclined SV-wave.

SUMMARY

1. The free-field-site response analysis represents the first and decisive step in any soil–structure interaction examination. Starting from the prescribed control motion acting in the selected control point and assuming the nature of the wave pattern, the free-field motion is determined on the line which subsequently will form the structure–soil interface.
2. As a broad-banded design-response spectrum or an historic earthquake record is used to define the control motion, it is meaningful to choose the control point either at the free surface of the site or at an assumed bedrock outcrop where the prescribed motion could conceivably exist. Under no circumstances can the control point be selected within the site at a specific depth.
3. In general, for a specific site, the mixture of the wave pattern is unknown because of the lack of data. Nor can it be determined analytically by modeling the source mechanism and the transmission path. Extreme cases of wave patterns thus have to be analyzed.
4. Using the familiar direct stiffness method and the substructure technique, a layered site's dynamic response caused by vertically incident and inclined body waves and surface waves is established. Assembling the dynamic-stiffness matrices of the layers and of the half-space leads to the left-hand side of the equations of motion of the site. The right-hand side is zero for the control point at the free surface. For the control point at the outcrop of the rock, the latter is regarded as the reference-soil system. The right-hand side is then equal to the product of the dynamic-stiffness matrix of the half-space and the vector of the control motion. Material damping is included in the formulation. Selecting the angle of incidence of the inclined body waves in the half-space determines the apparent velocity (and the wave number). For surface waves, the amplitudes of the incident waves in the half-space vanish, which results in a singular dynamic-stiffness matrix of the site. For a specific frequency, this condition is satisfied only for distinct phase velocities associated with the different modes. Surface waves are,

in general, dispersive. Inclined body waves and surface waves propagate horizontally across the site with an apparent velocity and attenuate. All variables, although in general complex, are interpreted physically.

5. At the free surface, an incident SH-wave is reflected as an SH-wave with the same amplitude. The horizontal out-of-plane displacement equals twice the amplitude of the SH-wave for all angles of incidence. An incident P-wave leads to a reflected P- and SV-wave for all angles of incidence. The resulting horizontal and vertical in-plane displacements are always 180° out of phase. An incident SV-wave leads to a reflected SV- and P-wave only for an angle of incidence larger than the critical one, with the horizontal and vertical in-plane displacements being in phase. For an angle of incidence of the incident SV-wave smaller than the critical angle, only an SV-wave is reflected and an additional plane wave is created which travels parallel to the free surface, decaying exponentially with depth. The in-plane motion at the free surface is prograde or retrograde for an angle of incidence of the SV-wave larger or smaller, respectively, than 45°.

6. For a half-space, no Love wave exists. There is only one (nondispersive) Rayleigh mode with a retrograde motion which propagates horizontally with an apparent velocity somewhat smaller than the shear-wave velocity and whose (frequency-independent) vertical displacement is larger than the horizontal one.

7. For an undamped layer built in at its base, the Love and Rayleigh modes start at the natural frequencies in the horizontal and in the horizontal/vertical directions, respectively. Thus, below the fundamental horizontal frequency, no surface waves exist. Introducing material damping does not change these properties from a practical point of view.

For a layer resting on a half-space, the dispersion curves of the Love modes start at the shear-wave velocity of the half-space (the first at zero frequency) and converge to the shear-wave velocity of the layer for increasing frequency. The first Rayleigh mode starts at zero frequency with the R-wave velocity of the half-space and converges to that of the layer. The higher R-modes exhibit cutoff frequencies, whereby the dispersion curves start with the shear-wave velocity of the half-space and converge to that of the layer. All surface waves decay exponentially with depth.

8. To identify the key features of the free-field response, a vast parametric study is performed, varying the location of the control point, the nature of the wave pattern, and the site properties. Harmonic and transient seismic excitations for a site consisting of a layer on bedrock and an actual soft site and a rock site are investigated.

9. The nature of the wave pattern producing the site response can be classified with respect to the apparent velocity of the motion propagating horizontally across the site as follows:

For the out-of-plane motion, the following applies:

(a) An infinite apparent velocity results from vertically incident SH-waves.
(b) A finite apparent velocity, which, however, is larger than the shear-wave velocity of the bedrock, arises from incident (inclined) body waves.
(c) A (frequency-dependent) apparent velocity which lies between the shear-wave velocity of the bedrock and that of the top layer occurs for Love waves. In contrast to (a) and (b), for a specific frequency, only distinct values of the apparent velocity associated with the different modes exist. The wave train is thus dispersive. In the horizontal direction, each Love mode attenuates more with increasing frequency.
(d) With an apparent velocity smaller than the shear-wave velocity of the top layer, no motion is possible.

For the in-plane motion, the following applies:

(a) An infinite apparent velocity results from vertically incident P- and SV-waves, which determine the vertical and horizontal motions, respectively.
(b) A finite apparent velocity which is larger than the compression-wave velocity of the bedrock arises from incident P-waves and from incident SV-waves (in the bedrock) with an angle of incidence larger than the critical angle of bedrock. For a layered site, the phase angle between the vertical and horizontal components depends on the frequency, but only weakly on the angle of incidence of the P-wave in the bedrock. For the SV-wave, the behavior is analogous. Arbitrary horizontal and vertical control motions can be interpreted as being caused by a combination of these two body waves. In contrast to vertically incident waves, the amplitudes of both waves depend as well on the horizontal as on the vertical motion.
(c) An apparent velocity bounded by the compression-wave velocity and the shear-wave velocity of the bedrock arises from incident SV-waves in the bedrock with an angle of incidence smaller than the critical angle of the bedrock. No incident P-waves exist in the bedrock. For a layered site, the phase angle is frequency dependent. The horizontal and vertical control motions cannot be independently prescribed.
(d) A (frequency-dependent) apparent velocity which lies between the shear-wave velocity of the bedrock and the Rayleigh-wave velocity of the top layer results from (generalized) Rayleigh waves. In contrast to body waves, for a specified frequency, only distinct values of the apparent velocity associated with the different modes exist. The wave train is thus dispersive. For a site without damping, the horizontal and vertical displacements are always 90° out of phase. The motion of the first Rayleigh mode is always retrograde in the low- and high-frequency ranges. In between, the motion can be prograde. Higher modes can exhibit retrograde and prograde motions. Introducing damping scarcely affects the ratio of the absolute values of the amplitudes of the motion

and the phase angle except when the damping ratio of the P-wave is different from that of the SV-wave. For a frequency below that at which the second mode starts, the horizontal and vertical motions cannot be independently prescribed by using only surface waves. For larger frequencies, higher modes exist which can be used to interpret arbitrary horizontal and vertical control motions.

(e) With an apparent velocity smaller than the Rayleigh-wave velocity of the top layer, no motion is possible.

10. Selecting the control point at the bedrock outcrop (which is meaningful only for body waves), the motion at the free surface can be calculated. The amplification of the motion from the outcropping bedrock to the free surface of the site depends on the properties of the whole site. The more the direction of the propagation of the SH-wave in the bedrock deviates from vertical incidence, the more this amplification of the out-of-plane motion (over the whole frequency range) decreases. The amplification of the vertical motion for incident P-waves depends only weakly (with a tendency of being somewhat larger compared to vertical incidence) on the angle of incidence in the bedrock. This applies for a large range of sites. The amplification of the horizontal motion for incident SV-waves is affected strongly by the angle of incidence, being larger or smaller than for vertical incidence. For combinations of P- and SV-waves, the amplification depends on the characteristic of the outcropping motion. For statistically independent outcropping horizontal and vertical control motions, the amplification in the vertical direction differs only slightly from that for vertical incidence, while that in the horizontal direction has a tendency to be smaller.

11. The variation of the motion with depth depends only on the properties of the site above the level where the motion is calculated. For the in-plane motion, this applies for a given ratio of the horizontal- and vertical-displacement amplitudes at the free surface and for a given apparent velocity (which, of course, have to be compatible with the assumed wave pattern). For the out-of-plane motion, the nature of the wave pattern is of no practical significance for the variation of the displacement with depth. For the in-plane motion, the fundamental frequencies in the horizontal and vertical directions of the site fixed at this level are important when discussing the variation with depth. Assuming that the motion results from incident P-waves in the bedrock, the amplification of the vertical motion within the site up to this vertical fundamental frequency is practically independent of the angle of incidence. This no longer holds for SV-waves for shallow angles of incidence for the amplification of the horizontal motion. Interpreting the two components of the motion at the free surface as arising from a combination of incident P- and SV-waves leads to amplifications within the site for the horizontal and vertical directions which do not depend, from a practical point of view, on the angle of incidence and on the prescribed motion up to the fundamental frequency in the horizontal and vertical directions, respectively. The variation of the motion for the first R-mode with depth in the two directions bears comparison, in general, with that of vertically incident waves up to the corresponding fundamental frequency. The higher R-modes lead to a completely different variation with depth. The agreements stated in this paragraph do not apply to an elastic half-space.

For inclined SH-waves and for Love waves, additional (horizontal) shear stresses acting on vertical planes arise. The variation of the maximum shear stress with depth is very similar to that for vertical incident SH-waves, with the exception of the immediate vicinity of the free surface. For the in-plane motion, horizontally propagating waves lead to significant normal stresses acting in the horizontal direction. Large maximum shear stresses appear close to the free surface. These will be more pronounced for sites approaching the elastic half-space.

12. For inclined body waves and surface waves, the motion varies in the horizontal direction.

(a) In contrast to surface waves, the attenuation of the motion for inclined body waves depends on the damping of the bedrock and not on that of of the layers. Surface waves thus attenuate more strongly than do body waves, especially for soft sites and in the higher-frequency range. Generally, higher surface modes attenuate less than the first one does.

(b) The apparent velocity depends only weakly on the damping. A finite apparent velocity results in a reduced horizontal and in an additional torsional seismic-input component for the out-of-plane motion and in a reduced horizontal and vertical and in an additional rocking seismic-input component for the in-plane motion.

13. For the out-of-plane motion, the control motion is matched with the component of the inclined body wave or, using Love waves, is associated with a certain mode up to the frequency where the next higher mode starts. This concept is appropriate, as each Love mode decays in the horizontal direction more for increasing frequency.

For the in-plane motion, if only one component of the control motion (e.g., the horizontal) is matched, it can be associated either with a body wave or with an R-wave. In the latter case, a specific mode is used up to the frequency at which the next higher mode starts, since for a given frequency, the higher modes attenuate less. The other component of the motion follows. If both components are prescribed, the motion can be interpreted as arising from a combination of a P- and an SV-wave (with a common apparent velocity). Surface waves alone cannot be used to match both components. A body wave has to be included, at least up to the frequency at which the second mode starts. As the motions of the shallow SV- and first Rayleigh mode are very similar in this range of frequency, a P-wave should be used. For the control motion specified at a fictitious rock outcrop, only body waves can be assumed.

14. For the soft site, the surface waves decay significantly, especially in the range of higher frequencies, where the apparent velocity is considerably smaller than the shear-wave velocity of the rock. For this site, it seems sufficient to examine only (extremely shallow) body waves. This does not apply to a structure–soil system whose fundamental frequency is small, as below 1 to 2 Hz the decay of the surface waves is small, even in soft soil. For a rock site, however, throughout the frequency range of interest, surface waves exist which attenuate little, leading to smaller apparent velocities than those of extremely shallow body waves.

PROBLEMS

6.1. A discrete model to calculate the free-field motion from vertically incident S-waves of an undamped site consisting of soil layers resting on elastic rock is shown in Fig. P6-1. The control motion v_o is assumed to act at an assumed rock outcrop. Derive expressions for all variables of the model.

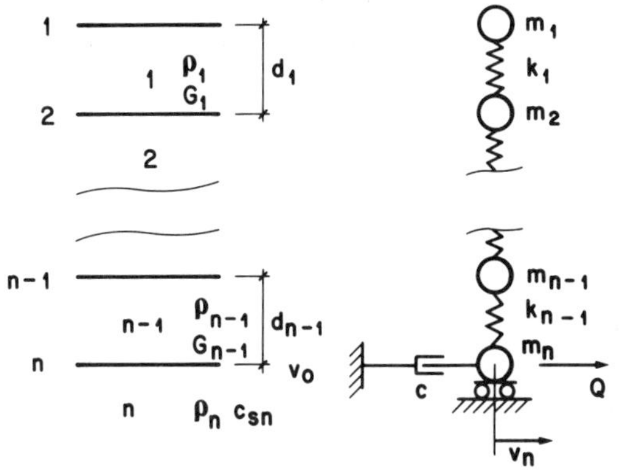

Figure P6-1 Discrete soil model for vertically incident S-waves.

Solution:

For vertically incident waves, the dynamic-stiffness matrices of a layer and of the half-space are specified in Eq. 5.106. Substituting $\sin \omega d/c_s \simeq \omega d/c_s$ and $\cos \omega d/c_s \simeq 1 - \omega^2 d^2/(2c_s^2)$ leads to

$$[S_{SH}^L] = \frac{G}{d}\begin{bmatrix} 1 & -1 \\ -1 & 1 \end{bmatrix} - \frac{\omega^2 \rho d}{2}\begin{bmatrix} 1 & \\ & 1 \end{bmatrix}$$

The spring constant is thus determined as $k_1 = G_1/d_1$, the masses $m_1 = \rho_1 d_1/2$, $m_2 = (\rho_1 d_1 + \rho_2 d_2)/2$, and so on. The rock's contribution is equal to

$$S^R_{SH} = iG_n \frac{\omega}{c_{sn}}$$

which is equal to a damper with a coefficient $c = \rho_n c_{sn}$. The equations of motion are formulated in Eq. 6.2, with the only load (at node n) specified in Eq. 6.3:

$$Q = iG_n \frac{\omega}{c_{sn}} v_o = \rho_n c_{sn} \dot{v}_o$$

The equation of motion of node n is thus formulated in the time domain as

$$\frac{\rho_{n-1} d_{n-1}}{2} \ddot{v}_n + \frac{G_{n-1}}{d_{n-1}} (v_n - v_{n-1}) + \rho_n c_{sn} \dot{v}_n = \rho_n c_{sn} \dot{v}_o$$

where v_n is the total displacement (as a function of time) at node n and $\dot{v}_o$ is the velocity of the outcropping motion.

6.2. Show that the Rayleigh-wave equation of the undamped half-space has three real roots for Poisson's ratio ν less than 0.26 and one real root plus a conjugate pair for $\nu > 0.26$. Verify that in both cases only one of the roots satisfies the condition that the corresponding motion decays with depth.

Hints:

Squaring Eq. 6.16 and factoring the quantity c^2/c_s^2 leads to the cubic equation

$$x^3 - 8x^2 + x\left(24 - 16\frac{c_s^2}{c_p^2}\right) + 16\left(\frac{c_s^2}{c_p^2} - 1\right) = 0$$

where $x = c^2/c_s^2$. (If the sign of one of the two terms in Eq. 6.16 is changed, this modified equation will also satisfy the cubic equation.) For the motion to correspond to a surface wave, the factors of B_P and B_{SV} must decay with depth (Eq. 5.130). This results in the imaginary ks and kt being negative, which is the case if $c < c_s$.

6.3. Using the rigorous expressions for the dynamic-stiffness matrices, the eigenvalue problem that has to be solved to calculate surface waves is transcendental. If the site is built in at its base (i.e., no half-space has to be modeled), and the dynamic-stiffness matrices are applied in discrete form (Problems 5.9 and 5.10), the resulting eigenvalue problem will be quadratic. From a computational point of view, this is an advantage.

For a single undamped homogeneous layer built in at its base, determine the equation of the dispersion curve (c/c_s versus $\omega d/c_s$) of the first Love mode using only one discrete element. Compare with the rigorous solution plotted in Fig. 6-11.

Solution:

The dynamic-stiffness coefficient of the single layer in discrete form (Problem 5.9) is as follows:

$$\left(\frac{G}{d} + \frac{k^2 dG}{3} - \omega^2 \frac{\rho d}{3}\right) v = 0$$

Setting the coefficient equal to zero leads to

$$\frac{c^2}{c_s^2} = \frac{1}{1 - 3c_s^2/\omega^2 d^2}$$

while the rigorous solution derived from Eq. 6.38 is equal to

$$\frac{c^2}{c_s^2} = \frac{1}{1 - (\pi^2/4)(c_s^2/\omega^2 d^2)}$$

7

MODELING OF SOIL

7.1 GENERAL CONSIDERATIONS

7.1.1 Dynamic-Stiffness Matrices of Soil

The soil medium is one of the two substructures in the analysis of soil–structure interaction. Its model will result in the force–displacement relationship of the soil defined at the structure–soil interface. This dynamic-stiffness matrix of the unbounded soil $[S^g_{bb}]$ in the nodes b is referred to the soil system g, as illustrated in Fig. 7-1 (ground with excavation). The matrix $[S^g_{bb}]$ enters directly in the coefficient matrix of the basic equation of motion (Eq. 3.9).

Modeling the excavated part of the soil (reference subsystem e) is straightforward, as it is a bounded domain. A standard finite-element discretization of the excavated part results in the dynamic-stiffness matrix $[S]$ (Eq. 2.15):

$$[S] = [K](1 + 2\zeta i) - \omega^2[M] \tag{7.1}$$

The matrix $[S]$ is decomposed into the submatrices $[S_{ii}]$, $[S_{ib}]$, and $[S_{bb}]$. The subscript b refers to the nodes on the structure–soil interface, the subscript i to the others (Fig. 7-1). Eliminating the degrees of freedom at nodes i leads to

$$[S^e_{bb}] = [S_{bb}] - [S_{bi}][S_{ii}]^{-1}[S_{ib}] \tag{7.2}$$

The superscript e is added to denote this dynamic-stiffness matrix of the total excavated soil referred to the nodes b.

Adding $[S^e_{bb}]$ to $[S^g_{bb}]$ results in the dynamic-stiffness matrix of the continuous soil $[S^f_{bb}]$, discretized in the same nodes b which will subsequently lie on the structure–soil interface (Eq. 3.6, Fig. 3-3). This matrix $[S^f_{bb}]$ appears in the load

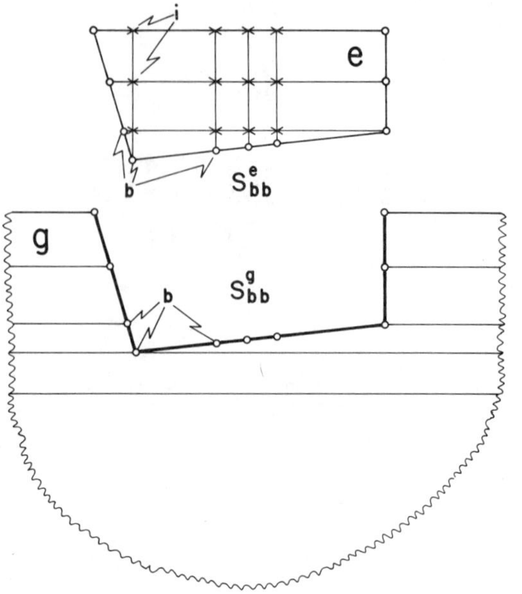

Figure 7-1 Dynamic-stiffness matrices of soil.

vector of the basic equation of motion (Eq. 3.9). The matrix $[S^g_{bb}]$ thus also enters indirectly on the right-hand side. Even to calculate the scattered motion $\{u^g_b\}$, no other matrix of the soil is needed (Eq. 3.12).

Similarly, for a rigid base, $[S^g_{bb}]$ captures all the dynamic properties of the soil. The corresponding dynamic-stiffness matrix of the soil $[S^g_{oo}]$ and the scattered motion $\{u^g_o\}$ follow from Eqs. 3.16c and 3.19, respectively, whereby for both, the rigid-body constraints are enforced.

An alternative procedure exists which from a computational point of view can be advantageous. As the soil subsystem f is a regular domain in contrast to g, calculating $[S^f_{bb}]$ will be simpler to perform than $[S^g_{bb}]$. This is demonstrated using the so-called boundary-element method in Section 7.5. The matrix $[S^g_{bb}]$ is then determined by subtracting $[S^e_{bb}]$ from $[S^f_{bb}]$.

The soil medium on the exterior of the line with the nodes b is assumed to consist of a layered half-space. If an irregular soil region surrounding the structure exists, this bounded region is regarded as part of the expanded structure. This concept is discussed in connection with Fig. 3-8.

To review the key aspects of modeling the unbounded soil, it is instructive to discuss briefly some procedures that model the soil region with, for example, finite elements. In the method developed later in this text, this is not necessary. As it is impossible to cover the unbounded domain with a finite number of elements with bounded dimensions, an artificial boundary has to be created for modeling purposes. This is illustrated in Fig. 7-2. Appropriate boundary conditions have to be formulated which must represent the missing soil. Besides

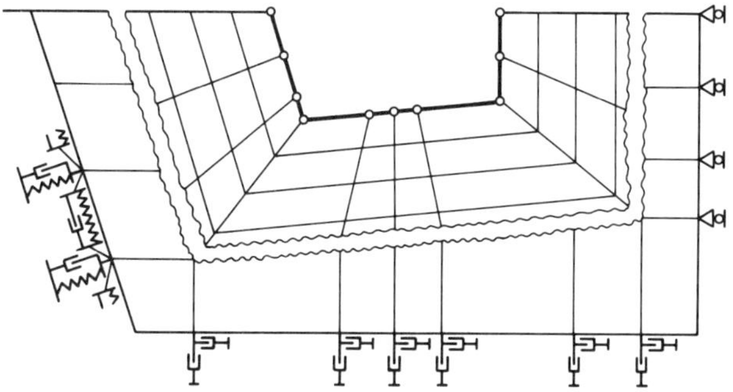

Figure 7-2 Finite-element mesh of soil terminated with artificial boundary: consistent, local (viscous), and elementary.

modeling its stiffness, reflections of the outwardly propagating waves at the artificial boundary have to be avoided. Conceptionally, the boundary conditions can be visualized as being determined from the elimination of all dynamic degrees of freedom not lying on the boundary of a mesh extending to infinity. Of course, they cannot really be calculated this way. Besides depending on the type of boundary condition enforced, the location of the artificial boundary is a function of the level of the material damping of the soil, the frequency range of interest, the wave velocity, and the duration of the excitation.

The following survey is perforce incomplete. The formulated boundary conditions can be classified into three groups.

7.1.2 Elementary Boundaries

The first contains the elementary boundary conditions. The soil mesh is brutally truncated at the artificial boundary, where either a zero displacement or a zero surface traction is enforced. They act as perfect reflectors for an impinging wave, not transmitting or absorbing any energy. The trapped energy in the dynamic system can lead to disastrous results. To improve the situation somewhat, the material damping of the modeled soil is artificially increased to ensure that the amplitudes of the reflected waves are significantly reduced before reaching the structure–soil interface. As discussed in Section 5.1.9, the use of unrealistically high damping will significantly influence the dynamic stiffness of the soil and will thus result in an erroneous structural response. In addition, the convergence of the dynamic stiffness of the finite domain to that of the unbounded one is slow and oscillatory in nature. One type of elementary boundary is indicated schematically on the vertical boundary in Fig. 7-2.

7.1.3 Local Boundaries

The second group contains the local boundaries. The boundary conditions do not couple the degrees of freedom of the nodes located on the artificial boundary. Usually, in each node and for each degree of freedom, they consist of a viscous dashpot, in more general cases of a (frequency-dependent) dynamic stiffness, which can be interpreted as a spring and a dashpot (analogous to the case of the rod with exponentially increasing area discussed in Section 5.1.6). To examine the essential features of these local boundaries, the out-of-plane motion of plane waves in an undamped layered half-space is used for illustration. The horizontal lower part of the artificial boundary (Fig. 7-2) is addressed first. An outgoing SH-wave with an angle of incidence ψ_{SH} and with an amplitude B_{SH} is assumed to propagate in the positive z-direction as indicated in Fig. 7-3a. At a given depth, an artificial boundary is introduced which completely absorbs the wave. (It is convenient to select the origin of the coordinate system on this boundary.) A reflected wave with an amplitude A_{SH} thus may not occur. The out-of-plane displacement with the amplitude v and the shear stress acting on a horizontal plane with the amplitude τ_{yz} are specified as functions of A_{SH} and B_{SH} in Eqs. 5.98 and 5.100. Solving these two equations at $z = 0$ for the amplitudes results in

$$A_{SH} = \frac{1}{2}\left(\frac{\tau_{yz}}{iktG} + v\right) \tag{7.3a}$$

$$B_{SH} = \frac{1}{2}\left(-\frac{\tau_{yz}}{iktG} + v\right) \tag{7.3b}$$

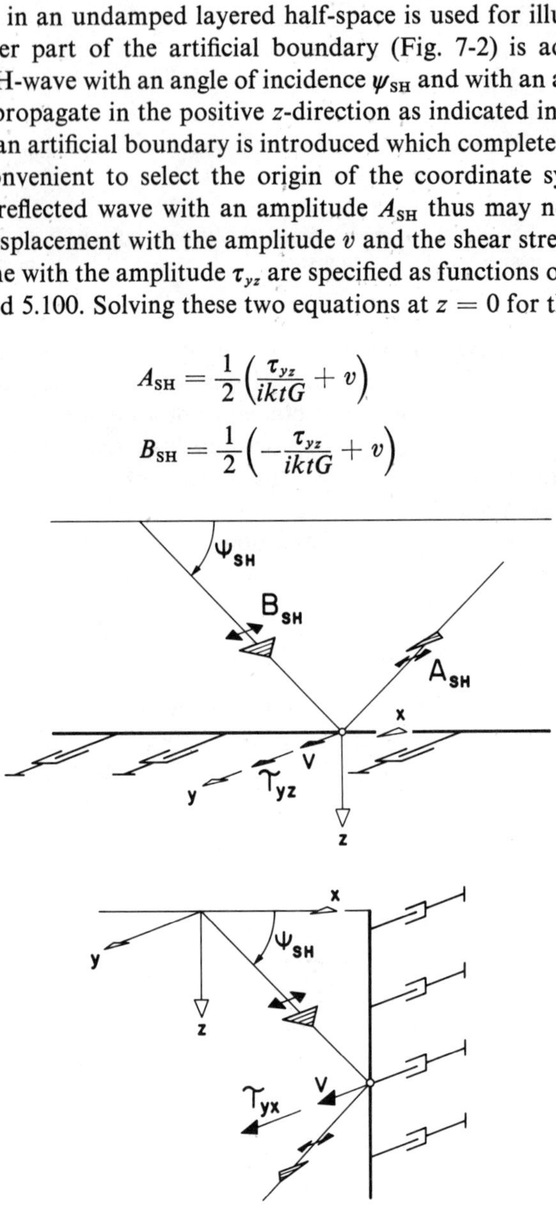

Figure 7-3 Viscous boundary for out-of-plane motion. (a) Horizontal; (b) vertical.

where the wave number k and the parameter t are specified in Eqs. 5.94 and 5.95b. The shear modulus is denoted by G.

Setting $A_{SH} = 0$ leads to

$$\tau_{yz} = -iktGv \tag{7.4}$$

Substituting Eqs. 5.94 and 5.120, Eq. 7.4 is reformulated as

$$\tau_{yz} = -\frac{t}{c}G\dot{v} \tag{7.5}$$

The value c denotes the phase velocity. This equation corresponds to the force–velocity relationship of a dashpot with a damping coefficient c_1:

$$c_1 = \frac{t}{c}G \tag{7.6}$$

For a positive $\dot{v}$, the damping force will act in the negative y-direction, as is expressed by the negative sign in Eq. 7.5. Formulating c and t as a function of $m_x = \cos \psi_{SH}$ using Eqs. 5.93 and 5.95 results in

$$c_1 = \sqrt{1 - m_x^2}\,\frac{G}{c_s} = \sin \psi_{SH}\frac{G}{c_s} \tag{7.7a}$$

or

$$c_1 = \sin \psi_{SH} \rho c_s \tag{7.7b}$$

The damping coefficient which achieves full absorption of an incident SH-wave is a function of the angle of incidence. Since along the lower artificial boundary, waves with varying ψ_{SH} will impinge, such a local boundary consisting of a specific dashpot will always reflect some part of the waves. For a vertically incident wave,

$$c_1 = \rho c_s \tag{7.8}$$

a result which is analogous to that of the prismatic rod in Section 5.1.6 (Eq. 5.35). Selecting the damping coefficient to absorb completely a vertically incident wave,

$$\tau_{yz} = -i\omega\rho c_s v \tag{7.9}$$

follows from Eqs. 7.8 and 5.120. Substituting Eq. 7.9 in Eq. 7.3, the ratio of the amplitudes of the reflected and incident waves results, using Eqs. 5.93 and 5.95, as

$$\frac{A_{SH}}{B_{SH}} = \frac{-1 + \sin \psi_{SH}}{1 + \sin \psi_{SH}} \tag{7.10}$$

A better measure for judging the effectiveness of this damper is the ratio of the absorbed rates of energy transmissions. From Eqs. 7.3 and 5.120, it is deduced that

$$\begin{aligned} \dot{v} &= i\omega(A_{SH} + B_{SH}) \\ \tau_{yz} &= iktG(A_{SH} - B_{SH}) \end{aligned} \tag{7.11}$$

It follows that $\dot{v}$ and τ_{yz} are imaginary. Equation 5.122 for the rate of energy transmission applies to a plane x = constant. In analogy to the case in which a plane z = constant, τ_{yx} is replaced by τ_{yz}. It follows that the absorbed rate of energy transmission in a point of the artificial boundary is proportional to the product of the imaginary parts of τ_{yz} and $\dot{v}$, that is, to $-(A_{SH} + B_{SH})(A_{SH} - B_{SH})$, which is equal to $B_{SH}^2 - A_{SH}^2$. As for full absorption $A_{SH} = 0$, A_{SH}^2/B_{SH}^2 represents the ratio of the rates of energy transmissions of the reflected and incident waves. Whereas for $\psi_{SH} = 45°$, for example, the amplitude of the reflected wave is significant ($A_{SH} = 0.172B_{SH}$), this no longer applies to its energy ($A_{SH}^2 = 0.030B_{SH}^2$).

The vertical boundary (Fig. 7-2) is examined next. The shear stress acting on a plane x = constant with the amplitude τ_{yx} is specified in Eq. 5.112b as (Fig. 7-3b)

$$\tau_{yx} = -ikGv = -\frac{G}{c}\dot{v} \tag{7.12}$$

This results in a damping coefficient c_1:

$$c_1 = \frac{G}{c} \tag{7.13}$$

For an inclined SH-wave, by using Eq. 5.93, c_1 is reformulated as

$$c_1 = \cos\psi_{SH}\frac{G}{c_s} \tag{7.14}$$

which is the same expression as in Eq. 7.7a for the horizontal boundary, taking the change in the orientation of the boundary into account. For a Love wave, it follows from Eq. 7.13 that c_1 is frequency dependent, as the phase velocity c depends on ω (dispersion). This means also that for a given frequency, a different damping coefficient is required for each mode to achieve full absorption.

Analogously, local boundaries can be constructed for the in-plane motion. These viscous dampers arranged at right angles to each other are shown on the horizontal boundary in Fig. 7-2 (see Problem 7.2).

Local boundaries can also be established using so-called infinite elements. These "finite" elements extend, as the name suggests, to infinity. Standard shape functions are multiplied with terms such as exp $(-s)$ and exp $(-iks)$, where s is the coordinate extending to infinity. As in the case of the viscous boundary described above, the form of the wave characterized by the wave number k has to be known a priori.

For a vertical boundary lying on the surface of a cylinder the following procedure can also be used. The adjacent soil layer is assumed to be composed of a series of infinitesimally thin independent layers. Based on the equations of motion formulated in cylindrical coordinates (Section 5.5), the local dynamic-stiffness coefficients of the independent layer extending to infinity can be established. These are complex and frequency dependent for all degrees of freedom (horizontal, vertical, torsional). See Problem 7.15. For this case of an embedded cylinder, the local boundary along the base is derived from an elastic half-space which is assumed to model the soil under the base of the cylinder.

7.1.4 Consistent Boundaries

The third group consists of the consistent boundaries. They are able to absorb perfectly all kinds of waves, that is, all types of body waves with varying angles of incidence and all surface waves. All impinging waves are thus transmitted fully without any reflections occurring. As is to be expected, all degrees of freedom on the artificial boundary are coupled. The force–displacement relationship on a consistent boundary is frequency dependent and can be visualized as a coupled spring–dashpot system with frequency-dependent coefficients. This is illustrated on the inclined boundary in Fig. 7-2. For each frequency, all possible wave types, characterized by the wave number k, are considered. As the consistent boundaries transmit perfectly all waves that can occur in the soil medium, the boundary can be placed directly on the structure–soil interface. No finite-element mesh of the (regular) soil is thus needed. In the following, the modeling of the soil with consistent boundaries is discussed in depth. The boundary integral-equation procedure, also called boundary-element method in discretized form, is used to develop the dynamic-stiffness matrix $[S_{bb}^{g}]$ for any shape of the structure–soil interface. The methods of the other two groups are not dealt with any further.

7.1.5 Sommerfeld's Radiation Condition

In Section 5.1.6 it is demonstrated that for the infinite rod with exponentially increasing area, it is not sufficient for the displacement to die out at infinity to determine a unique solution. Formulating the so-called radiation condition excludes the incoming wave, resulting in a unique solution. This boundary condition to be satisfied at infinity in an unbounded domain is further discussed for a spherically symmetric solution of the harmonic-wave equation in three dimensions. Material damping is disregarded. For polar symmetry, all variables depend on the radial coordinate r only. The amplitude of the only displacement, which arises in the radial direction, is denoted as u. The radial strain with amplitude ϵ_r and the normal strain in any direction perpendicular to r with amplitude ϵ_θ are specified as

$$\epsilon_r = u,_r \tag{7.15a}$$

$$\epsilon_\theta = \frac{u}{r} \tag{7.15b}$$

The dynamic-equilibrium equation for harmonic excitation is formulated as

$$\sigma_{r,r} + \frac{2}{r}(\sigma_r - \sigma_\theta) = -\rho\omega^2 u \tag{7.16}$$

where σ_r and σ_θ are the amplitudes of the stresses in the corresponding directions. Substituting Eq. 7.15 into Hooke's law and then introducing the stresses in Eq. 7.16 leads to the equation of motion

$$u,_{rr} + \frac{2}{r}u,_r - \frac{2u}{r^2} = -\frac{\omega^2}{c_p^2}u \tag{7.17}$$

where the dilatational-wave velocity c_p is defined in Eq. 5.72. Introducing the wave number k,

$$k = \frac{\omega}{c_p} \tag{7.18}$$

Eq. 7.17 is rewritten as

$$u_{,rr} + \frac{2}{r}u_{,r} - \frac{2u}{r^2} + k^2 u = 0 \tag{7.19}$$

Analogously to the case of cylindrical coordinates in Section 5.5.1, the displacement is expressed as a function of the potential φ as

$$u = \varphi_{,r} \tag{7.20}$$

Substituting Eq. 7.20 into Eq. 7.19 leads to the wave equation expressed in the potential. This equation is satisfied if the following equation applies for the product $r\varphi$:

$$(r\varphi)_{,rr} = -k^2 r\varphi \tag{7.21}$$

The solution of this one-dimensional wave equation is

$$\varphi = a\frac{\exp(-ikr)}{r} + b\frac{\exp(ikr)}{r} \tag{7.22}$$

The displacement amplitude u follows from Eq. 7.20 as

$$u = a\left(-\frac{1}{r^2} - \frac{ik}{r}\right)\exp(-ikr) + b\left(-\frac{1}{r^2} + \frac{ik}{r}\right)\exp(ikr) \tag{7.23}$$

It should be remembered that the displacement equals the product of the corresponding amplitude and the factor $\exp(+i\omega t)$. As the expression $\exp[i\omega(t - r/c_p)]$ describes a wave propagation in the positive radial direction with the velocity c_p, the first term in Eq. 7.23 corresponds to an outgoing spherical wave. Analogously, the second is associated with an incoming wave. Both waves, which are independent, attenuate with increasing radius.

Consider two points on the boundary of a bounded domain determined by r_1 and r_2. Enforcing a zero displacement in the two points will result in vanishing integration constants a and b. The solution is identical to zero throughout the domain. The solution for specified boundary conditions is thus unique. This no longer applies for an unbounded domain, as u dies out at infinity (Eq. 7.23). The solution

$$\frac{u}{a} = \left(-\frac{1}{r^2} - \frac{ik}{r}\right)\exp(-ikr) + \frac{1/r_1^2 + ik/r_1}{-1/r_1^2 + ik/r_1}\left(-\frac{1}{r^2} + \frac{ik}{r}\right)\exp[ik(r - 2r_1)] \tag{7.24}$$

vanishes at $r = r_1$ (and at $r_2 = \infty$), but is not identical to zero. To enforce a unique solution, the radiation condition, which excludes the incoming wave, but includes the outgoing one, has to be formulated at infinity. This so-called Sommerfeld radiation condition is derived as follows: As the term $1/r^2$ decays faster than $1/r$, only the second term has to be considered [i.e., $(1/r)\exp(-ikr)$

for the outgoing wave]. Taking the derivative with respect to r results in

$$\left(\frac{\exp(-ikr)}{r}\right)_{,r} = -\frac{ik}{r}\exp(-ikr) - \frac{\exp(-ikr)}{r^2} \tag{7.25}$$

Multiplying by r and observing that the second term on the right-hand side vanishes for large r compared to the first leads to

$$\lim_{r\to\infty} r\left[\left(\frac{\exp(-ikr)}{r}\right)_{,r} + ik\frac{\exp(-ikr)}{r}\right] = 0 \tag{7.26}$$

The displacement amplitude u of the outgoing wave will thus satisfy

$$\lim_{r\to\infty} r(u_{,r} + iku) = 0 \tag{7.27}$$

This radiation condition is violated by the displacement amplitude of the incoming wave, as can easily be verified by substituting $(1/r)\exp(ikr)$ for u. Formulating the radiation condition (Eq. 7.27) will thus lead in an unbounded domain to a boundary-value problem with a unique solution. Equation 7.27 applies to a three-dimensional situation. In two dimensions, an analogous relationship exists for the radiation condition which again excludes the incoming wave (see Problem 7.15).

7.1.6 Use of Analytical Solution

The exclusion of the incoming wave is virtually impossible to achieve numerically on the artificial boundary of a truncated finite-element mesh. In an analytical solution, it is, however, in general, quite straightforward. These analytical solutions, which satisfy all differential equations exactly in addition to the radiation condition, can be regarded as shape functions in a numerical scheme. Addressing specifically the task of calculating the dynamic-stiffness matrix $[S^g_{bb}]$ illustrated in Fig. 7-1, analytical solutions for the continuous layered half-space without the embedment (reference soil system f) can be constructed. These will also satisfy exactly the boundary conditions between two adjacent layers and at the free surface, but not those on the structure–soil interface. The latter can be enforced only in some average sense. On the exterior of the structure–soil interface (i.e., on that part of the system f that will subsequently be excavated), fictitious loads with unknown amplitudes are assumed to act. The corresponding amplitudes of the displacements on the line associated with the structure–soil interface can be calculated analytically. The amplitudes of the fictitious loads are then determined so as to satisfy in an average sense the prescribed displacement conditions of the definition of the stiffness matrix on the structure–soil interface. This represents the concept of the boundary integral-equation method, where the discretization is restricted to the irregular boundary (i.e., the structure–soil interface). This numerical scheme to model the unbounded domain will thus be accurate and efficient.

Surface foundations are addressed first: The computational procedure to calculate the dynamic-stiffness matrix in two- and three-dimensional situations is derived in Section 7.2. Parametric studies for a rigid strip and for a rigid circular basemat are discussed in Sections 7.3 and 7.4, respectively. A half-plane (space), a single layer resting on a half-plane (space), and a layer built in at its base are analyzed. The scattered motion is addressed in Section 7.4.3. Concepts of the boundary integral-equation method are contained in Section 7.5. Results of the dynamic-stiffness matrix for embedded foundations are presented in Section 7.6. Finally, the dynamic stiffnesses of two basemats of irregular shape on an actual site are calculated in Section 7.7, which are used for studying the through-soil-coupling effect of two adjacent structures.

7.2 DYNAMIC-STIFFNESS COEFFICIENTS OF SURFACE FOUNDATION

7.2.1 Weighted Residual Formulation

The calculation of the dynamic-stiffness coefficients [of a surface foundation (Fig. 7-4)] represents a mixed boundary-value problem. Over the structure–soil interface, the displacements are prescribed resulting from the definition of the

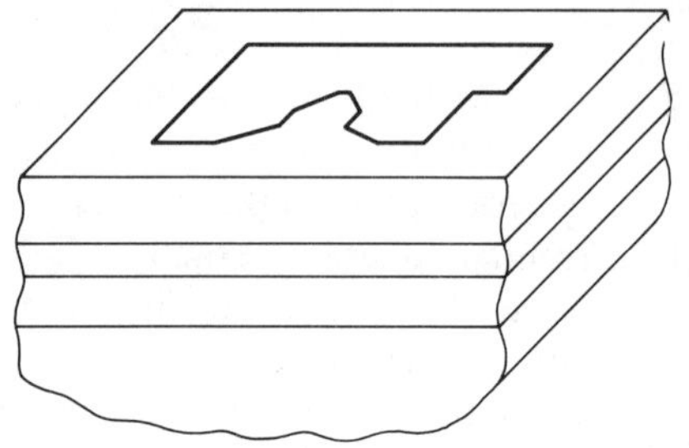

Figure 7-4 Surface foundation on layered half-space.

dynamic stiffness so as to achieve the continuity of displacements between structure and soil. Surface tractions (stresses) of zero value have to be enforced over the remainder of the surface. To reduce the problem to one where only surface tractions are prescribed on the surface, the structure–soil interface is discretized into elements (Fig. 7-5a). The corresponding surface tractions can then be expressed as a load to be used in the direct-stiffness approach of structural analysis applied to the site. The actual value of the surface traction, which is equal to the load arising from the base and acting on such an element of soil is unknown, but the distribution of the load expressed as a function of the unknown nodal values can be chosen. For the triangular element shown in Fig. 7-5b, a linear distribution of the loading is selected, which is defined by the nodal values in the corners. For the sake of clarity, only the vertical component is shown. In each node, three and two components are present in the three-dimensional and two-dimensional cases, respectively. For a quadrilateral element, a bilinear expansion can be chosen. As the nodal points are associated with the individual

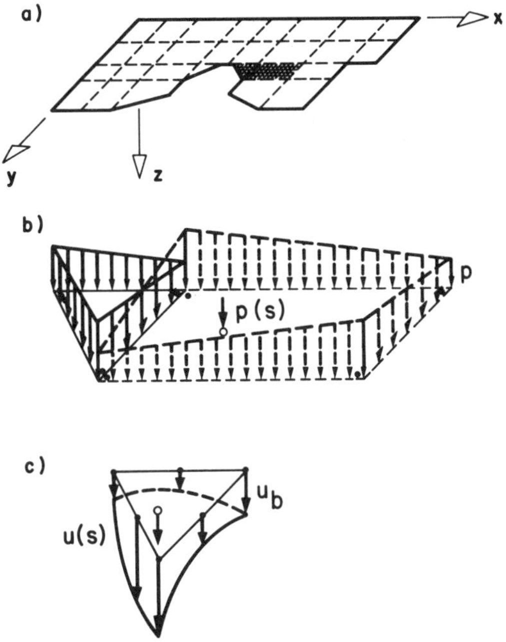

Figure 7-5 Discretized structure–soil interface. (a) Elements; (b) applied loads; (c) prescribed displacements.

elements, discontinuities in the loading across the sides of the element will, in general, arise with this definition. This is shown in Fig. 7-5b. For the examples discussed in Sections 7.3 and 7.4, a constant load is selected over each element. The unknown nodal values of the loading acting on the elements can be regarded as redundant quantities acting on a primary system. The latter, consisting of the soil, is statically indeterminate. The redundants are to be calculated such that the prescribed displacements on the structure–soil interface result. This concept is well known from the static analysis of simple frame structures. Denoting the unknown load intensities in all nodes as $\{p\}$, the load amplitudes on the structure–soil interface $\{p(s)\}$ are expressed as

$$\{p(s)\} = [L(s)]\{p\} \tag{7.28}$$

Here s denotes symbolically a point on the structure–soil interface and $[L(s)]$ represents the selected interpolation function. The amplitudes of the surface displacements in the three or two directions $\{u_p(s)\}$ are formulated as

$$\{u_p(s)\} = [g(s)]\{p\} \tag{7.29}$$

The calculation of the flexibility-influence functions of the layered half-space $[g(s)]$ is described further on in this section.

In Fig. 7-5c the nodes associated with the dynamic-stiffness matrix of the foundation are shown. In principle, they are independent of those of the load intensities. The corresponding shape functions $[N(s)]$ relate the nodal values of the displacement amplitudes $\{u_b\}$ to the amplitudes of the displacements of the structure–soil interface $\{u(s)\}$:

$$\{u(s)\} = [N(s)]\{u_b\} \tag{7.30}$$

The continuity and completeness requirements of the finite-element method have to be satisfied. In Fig. 7-5c, quadratic expansions are indicated. The nodes with subscript b are shown (for an embedded structure) in Fig. 7-1. For a rigid basemat, the rigid-body kinematics (as expressed by the transformation matrix $[A]$, Eq. 3.13) can be incorporated into $N(s)$. The vector $\{u_b\}$ is replaced by $\{u_o\}$, which denotes the rigid-body degrees of freedom (Fig. 3-5). In this case, $\{u(s)\}$ will vary linearly. For instance, for the vertical degree of freedom of a rigid basemat, $N(s)$ will be a constant of unity.

As only a finite number of load intensities can be introduced, the displacement-boundary condition on the structure–soil interface S cannot be satisfied exactly (i.e., in every point on S) but only in an average sense as

$$\int_S [W(s)]^T(\{u_p(s)\} - \{u(s)\})\, ds = \{0\} \tag{7.31}$$

This condition is used to determine the load intensities $\{p\}$. The matrix $[W(s)]$ denotes the weighting function, for which various choices are possible. In the following, $[W(s)] = [L(s)]$ is selected. For instance, for a constant distributed load acting on each element, the piecewise-constant interpolation function $L(s)$ will force the difference of $\{u_p(s)\}$ and $\{u(s)\}$, integrated over each element of the structure–soil interface, to vanish.

Substituting Eqs. 7.29 and 7.30 in Eq. 7.31 results in

$$[G]\{p\} = [T]\{u_b\} \tag{7.32}$$

where

$$[G] = \int_S [L(s)]^T[g(s)]\, ds \tag{7.33}$$

$$[T] = \int_S [L(s)]^T[N(s)]\, ds \tag{7.34}$$

The flexibility matrix $[G]$ is symmetric. This is easily verified by using Maxwell–Betti's reciprocity law, which states that the work of one set of loads with the displacements of another is equal to the work of the latter loads with the displacements of the former. Formulating this law for the two loading states corresponding to p_i and p_j results in

$$\int_S \{p(s)\}_i^T\{u_p(s)\}_j\, ds = \int_S \{p(s)\}_j^T\{u_p(s)\}_i\, ds \tag{7.35}$$

Substituting Eqs. 7.28 and 7.29 formulated for p_i and p_j into Eq. 7.35 leads to

$$p_i \int_S \{L(s)\}_i^T\{g(s)\}_j\, ds\, p_j = p_j \int_S \{L(s)\}_j^T\{g(s)\}_i\, ds\, p_i \tag{7.36}$$

where $\{L(s)\}_i$ and $\{g(s)\}_i$ represent the ith columns of $[L(s)]$ and $[g(s)]$, respectively. Using Eq. 7-33, Eq. 7.36 results in

$$G_{ij} = G_{ji} \tag{7.37}$$

As is well known from virtual-work considerations applied in finite-element analysis, the amplitudes of the concentrated loads $\{P_b\}$ are obtained as

$$\{P_b\} = \int_S [N(s)]^T \{p(s)\}\, ds \tag{7.38}$$

Solving Eq. 7.32 for $\{p\}$, substituting Eq. 7.28 in Eq. 7.38, and using Eq. 7.34 results in

$$\{P_b\} = [T]^T[G]^{-1}[T]\{u_b\} \tag{7.39}$$

This defines the dynamic-stiffness matrix $[S_{bb}]$ of the soil as

$$[S_{bb}] = [T]^T[G]^{-1}[T] \tag{7.40}$$

The superscript f or g is dropped for surface foundations. Obviously, $[S_{bb}]$ is symmetric.

7.2.2 Green's Influence Function for Two-Dimensional Case

The calculation of the flexibility-influence functions of the layered half-space (Green's functions) $[g(s)]$ is discussed next. At first, the two-dimensional case in the x-z plane is analyzed which is used to calculate strip foundations that are infinitely long in the y-direction. The coordinate on the structure–soil interface is thus equal to x. The load amplitudes $\{p(x)\}$ (arising from the load intensities $\{p\}$) can be expanded in the horizontal direction into a Fourier integral with terms $\exp(-ikx)$, k being the wave number. The following transforms (Eq. 2.17) apply, whereby use is made of Eq. 7.28.

$$\begin{aligned}\{p(k)\} &= \frac{1}{2\pi}\int_{-\infty}^{+\infty} \{p(x)\} \exp(ikx)\, dx \\ &= \frac{1}{2\pi}\left(\int_{-\infty}^{+\infty} [L(x)] \exp(ikx)\, dx\right)\{p\} = [L(k)]\{p\}\end{aligned} \tag{7.41a}$$

$$\begin{aligned}\{p(x)\} &= \int_{-\infty}^{+\infty} \{p(k)\} \exp(-ikx)\, dk \\ &= \left(\int_{-\infty}^{+\infty} [L(k)] \exp(-ikx)\, dk\right)\{p\} = [L(x)]\{p\}\end{aligned} \tag{7.41b}$$

The sign of the argument of the exponential function is the same as introduced in Sections 5.2, 5.3, and 5.4. As the wave number k runs from $-\infty$ to $+\infty$, all types of waves are treated. In this application k is a real variable, in contrast to the case when determining the free-field response, where for a damped site k is complex (Table 6-1).

The out-of-plane and in-plane motions decouple. The distributed $q(x)$ leads to $v(x)$, the loads $p(x)$ and $r(x)$ to $u(x)$ and $w(x)$ (Fig. 7-6). In the following, the

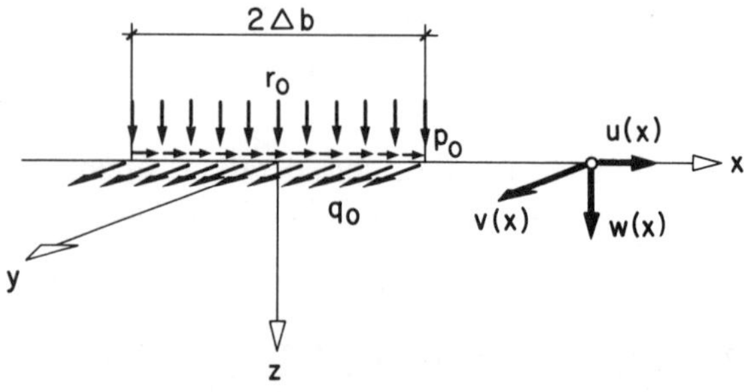

Figure 7-6 Substrip with selected constant load distribution.

distributed loads $p(x)$, $q(x)$, and $r(x)$ acting on the line element of width $2\,\Delta b$ (substrip) are selected to be constant (p_o, q_o, r_o). The out-of-plane motion is addressed first. Formulating Eq. 7.41a leads to the amplitudes of the load in the k-domain:

$$q(k) = \frac{q_o}{2\pi}\int_{-\Delta b}^{+\Delta b} \exp(ikx)\,dx = \frac{q_o}{\pi k}\sin k\Delta b \tag{7.42}$$

The corresponding displacement amplitude in the k-domain $v(k)$ is calculated applying the direct-stiffness approach to the site. As discussed in Section 6.2, assembling the dynamic-stiffness matrices of the individual layers $[S^L_{SH}]$ and of the half space S^R_{SH} results in the dynamic-equilibrium equation of the site (Eq. 6.2).

$$[S_{SH}]\{v\} = \{Q\} \tag{7.43}$$

The first element of the vector of the external load amplitudes $\{Q\}$ is equal to $q(k)$ of Eq. 7.42. All other elements are zero. Eliminating all displacement amplitudes $\{v\}$ with the exception of that corresponding to the surface $v(k)$ from Eq. 7.43, and then performing an inversion leads to

$$v(k) = F_{vv}(k)q(k) \tag{7.44}$$

The element $F_{vv}(k)$ is the flexibility coefficient (condensed at the surface) for the out-of-plane motion of the site. The inverse transform of Eq. 7.41b, but formulated for the displacement amplitude, is equal to

$$v(x) = \int_{-\infty}^{+\infty} v(k)\exp(-ikx)\,dk \tag{7.45}$$

The matrices $[S^L_{SH}]$ (Eq. 5.103) and S^R_{SH} (Eq. 5.104) depend on kt, which is an even function of k. The same thus also applies to $F_{vv}(k)$. Substituting Eqs. 7.42 and 7.44 in Eq. 7.45 leads to

$$v(x) = \frac{2}{\pi}\left[\int_0^{\infty} \frac{\sin k\Delta b}{k} F_{vv}(k)\cos kx\,dk\right] q_o \tag{7.46}$$

For a given ω, k now varies from 0 to infinity. The various special cases for

$[S_{SH}^L]$ and S_{SH}^R and thus also for $F_{vv}(k)$ are specified in Eqs. 5.106 to 5.108. In the static case ($\omega = 0$) for a site with a half-plane, the stiffness matrix of the site $[S_{SH}]$ will become singular for $k = 0$ (Eq. 5.108b). The corresponding flexibility coefficient $F_{vv}(k)$ is infinite.

For the in-plane motion, the load amplitudes $p(k)$ and $r(k)$ follow from Eq. 7.42, substituting the corresponding load amplitudes and intensities p_o and r_o. Analogously, the flexibility equation in the k-domain formulated at the surface is equal to

$$\begin{Bmatrix} u(k) \\ iw(k) \end{Bmatrix} = \begin{bmatrix} F_{uu}(k) & F_{uw}(k) \\ F_{wu}(k) & F_{ww}(k) \end{bmatrix} \begin{Bmatrix} p(k) \\ ir(k) \end{Bmatrix} \tag{7.47}$$

It should be remembered that to achieve symmetry of the dynamic-stiffness matrix $[S_{P\text{-}SV}]$ (Eqs. 5.134 and 5.135), the amplitudes corresponding to the vertical direction are multiplied by i. The elements $F_{uu}(k)$, $F_{ww}(k)$, and $F_{uw}(k)$ are even and odd functions of k, respectively. The inverse transformations leading to $u(x)$ and $w(x)$ follow from Eq. 7.45, replacing the variables. Performing the appropriate substitutions leads to

$$\begin{Bmatrix} u(x) \\ w(x) \end{Bmatrix} = \frac{2}{\pi}\left(\int_0^\infty \frac{\sin k\Delta b}{k} \begin{bmatrix} F_{uu}(k)\cos kx & F_{uw}(k)\sin kx \\ -F_{wu}(k)\sin kx & F_{ww}(k)\cos kx \end{bmatrix} dk\right) \begin{Bmatrix} p_o \\ r_o \end{Bmatrix} \tag{7.48}$$

The various special cases for $[S_{P\text{-}SV}^L]$ and $[S_{P\text{-}SV}^R]$ which affect $F_{uu}(k)$, $F_{uw}(k)$, and $F_{ww}(k)$ are described in Eqs. 5.136 to 5.138. Again, the flexibility coefficients will become infinite for the static case with $k = 0$, if a half-plane is present.

The $[g(s)]$ matrix introduced in Eq. 7.29 has as elements the coefficients in Eqs. 7.46 and 7.48.

7.2.3 Green's Influence Function for Axisymmetric Case

Next, the three-dimensional case is addressed. The load amplitude $\{p(x, y)\}$ could be represented by a Fourier integral with terms $\exp[-i(kx + ly)]$, k and l being the wave numbers. For computational efficiency, the loading is assumed to act on a (circular) subdisk. This allows cylindrical coordinates to be introduced. The corresponding loading is expanded in a Fourier series in the circumferential direction θ (integer $n = 0, 1, 2, \ldots$) and into Bessel functions involving the wave number k in the radial direction r. As k runs from 0 to infinity, all types of waves are captured. The corresponding load amplitudes are related by the following Bessel transform pair

$$\begin{aligned} \{p(k, n)\} &= a_n \int_{r=0}^{\infty} r[C_n(kr)] \int_{\theta=0}^{2\pi} [D(n\theta)]\{p(r, \theta)\}\, d\theta\, dr \\ &= a_n \left(\int_{r=0}^{\infty} r[C_n(kr)] \int_{\theta=0}^{2\pi} [D(n\theta)][L(r, \theta)]\, d\theta\, dr\right)\{p\} \\ &= [L(k, n)]\{p\} \end{aligned} \tag{7.49a}$$

$$
\begin{aligned}
\{p(r,\theta)\} &= \sum_{n=0}^{\infty}[D(n\theta)]\int_{k=0}^{\infty} k[C_n(kr)]\{p(k,n)\}\,dk \\
&= \left(\sum_{n=0}^{\infty}[D(n\theta)]\int_{k=0}^{\infty} k[C_n(kr)][L(k,n)]\,dk\right)\{p\} \\
&= [L(r,\theta)]\{p\}
\end{aligned}
\tag{7.49b}
$$

The matrix $[C_n(kr)]$ contains the Bessel functions, $[D(n\theta)]$ sine and cosine functions of $n\theta$ on the diagonal. The scalar a_n is the normalization factor, which is equal to $1/2\pi$ for $n = 0$ and $1/\pi$ for $n \neq 0$.

The transformation of Eq. 7.49b is introduced for the amplitudes of the displacements and stresses in Section 5.5.2. For instance, for the vector of the amplitudes of the displacements in the radial, circumferential, and vertical directions $u(r, \theta)$, $v(r, \theta)$, and $w(r, \theta)$, the diagonal matrix $[D(n\theta)]$ consisting of $\cos n\theta$, $-\sin n\theta$, and $\cos n\theta$ for the symmetric case and $\sin n\theta$, $\cos n\theta$, and $\sin n\theta$ for the antimetric case is specified in Eq. 5.162. From Eq. 5.180, the matrix $[C_n(kr)]$ is identified as

$$
[C_n(kr)] = \begin{bmatrix} \frac{1}{k}J_n(kr)_{,r} & \frac{n}{kr}J_n(kr) & \\ \frac{n}{kr}J_n(kr) & \frac{1}{k}J_n(kr)_{,r} & \\ & & -J_n(kr) \end{bmatrix}
\tag{7.50}
$$

where $J_n(kr)$ denotes the Bessel function of order n of the first kind. This matrix is postmultiplied by the vector of the amplitudes of the displacements in the k-domain $u(k, n)$, $v(k, n)$, $w(k, n)$ (as defined in Eq. 5.180) divided by k. The latter occurs because in Eq. 7.49b a factor k is present in the integrand. The same transformation exists relating $\tau_{rz}(k, n)$, $\tau_{\theta z}(k, n)$, and $\sigma_z(k, n)$ (divided by k) to $\tau_{rz}(r, \theta)$, $\tau_{\theta z}(r, \theta)$, and $\sigma_z(r, \theta)$ (see Eq. 5.182).

The vertical distributed load with amplitude $r(s)$, which acts on a subdisk of radius Δa and which is selected to have a constant value r_o, is examined first (Fig. 7-7a). Only the zeroth symmetric Fourier term ($n = 0$) resulting in

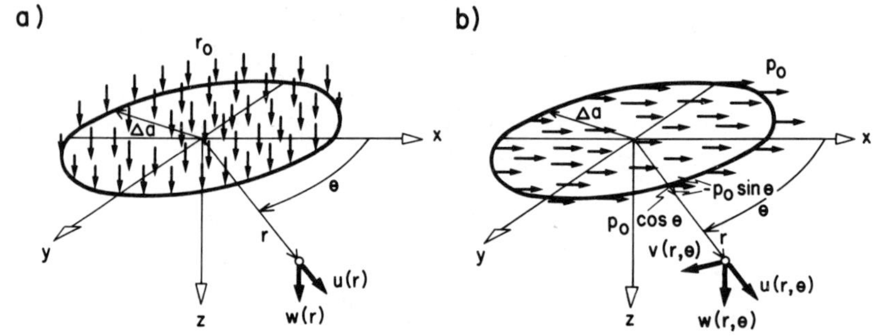

Figure 7-7 Subdisk with selected constant load distribution. (a) Vertical; (b) horizontal.

constant values around the circumference arises. Formulating Eq. 7.49a and making use of Eq. 7.50 leads to the amplitude of the load in the k-domain (divided by k)

$$r(k) = -\frac{1}{2\pi}\int_{r=0}^{\Delta a} rJ_o(kr)\int_{\theta=0}^{2\pi} r_o\, d\theta\, dr = -r_o\int_{r=0}^{\Delta a} rJ_o(kr)\, dr \tag{7.51}$$

The following two identities of the Bessel functions are used:

$$[x^n J_n(x)]_{,x} = x^n J_{n-1}(x) \tag{7.52a}$$

which in integral form is equal to

$$\int x^n J_{n-1}(x)\, dx = x^n J_n(x) + \text{constant} \tag{7.52b}$$

and

$$[x^{-n} J_n(x)]_{,x} = -x^{-n} J_{n+1}(x) \tag{7.53}$$

Using Eq. 7.52b, Eq. 7.51 is formulated as

$$r(k) = -\frac{r_o \Delta a}{k} J_1(k\Delta a) \tag{7.54}$$

noting that $J_1(0) = 0$.

As expressed in Eq. 5.183, the same dynamic-stiffness matrix of the site (independent of the Fourier index n) applies for cylindrical coordinates as for the case of plane waves expressed in Cartesian coordinates. Thus the same flexibility matrix (condensed at the surface) applies. Analogously to Eq. 7.47, this results in displacement amplitudes in the k-domain (divided by k)

$$\begin{Bmatrix} u(k) \\ w(k) \end{Bmatrix} = \begin{Bmatrix} F_{uw}(k) \\ F_{ww}(k) \end{Bmatrix} r(k) \tag{7.55}$$

The inverse transformation leading to $u(r)$ and $w(r)$ follows from Eq. 7.49b, formulated for the displacement amplitudes.

$$\begin{Bmatrix} u(r) \\ w(r) \end{Bmatrix} = \int_{k=0}^{\infty} \begin{bmatrix} J_o(kr)_{,r} & \\ & -kJ_o(kr) \end{bmatrix} \begin{Bmatrix} u(k) \\ w(k) \end{Bmatrix} dk \tag{7.56}$$

Making use of the identity of Eq. 7.53 and substituting Eqs. 7.54 and 7.55 leads to

$$\begin{Bmatrix} u(r) \\ w(r) \end{Bmatrix} = \Delta a \left[\int_{k=0}^{\infty} J_1(k\Delta a) \begin{Bmatrix} F_{uw}(k) J_1(kr) \\ F_{ww}(k) J_o(kr) \end{Bmatrix} dk \right] r_o \tag{7.57}$$

The horizontal distributed load with amplitude $p(s)$ (acting in the x-direction), which is assumed to have the constant value p_o, is discussed next (Fig. 7-7b). The radial and circumferential distributions vary as $p_o \cos\theta$ and $-p_o \sin\theta$, respectively. The first symmetric Fourier term is thus involved. The corresponding amplitudes of the load in the k-domain (divided by k) are calculated using Eq. 7.49a.

$$\begin{Bmatrix} p(k) \\ q(k) \end{Bmatrix} = \frac{p_o}{k}\int_{r=0}^{\Delta a} \begin{Bmatrix} rJ_1(kr)_{,r} + J_1(kr) \\ J_1(kr) + rJ_1(kr)_{,r} \end{Bmatrix} dr \tag{7.58}$$

Applying Eq. 7.52, Eq. 7.58 is reformulated as

$$p(k) = q(k) = \frac{p_o \Delta a}{k} J_1(k\,\Delta a) \tag{7.59}$$

The corresponding displacement amplitudes in the k-domain (divided by k) follow from the flexibility matrix (condensed at the surface)

$$\begin{Bmatrix} u(k) \\ v(k) \\ w(k) \end{Bmatrix} = \begin{bmatrix} F_{uu}(k) & & \\ & F_{vv}(k) & \\ F_{wu}(k) & & \end{bmatrix} \begin{Bmatrix} p(k) \\ \\ q(k) \end{Bmatrix} \tag{7.60}$$

Applying the inverse transform of Eq. 7.49b to the displacements leads to the displacement amplitudes $u(r, \theta)$, $v(r, \theta)$, and $w(r, \theta)$. Substituting Eqs. 7.59 and 7.60,

$$\begin{Bmatrix} u(r,\theta) \\ v(r,\theta) \\ w(r,\theta) \end{Bmatrix} = \frac{\Delta a}{2} \begin{bmatrix} \cos\theta & & \\ & -\sin\theta & \\ & & \cos\theta \end{bmatrix}$$

$$\times \left(\int_{k=0}^{\infty} J_1(k\Delta a) \begin{bmatrix} J_o(kr) - J_2(kr) & J_o(kr) + J_2(kr) & \\ J_o(kr) + J_2(kr) & J_o(kr) - J_2(kr) & \\ & & -2J_1(kr) \end{bmatrix} \begin{Bmatrix} F_{uu}(k) \\ F_{vv}(k) \\ F_{wu}(k) \end{Bmatrix} dk \right) p_o \tag{7.61}$$

results, whereby the following identity, which follows from Eqs. 7.52a and 7.53, is used:

$$[J_n(x)]_{,x} = J_{n-1}(x) - \frac{n}{x} J_n(x) = \frac{n}{x} J_n(x) - J_{n+1}(x) \tag{7.62}$$

Applying the last identity, $J_2(kr)$ could be eliminated in Eq. 7.61. For the distributed horizontal load in the y-direction with amplitude q_o, the same equation (7.61) applies, with $\cos\theta$ and $-\sin\theta$ replaced by $\sin\theta$ and $\cos\theta$, respectively.

The coefficients in Eqs. 7.57 and 7.61 determine the influence matrix $[g(s)]$ of Eq. 7.29.

The influence functions for the disk loaded by a torsion moment and a rocking moment are specified in Problems 7.12 and 7.13, respectively.

For the sake of illustration, the vertical flexibility-influence function $w(s)$ is presented. A site consisting of a layer of thickness d resting on a half-space is used (Fig. 6-13). The shear-wave velocity of the half space c_s^R (R stands for rock) is twice that of the layer c_s^L (L denotes the layer). Poisson's ratio equals 0.33. This site is also investigated in the Sections 7.3 and 7.4. The dimensionless frequency $\omega d/c_s^L$ is selected as 7. A damping ratio of $\zeta = 0.001$ is introduced. The vertical-flexibility coefficient in the k-domain $F_{ww}(k)$ (which is used in Eqs. 7.47 and 7.55) is plotted as a function of the dimensionless wave number kd in Fig. 7-8. For four distinct wave numbers, the real and imaginary parts exhibit sharp peaks, whereby the real part changes sign. For $\zeta = 0$, the values of the

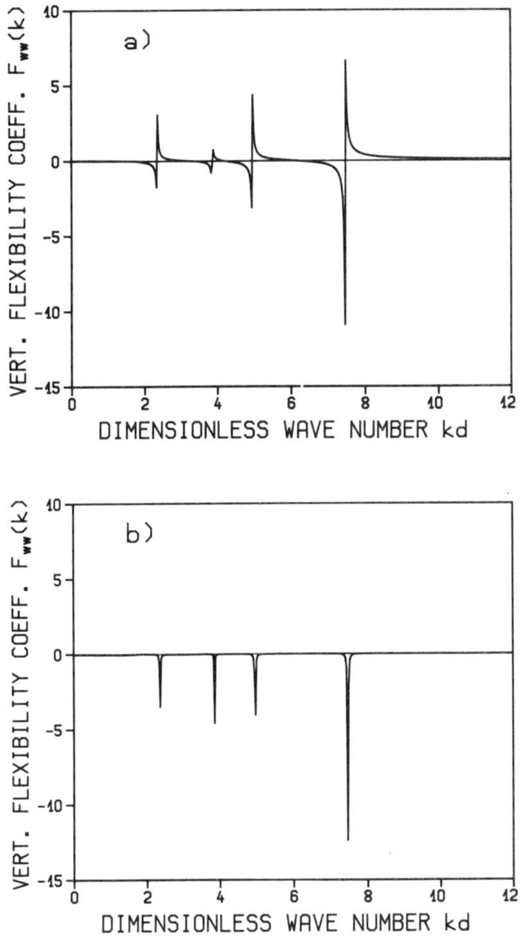

Figure 7-8 Vertical flexibility coefficient in the wave-number domain (Ref. [8]). (a) Real part; (b) imaginary part.

second, third, and fourth peaks are infinite. For the selected small value of ζ, these peaks remain finite. The dispersion curve representing the dimensionless-apparent velocity $\bar{c}_a = c_a/c_s^L$ as function of $\omega d/c_s^L$ is shown in Fig. 6-44. The determinant of the inverse of the flexibility matrix vanishes along these curves. For zero damping, the phase velocity (which is equal to c_a) is real, and thus also the wave number $k = \omega/c$. The phase velocities of the three Rayleigh modes for $\omega d/c_s^L = 7$ correspond to the value of kd for which the peaks of $F_{ww}(k)$ are infinite. The vertical flexibility-influence function $w(s)$ follows from the lower equation in Eq. 7.57 for the three-dimensional case and from the equation involving the lower diagonal element in Eq. 7.48 for two dimensions. The integrations are performed numerically. No problems arise if a fictitious

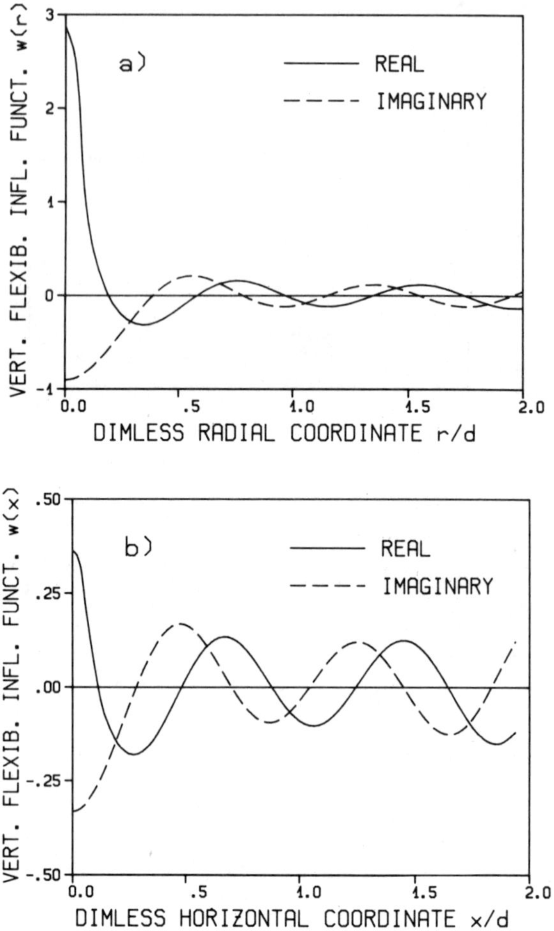

Figure 7-9 Vertical flexibility-influence function (Ref. [8]). (a) Three-dimensional case; (b) two-dimensional case.

damping ratio $\zeta = 0.001$ is introduced for an undamped site. For a subdisk with a radius $\Delta a = 0.0705d$, $w(r)$ is shown as a function of r/d in Fig. 7-9a. For the two-dimensional case with a half-width of a substrip $\Delta b = 0.0625d$, $w(x)$ is shown in Fig. 7-9b. As expected, $w(s)$ decays more rapidly with s in the three-dimensional than in the two-dimensional case.

7.2.4 Element Size and Numerical Integration

The structure–soil interface is discretized into elements that do not have an extreme aspect ratio. The latter are then replaced by subdisks with the same area, as the corresponding elements. The number of elements will depend on the (highest) frequency which is to be transmitted properly. Experience shows that at least six points are needed to represent accurately the wavelength $2\pi c_s/\omega$. The maximum length $2\,\Delta l$ of an element will thus equal $\pi c_s/(3\omega)$. If on a line running across a basemat of length $2l$ (e.g., the diameter), m elements are chosen, the dimensionless frequency $a_o = \omega l/c_s$ up to the value $\pi m/6$ is adequately modeled using this criterion.

The integrations in $[G]$ (Eq. 7.33) are performed numerically. For the elements on the diagonal, 5×5 Gaussian-integration points are selected. For the other elements, 2×2 points are chosen. In general, a horizontal load with amplitude p_o will lead to horizontal and vertical flexibility-influence functions $u(s)$, $v(s)$, and $w(s)$. The same applies to a vertical load. Rigorously taking account of all components in $[G]$ leads to the dynamic-stiffness matrix $[S_{bb}]$ for welded contact. As an approximation, the vertical component due to the horizontal load and the horizontal components due to the vertical load can be omitted ($F_{uw} = F_{wu} = 0$ in Eqs. 7.48, 7.57, and 7.61). This represents the condition of relaxed contact.

7.2.5 Nondimensionalized Spring and Damping Coefficients

The dynamic-stiffness matrix $[S_{bb}]$ in Eq. 7.40 is decomposed for the three-dimensional case as follows:

$$[S_{bb}] = [K_{bb}]([k] + ia_o[c]) \tag{7.63}$$

The matrix $[K_{bb}]$ contains the static-stiffness coefficients, $[k]$ and $[c]$ the (nondimensionalized) spring and damping coefficients, respectively. The variable a_o represents the dimensionless frequency

$$a_o = \frac{\omega l}{c_s} \tag{7.64}$$

where c_s is the shear-wave velocity of the medium at the surface. For a rigid basemat, the Eq. 7.63 still applies, replacing b by o. For a symmetric basemat and for relaxed contact of a surface foundation, $[K_{oo}]$ will be a diagonal matrix. As discussed in connection with Eqs. 7.46 and 7.48, the translational static-flexibility coefficients for a site with a half-plane are infinite. The static-stiffness coefficients $[K_{bb}]$, which vanish, can thus not be used to nondimensionalize $[S_{bb}]$ for the two-dimensional case. Quite arbitrarily, values πG for the translational degrees of freedom and $\pi G l^2$ for the rotational degrees of freedom can be chosen (G is the shear modulus of the medium at the surface).

7.3 TWO-DIMENSIONAL RIGID BASEMAT (STRIP FOUNDATION)

To be able to reach conclusions applicable to a wide range of sites, the following three cases are investigated (Fig. 7-10):

1. The viscoelastic half-plane with Poisson's ratio $\nu^R = 0.33$;
2. A single layer of depth d and $\nu^L = 0.33$ built in at its base;

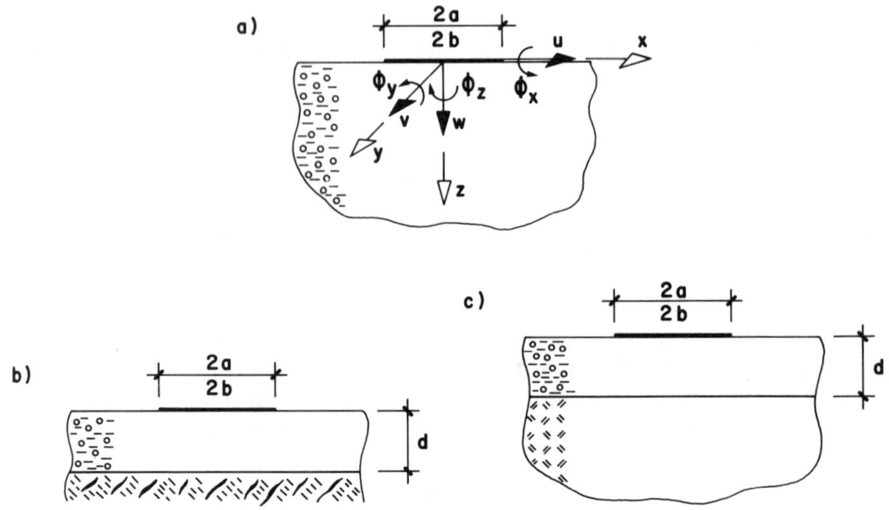

Figure 7-10 Investigated sites (Ref. [8]).

3. A single layer of depth d and of Poisson's ratio $\nu^L = 0.33$ resting on a half-plane with $\nu^R = 0.33$. The ratios of the shear-wave velocities $\bar{c}_s = c_s^R/c_s^L$ and of the mass densities $\bar{\rho} = \rho^R/\rho^L$ equal 2 and 1, respectively.

The damping ratios ζ^L and ζ^R are either equal to 0.05 or 0.001, the latter value representing the undamped case. The same sites are investigated for the free-field response in Chapter 6. The half-plane and the layer built in at its base represent limiting cases. A rigid strip foundation with half-width b (two-dimensional case) is examined. In Section 7.4 a rigid circular foundation (three-dimensional case) is addressed. For the two sites involving a layer, the depth d is selected equal to the half-width b. The half-width is subdivided into eight elements (substrips) of equal length. This results in a total of 16 elements for the strip. Applying the criterion developed in Section 7.2.4 leads to dependable results up to the value of the dimensionless frequency $a_o = \omega b/c_s$ of over 8. The spring and damping coefficients are presented up to $a_o = 10$. To nondimensionalize the dynamic-stiffness matrix, the values πG and $\pi G b^2$ are used as elements for the translational and rotational degrees of freedom, respectively, in $[K_{oo}]$, which appears in the analogous relation shown in Eq. 7.63 formulated for a rigid basemat.

In Fig. 7-11a, the spring coefficients k_x in the horizontal direction in the plane, k_z in the vertical direction, k_{ϕ_y} for rocking, $k_{x\phi_y}$, representing the coupling between the horizontal direction and rocking and finally k_y in the horizontal direction but out-of-plane, are plotted for the rigid strip on the undamped half-plane. The corresponding damping coefficients are indicated in Fig. 7-11b. Welded contact is enforced. The dynamic-stiffness coefficients depend, to a certain degree, on the frequency of excitation, especially for the rocking and the

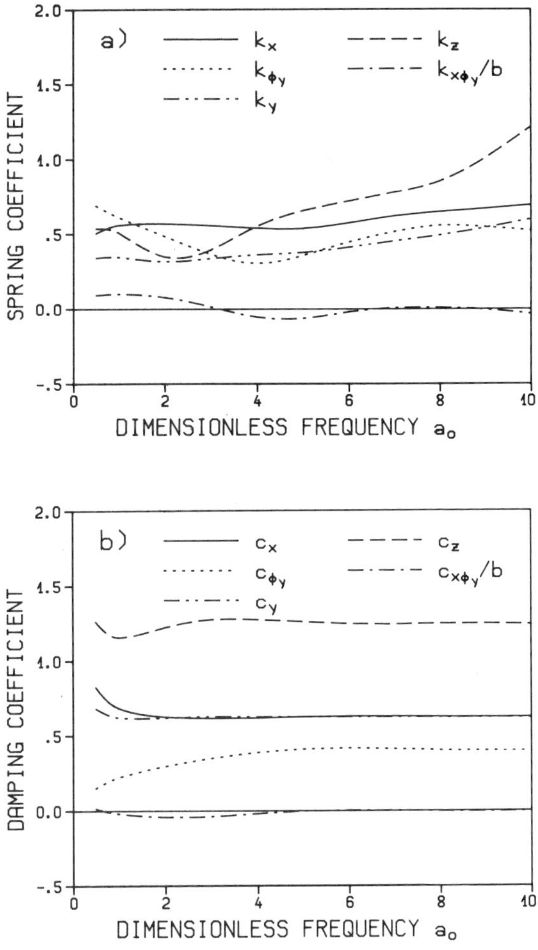

Figure 7-11 Dynamic stiffness of strip, half-plane, no damping, welded contact (Ref. [8]).

vertical degrees of freedom. Introducing damping reduces the spring coefficients in the higher-frequency range (Fig. 7-12a) and increases the damping coefficients in the lower-frequency range (Fig. 7-12b). In addition, the dependency on a_o is reduced.

For the other limiting site (i.e., the layer built in at its base), the dynamic-stiffness coefficients without and with damping are shown in Figs. 7-13 and 7-14, respectively. Especially in the high-frequency range, the dynamic-stiffness coefficients exhibit large oscillations. Even negative spring coefficients arise. In contrast to the elastic half-plane, the built-in layer exhibits a cutoff frequency below which, for the case of no damping, no radiation of energy occurs. The damping coefficients c_x and c_y are zero below the horizontal fundamental frequency of the

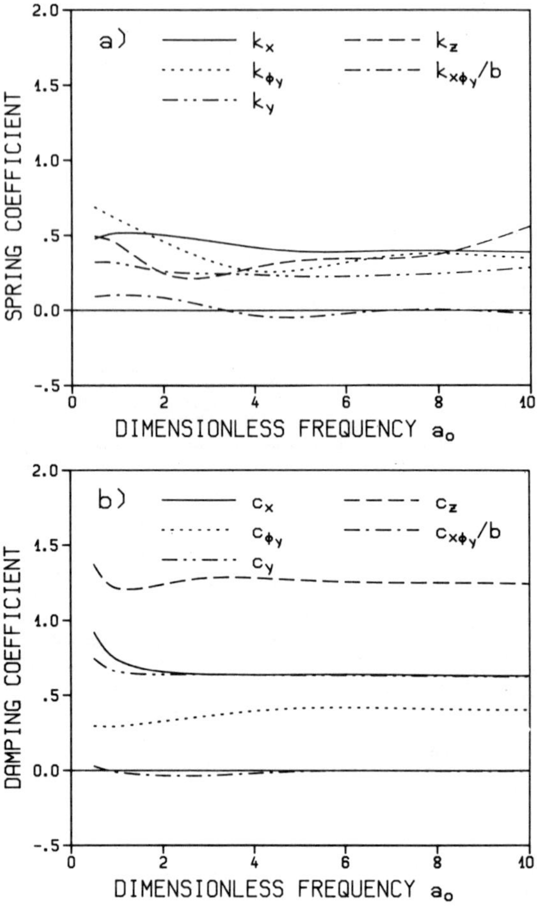

Figure 7-12 Dynamic stiffness of strip, half-plane, with damping, welded contact (Ref. [8]).

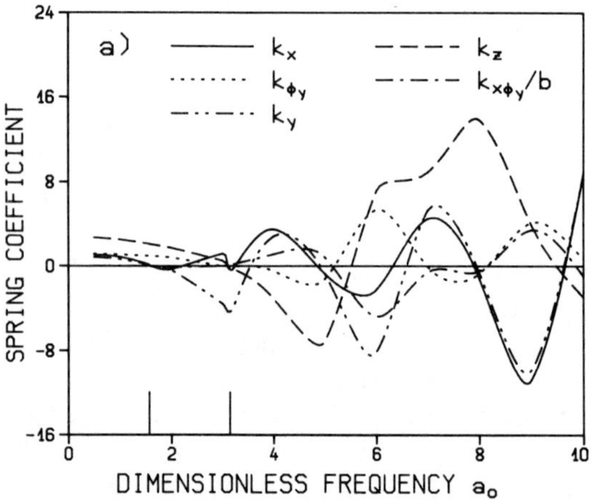

Figure 7-13 Dynamic stiffness of strip, layer built in at its base, no damping, welded contact (Ref. [8]).

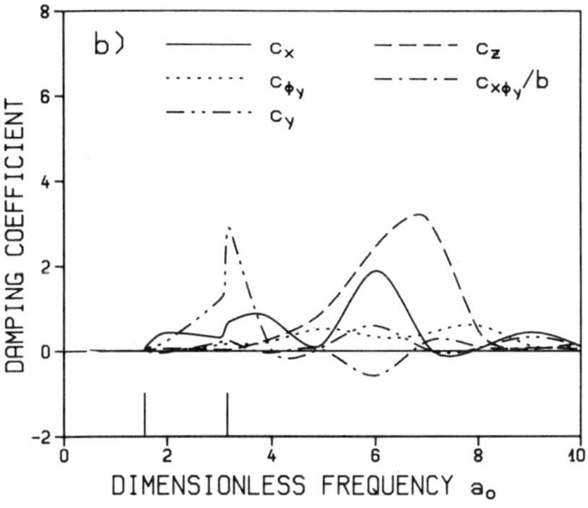

Figure 7-13 (Continued)

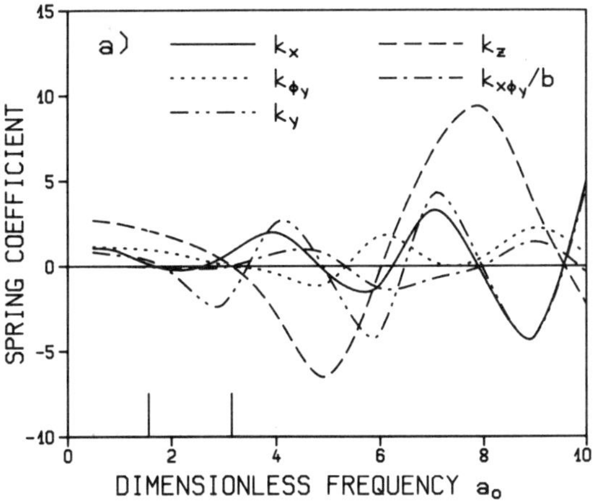

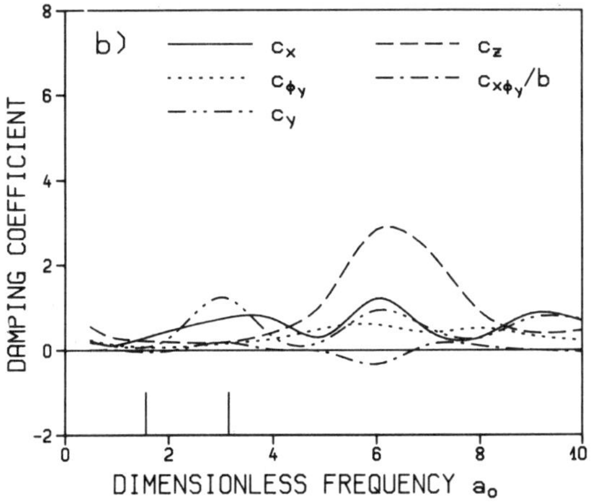

Figure 7-14 Dynamic stiffness of strip, layer built in at its base, with damping, welded contact (Ref. [8]).

layer $[\omega/2\pi = c_s/(4d)]$, which corresponds to $a_o = \pi/2$. For larger frequencies significant values arise. The values c_z and c_{ϕ_y} are also zero for $a_o < \pi/2$. In the range up to the vertical fundamental frequency $a_o = \pi$, c_z and c_{ϕ_y} remain small. For $a_o > \pi$ they increase significantly. With damping, a cutoff frequency no longer exists. Below the fundamental frequencies, however, the imaginary parts of the dynamic-stiffness coefficients $a_o c$ remain small. In addition, the dependency on frequency is reduced, and the values of the spring coefficients are diminished. As expected, the corresponding dynamic-stiffness coefficients of the layer on half-plane (Figs. 7-15 and 7-16) exhibit characteristics lying between those of the two extreme sites. For the undamped

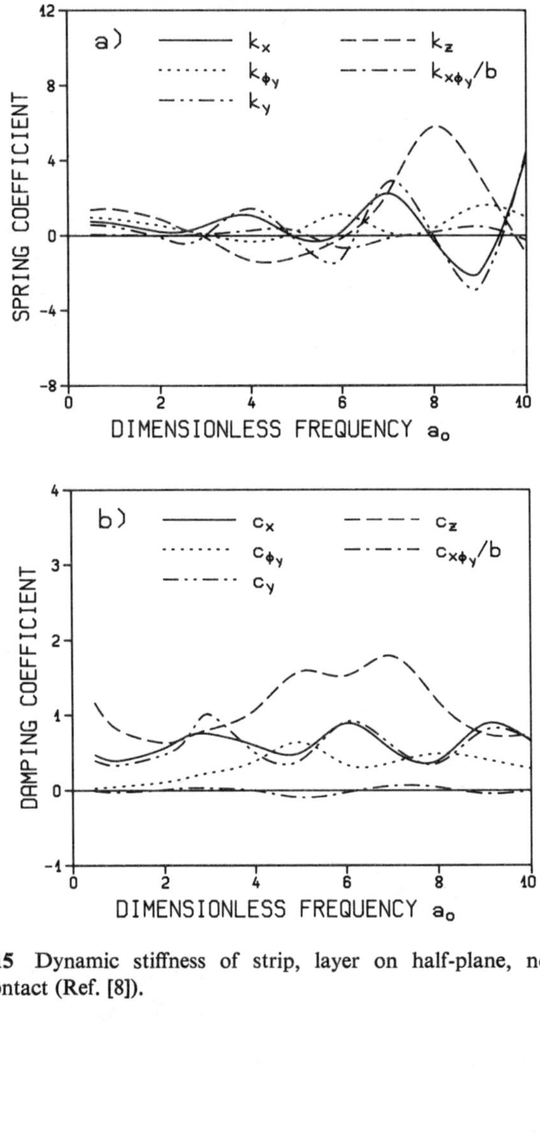

Figure 7-15 Dynamic stiffness of strip, layer on half-plane, no damping, welded contact (Ref. [8]).

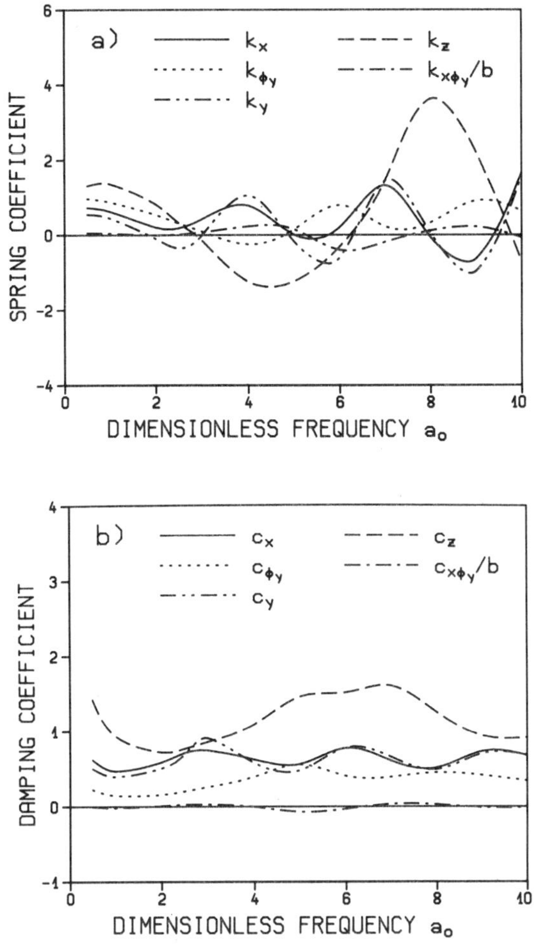

Figure 7-16 Dynamic stiffness of strip, layer on half-plane, with damping, welded contact (Ref. [8]).

case a cutoff frequency no longer exists. The dependency on frequency is quite large.

Modifying the contact condition from welded to relaxed hardly affects any dynamic-stiffness coefficients for the half-plane (results not shown) or those in the two horizontal directions for the layer built in at its base. For this site, however, the vertical and rocking coefficients are affected, especially in the high-frequency range (Fig. 7-17). For the damped site the values agree better (not shown).

Integrating in the wave-number domain takes all types of waves into account. Assuming only vertically incident waves corresponds to evaluating the

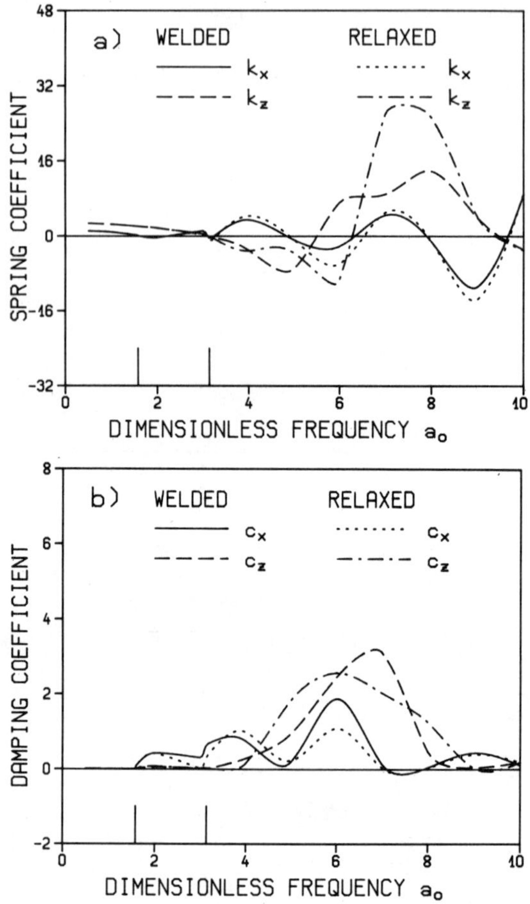

Figure 7-17 Dynamic stiffness of strip, layer built in at its base, no damping, relaxed versus welded contact (Ref. [8]).

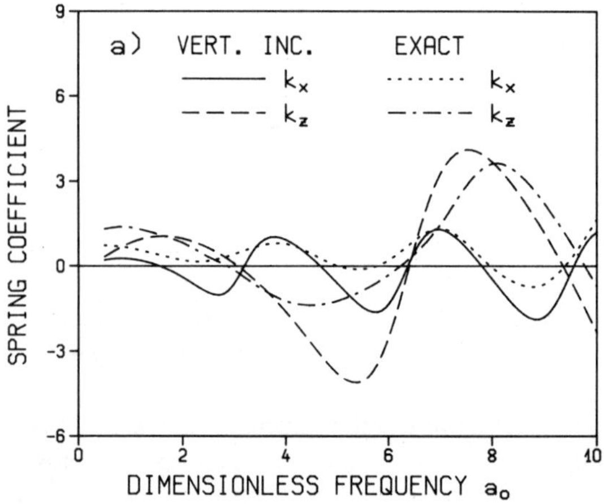

Figure 7-18 Dynamic stiffness of strip, layer on half-plane, with damping, vertically incident waves versus all wave types (Ref. [8]).

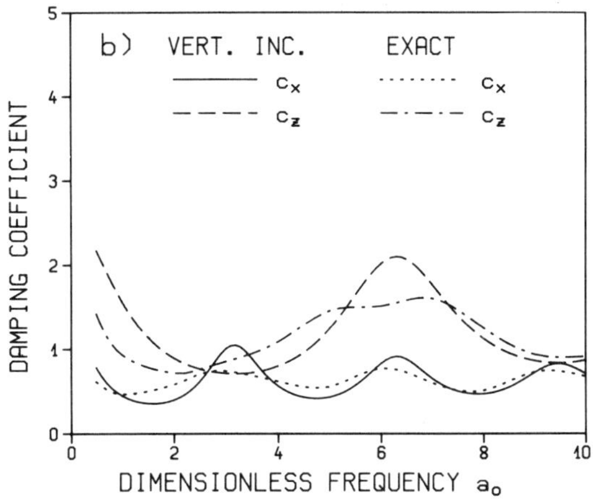

Figure 7-18 (Continued)

equations for $k = 0$ only, as the phase velocity c is infinite for this case ($k = \omega/c$). The dynamic-stiffness coefficients resulting from this stringent assumption are compared with the exact values for the strip on the layer on half-plane in Fig. 7-18. For k_x and k_z only vertically incident S- and P-waves, respectively, occur. Even in the higher-frequency range, where Rayleigh waves arise, the corresponding curves still bear comparison.

7.4 THREE-DIMENSIONAL RIGID BASEMAT (DISK FOUNDATION)

A rigid circular foundation of radius a is examined in this section. The same three sites as defined in Section 7.3 are analyzed (Fig. 7-10). For the two sites involving a layer, the depth d is chosen equal to the radius a. The results will allow the analyst to avoid many calculations or at least the plots for a large range of frequencies can be used to check the results for cases that are not covered accurately.

7.4.1 Dynamic-Stiffness Coefficients

The radius is subdivided into eight elements (subdisks) of equal radius. This results in a total of 200 subdisks. (Obviously, a discretization into annular rings with a finite width would be more efficient; see Problem 7.11.) The spring and damping coefficients are shown in the frequency range of $a_o = \omega a/c_s$ up to 10. The static-stiffness coefficients for the three sites, assuming welded and relaxed contact, are presented in Table 7-1. The values K_x and K_z denote the static stiffnesses in the horizontal and vertical directions, respectively (Fig. 7-10). The value K_{ϕ_y} is the rocking stiffness and $K_{x\phi_y}$ represents the coupling term involving the horizontal direction and rocking. Finally, the twisting stiffness is

TABLE 7-1 Static-Stiffness Coefficients

	Welded Contact			Relaxed Contact			
Coefficient	Half-space	Single Layer on Half-space	Single Layer Built in	Half-space	Single Layer on Half-space	Single Layer Built in	Multiplier
K_x	1.01	1.32	1.56	1.00	1.32	1.55	$\frac{8Ga}{2-\nu}$
K_z	1.02	1.82	2.56	1.00	1.80	2.55	$\frac{4Ga}{1-\nu}$
K_{ϕ_y}	1.03	1.19	1.28	0.99	1.17	1.26	$\frac{8Ga^3}{3(1-\nu)}$
$K_{x\phi_y}$	1.03	0.63	0.25	0	0	0	$\frac{4(1-2\nu)Ga^2}{\pi(2-\nu)(1-\nu)}$
K_{ϕ_z}	1.00	1.04	1.06	1.00	1.04	1.06	$\frac{16Ga^3}{3}$

denoted as K_{ϕ_z}. The selected multipliers correspond to the well-known static-stiffness coefficients of a disk resting on an elastic half-space. These static-stiffness coefficients are used to nondimensionalize the dynamic-stiffness coefficients. All spring coefficients (with the exception of the coupling term) are equal to 1 for $a_o = 0$.

For the disk resting on top of the undamped half-space, the spring coefficients and damping coefficients are plotted as a function of a_o in Fig. 7-19a and b, respectively. The degrees of freedom are indicated in the figures. The corresponding values for the damped half-space are shown in Fig. 7-20. Figures

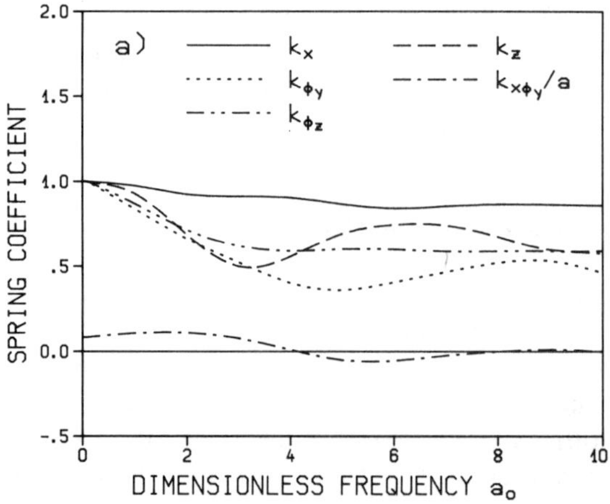

Figure 7-19 Dynamic stiffness of disk, half-space, no damping, welded contact (Ref. [8]).

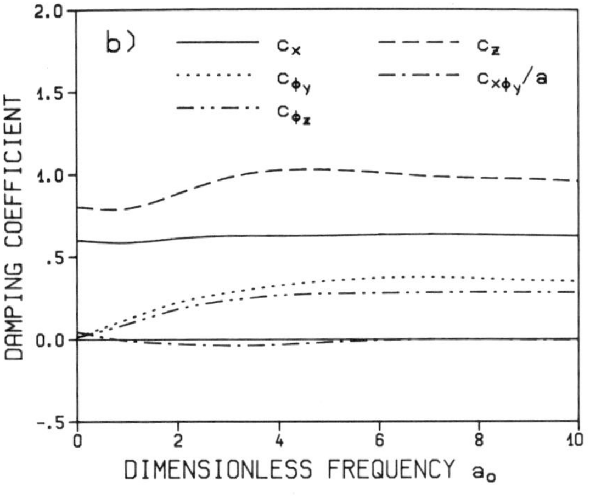

Figure 7-19 (Continued)

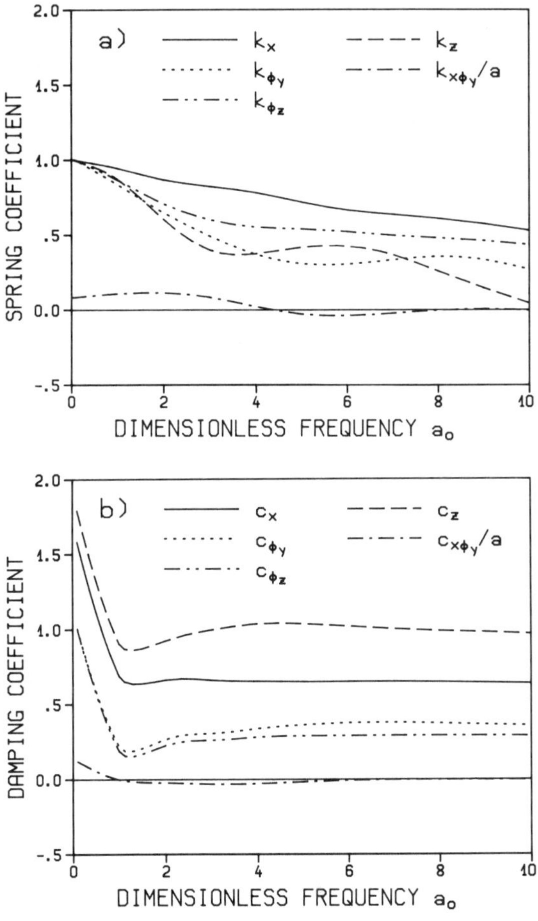

Figure 7-20 Dynamic stiffness of disk, half-space, with damping, welded contact (Ref. [8]).

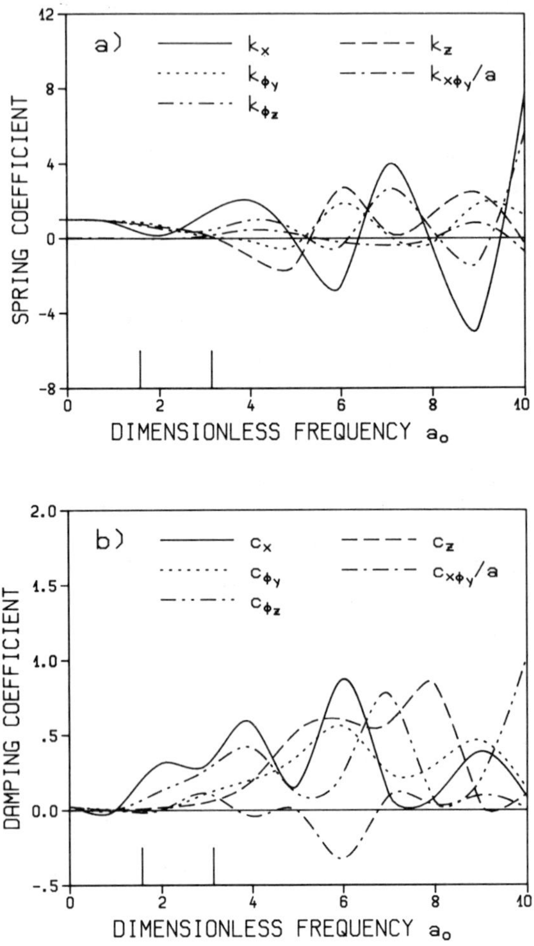

Figure 7-21 Dynamic stiffness of disk, layer built in at its base, no damping, welded contact (Ref. [8]).

7-21 to 7-24 contain the dynamic-stiffness coefficients for the layer built in at its base and for the layer on half-space. The same characteristics as in the two-dimensional case are visible (Section 7.3). In particular, below the fundamental frequencies of the layer in the horizontal and vertical directions the damping coefficients c_x, c_{ϕ_z} and c_z, c_{ϕ_y}, respectively, vanish for the undamped layer built in at its base (Fig. 7-21b). The damping coefficient of rocking, c_{ϕ_y}, which generally increases for increasing a_o, is significantly smaller than those of the translational degrees of freedom.

The influence of changing the contact condition from welded to relaxed is examined next. For the elastic half-space all dynamic-stiffness coefficients are

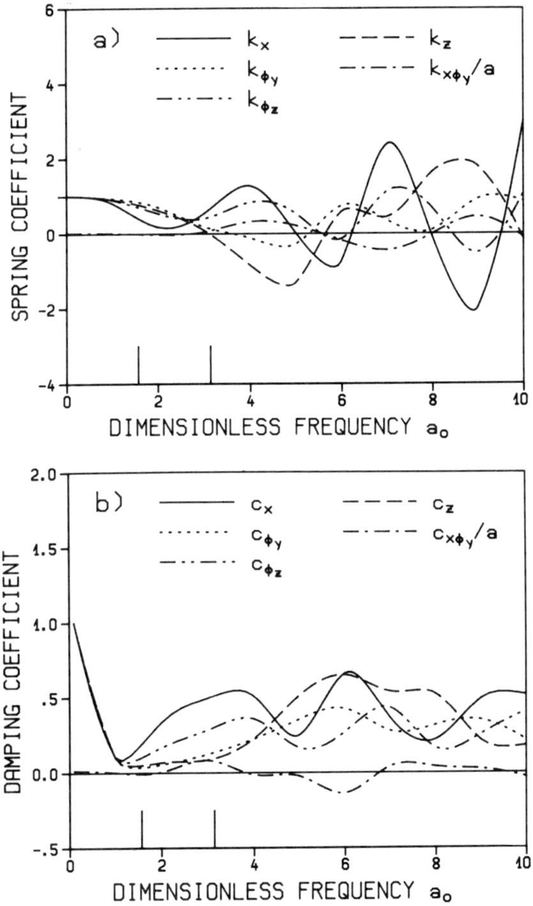

Figure 7-22 Dynamic stiffness of disk, layer built in at its base, with damping, welded contact (Ref. [8]).

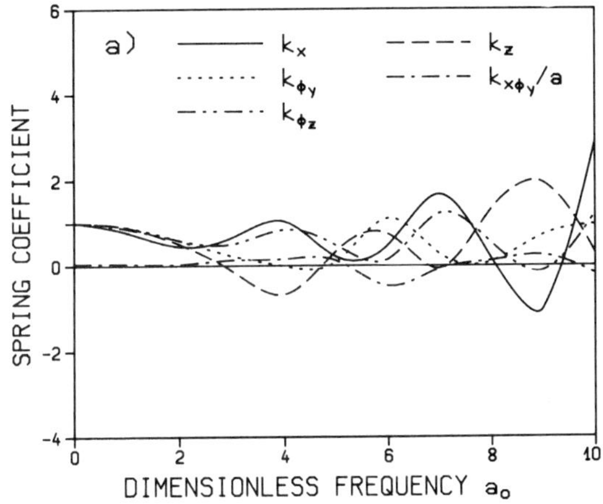

Figure 7-23 Dynamic stiffness of disk, layer on half-space, no damping, welded contact (Ref. [8]).

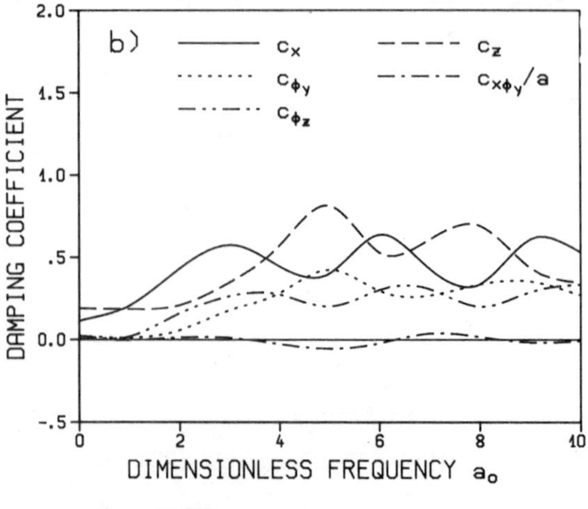

Figure 7-23 (Continued)

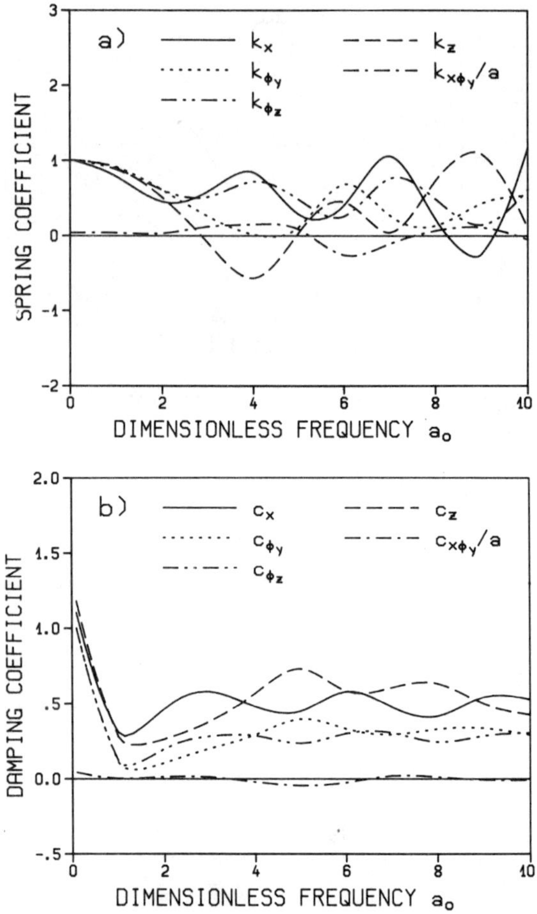

Figure 7-24 Dynamic stiffness of disk, layer on half-space, with damping, welded contact (Ref. [8]).

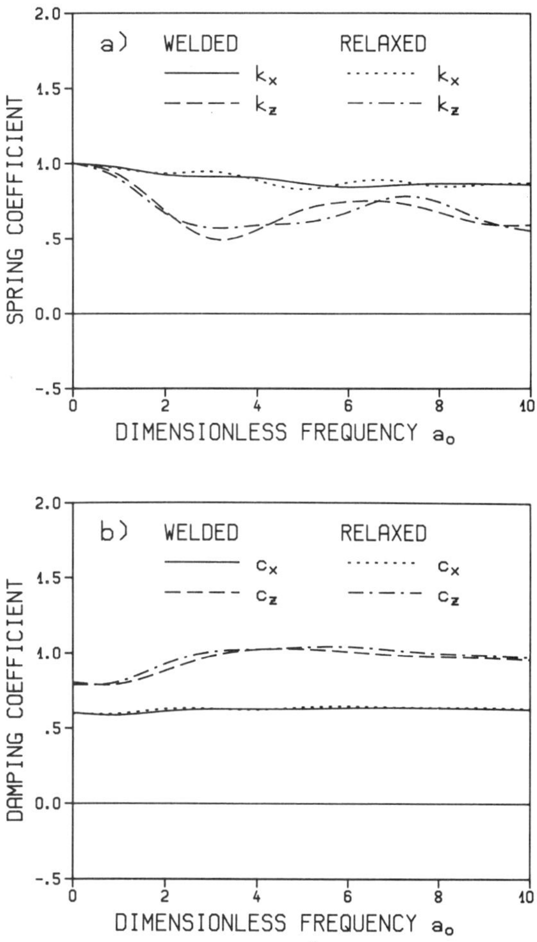

Figure 7-25 Dynamic stiffness of disk, half-space, no damping, relaxed versus welded contact (Ref. [8]).

hardly affected. This is shown in Fig. 7-25 for the horizontal and vertical directions. For the layer built in at its base the dynamic-stiffness coefficients of the vertical and rocking degrees of freedom are quite sensitive to the formulated contact condition. This is illustrated in Fig. 7-26. For the damped site the values agree better (not shown). The coefficients of the other degrees of freedom agree well even in the undamped case.

7.4.2 Two-Dimensional versus Three-Dimensional Modeling

Finally, the controversial matter of modeling a three-dimensional problem with a two-dimensional model is briefly addressed. Many aspects exist; here only that one affecting the dynamic-stiffness matrix of the soil is examined. A disk is selected as the three-dimensional problem. Besides the strip's depth,

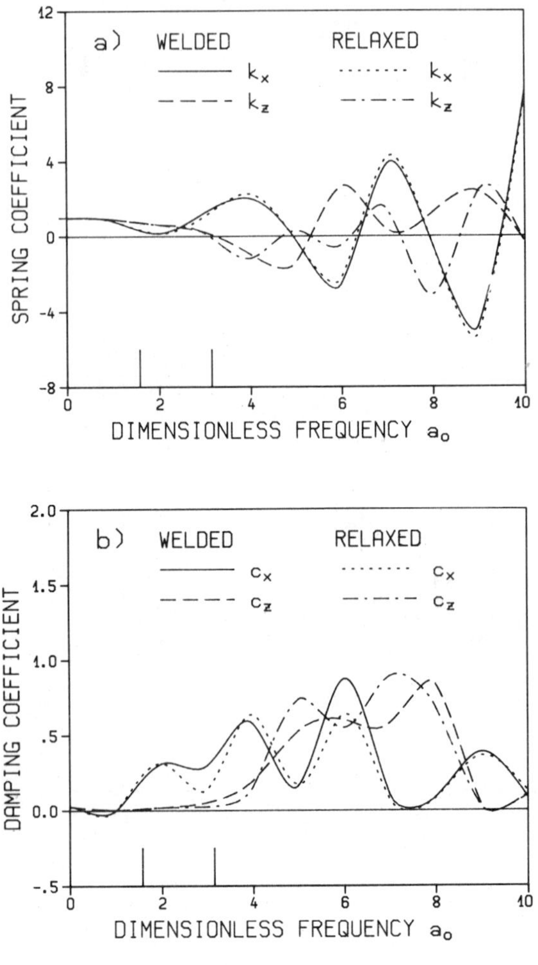

Figure 7-26 Dynamic stiffness of disk, layer built in at its base, no damping, relaxed versus welded contact (Ref. [8]).

which enters only into the static-stiffness matrix $[K_{oo}]$, the half-width b of the equivalent two-dimensional model has to be selected. Arbitrarily choosing $b = a$ when modeling the disk, it is instructive to examine the ratios of the damping coefficient to the spring coefficient for the two- and three-dimensional cases. For a given ω this ratio specifies the damping ratio (up to a factor of 2), as can be concluded from, for example, Eq. 3.54. In Fig. 7-27, $(c/k)_{2-D}/(c/k)_{3-D}$ is plotted for the various degrees of freedom as a function of a_o for the half-plane/half-space. In the vertical direction this ratio is > 1 for $a_o < 8$. Especially in the lower-frequency range a large ratio results, suggesting that the two-dimensional model will significantly underestimate the true three-dimensional dynamic response. For the other degrees of freedom, the

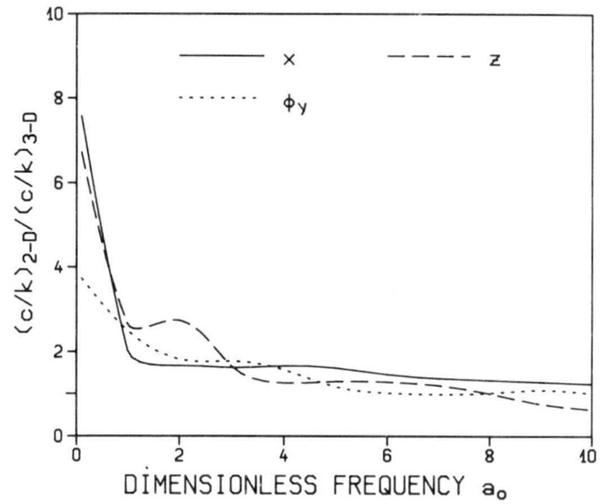

Figure 7-27 Ratio of damping to spring coefficients, half-plane/half-space, no damping, welded contact, two-dimensional versus three-dimensional (Ref. [8]).

ratio is also > 1. However, the horizontal and rocking degrees of freedom are generally coupled, making it more difficult to determine the influence. For another choice of b (i.e., $b \neq a$), the conclusions still apply. The corresponding results are shown in Fig. 7-28 for the layer built in at its base. For this other limiting case of a site, no generally applicable statements can be made, except perhaps the strong dependency on a_o (and thus on the equivalent width of the two-dimensional strip). It is fair to state that a two-dimensional model cannot be used to represent a truly three-dimensional situation.

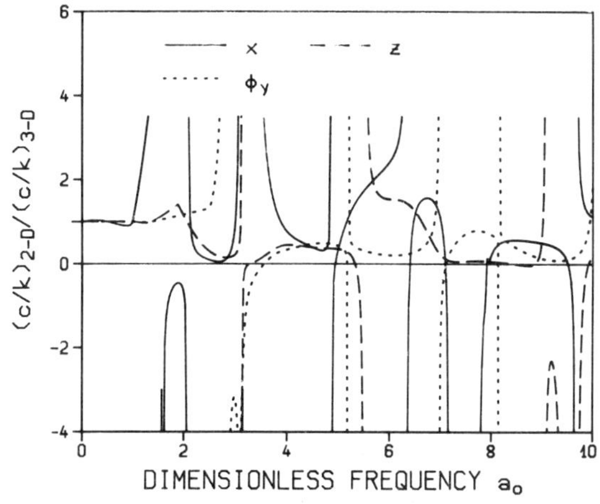

Figure 7-28 Ratio of damping to spring coefficients, layer built in at its base, no damping, welded contact, two-dimensional versus three-dimensional.

7.4.3 Scattered Motion

Finally, the scattered motion $\{u_o^g\}$ for a circular rigid basemat with radius a on the surface of an undamped half-space is calculated. The Poisson's ratio is selected as 0.33. The scattered motion $\{u_o^g\}$ is determined as a function of that of the free-field $\{u_b^f\}$ using Eq. 3.19. The same number of subdisks as described in Section 7.4.1 is used. As already discussed in Section 6.2.5, the effects of the horizontally propagating waves can be characterized essentially by the ratio $\omega a/c_a$, where c_a denotes the apparent velocity (of the motion in the positive x-direction, Fig. 7-29). In the following, relaxed contact is assumed. The free-field

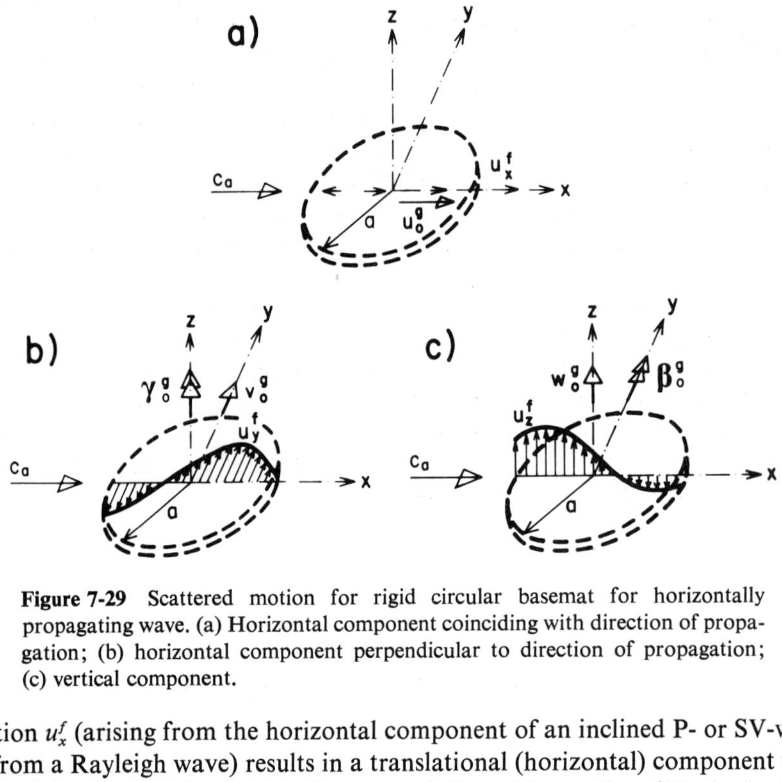

Figure 7-29 Scattered motion for rigid circular basemat for horizontally propagating wave. (a) Horizontal component coinciding with direction of propagation; (b) horizontal component perpendicular to direction of propagation; (c) vertical component.

motion u_x^f (arising from the horizontal component of an inclined P- or SV-wave or from a Rayleigh wave) results in a translational (horizontal) component with the amplitude u_o^g (Fig. 7-29a). Because of the self-canceling (filtering) effect, $|u_o^g|$ will be smaller than $|u_x^f|$ for all $\omega a/c_a$. The real and imaginary parts of u_o^g are plotted versus $\omega a/c_a$ in Fig. 7-30. The out-of-plane motion u_y^f (occurring from an inclined SH-wave or from a Love wave) leads to a translational (horizontal) component with the amplitude v_o^g and to a torsional component with the amplitude γ_o^g (Fig. 7-29b). The latter, which is negligible for small and large ratios $\omega a/c_a$, reaches a maximum at approximately $\omega a/c_a = 2$ (Fig. 7-31). The free-field motion u_z^f (from the vertical component of an inclined P- or SV-wave or from a Rayleigh wave) gives rise to a translational (vertical) component

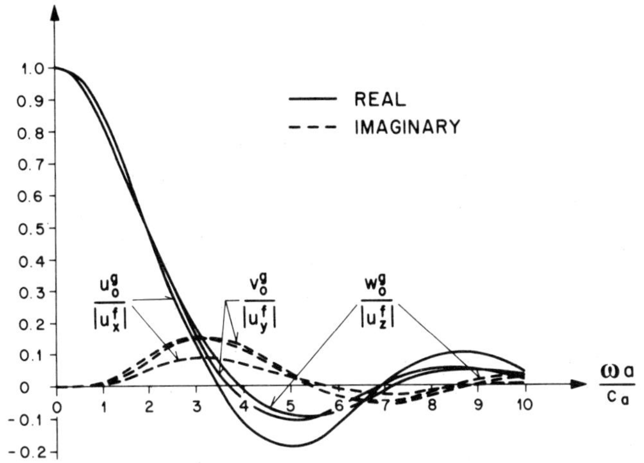

Figure 7-30 Translational components of scattered motion of rigid circular basemat for horizontally propagating wave.

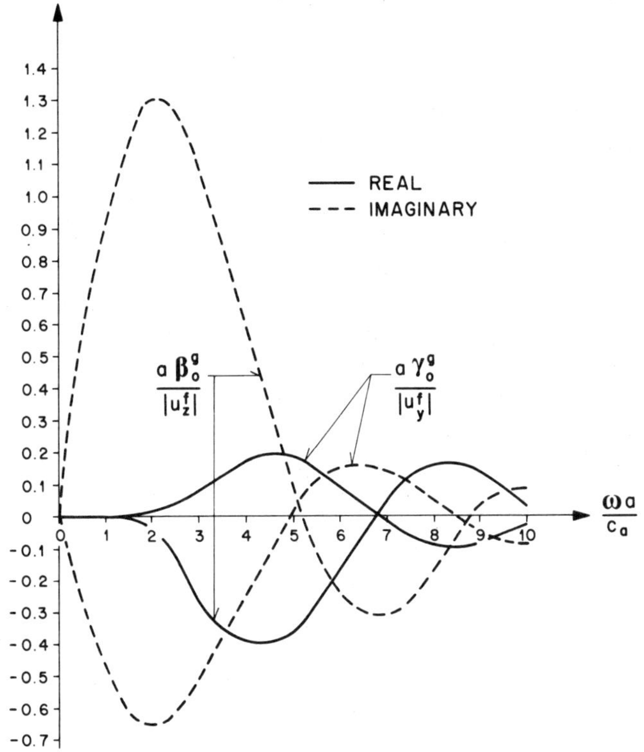

Figure 7-31 Rotational components of scattered motion of rigid circular basemat for horizontally propagating wave.

with the amplitude w_o^g and to a rocking component with the amplitude β_o^g (Fig. 7-29c). While the three translational amplitudes $|u_o^g|$, $|v_o^g|$, and $|w_o^g|$ exhibit a similar variation with $\omega a/c_a$ (Fig. 7-30), the rocking amplitude $|\beta_o^g|$ is roughly twice as large as the torsional one $|\gamma_o^g|$ (Fig. 7-31). The corresponding applied seismic loading is discussed in Section 4.2.2 (Fig. 4-3).

If welded contact were assumed, additional terms would arise in $\{u_o^g\}$. These could, however, be neglected in actual practice. For the free-field motion u_y^f, an additional rocking component with the amplitude α_o^g would occur, for u_x^f additional amplitudes w_o^g, β_o^g, and for u_z^f an additional u_o^g. Furthermore, strictly speaking, the wave effects could no longer be characterized only by c_a. For instance, to be able to evaluate the effect of the horizontally propagating component u_z^f, the other component u_x^f would also have to be considered, which would mean that the type of wave (including its angle of incidence) would have to be specified.

An approximate procedure is described in Problems 4.2, 4.3, and 4.4.

7.5 DYNAMIC-STIFFNESS COEFFICIENTS OF EMBEDDED FOUNDATION

7.5.1 Significance of Boundary-Element Method

The boundary-integral-equation method is well suited for modeling the unbounded soil domain. As analytical solutions exactly satisfying all field equations and the radiation condition at infinity are used as fundamental solutions, only the boundary (i.e., the structure–soil interface) needs to be discretized. Well-established procedures exist for calculating, for a layered half-space, these analytical solutions, which will, in addition, rigorously enforce the conditions at the interface of two adjacent layers and at the free surface. The latter results in only the structure–soil interface having to be modeled by this formulation of the boundary-element method. This term is used to denote the discretized form of the boundary-integral-equation method. The discretization effort and the number of unknowns are thus strongly reduced. For instance, for a three-dimensional case, only a two-dimensional surface has to be addressed. The resulting equations will turn out to be coupled, in contrast to the narrow-banded structure of the finite-element method. Analytically available solutions are used to supplement numerical procedures in this boundary-element method.

Only the so-called indirect boundary-element method is addressed in this text. Other formulations with the same base exist, for example, the direct boundary-element method and other weighted-residual techniques. These are not examined, as the accuracy of the results is, in general, inferior to that of the indirect method for the calculation of the dynamic-stiffness matrix, and the symmetry of this matrix cannot be guaranteed. The direct boundary-element method also does not permit a comparison of the displacement associated with the discretization with the prescribed displacement along the structure–soil interface.

A very simple example, the homogeneous built-in layer in out-of-plane motion, is used for demonstration in Section 7.5.2. The base and the features of the indirect boundary-element method are discussed. As the analytical solution is available, the accuracy can easily be evaluated. The formulations are generalized to the three-dimensional case in Section 7.5.3. In particular, the procedure for calculating $[S^g_{bb}]$ and $[S^f_{bb}]$ is discussed, whereby the latter is easier to obtain, because a regular domain (i.e., the continuous soil without an excavation) is examined. The procedure to calculate Green's influence functions (which are used as fundamental solutions) for distributed loads acting on a layered half-plane is described for out-of-plane and in-plane motions in Section 7.5.4. As an example of a three-dimensional application, the horizontal dynamic-stiffness coefficient of a 2×2 pile foundation is examined in Section 7.5.5.

7.5.2 Example to Illustrate Basic Concepts of Boundary-Element Method

Analytical solution of illustrative example. The static-stiffness coefficient at the free surface for the out-of-plane motion of the semi-infinite homogeneous layer ($x > 0$) built in at its base is calculated (Fig. 7-32). This corresponds to $[S^g_{bb}]$ in the general formulation with $[S^f_{bb}]$ being equal to $2[S^g_{bb}]$. For the sake of conciseness, the superscript g is deleted in this section. The depth is indicated as d, and G denotes the shear modulus. A linear variation of the displacement $v(z)$ associated with the stiffness coefficient is selected at $x = 0$. As the displacement is zero at the base, only one node at the free surface with the displacement v_b is introduced:

$$v(z) = N(z)v_b \tag{7.65}$$

where $N(z)$ is the shape function:

$$N(z) = 1 - \frac{z}{d} \tag{7.66}$$

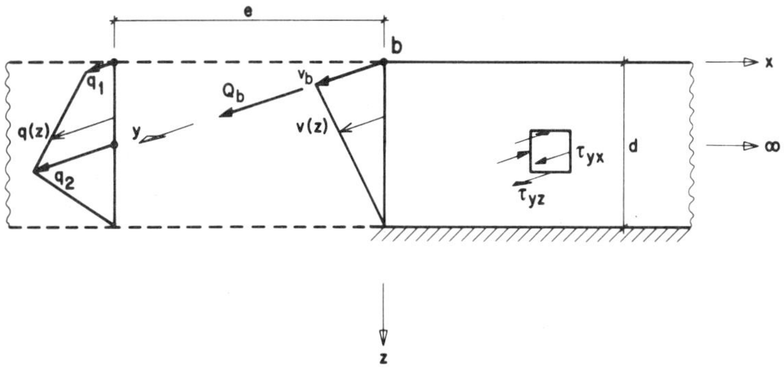

Figure 7-32 Investigated layer.

The static-stiffness coefficient is denoted by S_{bb}. The governing differential equation

$$v_{,xx} + v_{,zz} = 0 \tag{7.67}$$

is transformed by the separation of variables

$$v(x, z) = X(x)Z(z) \tag{7.68}$$

into

$$\frac{X_{,xx}}{X} = -\frac{Z_{,zz}}{Z} = k^2 \tag{7.69}$$

The corresponding solution equals

$$v(x, z) = (a \cos kz + b \sin kz) \exp (\pm kx) \tag{7.70}$$

For a bounded solution at $x \longrightarrow + \infty$, only the negative sign in the argument of the exponential function remains. Enforcing the traction-free condition at $z = 0$,

$$\tau_{yz}(x, 0) = Gv_{,z}(x, 0) = 0 \tag{7.71}$$

leads to $b = 0$. The other boundary condition at the fixed base

$$v(x, d) = 0 \tag{7.72}$$

results in the characteristic equation

$$\cos kd = 0 \tag{7.73}$$

which is satisfied for the discrete values of k_j:

$$k_j d = \frac{2j - 1}{2}\pi \qquad j = 1, 2, 3, \ldots \tag{7.74}$$

The complete solution follows as

$$v(x, z) = \sum_j a_j \cos k_j z \exp (-k_j x) \tag{7.75}$$

The corresponding shear stress $\tau_{yx}(x, z) = Gv_{,x}(x, z)$ is specified as

$$\tau_{yx}(x, z) = -G \sum_j a_j k_j \cos k_j z \exp (-k_j x) \tag{7.76}$$

The analytical solution for the prescribed (linear) displacement at $x = 0$ is derived by expanding $v(z)$ in Eq. 7.65 into the Fourier series with terms $\cos k_j z$:

$$v(z) = \sum_j \frac{2}{d}\left(\int_{z=0}^{d} \left(1 - \frac{z}{d}\right) \cos k_j z \, dz\right) v_b \cos k_j z = \sum_j \frac{2v_b}{(k_j d)^2} \cos k_j z \tag{7.77}$$

This determines the coefficients a_j in Eq. 7.75, and thus also $\tau_{yx}(0, z)$ as a function of v_b. Integrating the surface traction, which is equal to the negative value of $\tau_{yx}(0, z)$, with the shape function $N(z)$ results in the nodal force Q_b:

$$Q_b = -\int_{z=0}^{d} N(z)^T \tau_{yx}(0, z) \, dz \tag{7.78a}$$

which can be rewritten as

$$Q_b = S_{bb} v_b \tag{7.78b}$$

Obviously, $N(z)^T = N(z)$ for a scalar. However, the equation is specified in the form applicable to more than one degree of freedom. The stiffness coefficient S_{bb} is given by

$$S_{bb} = 2G \sum_j \frac{1}{(k_j d)^3} \tag{7.79}$$

This results in the analytical value $S_{bb} = 0.543G$.

As will be demonstrated later, the response for a load acting on the infinite system is used, that is, on the layer extending to infinity in the positive and negative x-directions. This so-called fundamental solution is calculated for a load acting on a line defined by x equals the constant $-e$:

$$q(z) = [L(z)]\{q\} \tag{7.80}$$

where $[L(z)]$ represents the interpolation function and $\{q\}$ the load parameters. In Fig. 7-32, a piecewise linear applied load $q(z)$ with two load parameters q_1 and q_2 is illustrated as an example. Point loads can also be represented by choosing delta Dirac functions for $[L(z)]$. Also for these loading cases, the calculation can be restricted to the semi-infinite system adjacent to the applied load and on which only half of the load acts. The boundary condition on the shear stress for the semi-infinite layer extending in the positive x-direction is formulated as

$$\tau_{yx}(-e, z) = -\tfrac{1}{2}q(z) \tag{7.81}$$

Expanding the load in a Fourier series and equating Eqs. 7.81 and 7.76 determines the coefficients a_j as

$$a_j = \frac{1}{G}\frac{1}{k_j d}\exp(-k_j e)\int_{z=0}^{d} \cos k_j z\,[L(z)]\,dz\,\{q\} \tag{7.82}$$

The corresponding displacement and surface traction along $x = 0$ are functions of $\{q\}$ and are denoted symbolically as

$$v(0, z) = [g_u(z)]\{q\} \tag{7.83}$$

$$-\tau_{yx}(0, z) = [g_t(z)]\{q\} \tag{7.84}$$

where, using Eqs. 7.75, 7.76, and 7.82,

$$[g_u(z)] = \frac{1}{G}\sum_j \frac{1}{k_j d}\exp(-k_j e)\cos k_j z \int_{z=0}^{d} \cos k_j z\,[L(z)]\,dz \tag{7.85}$$

$$[g_t(z)] = \frac{1}{d}\sum_j \exp(-k_j e)\cos k_j z \int_{z=0}^{d} \cos k_j z\,[L(z)]\,dz \tag{7.86}$$

Equations 7.85 and 7.86 apply only to the semi-infinite layer extending to the right of the applied load ($e > 0$), that is, the load acts on the exterior of the investigated domain. For loads applied directly to the semi-infinite layer ($e < 0$), the following equations for $[g_u(z)]$ and $[g_t(z)]$ follow:

$$[g_u(z)] = \frac{1}{G}\sum_j \frac{1}{k_j d}\exp(k_j e)\cos k_j z \int_{z=0}^{d} \cos k_j z\,[L(z)]\,dz \tag{7.87}$$

$$[g_t(z)] = -\frac{1}{d} \sum_j \exp(k_j e) \cos k_j z \int_{z=0}^{d} \cos k_j z \,[L(z)]\, dz \tag{7.88}$$

As expected, when moving the load across the interface $x = 0$ (i.e., from $e = +\epsilon$ to $e = -\epsilon$), $[g_u(z)]$ is continuous while $[g_t(z)]$ shows a discontinuity equivalent to $[L(z)]$, which represents the load for unit-load parameters.

Weighted-residual technique. A physical explanation of the indirect boundary-element method is given first. Assume that a loading pattern exists acting on the part $x < 0$ of the infinite layer and that these loads result along the line $x = 0$ of the infinite layer in the prescribed displacement of unit nodal value associated with the definition of the stiffness coefficient. In the actual formulation this can be achieved by selecting the location of the loads and adjusting their intensities to satisfy this condition. (The line $x = 0$ will subsequently form the structure–soil interface of the semi-infinite layer.) Integrating the corresponding surface traction $-\tau_{yx}$ acting on the line $x = 0$ with the above-mentioned prescribed displacement will lead to the stiffness coefficient. It is important to note that all calculations for the stiffness coefficient of the semi-infinite layer are performed for the infinite layer using the corresponding analytical solution derived above. In general, the results for an infinite system are readily available, in contrast to those of the semi-infinite system with an arbitrary structure–soil interface. As in the actual formulation only a finite number of loads can be chosen, an approximate solution results.

The actual formulation proceeds as follows: Distributed loads and point loads with initially unknown intensities are assumed to act on a source line S' of the infinite system (Fig. 7-33). The line S' is offset toward that part of the infinite system which lies outside the region for which the stiffness coefficient is calculated (i.e., in the negative x-direction). In the limit it can be moved toward S up to a distance ϵ (i.e., $x = 0 - \epsilon$). Selecting an interpolation function $[L(z')]$, the load

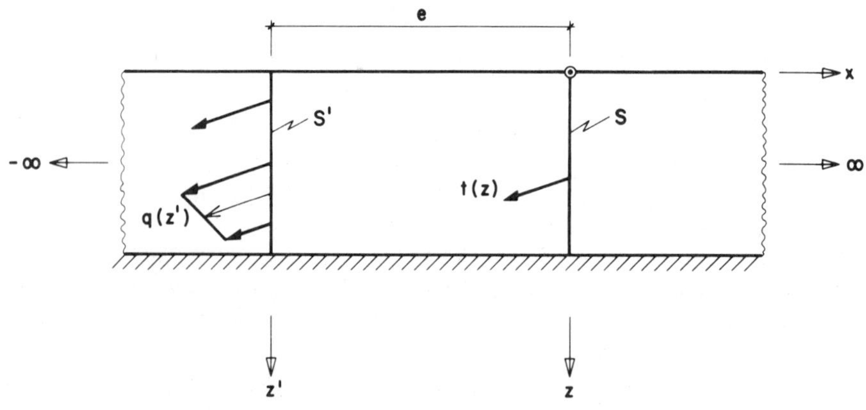

Figure 7-33 Source line and structure–soil interface.

$q(z')$ is expressed as a function of the unknown source parameters $\{q\}$ associated with the values at all nodes:

$$q(z') = [L(z')]\{q\} \tag{7.89}$$

The coordinate z' denotes symbolically a point on the source line. An example of such a load distribution is illustrated on the left-hand side of Fig. 7-33. The displacement $v_p(z)$ and the surface traction $t_p(z)$ on the line S calculated for the infinite layer can be formulated as

$$v_p(z) = [g_u(z)]\{q\} \tag{7.90}$$

$$t_p(z) = [g_t(z)]\{q\} \qquad [= -\tau_{yx}(0, z)] \tag{7.91}$$

The matrices $[g_u(z)]$ and $[g_t(z)]$ contain the influence functions (Green's functions) and are specified in Eqs. 7.85 and 7.86. The prescribed displacement $v(z)$ associated with the stiffness coefficient is specified in Eq. 7.65.

The displacement-boundary condition on the structure–soil interface S can be satisfied only in an average sense as

$$\int_{z=0}^{d} [W(z)]^T(v_p(z) - v(z))\,dz = \{0\} \tag{7.92}$$

which is used to determine the source parameters $\{q\}$. The matrix $[W(z)]$ contains the weighting functions, for which various choices are possible. The number of weighting functions must be equal to that of the source parameters. In the indirect boundary-element method, the matrix of the weighting functions $[W(z]$ is chosen to be equal to the matrix of the Green's functions $[g_t(z)]$.

Substituting Eqs. 7.90 and 7.65 in Eq. 7.92 results in

$$[G]\{q\} = [T]v_b \tag{7.93}$$

where

$$[G] = \int_{z=0}^{d} [g_t(z)]^T[g_u(z)]\,dz \tag{7.94}$$

$$[T] = \int_{z=0}^{d} [g_t(z)]^T N(z)\,dz \tag{7.95}$$

The flexibility matrix $[G]$ is symmetric, as is easily verified using Maxwell–Betti's reciprocity law. For the semi-infinite system, the surface tractions represent the load. Formulating the reciprocity law for the two loading states corresponding to q_i and q_j results in

$$\int_{z=0}^{d} t_p(z)_i^T v_p(z)_j\,dz = \int_{z=0}^{d} t_p(z)_j^T v_p(z)_i\,dz \tag{7.96}$$

where $v_p(z)_j$ and $t_p(z)_i$ are the jth and ith elements of the products $[g_u(z)]\{q\} = g_u(z)_j q_j$ and $[g_t(z)]\{q\} = g_t(z)_i q_i$, respectively. Substituting these relations (Eqs. 7.90 and 7.91) into Eq. 7.96 leads to

$$q_i \int_{z=0}^{d} g_t(z)_i^T g_u(z)_j\,dz\,q_j = q_j \int_{z=0}^{d} g_t(z)_j^T g_u(z)_i\,dz\,q_i \tag{7.97}$$

or, using Eq. 7.94,

$$G_{ij} = G_{ji} \tag{7.98}$$

The concentrated nodal force is obtained as

$$Q_b = \int_{z=0}^{d} N(z)^T t_p(z)\, dz \tag{7.99}$$

Solving Eq. 7.93 for $\{q\}$ and substituting Eq. 7.91 in Eq. 7.99 leads to

$$Q_b = [T]^T[G]^{-1}[T]v_b \tag{7.100}$$

The stiffness coefficient S_{bb} is equal to

$$S_{bb} = [T]^T[G]^{-1}[T] \tag{7.101}$$

For a number of degrees of freedom greater than one, it is obvious that the matrix $[S_{bb}]$ is always symmetric. The choice of the weighting function $[W] = [g_t]$ thus guarantees the symmetry.

As an example, the piecewise-linear load variation with two load parameters q_1 and q_2 shown in Fig. 7-32 acting at $e = d$ is used. The calculated stiffness coefficient equals $0.535G$, which is 98.5% of the exact value. Moving the source line toward the structure–soil interface up to an infinitesimal distance leads to a coefficient which is equal to 99.3% of the exact value. Solving Eq. 7.93 for $\{q\}$ and substituting in Eq. 7.90, the displacement $v_p(z)$ is determined, which can then be compared with the prescribed (linear) displacement $v(z)$.

The same stiffness coefficient can also be derived by starting with the reciprocity law formulated for the loading state of the sources (subscript p) and that corresponding to the enforced displacement $v(z)$ (no subscript):

$$\int_{z=0}^{d} t(z)^T v_p(z)\, dz = \int_{z=0}^{d} t_p(z)^T v(z)\, dz \tag{7.102}$$

Assuming that the surface traction $t(z)$ corresponding to $v(z)$ can be expressed as a function of the same source parameters $\{q\}$,

$$t(z) = [g_t(z)]\{q\} \tag{7.103}$$

and making the appropriate substitutions in Eq. 7.102 leads to Eq. 7.101.

7.5.3 Dynamic-Stiffness Coefficients of Soil Domains Calculated by Boundary-Element Method

Reference soil systems. The three reference soil systems are shown in Fig. 7-34: the continuous free-field system f, the soil with excavation (system ground g) and the excavated part of the soil (system e). For instance, the matrix $[S^g_{bb}]$ specifies the amplitudes of the forces $\{P_b\}$ due to displacements of unit amplitude $\{u_b\}$ that are applied at the nodes b of the structure–soil interface for harmonic motion with the excitation frequency ω:

$$\{P_b\} = [S^g_{bb}]\{u_b\} \tag{7.104}$$

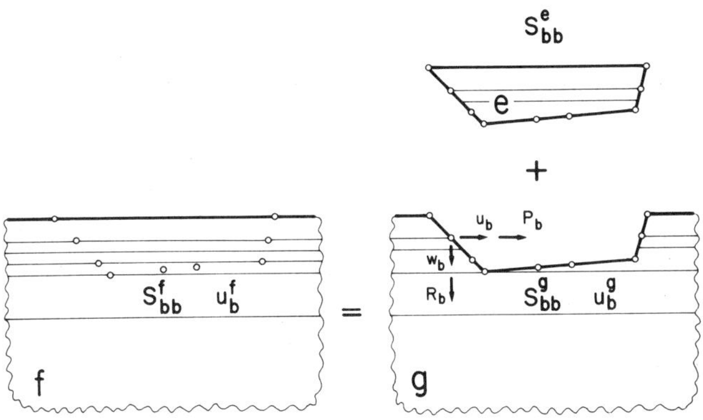

Figure 7-34 Reference soil systems (Ref. [9]).

The vectors $\{P_b\}$ and $\{u_b\}$ contain the amplitudes of the forces P_b, Q_b, and R_b and of the displacements u_b, v_b, and w_b, respectively, in all nodes. This is illustrated in Fig. 7-34 for the two-dimensional case.

Generalization to three dimensions for system ground. The formulation specified for the out-of-plane motion in Section 7.5.2 is generalized to three dimensions. At the same time, the harmonic response is addressed. In Fig. 7-35, the nomenclature is illustrated for the calculation of $[S^g_{bb}]$ for the in-plane motion (i.e., for two dimensions). The prescribed displacement amplitudes on the structure–soil interface S are specified as

$$\{u(s)\} = [N(s)]\{u_b\} \tag{7.105}$$

where $\{u(s)\}$ contains the three elements $u(s)$, $v(s)$, and $w(s)$ in the directions of the three coordinate axes x, y, and z. For a rigid foundation, $[N(s)]$ expresses rigid-body kinematics. The vector $\{u_b\}$ (which could also be denoted as $\{u_o\}$) then represents the rigid-body degrees of freedom.

The variation of load amplitudes $\{p(s')\}$ with the three components $p(s')$, $q(s')$, and $r(s')$ with initially unknown intensities $\{p\}$ is assumed to act on a source line S'. The line S' is always offset toward the soil region to be excavated, in the limit by an infinitesimal amount. These so-called source loads act on the dynamic system consisting of the continuous soil (i.e., on the layered half-space without excavation). The vector $\{p(s')\}$ is expressed as a function of the source parameters $\{p\}$ as

$$\{p(s')\} = [L(s')]\{p\} \tag{7.106}$$

Discontinuities can be introduced as shown in Fig. 7-35c. The number of source parameters has to be larger than or equal to the order of $\{u_b\}$. The amplitudes of the displacement $\{u_p(s)\}$ with the components $u_p(s)$, $v_p(s)$, and $w_p(s)$ and of the surface traction $\{t_p(s)\}$ with the components $t_{px}(s)$, $t_{py}(s)$, and $t_{pz}(s)$ on the

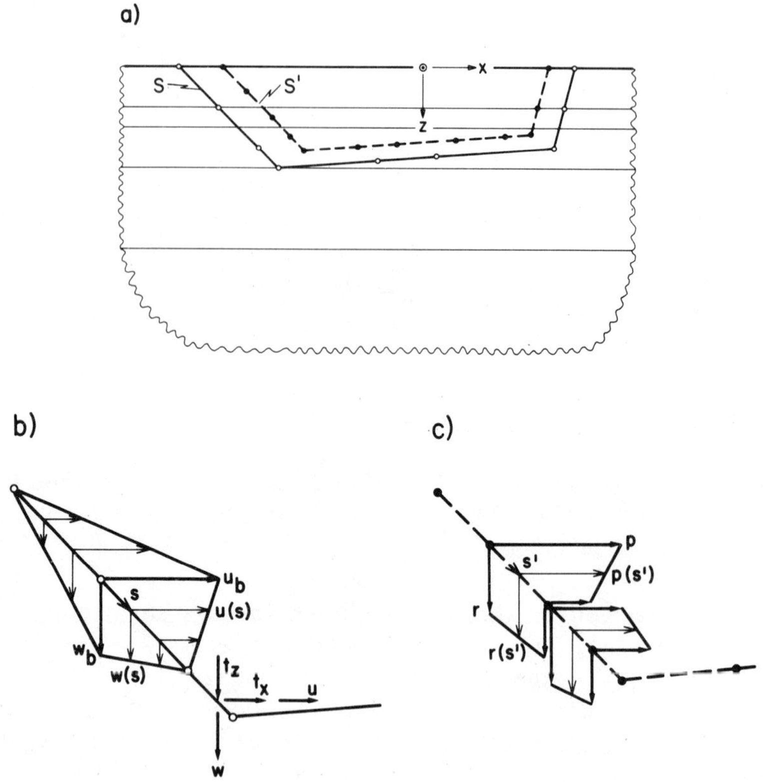

Figure 7-35 Elements of discretization (Ref. [9]). (a) Source line and structure–soil interface; (b) prescribed displacement; (c) selected load distribution.

surface S, which subsequently will form the structure–soil interface, are formulated as

$$\{u_p(s)\} = [g_u(s)]\{p\} \tag{7.107}$$

$$\{t_p(s)\} = [g_t(s)]\{p\} \tag{7.108}$$

The calculation of the Green's functions $[g_u(s)]$ and $[g_t(s)]$ of the continuous layered half-space is discussed in Section 7.5.4. One may note that identical Green's functions arise for all the elements of the source line S' which differ by a horizontal translation. This results in a saving in the computational effort associated with the calculation of the Green's functions. The generalized strain-displacement matrix is specified as

$$[T] = \int_S [g_t(s)]^T[N(s)]\, ds \tag{7.109}$$

and the flexibility matrix as

$$[G] = \int_S [g_t(s)]^T[g_u(s)]\, ds \tag{7.110}$$

The generalized stress-displacement relationship is equal to

$$[G]\{p\} = [T]\{u_b\} \tag{7.111}$$

The coefficient matrix of the equilibrium equation is given by $[T]^T$. Finally, the dynamic-stiffness matrix follows as

$$[S^g_{bb}] = [T]^T[G]^{-1}[T] \tag{7.112}$$

The procedure discussed in Section 7.2 for a surface foundation is obviously a special case of this indirect boundary-element method, as for this case $[g_t(s)] = [L(s)]$.

For an undamped system, the method breaks down for a frequency equal to one of the natural frequencies of the soil region bounded by the source line S' and the free surface, built in along the former and free to move along the latter (Fig. 7-35a). The motion is limited to this soil region, and no response arises along the line S. If there is no offset, this will thus arise for the natural frequencies of the excavated part of the soil fixed along the structure–soil interface. The easiest way to circumvent this problem is to calculate the dynamic-stiffness coefficients just below and above these frequencies and then to interpolate the results.

System excavated part. To calculate the dynamic-stiffness matrix of the excavated part $[S^e_{bb}]$, the same equations apply. As the source line S' always has to be selected as lying outside the investigated domain, it is placed on the other side of the structure–soil interface S than indicated in Fig. 7-35a.

In practice, $[S^e_{bb}]$ can also be calculated using finite elements (Section 7.1.1). This procedure, which will involve the elimination of all nodes not lying on S, is straightforward, as the excavated part represents a bounded domain.

System free field. Although for the continuous system f the line S is not a boundary, the concept of the boundary-element method can still be applied. The loaded source line S' has to coincide with the line S to be able to determine $[S^f_{bb}]$. These applied loads $\{p(s)\}$ specified in Eq. 7.106 are integrated with the shape function $[N(s)]$ to determine the concentrated forces $\{P_b\}$ for the indirect boundary-element method. The matrix $[g_t(s)]$ in Eq. 7.108 is thus replaced by $[L(s)]$, which affects the $[T]$ matrix (Eq. 7.109) and the $[G]$ matrix (Eq. 7.110). The surface tractions $\{t_p(s)\}$ appearing in Eq. 7.108 are thus replaced by the load $\{p(s)\}$.

It is important to realize that for the calculation of $[S^f_{bb}]$, $[g_t(s)]$ is not needed and thus does not have to be calculated. Only one Green's function ($[g_u(s)]$) is determined in this case.

As the calculation of $[S^f_{bb}]$ is simpler than $[S^g_{bb}]$, using boundary-element methods, it can be advantageous to calculate $[S^e_{bb}]$ by the finite-element method and substract it from $[S^f_{bb}]$ (Eq. 3.6) to get $[S^g_{bb}]$. In the vicinity of the natural frequencies of the excavated part of the soil built in along the line S, the dynamic-stiffness matrix $[S^e_{bb}]$ will be very large, especially for a soil with small material damping. The same also applies to $[S^f_{bb}]$. As $[S^g_{bb}]$ is calculated as the difference of two large numbers, care must be taken for this range of frequencies when discretizing, especially as $[S^e_{bb}]$ and $[S^f_{bb}]$ are calculated by two different methods (see Section 7.6.4).

Finally, it should be noted that $[S^g_{bb}]$ and $[S^e_{bb}]$ can be calculated using the surface tractions $\{t^g_p(s)\}$ and $\{t^e_p(s)\}$, respectively, resulting from the load $\{p(s)\}$ applied on the line S. The vector $\{t^e_p(s)\}$ denotes the surface traction on that side which subsequently will be excavated and $\{t^g_p(s)\}$ the one on the other side. The following relationship exists:

$$\{t^g_p(s)\} + \{t^e_p(s)\} = \{p(s)\} \tag{7.113}$$

Using Eqs. 7.108 and 7.106 leads to

$$[g^g_t(s)] + [g^e_t(s)] = [L(s)] \tag{7.114}$$

7.5.4 Green's Influence Functions

In the boundary-element method, the so-called fundamental solution (i.e., the displacement and surface-traction amplitudes on the line S which subsequently will form the structure–soil interface), is needed for applied distributed and point loads acting on the source line S' of the continuous system (i.e., of the soil without excavation). These Green's functions are thus determined for the free-field system (Fig. 7-34).

In the following, the equations are summarized for a Cartesian coordinate system for out-of-plane and in-plane motions. The former occurs along the y-axis with the displacement amplitudes v, the latter in the x-z plane with the amplitudes u and w. The corresponding amplitudes of the distributed loads are denoted by q, and p, r, respectively.

As an example, the linearly distributed load $p(s')$ acting in the x-direction with nodal values p_1 and p_2 is shown in the upper part of Fig. 7-36. As only part of the layer is loaded, an additional interface is introduced in node 1. The layer on which the distributed load acts is first fixed at the two interfaces. The corresponding reaction forces (external loads) P_1, R_1, P_2, and R_2 are calculated to achieve this condition, whereby the analysis is restricted to the loaded layer. They are then applied with the opposite sign to the total system as shown in the lower part of this figure. To this global response, the analysis of the fixed layer has to be added to calculate the total one. The analysis of the total system for loads acting at the interfaces based on the direct stiffness approach is described in depth in Section 6.2.1. The analysis of the fixed layer is performed for a distributed load acting on a line $x =$ constant in Sections 5.3.4 and 5.4.4 for the out-of-plane and in-plane motions, respectively. Thus only the modifications for the inclined line are specified in the following.

It is appropriate to discuss the out-of-plane case first. The load amplitude

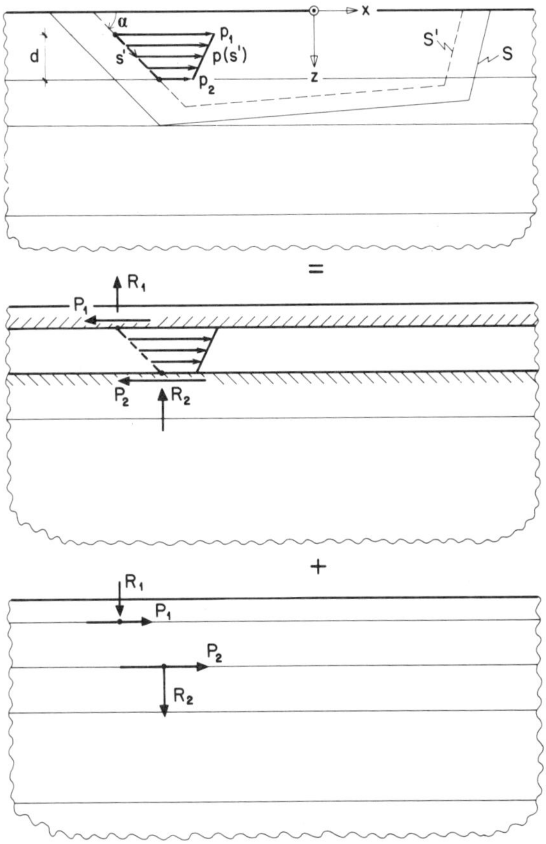

Figure 7-36 Distributed load acting on inclined line over part of layer (Ref. [10]).

$q(s')$ arising from the nodal values q_1 and q_2 is formulated as

$$q(s') = \left[q_1 + (q_2 - q_1)\frac{z}{d} \right] \delta(z - x \tan \alpha) \tag{7.115}$$

where δ is the Dirac delta function, d the depth of the loaded layer, and α the angle of the line S' measured from the horizontal (see Fig. 7-36). The load is expanded in the horizontal direction into a Fourier integral with terms $\exp(-ikx)$, k being the wave number, as (Eq. 7.41a)

$$\begin{aligned} q(k, z) &= \frac{1}{2\pi} \int_{-\infty}^{\infty} q(x, z) \exp(ikx)\, dx \\ &= \frac{1}{2\pi} \left[q_1 + (q_2 - q_1)\frac{z}{d} \right] \exp(ik \cot \alpha\, z) \end{aligned} \tag{7.116}$$

The dynamic-equilibrium equation for harmonic motion expressed in terms of $v(k, z)$ with $v(k, z, x) = v(k, z) \exp(-ikx)$ is equal to (Eq. 5.113)

$$G^*[-k^2 v(k, z) + v_{,zz}(k, z)] = -\rho\omega^2 v(k, z) - q(k, z) \tag{7.117}$$

where G^* and ρ denote the (complex) shear modulus and the mass density. By inspection, a particular solution equals

$$v^p(k, z) = \frac{1}{2\pi}\left(b_1 + b_2\frac{z}{d}\right) \exp(ik \cot\alpha\, z) \tag{7.118}$$

with

$$b_1 = \frac{d^2}{G^*}\frac{1}{(kd)^2(\cot^2\alpha - t^2)}\left[q_1 + \frac{2i\cot\alpha}{kd(\cot^2\alpha - t^2)}(q_2 - q_1)\right] \tag{7.119}$$

$$b_2 = \frac{d^2}{G^*}\frac{1}{(kd)^2(\cot^2\alpha - t^2)}(q_2 - q_1) \tag{7.120}$$

where t is specified in Eq. 5.95b.

With this particular solution specified, the method proceeds as in Section 5.3.4. Then formulating the inverse Fourier transform leads to the Green's influence functions in the space domain (Eq. 7.41b).

The procedure is analogous for the in-plane motion. For a linear variation of the loads in the x- and z-directions, the amplitudes in the k-domain are equal to (see Eq. 7.116)

$$p(k, z) = \frac{1}{2\pi}\left[p_1 + (p_2 - p_1)\frac{z}{d}\right] \exp(ik \cot\alpha\, z) \tag{7.121}$$

$$r(k, z) = \frac{1}{2\pi}\left[r_1 + (r_2 - r_1)\frac{z}{d}\right] \exp(ik \cot\alpha\, z) \tag{7.122}$$

The dynamic-equilibrium equations formulated in terms of $u(k, z)$ and $w(k, z)$ with the variation in x-direction $\exp(-ikx)$ being implied are equal to (Eq. 5.142)

$$\begin{aligned} -(\lambda^* + 2G^*)k^2 u(k, z) + G^* u_{,zz}(k, z) - ik(\lambda^* + G^*)w_{,z}(k, z) \\ = -\rho\omega^2 u(k, z) - p(k, z) \end{aligned} \tag{7.123a}$$

$$\begin{aligned} -ik(\lambda^* + G^*)u_{,z}(k, z) - k^2 G^* w(k, z) + (\lambda^* + 2G^*)w_{,zz}(k, z) \\ = -\rho\omega^2 w(k, z) - r(k, z) \end{aligned} \tag{7.123b}$$

where λ^* denotes one of the Lamé's constants. A particular solution is specified as

$$u^p(k, z) = \frac{1}{2\pi}\left(a_1 + a_2\frac{z}{d}\right) \exp(ik \cot\alpha\, z) \tag{7.124a}$$

$$w^p(k, z) = \frac{1}{2\pi}\left(c_1 + c_2\frac{z}{d}\right) \exp(ik \cot\alpha\, z) \tag{7.124b}$$

Substituting Eq. 7.124 in Eq. 7.123 and identifying the constant and linear terms leads to the following system of equations for the four constants:

$$\begin{bmatrix} \frac{c_p^{*2}}{c_s^{*2}} s^2 - \cot^2 \alpha & \frac{2i}{kd} \cot \alpha & \left(\frac{c_p^{*2}}{c_s^{*2}} - 1\right) \cot \alpha & -\frac{i}{kd}\left(\frac{c_p^{*2}}{c_s^{*2}} - 1\right) \\ 0 & \frac{c_p^{*2}}{c_s^{*2}} s^2 - \cot^2 \alpha & 0 & \left(\frac{c_p^{*2}}{c_s^{*2}} - 1\right) \cot \alpha \\ \left(\frac{c_p^{*2}}{c_s^{*2}} - 1\right) \cot \alpha & -\frac{i}{kd}\left(\frac{c_p^{*2}}{c_s^{*2}} - 1\right) & t^2 - \frac{c_p^{*2}}{c_s^{*2}} \cot^2 \alpha & \frac{2i}{kd} \frac{c_p^{*2}}{c_s^{*2}} \cot \alpha \\ 0 & \left(\frac{c_p^{*2}}{c_s^{*2}} - 1\right) \cot \alpha & 0 & t^2 - \frac{c_p^{*2}}{c_s^{*2}} \cot^2 \alpha \end{bmatrix} \begin{Bmatrix} a_1 \\ a_2 \\ c_1 \\ c_2 \end{Bmatrix}$$

$$= -\frac{d^2}{G^*} \frac{1}{(kd)^2} \begin{Bmatrix} p_1 \\ p_2 - p_1 \\ r_1 \\ r_2 - r_1 \end{Bmatrix} \qquad (7.125)$$

where s and t are specified in Eq. 5.128. From here on, the procedure is straightforward and no details will be specified.

7.5.5 Pile Foundation

Three-dimensional cases can also be calculated by the boundary-element method. In particular, cylindrical coordinates can be introduced. This topic is not treated in depth in this text. Only the principle is illustrated.

Three-dimensional loads acting on a layered half-space (Fig. 7-37) can also be processed by choosing a Fourier series in the circumferential direction θ and Bessel functions involving the wave number k in the radial direction r (Sections 5.5.2 and 7.2.3). This allows the dynamic-stiffness matrix of three-dimensional embedded foundations and of pile groups, including pile–soil–pile interaction, to be calculated. The source loads can be selected to act along the axes of the piles. The computational procedure is outlined schematically in Fig. 7-37.

As an example, a foundation with 2×2 end-bearing piles in a homogeneous layer of soil of depth d (built in at its base) is investigated (Fig. 7-38). The radius a of a pile and the distance l equal $d/40$ and $d/6.67$, respectively. Poisson's ratio $= 0.33$. The ratio of the modulus of elasticity of the pile to that of the soil equals 200, the ratio of the mass densities 1.67. The damping ratio equals 0.05. The horizontal dynamic-stiffness coefficient nondimensionalized with the static value, is decomposed into k_x and $ia_o c_x$, whereby the dimensionless frequency $a_o = \omega a / c_s$ is referred to the radius of the pile. The spring and damping coefficients are shown in Fig. 7-39. The influence of the cutoff frequency of the layer at $\pi/80$ is clearly visible.

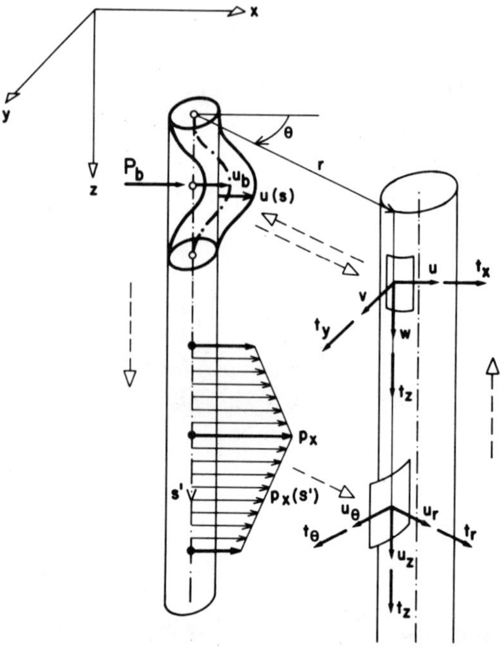

Figure 7-37 Computational procedure for dynamic-stiffness coefficients of pile groups: prescribed displacement, source load with corresponding surface traction and displacement (Ref. [11]).

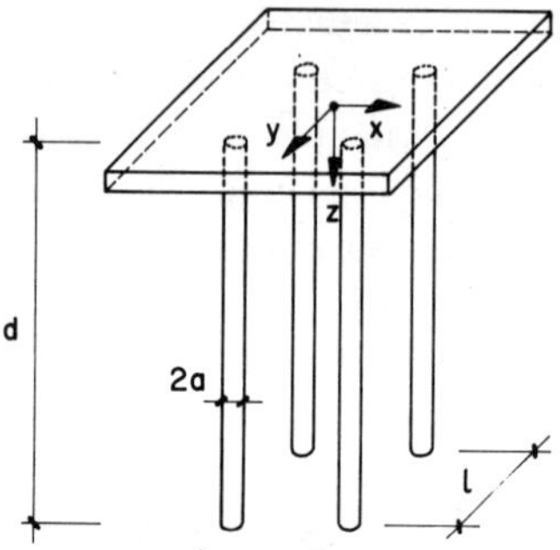

Figure 7-38 Foundation with 2 × 2 piles (after Ref. [12]).

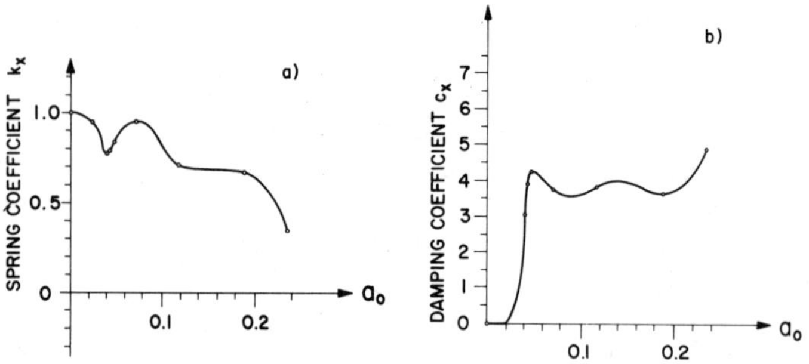

Figure 7-39 Horizontal dynamic-stiffness coefficient of pile foundation (after Ref. [12]).

7.6 EMBEDDED RECTANGULAR FOUNDATION

7.6.1 Scope of Investigation

To evaluate the influence of embedment on the dynamic-stiffness matrix of the soil, the two extreme cases of a two-dimensional site, the half-plane and the homogeneous layer with depth d built in at its base, are examined (Fig. 7-40). The rectangular soil–structure interface, along which welded contact is assumed, is regarded as rigid. The ratio of embedment to half-width h/b is varied parametrically. Poisson's ratio equals 0.33. The damping ratio equals 0.001, which represents the undamped case. For the layer, a damping ratio of 0.05 is also investigated. The various dynamic-stiffness matrices $[S_{oo}]$ are decomposed as follows:

$$[S_{oo}] = [K_{oo}]([k] + ia_o[c]) \tag{7.126}$$

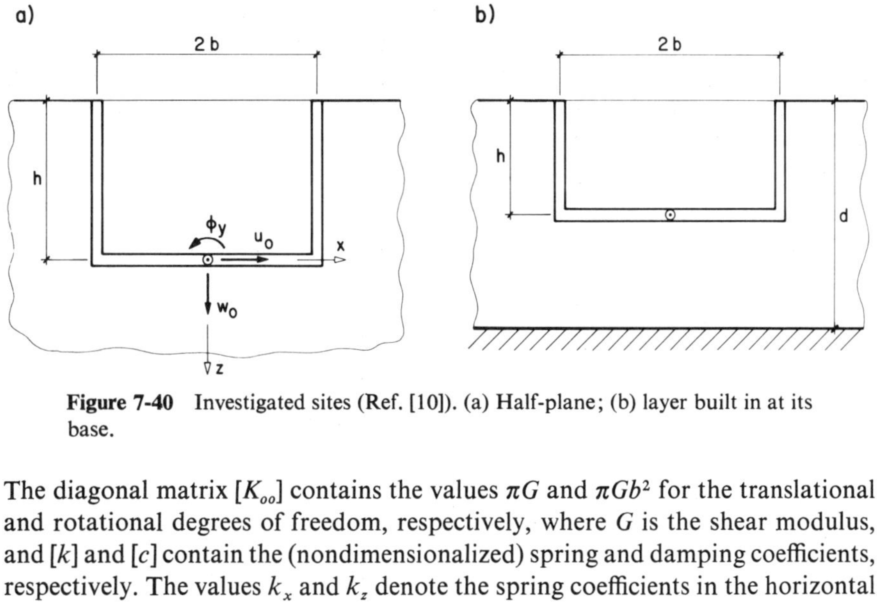

Figure 7-40 Investigated sites (Ref. [10]). (a) Half-plane; (b) layer built in at its base.

The diagonal matrix $[K_{oo}]$ contains the values πG and $\pi G b^2$ for the translational and rotational degrees of freedom, respectively, where G is the shear modulus, and $[k]$ and $[c]$ contain the (nondimensionalized) spring and damping coefficients, respectively. The values k_x and k_z denote the spring coefficients in the horizontal and vertical directions; k_{ϕ_y} is used for rocking (referred to the center of the rigid basemat, Fig. 7-40a), and $k_{x\phi_y}$ represents the coupling between the horizontal direction and rocking. The dimensionless frequency a_o is defined as

$$a_o = \frac{\omega b}{c_s} \tag{7.127}$$

where c_s is the shear-wave velocity.

In all calculations of the indirect boundary-element method, the source line S' coincides with or is offset by an infinitesimal amount from the structure–soil interface. If a finite offset were selected, the accuracy of the results would be reduced. Piecewise linearly distributed source loads are used. On the horizontal part of the source line four elements are used per half-width. In the vertical direction, three elements are chosen for $h/b = 0.5$, four elements for $h/b = 1$, and six elements for $h/b = 2$. The load of the source varies linearly along an element with continuity enforced at all nodes except the corner one. This results in a total of 18, 20, and 24 source parameters per half-width, respectively, for the three cases. If point loads instead of distributed loads were applied, the quality of the results would be reduced.

Before presenting the dynamic-stiffness coefficients of the parametric study, typical Green's functions are discussed. In addition, for one case all results are examined, that is, the loads acting on the source line, the surface tractions, and the displacements on the structure–soil interface. The properties of the dynamic-stiffness coefficients of the three reference soil systems are addressed and the relationships between them investigated.

7.6.2 Green's Influence Function

The foundation with $h/b = 0.5$ embedded in a half-plane for $a_o = 1$ is used to present Green's functions. The influence functions for a horizontal load whose amplitude varies linearly from unity at the free surface to zero at one-third of the embedment and which acts on a line in the continuous soil, which will subsequently form the vertical part of the structure–soil interface shown in Fig. 7-41a, are investigated. This loading distribution corresponds to a source parameter of unity at the node on the free surface. The source line is selected to coincide with the structure–soil interface. The corresponding amplitudes of the displacements in the horizontal and vertical directions u_p and w_p, which are the two elements of the column of $[g_u(s)]$ (Eq. 7.107) associated with the source parameter, are shown in Fig. 7-41a and b. Figure 7-41c and d shows the corresponding amplitudes of the line tractions t_{px} and t_{pz}, which are the elements of a column of $[g_t(s)]$ (Eq. 7.108).

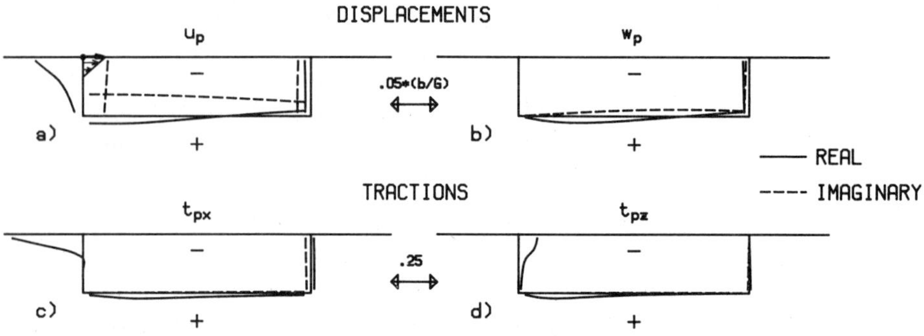

Figure 7-41 Green's functions for source load of unit amplitude (Ref. [11]).

7.6.3 Complete Set of Results

For the same case, all results are presented, that is, the amplitudes of the loads $\{p(s')\}$ acting on the source line (Eqs. 7.106 and 7.111), the amplitudes of the surface tractions $\{t_p(s)\}$ (Eq. 7.108), and the corresponding amplitudes of the surface displacements $\{u_p(s)\}$ (Eq. 7.107) on the structure–soil interface. Figure 7-42a, b, and c corresponds to the prescribed displacement amplitudes u_o, w_o, and ϕ_y, respectively (Fig. 7-40). It should be noted how well $\{u_p(s)\}$ (Eq. 7.107) agrees with the prescribed values $\{u(s)\}$ (Eq. 7.105).

7.6.4 Properties of Dynamic-Stiffness Coefficients of Excavated Part and System Free Field

The calculation of the dynamic-stiffness matrix of the free field $[S^f_{bb}]$ is computationally simpler than that of the reference system ground $[S^g_{bb}]$. The two matrices differ by the dynamic-stiffness matrix of the excavated part of the soil $[S^e_{bb}]$, whose properties are first examined.

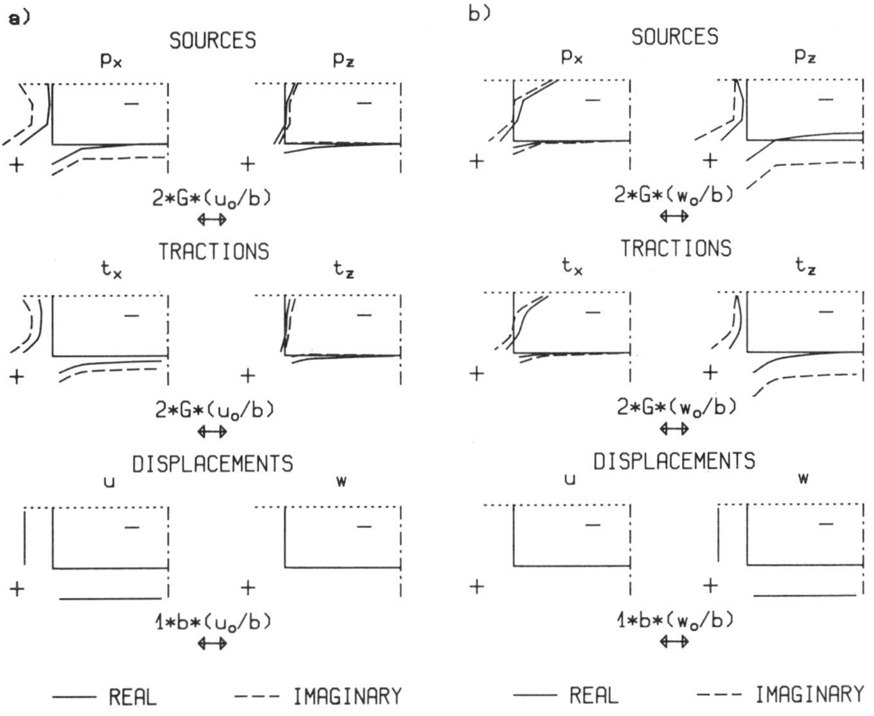

Figure 7-42 Variation of sources, tractions, and displacements for (Ref. [11]): (a) Prescribed horizontal displacement; (b) prescribed vertical displacement;

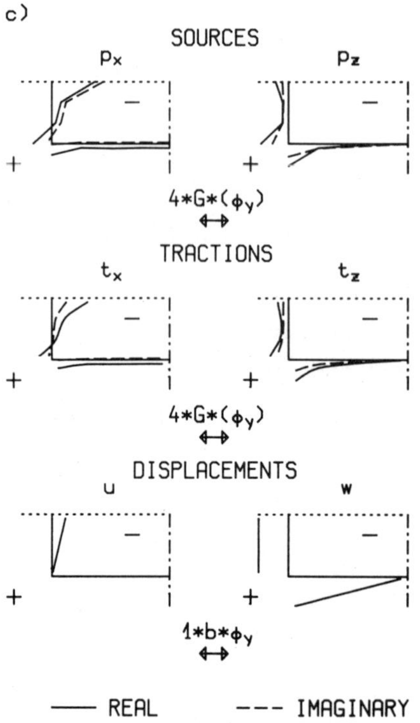

Figure 7-42 (Continued) (c) prescribed rocking displacement.

The spring coefficients of the undamped system *e* with $h/b = 2$ are shown in Fig. 7-43. The corresponding damping coefficient is identically equal to zero. As the excavated part is a bounded domain, vibrational modes associated with natural frequencies can be calculated. Using a very fine finite-element discretization (with 50 elements per half-width of the foundation) of the undamped excavated part built in along the structure–soil interface, the four lowest dimen-

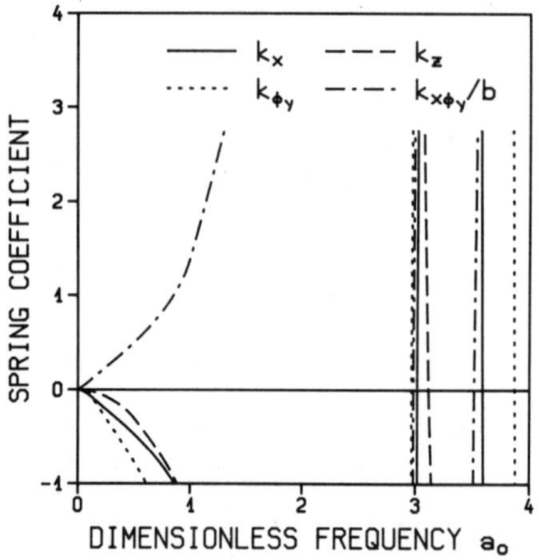

Figure 7-43 Spring coefficient of excavated part of soil, $h/b = 2$, no damping (Ref. [10]).

TABLE 7-2 Dimensionless Natural Frequencies of Excavated Part Built in along the Structure–Soil Interface

Symmetric Mode	Antimetric Mode
2.11	2.93
4.56	3.04
4.77	4.05
5.67	4.66

sionless natural frequencies are determined and presented in Table 7-2. The symmetric and antimetric modes apply to the vertical degree of freedom w_o and the horizontal and rocking degrees of freedom u_o and ϕ_y, respectively. It is easily verified that at these dimensionless natural frequencies, the corresponding spring coefficients are infinite. The zero values of k_x, k_z, and $k_{\phi v}$ in Fig. 7-43 correspond to the natural frequencies of the excavated part having a rigid-body constraint enforced along the structure–soil interface, and being supported as specified in Fig. 7-44a, b and c, respectively. The corresponding frequencies for the excavated part with these released degrees of freedom are specified in Table 7-3.

The corresponding dynamic-stiffness coefficients of the reference system ground of the undamped half-plane are shown in Fig. 7-45. As is to be expected,

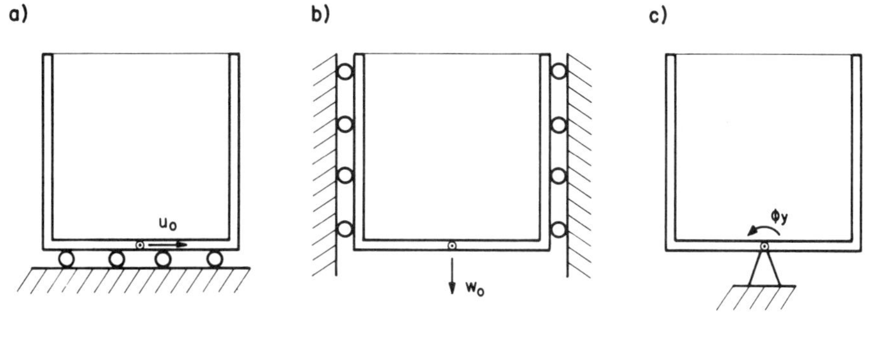

Figure 7-44 Support conditions of excavated part of soil resulting in zero spring coefficients (Ref. [10]). (a) Horizontal; (b) vertical; (c) rocking.

TABLE 7-3 Dimensionless Natural Frequencies of Excavated Part with Released Degrees of Freedom

u_o	w_o	ϕ_y
0	0	0
3.02	3.12	2.97
3.59	4.74	3.87
4.44	5.18	4.59

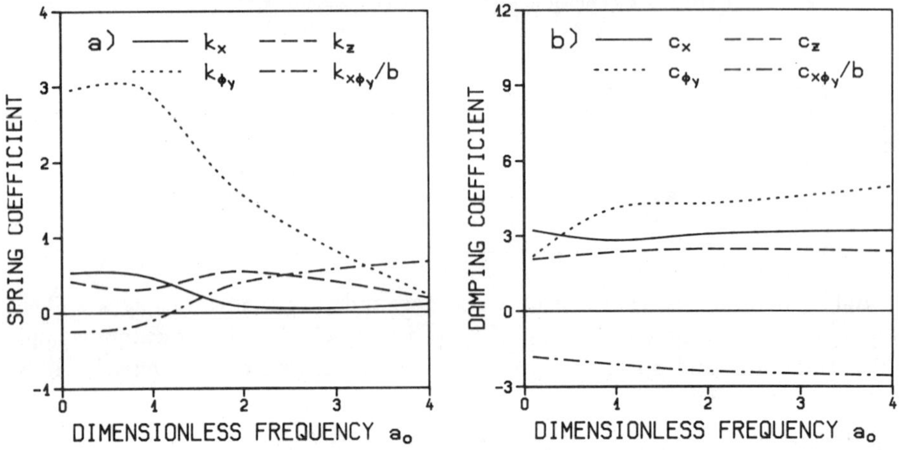

Figure 7-45 Dynamic stiffness of system ground, half-plane, $h/b = 2$, no damping (Ref. [11]).

the spring and damping coefficients are smooth curves for this unbounded domain. Adding $[S^e_{bb}]$ (Fig. 7-43) to these values $[S^g_{bb}]$ (Fig. 7-45) results in the dynamic-stiffness matrix of the free field $[S^f_{bb}]$. The corresponding spring coefficient is shown in Fig. 7-46. Infinite values for the spring coefficients at the same frequencies as for the excavated part (Table 7-2) appear. The damping coefficients of the free field are equal to those of the ground (Fig. 7-45b) for this undamped system.

In Section 7.5.3 it is suggested to calculate $[S^g_{bb}]$ by subtracting $[S^e_{bb}]$ from $[S^f_{bb}]$. In the vicinity of the natural frequencies of the excavated part built in along the structure–soil interface (Table 7-2), a small difference of two large numbers will occur for the spring coefficient. This can be problematic if the two systems f and e are discretized using different methods.

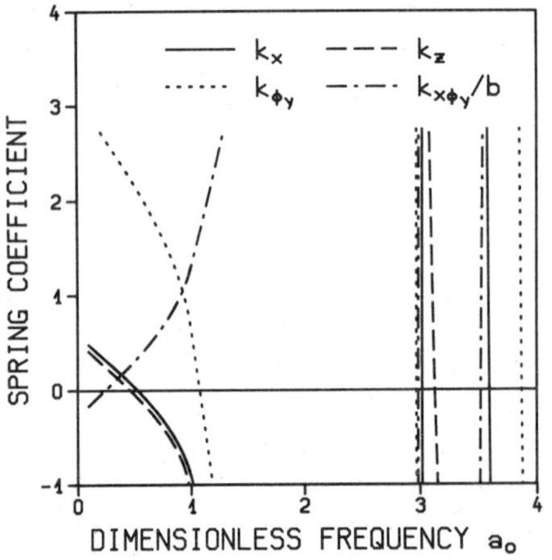

Figure 7-46 Spring coefficient of free field, half-plane, $h/b = 2$, no damping (Ref. [10]).

7.6.5 Parametric Study System Ground

The results for the (undamped) half-plane (Fig. 7-40a) are discussed first. For comparison the dynamic-stiffness coefficients are presented for the surface foundation ($h/b = 0$) in Fig. 7-11. For the embedded foundation, the results are shown in Fig. 7-47 for $h/b = 0.5$, in Fig. 7-48 for $h/b = 1$, and in Fig. 7-45 for $h/b = 2$. Quite surprisingly, the two translational spring coefficients are not strongly affected by embedment, while the increase of the corresponding damping coefficients is approximately proportional to the embedment, the increase in the

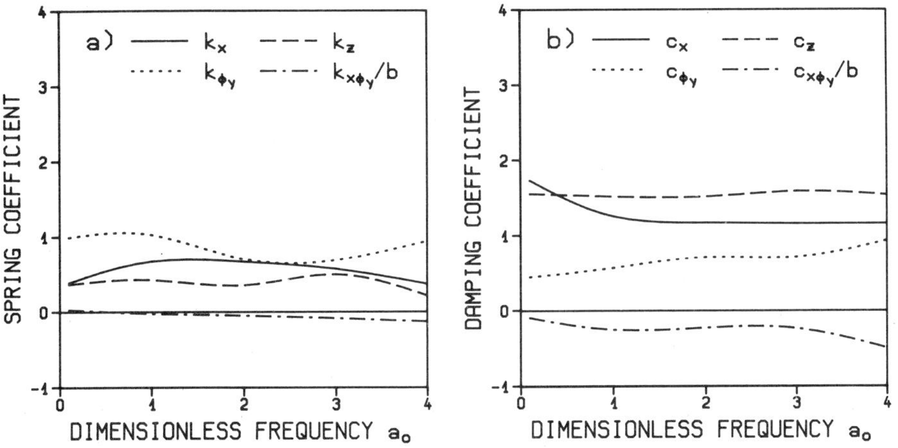

Figure 7-47 Dynamic stiffness of system ground, half-plane, $h/b = 0.5$, no damping (Ref. [11]).

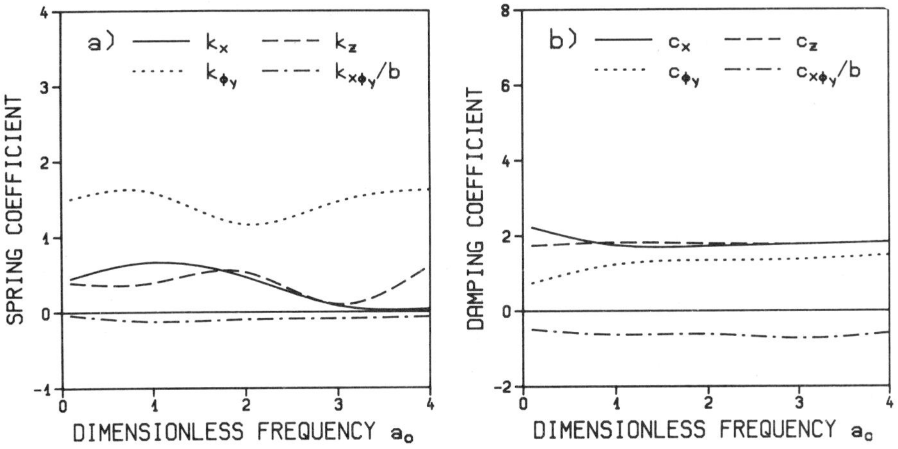

Figure 7-48 Dynamic stiffness of system ground, half-plane, $h/b = 1$, no damping (Ref. [11]).

horizontal direction being more marked. In the rocking direction, both the spring and damping coefficients increase with embedment. Augmenting the embedment results in stronger frequency dependence of the dynamic-stiffness coefficients.

Turning to the foundation embedded in a homogeneous layer (Fig. 7-40b), the dynamic-stiffness matrix for $h/b = 1$ and $d/b = 2$ is examined. For the undamped case, the results are presented in Fig. 7-49. Especially in the high-frequency range, large oscillations exist. Even negative spring coefficients arise.

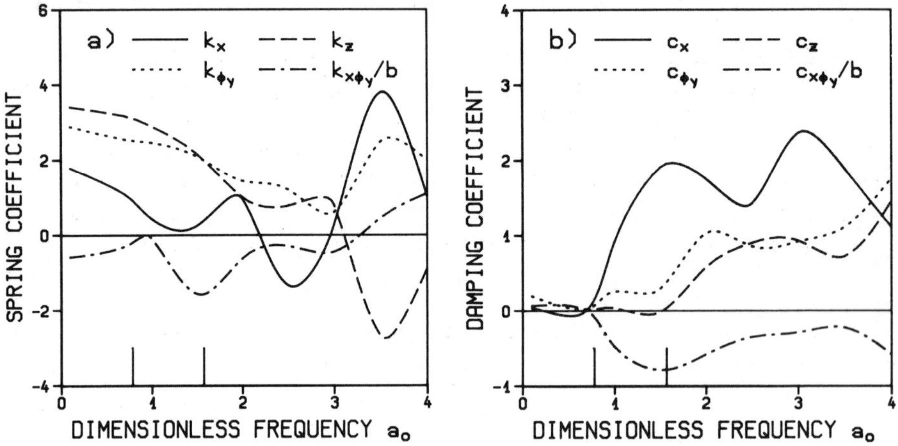

Figure 7-49 Dynamic stiffness of system ground, layer built in at its base, $h/b = 1$, $d/b = 2$, no damping (Ref. [11]).

In contrast to the elastic half-plane, the built-in layer exhibits a cutoff frequency, below which, for the case of no damping, no radiation of energy occurs. The damping coefficient c_x is zero below the horizontal fundamental frequency of the layer which corresponds to $a_o = \pi/4$. For larger frequencies, significant values arise. The coefficients c_z and c_{ϕ_y} are almost zero below the vertical fundamental frequency $a_o = \pi/2$. For the damped case (Fig. 7-50), a cutoff frequency no longer exists. Below the fundamental frequencies, the imaginary part of the dynamic-stiffness coefficients $a_o c$ remains small. In addition, the dependency on frequency is reduced. The dynamic-stiffness coefficients for a surface foundation ($h/b = 0$) resting on the same layer are presented in Fig. 7-14 as comparison. As expected, the translational spring coefficients depend strongly on the embedment for the layer built in at its base (Figs. 7-50a and 7-14a).

7.6.6 Parametric Study System Free Field

The undamped half-plane (Fig. 7-40a) is addressed first. The results are shown in Fig. 7-51 for $h/b = 0.5$, in Fig. 7-52 for $h/b = 1$, and in Fig. 7-46 for $h/b = 2$. The dominant effect of $[S^e_{bb}]$ on the spring coefficients is clearly visible. For a_o approaching zero, $[S^e_{bb}]$ tends toward zero. In the low-frequency range, $[S^f_{bb}]$ thus bears comparison with $[S^g_{bb}]$.

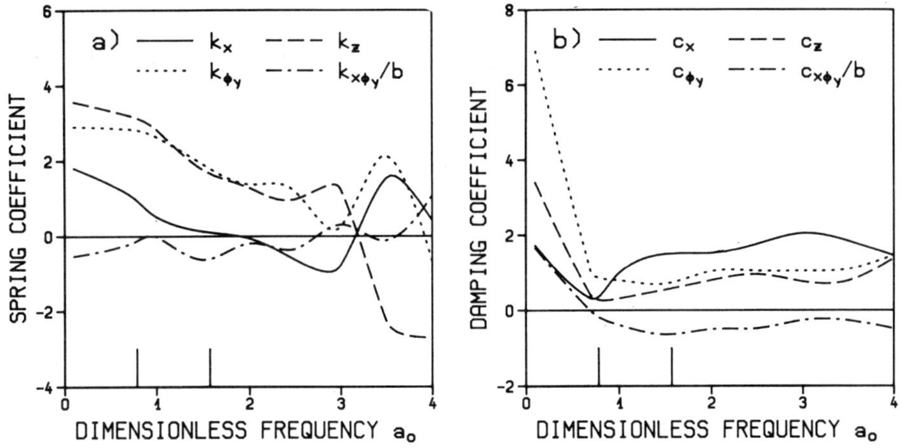

Figure 7-50 Dynamic stiffness of system ground, layer built in at its base, $h/b = 1$, $d/b = 2$, with damping (Ref. [11]).

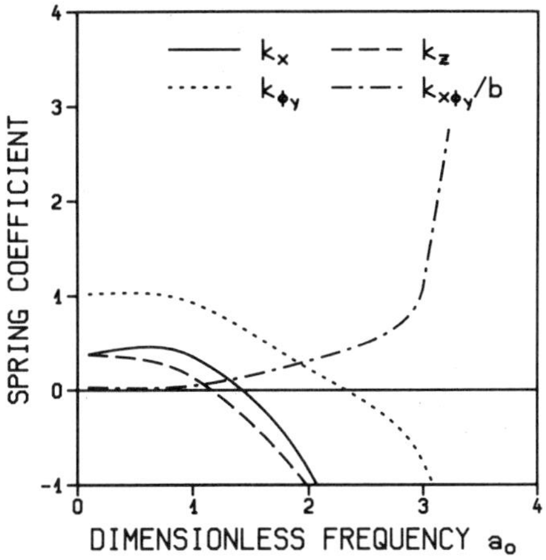

Figure 7-51 Spring coefficient of free field, half-plane, $h/b = 0.5$, no damping (Ref. [10]).

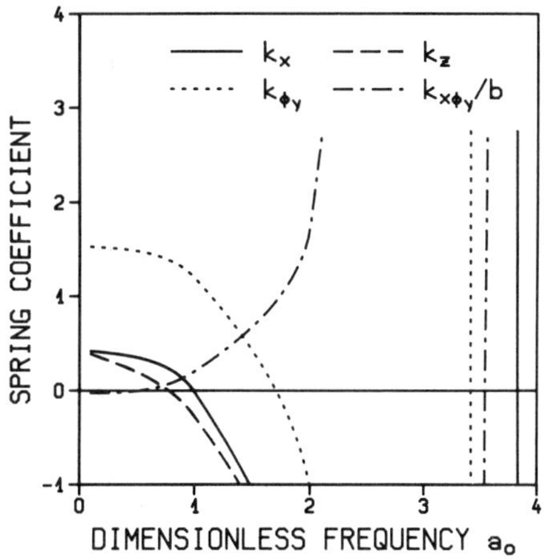

Figure 7-52 Spring coefficient of free field, half-plane, $h/b = 1$, no damping (Ref. [10]).

Turning to the foundation embedded in a homogeneous layer (Fig. 7-40b), the spring coefficients for $h/b = 1$ and $d/b = 2$ are examined. For the undamped case, the results are presented in Fig. 7-53. It is worth mentioning that the difference of $[S^f_{bb}]$ and $[S^g_{bb}]$ (which corresponds to $[S^e_{bb}]$) is the same for the layer and for the half-plane for the same aspect ratio. For the damped case (Fig. 7-54), the spring coefficients remain finite for all frequencies.

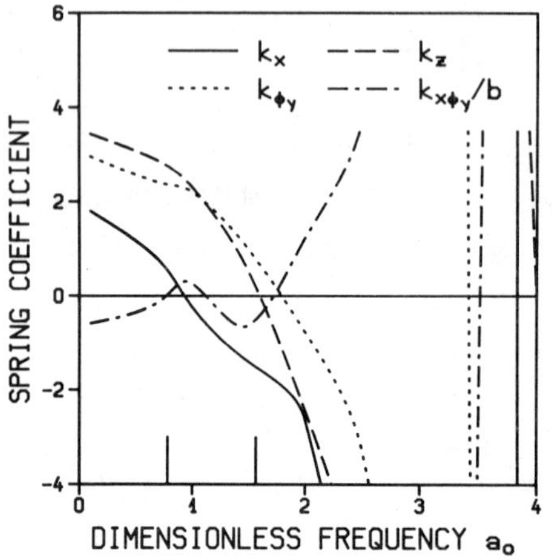

Figure 7-53 Spring coefficient of free field, layer built in at its base, $h/b = 1$, $d/b = 2$, no damping (Ref. [10]).

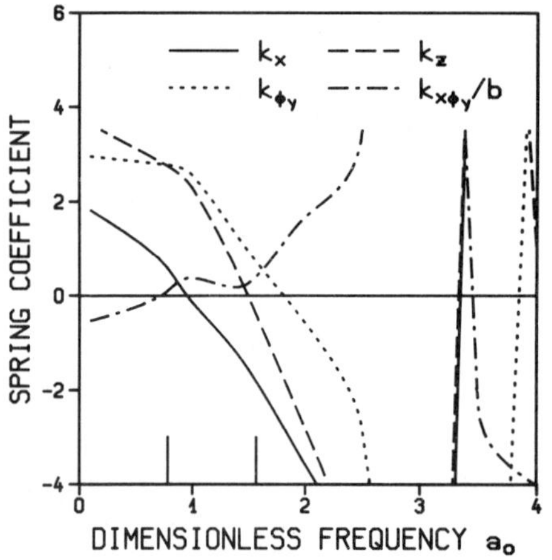

Figure 7-54 Spring coefficient of free field, layer built in at its base, $h/b = 1$, $d/b = 2$, with damping (Ref. [10]).

7.7 DYNAMIC-STIFFNESS COEFFICIENTS OF ADJACENT FOUNDATIONS

To be able to analyze the through-soil coupling for seismic excitation of the reactor building and of the combined reactor-auxiliary and fuel-handling building shown in Figs. 4-15 and 4-16, the dynamic-stiffness matrix of the coupled system of the two adjacent foundations is calculated. The results of the dynamic calculation of the total system are presented in Section 9.3.1. The two structures are founded at a depth of 10 m on the soft site introduced in Section 6.7 (Table 6-4). As other structures surround these two buildings, the latter can be assumed to be founded on the surface of a modified site which actually starts at a depth of 10 m. In this actual practical case, the relative displacements between the reactor building and the other structure, hereinafter called reactor-auxiliary building, is to be calculated. To be able to determine values which are sufficiently conservative for design, the shear-wave velocities c_s shown in Table 6-4 (from a depth of 10 m onward) are reduced by an average of 0.67 by modifying the dynamic-shear modulus G. The damping ratio of the soil layer is selected as 0.07. The site consists of four layer of soil with $c_s = 240$ m/s at the top resting on bedrock with $c_s = 1000$ m/s. Assuming the soil layer to be built in at the top of the bedrock results in fundamental frequencies of 3.2 Hz and 6.8 Hz in the horizontal and vertical directions, respectively.

The discretization into subdisks of the two rigid basemats of the reactor and of the reactor-auxiliary building is shown in Fig. 7-55. Quite a coarse mesh with 149 subdisks is selected, as only displacements are of interest (see Section 4.1).

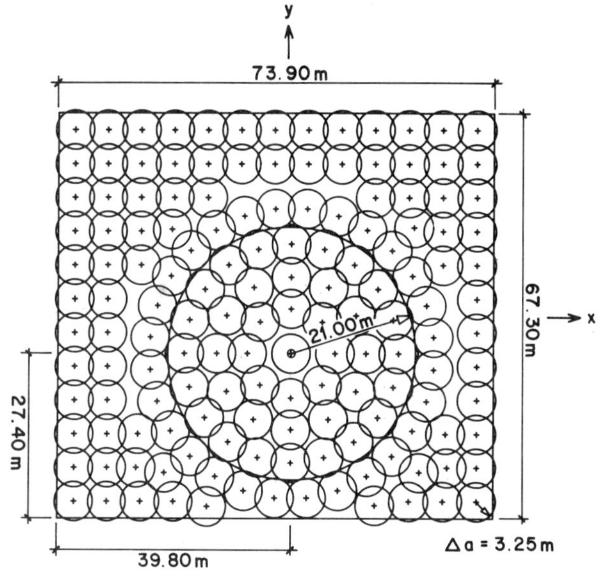

Figure 7-55 Plan view of two basemats with subdisk discretization.

After enforcing the geometric constraints of the two rigid basemats, the dynamic-stiffness matrix of the two foundations will be of order 12 × 12. The spring and damping coefficients k_x and c_x of the center of the reactor building in the horizontal direction are plotted in Fig. 7-56 ($k_x + i\omega c_x$). The values are shown as a solid line when the reactor-auxiliary building is fixed, and as a

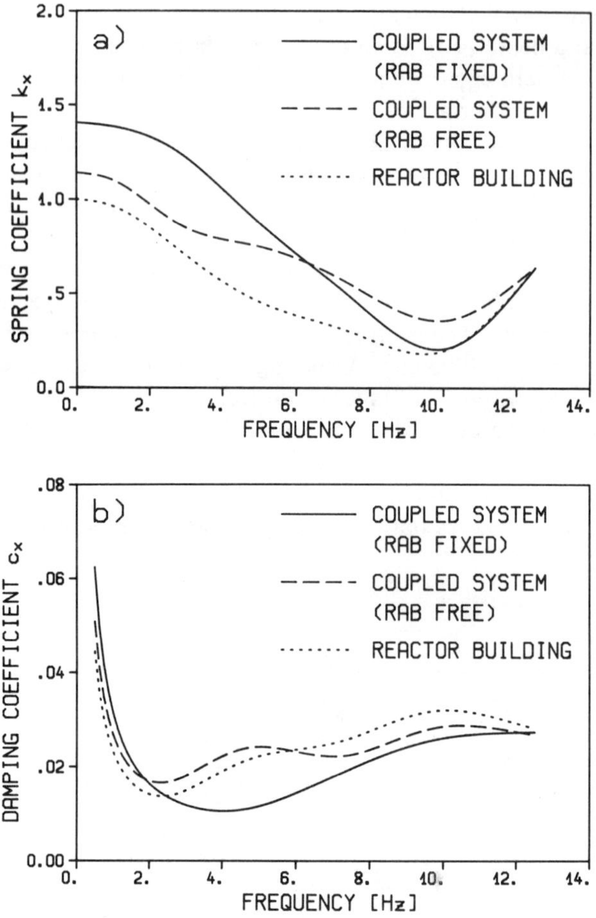

Figure 7-56 Horizontal dynamic-stiffness coefficient of reactor building.

dashed line when the reactor-auxiliary building is free to displace as a whole (external load vanishes). The former are equal to the corresponding elements of the dynamic-stiffness matrix of the two foundations. The latter are calculated performing a partial inversion. The values for the reactor building alone are indicated as a dotted line. As expected, the latter values for the spring coefficients are the smallest. All values are normalized with the static stiffness corresponding to the reactor building alone. Figure 7-57 shows the corresponding values k_{ϕ_v} and c_{ϕ_v} for the rocking degree of freedom. The shape of the fundamental mode of the reactor building (being a tall structure) will consist mostly of the rocking motion. Calculating the corresponding damping ratio $\omega c_{\phi_v}/2k_{\phi_v}$ (see, e.g., Eq. 3.54) using the information shown for the reactor building alone in Fig. 7-57 indicates that, from a practical point of view, only the material damping of the soil (0.07) contributes. The radiation damping is thus not effective in the frequency range ($\sim$2 Hz) of the fundamental mode (Fig. 7-58).

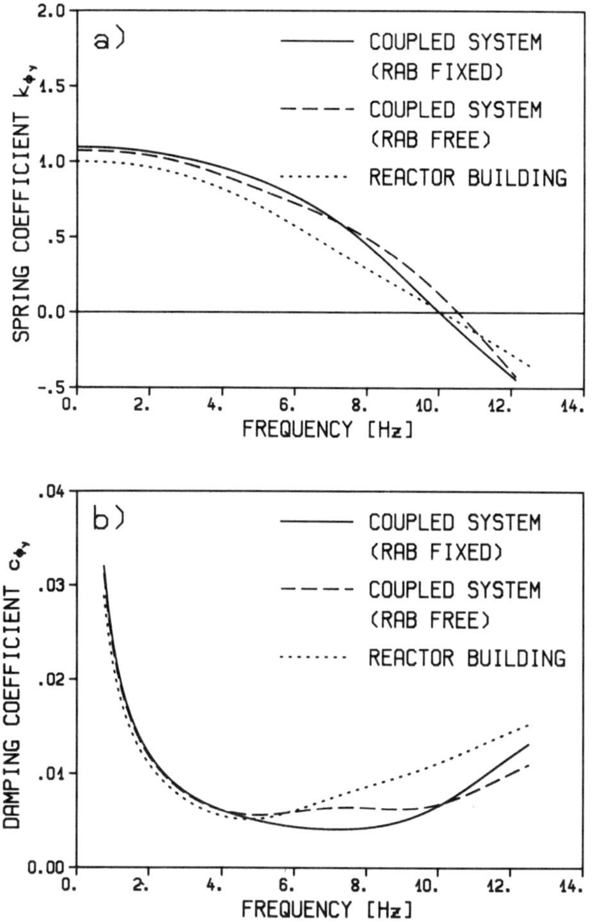

Figure 7-57 Dynamic-stiffness coefficient for rocking of reactor building.

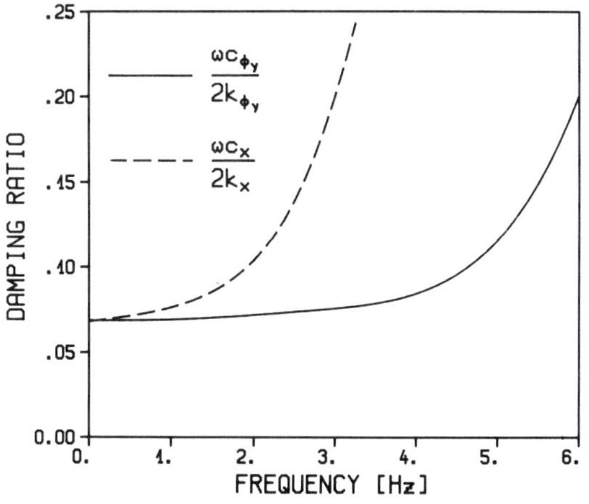

Figure 7-58 Equivalent damping ratios.

This is confirmed in Section 9.3.1. For a squat structure for which the horizontal motion dominates the radiation damping could contribute significantly.

SUMMARY

1. The dynamic-stiffness matrix of the soil with excavation discretized in the nodes located on the structure–soil interface captures all dynamic properties of the unbounded medium.
2. On the artificial boundary, which terminates the finite-element mesh of the soil, three groups of boundary conditions can be identified which model the missing unbounded soil:
   (a) Elementary boundaries, which are perfect reflectors of all impinging waves, not transmitting or absorbing any energy.
   (b) Local (viscous) boundaries, which can be made to act as perfect absorbers only if the type of the impinging wave is known.
   (c) Consistent boundaries, which are perfect absorbers transmitting all types of impinging waves which can occur in the soil without any reflections occurring. For any shape of the structure–soil interface, the dynamic-stiffness matrix of the consistent boundary can be constructed with the boundary integral-equation procedure (boundary-element method).
3. In an unbounded domain, it is not sufficient for the displacement to die out at infinity to determine a unique solution. The latter is achieved by enforcing the radiation condition (at infinity), which excludes the incoming wave.
4. To calculate the dynamic-stiffness matrix of a surface foundation, the structure–soil interface is discretized into elements. Over each element, a distribution of the load is assumed. The intensities of these loads (redundants) are determined such that the corresponding displacements of the site at the surface (primary system) match the prescribed ones associated with the dynamic-stiffness matrix in an integral sense. Choosing the weighting function appropriately, a symmetric dynamic-stiffness matrix results.
5. To be able to determine the flexibility-influence functions of the layered half-space (Green's functions), the specified distribution of the load of an element is transformed in the two-dimensional case into a Fourier integral in the wave-number domain, and in the three-dimensional case it is expanded in a Fourier series in the circumferential direction, and into Bessel functions involving the wave-number domain in the radial direction. From the flexibility equation in the wave-number domain of the site (condensed at its surface), the corresponding displacements follow. Applying the inverse transformation leads to the flexibility-influence functions. As this involves integrals over all possible wave numbers, all types of waves are taken into

account. All integrals can be performed numerically; to keep poles from occurring, a fictitious small damping is introduced.

6. Based on a parametric study, the following conclusions concerning the dynamic stiffness of a surface foundation (strip and disk) apply:
   - **(a)** The dynamic-stiffness coefficients for a half-space in the horizontal direction depend only weakly on the dimensionless frequency.
   - **(b)** The damping coefficients for a half-space for rocking and twisting increase with increasing dimensionless frequency and are significantly smaller than those for the translational degrees of freedom.
   - **(c)** The coupling term between the horizontal and rocking motion is negligible.
   - **(d)** Changing the condition of contact hardly affects the dynamic-stiffness coefficients of the half-space and only somewhat those of the vertical and rocking motions of the layer built in at its base.
   - **(e)** Introducing damping reduces the spring coefficients in the higher-frequency range and increases the damping coefficients in the lower-frequency range. In addition, the dependency on the dimensionless frequency is reduced.
   - **(f)** For the layer built in at its base, large oscillations arise, especially in the higher-frequency range. A cutoff frequency exists below which, for the case of zero material damping, no radiation of energy occurs. Below the horizontal fundamental frequency of the layer, the damping coefficients of the horizontal and of the torsional motions vanish; below the vertical fundamental frequency, the damping coefficients of the vertical and rocking motions remain small. Introducing damping, the cutoff frequency disappears. The damping coefficients, however, remain small.
7. A truly three-dimensional situation cannot be modeled with a two-dimensional model.
8. For a surface foundation, the three translational components of the scattered-wave motion (which are always smaller than the corresponding horizontally propagating free-field motions) exhibit approximately the same decrease (starting from the free-field values) with increasing ratio of the product of frequency and radius to the apparent velocity. The corresponding rotational components are zero for very small and very large ratios; the maximum is reached at a ratio of about 2, the rocking component being approximately twice as large as the torsional one.
9. The boundary-integral-equation method, called boundary-element method in discretized form, is well suited for modeling the unbounded domain. Only the structure–soil interface has to be modeled. Analytical solutions available for a regular domain are used to supplement numerical procedures.
10. In the indirect boundary-element method, fictitious loads with initially unknown intensities are applied on a source line located outside the soil

region to be investigated (excavated part). These loads act on the dynamic system consisting of the continuous soil, that is, the layered half-space without excavation (free field), a system for which the displacement and surface tractions (Green's influence functions) on the line which subsequently will form the structure–soil interface can be calculated analytically. The amplitudes of the source loads are then determined so as to satisfy in an average sense the prescribed displacement condition of the stiffness matrix. Selecting the weighting functions equal to the Green's functions for the surface tractions guarantees the symmetry of the dynamic-stiffness matrix. The best results occur for: (a) no offset, that is, placing the load at an infinitestimal distance from the structure–soil interface, and (b) linearly distributed loads (in contrast to using concentrated loads). The displacements arising from the applied loads can easily be calculated and compared to the prescribed displacements.

11. Green's influence functions, which are used as fundamental solutions in the boundary-element method, are derived for a linearly distributed load acting on part of a layered half-plane on a line that is inclined from the horizontal.
12. The calculation of the dynamic-stiffness matrix of the free field does not require the Green's functions for the surface tractions.
13. If the dynamic-stiffness matrix of the soil with excavation (ground) is calculated as the difference between that of the free field and that of the excavated part, the behavior of the latter, being a bounded domain, has to be examined in detail. At the natural frequencies of the undamped excavated part built in along the structure–soil interface, the spring coefficients associated with the dynamic-stiffness matrices of the excavated part and of the free field will become infinite. In the vicinity of these frequencies, the dynamic-stiffness matrix of the soil with excavation will thus follow as the difference of two large numbers. A consistent discretization must therefore be used. In particular, the dynamic-stiffness matrix of the embedded part cannot be determined by the finite-element method close to these frequencies.
14. A parametric study is performed for the dynamic-stiffness matrix of the soil with excavation (ground) for a rectangular foundation embedded in a half-plane and in a layer built in at its base, varying the aspect ratio and the damping of the soil. The two translational spring coefficients are not strongly affected by embedment for the half-plane, while the increase of the corresponding damping coefficients is approximately proportional to the embedment, the increase in the horizontal direction being more marked. In the rocking direction both the damping and spring coefficients increase with embedment. Augmenting the embedment results in strong frequency dependence of the dynamic-stiffness coefficients. In contrast to the elastic half-plane, the built-in layer exhibits a cutoff frequency below which, for the case of no damping, no radiation of energy occurs.

PROBLEMS

7.1. For a structure (with or without an irregular soil region) embedded in a single layer built in at its base, a vertical soil–structure interface can be selected (Fig. P7-1a). Selecting only one node b at the free surface, calculate the stiffness coefficient S_y of the semi-infinite layer for the out-of-plane motion, assuming a linear

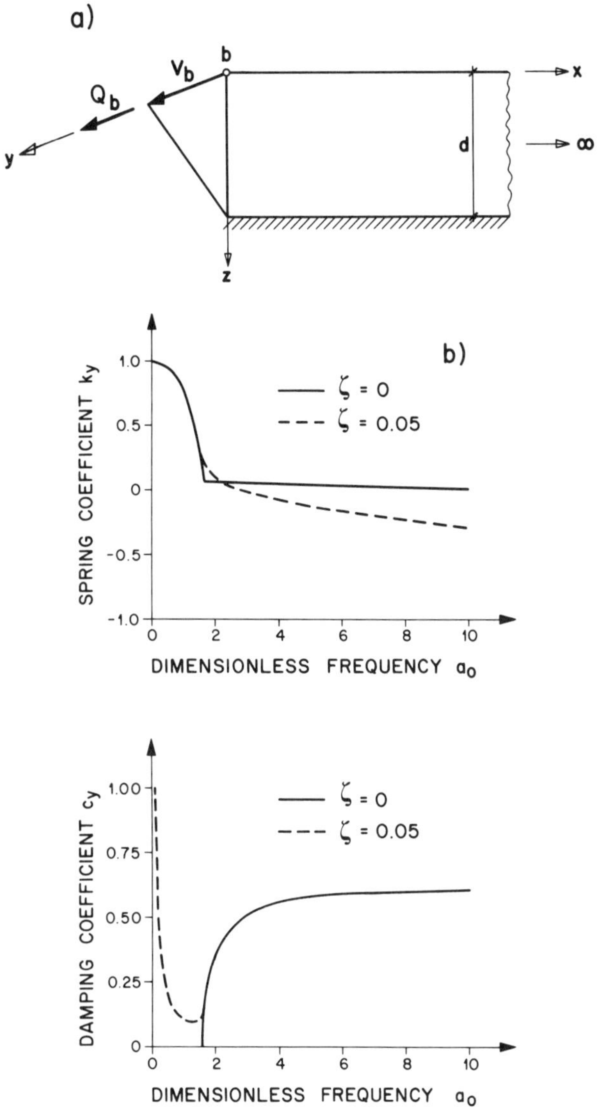

Figure P7-1 Out-of-plane motion of layer built in at base. (a) Nomenclature; (b) dynamic-stiffness coefficient.

variation of the displacement. Decompose the nondimensionalized dynamic-stiffness coefficient into k_y and $ia_o c_y$, where $a_o = \omega d/c_s$. Plot k_y and c_y for the undamped and damped cases ($\zeta = 0.05$) as a function of a_o.

Solution:

The amplitude $v(z)$ corresponding to the wave number k is specified in Eq. 5.98. Enforcing the boundary condition at the free surface (Eq. 5.100)

$$\tau_{yz}(z=0) = 0$$

results in

$$v(z) = A_{\text{SH}}[\exp(iktz) + \exp(-iktz)] = a \cos ktz$$

where (Eq. 5.95b)

$$k^2 t^2 = \frac{\omega^2}{c_s^{*2}} - k^2$$

The other boundary condition at the fixed base $z = d$ leads to the characteristic equation

$$\cos ktd = 0$$

which is satisfied for the discrete values of $k_j td$,

$$k_j td = \frac{(2j-1)\pi}{2} \qquad j = 1, 2, \ldots$$

The complete solution for all k_j equals

$$v(z) = \sum_j a_j \cos k_j tz$$

The prescribed displacement for $x = 0$ is formulated as $N(z)v_b$ with

$$N(z) = 1 - \frac{z}{d}$$

Expanding $N(z)v_b$ into a Fourier series with terms $\cos k_j tz$ determines the coefficients a_j as

$$a_j = \frac{2}{d}\int_0^d \left(1 - \frac{z}{d}\right) \cos k_j tz \, dz \, v_b = \frac{2}{(k_j td)^2} v_b$$

The amplitudes of the shear stress $\tau_{yx}(z)$ at $x = 0$ are determined from Eq. 5.112a as

$$\tau_{yx}(z) = -iG^* \sum_j a_j k_j \cos k_j tz$$

The concentrated nodal load Q_b is calculated as

$$Q_b = -\int_0^d N(z)^T \tau_{yx}(z)\, dz = -2iG^* \sum_j \frac{k_j d}{(k_j td)^4} v_b$$

The dynamic-stiffness coefficient S_y thus is equal to

$$S_y = -2iG^* \sum_j \frac{k_j d}{(k_j td)^4}$$

Substituting

$$t = i\sqrt{\frac{1}{1 - a_o^{*2}/k_j^2 t^2 d^2}}$$

leads to

$$S_y = 2G^* \sum_j \left[\sqrt{1 - \frac{4a_o^{*2}}{(2j-1)^2\pi^2}} \frac{8}{(2j-1)^3\pi^3} \right]$$

The static value K_y equals $0.543G$. Defining

$$S_y = K_y(k_y + ia_o c_y)$$

k_y and c_y are plotted versus a_o in Fig. P7-1b. For the undamped case, the cutoff frequency equals $\pi/2$.

7.2. On a horizontal artificial boundary, which truncates a finite-element discretization, local horizontal and vertical dashpots with coefficients ρc_s and ρc_p are introduced, which completely absorb vertically incident SV- and P-waves (analogous to Eq. 7.8). For an incident P-wave propagating with an angle of incidence $\psi_P = 30°$ (measured from the horizontal) toward this boundary, determine the ratio of the reflected and incident amplitudes A_P/B_P of the P-wave and the ratio A_{SV}/B_P of the created SV-wave for Poisson's ratio $\nu = 0.33$.

Solution:

Horizontal boundary at $z = 0$ (Eqs. 5.130 and 5.132):

$$u = l_x(A_P + B_P) - m_x t(A_{SV} - B_{SV})$$

$$w = -l_x s(A_P - B_P) - m_x(A_{SV} + B_{SV})$$

$$\tau_{xz} = i2kl_x sG(A_P - B_P) + ikm_x(1 - t^2)G(A_{SV} + B_{SV})$$

$$\sigma_z = ikl_x(1 - t^2)G(A_P + B_P) - i2km_x tG(A_{SV} - B_{SV})$$

Substituting

$$B_{SV} = 0$$

$$l_x = \cos\psi_P = c_p\frac{k}{\omega} = 2c_s\frac{k}{\omega}$$

$$s = \tan\psi_P$$

$$m_x = \cos\psi_{SV} = c_s\frac{k}{\omega}$$

$$t = \tan\psi_{SV}$$

and

$$\tau_{xz} = -i\omega\rho c_s u$$

$$\sigma_z = -i\omega\rho c_p w$$

into these equations leads to

$$(2\sin\psi_P\cos\psi_{SV} + \cos\psi_P)A_P + (\cos^2\psi_{SV} - \sin^2\psi_{SV} - \sin\psi_{SV})A_{SV} = (-\cos\psi_P + 2\sin\psi_P\cos\psi_{SV})B_P$$

$$(-\sin\psi_P + \cos^2\psi_{SV}(1 - \tan^2\psi_{SV}))A_P + \left(-\frac{2\cos^2\psi_{SV}}{\cos\psi_P}\sin\psi_{SV} - \cos\psi_{SV}\right)A_{SV} = (-\cos^2\psi_{SV}(1 - \tan^2\psi_{SV}) - \sin\psi_P)B_P$$

For $\sin\psi_P = 0.5$, this results in

$$A_{SV} = 0.117B_P$$

$$A_P = -0.196B_P$$

7.3. The dynamic-stiffness coefficients of a damped system are straightforwardly calculated from those of the elastic system using the correspondence principle. The Lamé constants $\lambda + 2G$ and G are replaced by their corresponding complex values (Eq. 5.87). If an analytical expression exists for the dynamic-stiffness coefficients, this is extremely easy to perform. In other cases, the calculation has to be repeated from the start, using the complex values. The question arises whether it is possible to approximate the stiffness coefficients of the damped system using those of the undamped system (at the same frequency or at neighboring frequencies) which are available as tabulated values or as curves. (In passing, it is worth mentioning that a rigorous procedure exists, which evaluates an infinite integral over a_o. The undamped solution thus has to be known over a large range and has to decay rapidly with increasing a_o.)

One specific dynamic-stiffness coefficient $S(a_o)$ is addressed, which can be decomposed for the undamped unbounded domain as follows:

$$S(a_o) = K[k(a_o) + ia_o c(a_o)]$$

where K is the static value, k and c denote the spring and damping coefficients, and a_o is equal to the dimensionless frequency. For a damped system with a damping ratio ζ, a subscript ζ is added.

$$S_\zeta(a_o) = K[k_\zeta(a_o) + ia_o c_\zeta(a_o)]$$

Applying the correspondence principle to the expression of the undamped case results in

$$S(a_o^*) = K(1 + 2\zeta i)[k(a_o^*) + ia_o^* c(a_o^*)] \qquad [= S_\zeta(a_o)]$$

where

$$a_o^* = \frac{a_o}{\sqrt{1 + 2\zeta i}}$$

The damping ratio thus affects the dynamic-stiffness coefficient in three ways. First, the static-stiffness coefficient used to nondimensionalize the expression is multiplied by $(1 + 2\zeta i)$. Second, a_o^* is substituted for a_o and third, $k(a_o)$ and $c(a_o)$ are replaced by their damped (complex) counterparts $k(a_o^*)$ and $c(a_o^*)$. The following three possibilities to approximate $k_\zeta(a_o)$ and $c_\zeta(a_o)$ are discussed.

(a) The unbounded domain is treated as in the case of a finite body; that is, only the factor $(1 + 2\zeta i)$ is considered.

$$S_\zeta(a_o) = K(1 + 2\zeta i)[k(a_o) + ia_o c(a_o)]$$

This results in

$$k_\zeta(a_o) = k(a_o) - 2\zeta a_o c(a_o)$$

$$c_\zeta(a_o) = c(a_o) + 2\frac{\zeta}{a_o}k(a_o)$$

(b) The damped values $k(a_o^*)$ and $c(a_o^*)$ are set equal to their undamped (real) values $k(a_o)$ and $c(a_o)$, respectively. With $a_o^* \simeq a_o(1 - \zeta i)$, this leads to

$$S_\zeta(a_o) = K(1 + 2\zeta i)[k(a_o) + ia_o(1 - \zeta i)c(a_o)]$$

or

$$k_\zeta(a_o) = k(a_o) - \zeta a_o c(a_o)$$

$$c_\zeta(a_o) = c(a_o) + 2\frac{\zeta}{a_o}k(a_o)$$

(c) The damped values $k(a_o^*)$ and $c(a_o^*)$ are calculated using a Taylor expansion at a_o. For $k(a_o^*)$, this leads to

$$k(a_o^*) = k(a_o) + k(a_o),_{a_o}(a_o^* - a_o)$$

With $a_o^* - a_o \simeq -i\zeta a_o$,

$$k(a_o^*) = k(a_o) - i\zeta a_o k(a_o),_{a_o}$$

results. Analogously,

$$c(a_o^*) = c(a_o) - i\zeta a_o c(a_o),_{a_o}$$

Substituting leads to

$$S_\zeta(a_o) = K(1 + 2\zeta i)[k(a_o) - i\zeta a_o k(a_o),_{a_o} + ia_o(1 - \zeta i)(c(a_o) - i\zeta a_o c(a_o),_{a_o})]$$

or

$$k_\zeta(a_o) = k(a_o) - \zeta a_o c(a_o) + a_o^2\zeta c(a_o),_{a_o}$$

$$c_\zeta(a_o) = c(a_o) + 2\frac{\zeta}{a_o}k(a_o) - \zeta k(a_o),_{a_o}$$

Generally, because no analytical expression exists, the derivatives are determined by using a numerical-difference formula, for example, that based on the central-difference equation. For the conical shear beam used to determine approximately the dynamic-stiffness coefficient for twisting (Problem 5.6), calculate for the three methods $k_\zeta(a_o)$ and $c_\zeta(a_o)$, starting from the solution $k(a_o)$ and $c(a_o)$ for $\zeta = 0.20$. Compare the results in a plot with the exact solution ($0 < a_o < 6$).

Results:

The comparison is shown in Fig. P7-3. Method (b) results in the best approximation.

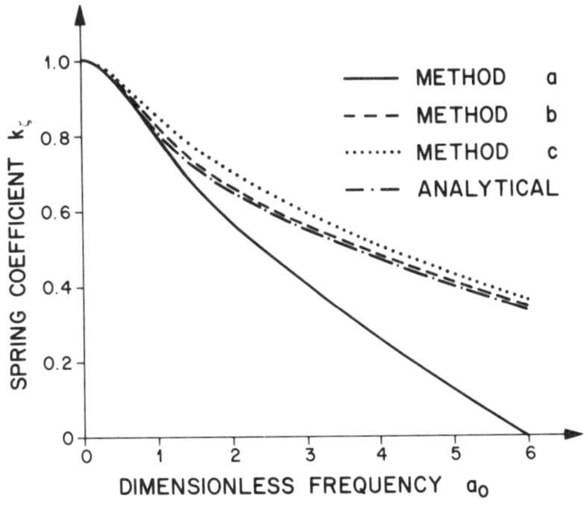

Figure P7-3 Approximate twisting dynamic-stiffness coefficient of damped conical shear beam.

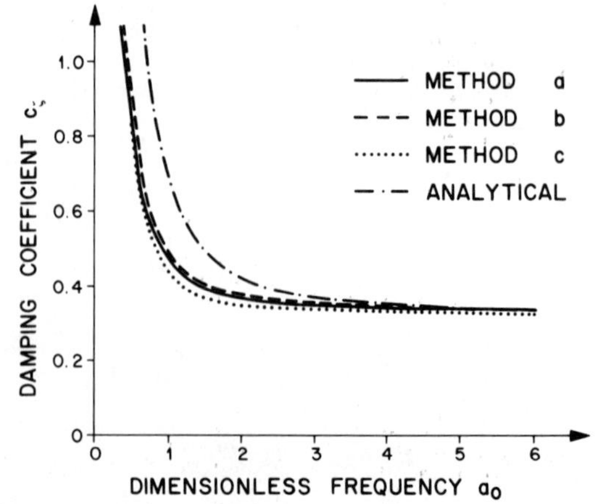

Figure P7-3 (Continued)

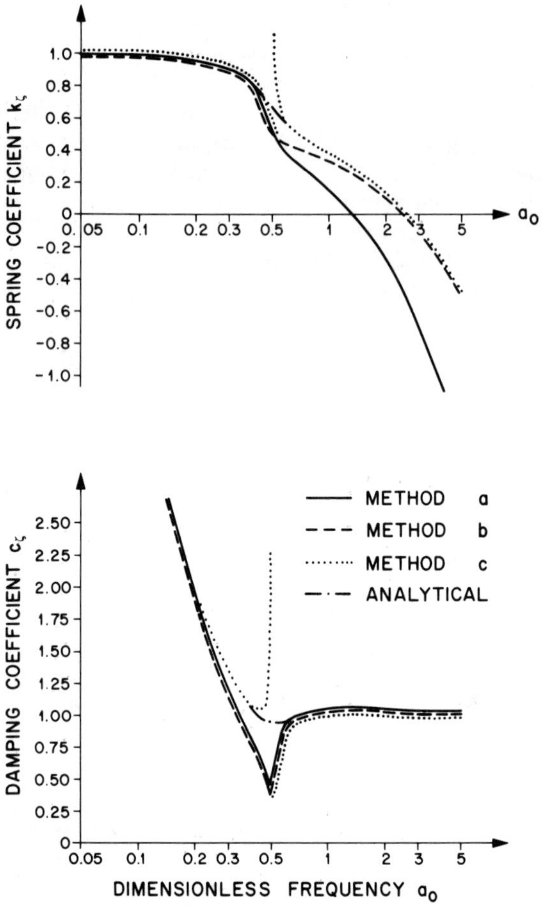

Figure P7-4 Approximate dynamic-stiffness coefficient of damped rod with exponentially increasing area.

7.4. For the infinite rod with exponentially varying area discussed in Section 5.1, calculate for the three methods developed in Problem 7.3 $k_\zeta(a_o)$ and $c_\zeta(a_o)$, starting from the solution $k(a_o)$ and $c(a_o)$ for $\zeta = 0.20$. Compare the results in a plot with the exact solution ($0.05 < a_o < 5$, Fig. 5-6).

Results:

As the derivative of the spring and damping coefficient is discontinuous at the cutoff frequency $a_o = 0.5$, method (c) leads to unreliable results in the vicinity of this frequency. The comparison is shown in Fig. P7-4.

7.5. For a more restrictive class of sites than dealt with in this chapter, other innovative procedures exist for calculating the dynamic-stiffness coefficients. For instance, for a horizontally stratified layer built in at its base, the dynamic-stiffness coefficients associated with a vertical structure–soil interface extending over the entire depth (two-dimensional problem) can be determined by using the so-called cloning algorithm, which is based completely on the finite-element method. This formulation captures the essential notion of infinity by stating that adding a finite part to an infinite quantity does not change its value. The fundamental idea of cloning is illustrated in Fig. P7-5a for the semi-infinite soil taking the embedment into account. Adding the bounded cell of finite elements to the semi-infinite domain with the characteristic length l_e results in a similar semi-infinite domain with length l_i. The concept can be applied to their dynamic-stiffness matrices by assembling the known dynamic-stiffness matrix of the cell and the unknown matrix of the unbounded soil referenced by the length l_e, which results in the unknown dynamic-stiffness matrix of the unbounded soil with length l_i. As a relationship for the dynamic-stiffness matrices referenced by different lengths exists, the cloning algorithm leads to an expression for the dynamic-stiffness matrix of the unbounded soil as a function of that of the cell.

To discuss the fundamental features of the cloning algorithm, the simple one-dimensional rod with exponentially increasing cross-sectional area is used by way of illustration (Fig. P7-5b). This example is also selected in Section 5.1.

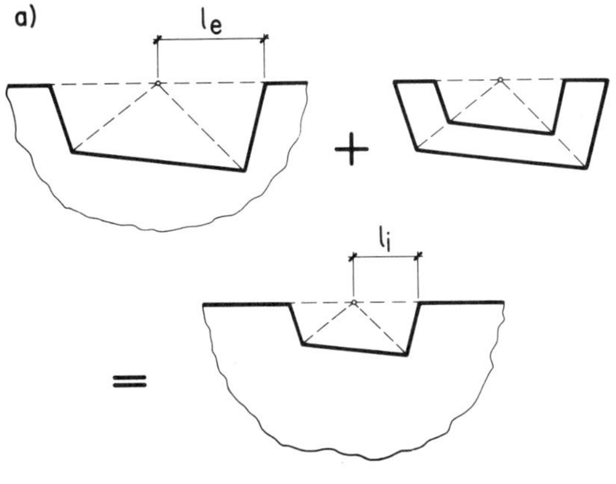

Figure P7-5 Cloning algorithm (after Ref. [13]). (a) Fundamental concept;

The area $A(z)$ is specified as

$$A(z) = A_o \exp\left(\frac{z}{f}\right)$$

where A_o is the area at the cross section at $z = 0$ and f is a specified parameter which can be interpreted as a length. The dynamic-stiffness coefficient S of the unbounded rod at the end of the rod at $z = 0$ is to be determined for a specific frequency ω.

For a cross section at the location z, the dynamic-stiffness coefficient of the infinite rod is formulated as

$$S(z) = \frac{EA(z)}{f}\bar{S}$$

where $\bar{S}$ is the dimensionless dynamic-stiffness coefficient, independent of z, and E denotes the modulus of elasticity. Specifically, for the two sections denoted with subscripts i and e, as shown in Fig. P7-5b, we have

$$S_i = \frac{EA_o}{f}\bar{S} \qquad \text{and} \qquad S_e = \frac{EA_o e^{\alpha}}{f}\bar{S}$$

with $\alpha = l/f$ and e denoting the exponential function.

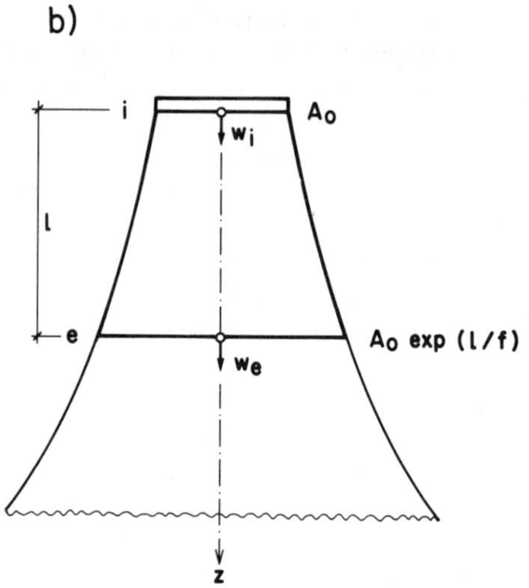

Figure P7-5 (Continued) (b) rod with exponentially increasing area.

The cell bounded by the interior section i and the exterior section e consists of one finite element. Its dynamic-stiffness matrix $[S]$ is nondimensionalized as

$$[S] = \frac{EA_o}{f}[D]$$

For linear shape functions of the amplitude of the axial displacement,

$$w(z) = \left(1 - \frac{z}{l}\right)w_i + \frac{z}{l}w_e$$

$$[D] = \frac{e^{\alpha} - 1}{\alpha^2}\begin{bmatrix} 1 & -1 \\ -1 & 1 \end{bmatrix}$$

$$+ a_o^2 \left[\begin{array}{c|c} 1 + \frac{2}{\alpha}\left(\frac{1 - e^{\alpha}}{\alpha} + 1\right) & -\frac{1 - e^{\alpha}}{\alpha} - \frac{2}{\alpha}\left(\frac{1 - e^{\alpha}}{\alpha} + e^{\alpha}\right) \\ -\frac{1 - e^{\alpha}}{\alpha} - \frac{2}{\alpha}\left(\frac{1 - e^{\alpha}}{\alpha} + e^{\alpha}\right) & -e^{\alpha} + \frac{2}{\alpha}\left(\frac{1 - e^{\alpha}}{\alpha} + e^{\alpha}\right) \end{array}\right]$$

results, where the dimensionless frequency is defined as

$$a_o = \frac{\omega f}{\sqrt{E/\rho}}$$

The mass density is denoted as ρ.

The actual cloning algorithm proceeds as follows. Formulating the dynamic-equilibrium equation at section e results in

$$S_e u_e + S_{ee} u_e + S_{ei} u_i = 0$$

The elements of $[S]$ and $[D]$ are denoted with subscripts i and e. Analogously for the interior boundary

$$S_i u_i = S_{ii} u_i + S_{ie} u_e$$

Eliminating u_e leads to

$$\bar{S}^2 - (D_{ii} - D_{ee}e^{-\alpha})\bar{S} - e^{-\alpha}(D_{ii}D_{ee} - D_{ie}^2) = 0$$

The solution of this quadratic equation is

$$\bar{S} = \tfrac{1}{2}[D_{ii} - D_{ee}e^{-\alpha} \pm \sqrt{(D_{ii} + D_{ee}e^{-\alpha})^2 - 4D_{ie}^2 e^{-\alpha}}]$$

or

$$\bar{S} = \frac{\beta}{2}\left[1 + 2a_o^2\left(\frac{1}{\beta} - 1\right)_{(}\pm_{)}\sqrt{1 - 4a_o^2 + 4a_o^4(\beta - 1)}\right]$$

with

$$\beta = \left(\frac{1 - e^{-\alpha}}{\alpha}\right)^2 e^{\alpha}$$

The positive sign in front of the square root is to be used. This can easily be verified by considering the static case ($a_o = 0$), where the negative sign would result in $\bar{S} = 0$.

Introducing material damping, characterized by the constant hysteretic-damping ratio ζ in the cloning algorithm, affects only the dynamic-stiffness matrix of the finite-element cell. Its static stiffness is multiplied by $(1 + 2\zeta i)$. Proceeding analogously, the dynamic-stiffness coefficient $\bar{S}$, which is nondimensionalized with the same (undamped) static-stiffness coefficient EA_o/f, follows.

The nondimensionalized dynamic-stiffness coefficient can be split up into its real and imaginary parts as

$$\bar{S} = k_1 + ia_o c_1$$

To establish the accuracy of the cloning algorithm, plot k_1 and c_1 for the undamped case, as a function of a_o for two different element lengths $\alpha = 0.2$ and 0.5 ($\alpha = l/f$). In addition, for a damping ratio $\zeta = 0.2$, plot k_1 and c_1 versus a_o using $\alpha = 0.5$.

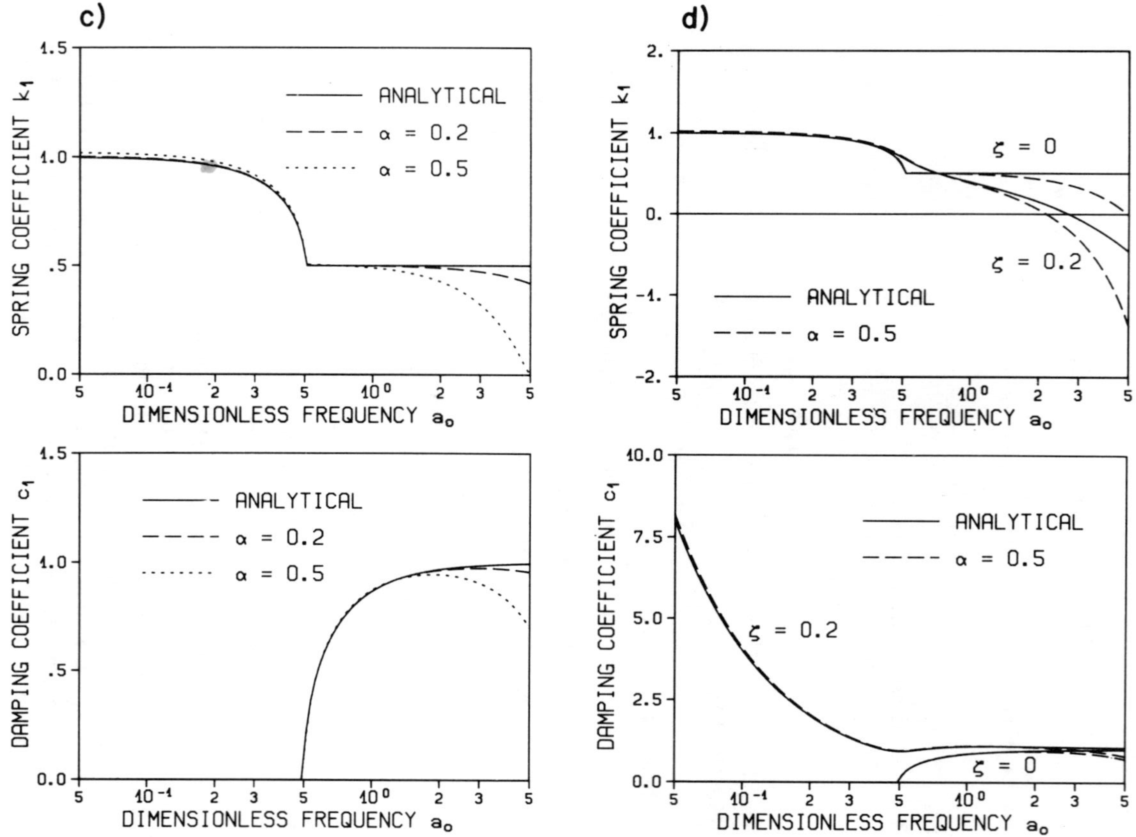

Figure P7-5 (Continued) (c) dynamic-stiffness coefficient, varying finite-element length; (d) dynamic-stiffness coefficient, damped case, varying damping.

Solution:

The plots are shown in Fig. P7-5c and d. Excellent agreement exists with the analytical solution specified in Eqs. 5.33 and 5.34 for $\zeta = 0$ and in Eq. 5.53 for the damped case, with the exception of the range of highest frequencies, for which the wavelength starts to approach twice the length of the finite-element cell. In particular, the cutoff frequency is well captured. The cloning algorithm can thus determine a complex dynamic-stiffness coefficient, starting from the real dynamic-stiffness matrix of the cell.

7.6. Another innovative method, called infinite substructuring, is briefly discussed. In this procedure, the outer boundary of a discretized region of finite elements is moved outward from the structure–soil interface by a recursive technique. A certain connection with the cloning algorithm does exist. The application is again limited to certain special cases. For instance, the dynamic-stiffness coefficients associated with the vertical structure–soil interface of a horizontally stratified layer can be determined (Fig. P7-6a). The static case is addressed first.

A cell of finite elements is placed with the interior boundary (subscript i) coinciding with the structure–soil interface. The exterior boundary (subscript e) is similar in shape to the interior one. The static-stiffness matrix of the cell is partitioned as $[S_{ii}]$, $[S_{ie}]$, $[S_{ei}]$, and $[S_{ee}]$. Another cell of finite elements with the same proportions, but their dimensions increasing with distance by a factor γ, is placed adjacent to the first (with its interior boundary coinciding with the exterior boundary of the first cell). Its static-stiffness matrix is formulated as $\gamma^{n-2}[S]$, where n is equal to 2 for the two-dimensional and 3 for the three-dimensional case. Assembling the two cells results in the following static-stiffness matrix:

$$\left[\begin{array}{c|c|c} [S_{ii}] & [S_{ie}] & \\ [S_{ei}] & [S_{ee}] + \gamma^{n-2}[S_{ii}] & \gamma^{n-2}[S_{ie}] \\ & \gamma^{n-2}[S_{ei}] & \gamma^{n-2}[S_{ee}] \end{array}\right]$$

Eliminating the displacements at the intermediate nodes leads to the condensed static-stiffness matrix $[S^1]$ after one step:

$$[S_{ii}^1] = [S_{ii}] - [S_{ie}]([S_{ee}] + \gamma^{n-2}[S_{ii}])^{-1}[S_{ei}]$$

$$[S_{ie}^1] = -\gamma^{n-2}[S_{ie}]([S_{ee}] + \gamma^{n-2}[S_{ii}])^{-1}[S_{ie}] = [S_{ei}^1]^T$$

$$[S_{ee}^1] = \gamma^{n-2}[S_{ee}] - \gamma^{2(n-2)}[S_{ei}]([S_{ee}] + \gamma^{n-2}[S_{ii}])^{-1}[S_{ie}]$$

After j analogous steps, this leads to

$$[S_{ii}^j] = [S_{ii}^{j-1}] - [S_{ie}^{j-1}]([S_{ee}^{j-1}] + (\gamma^{n-2})^{\kappa}[S_{ii}^{j-1}])^{-1}[S_{ei}^{j-1}]$$

and so on, where $\kappa = 2^{j-1}$. The generalization to the dynamic case is not straightforward, as the dynamic-stiffness matrix of a cell is proportional to $(\gamma^{n-2})^{\kappa}$ only if the characteristic length of the foundation used to define the dimensionless frequency a_o is independent of γ. This is, for example, the case for the built-in layer, for which its depth can be used to define a_o. (The same restriction also holds for the cloning algorithm as described in Problem 7.5). If this applies, then a certain amount of damping can be included in the analysis to avoid the fact that the reflected waves reach the structure-soil interface. $[S_{ii}^j]$ is a good approximation of the dynamic-stiffness matrix. This infinite-

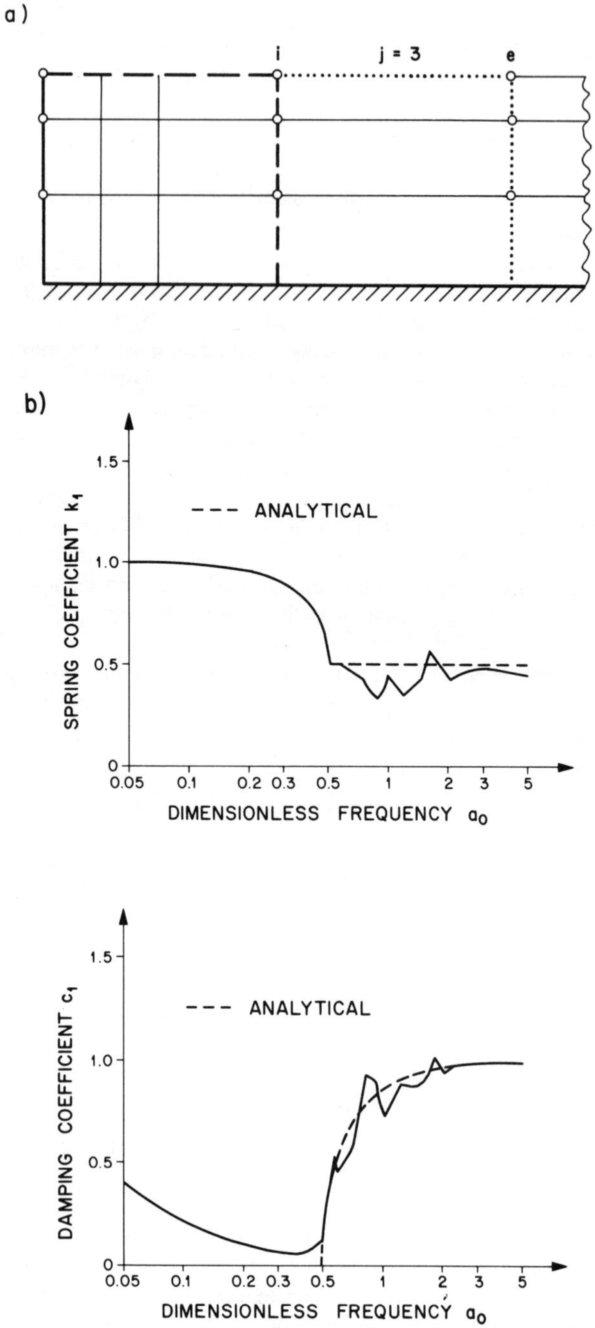

Figure P7-6 Infinite substructuring. (a) Fundamental concept; (b) dynamic-stiffness coefficient of rod with exponentially increasing area.

substructuring procedure is illustrated in Fig. P7-6a, where the third step to calculate the dynamic-stiffness matrix of a stratified layer built in at its base is explained. In this case $\gamma = 1$.

Use the concept of infinite substructuring to calculate the dynamic-stiffness coefficient of the undamped rod with exponentially increasing area discussed in Section 5.1 and Problem 7.5. Employ cells of constant length l consisting of one rod element. Plot the dimensionless dynamic-stiffness coefficient versus a_o. Investigate the influence of the number j of steps, of $\alpha = l/f$ (see Fig. P7-5b) and of the fictitious damping ratio ζ which has to be introduced.

Solution:

For the jth step, the assembled dynamic-stiffness matrix is equal to

$$\left[\begin{array}{c|c|c} \bar{S}_{ii}^{j-1} & \bar{S}_{ie}^{j-1} & \\ \bar{S}_{ei}^{j-1} & \bar{S}_{ee}^{j-1} + \exp(j\alpha)D_{ii} & \exp(j\alpha)D_{ie} \\ & \exp(j\alpha)D_{ei} & \exp(j\alpha)D_{ee} \end{array}\right]$$

where

$$S_{ii}^{j-1} = \frac{EA_o}{f}\bar{S}_{ii}^{j-1}$$

and so on. The results are plotted in Fig. P7-6b for $\alpha = 0.01$, $\zeta = 0.01$ and after 10000 steps. The analytical solution for the undamped case is also indicated. This example is not well suited for infinite substructuring, as the length l of the element has to be constant to be able to use the same stiffness matrix.

7.7. The dynamic-stiffness coefficients of the soil are frequency dependent, those of the rotational degrees of freedom to a larger extent than those of the translational ones (especially in the horizontal direction). Various approximations exist. Besides using the frequency-independent coefficients of Eq. 3.65, a fictitious added mass can be introduced to represent the dependence on frequency of the soil. For a circular rigid basemat with radius a resting on the surface of an elastic half-space with Poisson's ratio ν and mass density ρ, the following added mass Δm can be introduced for the horizontal direction:

$$\Delta m = 0.095\frac{m}{\bar{m}}$$

where m denotes the mass of the basemat and $\bar{m}$ the mass ratio defined as

$$\bar{m} = \frac{m(2-\nu)}{8\rho a^3}$$

For rocking, the nondimensional damping coefficient c_ϕ is introduced as a function of $\bar{I}$, the ratio of the mass moments of inertia,

$$c_\phi = \frac{0.30}{1+\bar{I}}$$

where

$$\bar{I} = \frac{I3(1-\nu)}{8\rho a^5}$$

with I denoting the mass moment of inertia of the basemat. The added mass moment of inertia ΔI is equal to

$$\Delta I = \frac{0.24I}{\bar{I}}$$

For a horizontal excitation with amplitude u_g plot the relative displacement with amplitude u of a rigid circular basemat with mass ($\bar{m} = 0.5$) as a function of a_o using the following three methods: **(a)** with the frequency-dependent dynamic-stiffness coefficients of Fig. 7-19 (analytical), **(b)** with the frequency-independent values of Eq. 3.65a, and 3.65b without the added mass, and **(c)** with the added mass. For a rocking excitation with amplitude ϕ_g, plot the relative rotation with amplitude ϕ of a rigid circular basemat with mass moment of inertia ($\bar{I} = 0.5$) as a function of a_o using the following four methods: **(a)** with the frequency-dependent dynamic-stiffness coefficients of Fig. 7-19 (analytical), **(b)** with the

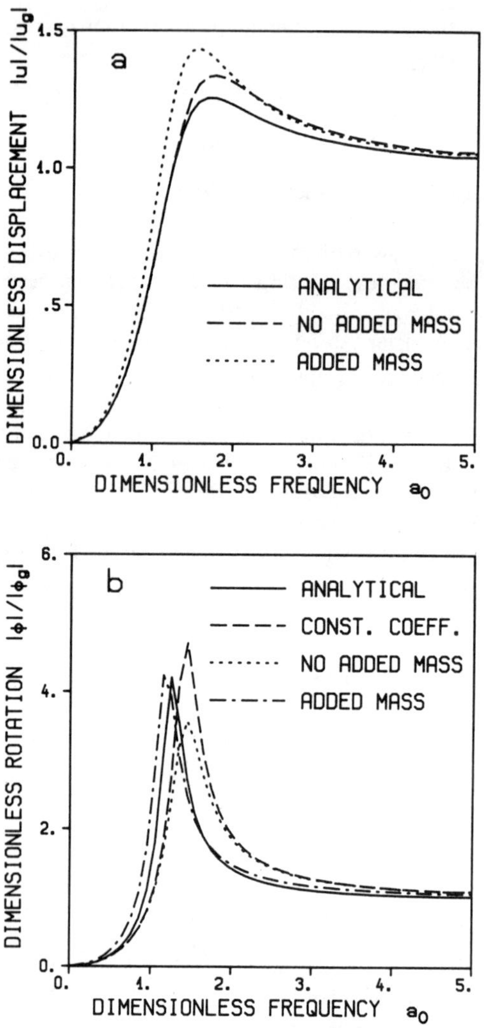

Figure P7-7 Amplification of rigid basemat with mass and mass moment of inertia. (a) Horizontal; (b) rocking.

frequency-independent constant coefficients of Eq. 3.65c and 3.65d, **(c)** with the spring coefficient of Eq. 3.65c and the damping coefficient specified above without, and **(d)** with the added mass moment of inertia. Note that the damping coefficients in Eq. 3.65b and 3.65d have a dimension.

Solution:

Horizontal (Fig. P7-7a):

$$\left[-1 + \frac{k_x + ia_o c_x}{\bar{m}(1 + \Delta m/m)a_o^2}\right]u = u_g$$

(a) $k_x(a_o)$, $c_x(a_o)$, $\Delta m = 0$
(b) $k_x = 1$, $c_x = 0.575$, $\Delta m = 0$
(c) $k_x = 1$, $c_x = 0.575$, $\Delta m = 0.095\dfrac{m}{\bar{m}}$

Rocking (Fig. P7-7b):

$$\left[-1 + \frac{k_\phi + ia_o c_\phi}{\bar{I}(1 + \Delta I/I)a_o^2}\right]\phi = \phi_g$$

(a) $k_\phi(a_o)$, $c_\phi(a_o)$, $\Delta I = 0$
(b) $k_\phi = 1$, $c_\phi = 0.15$, $\Delta I = 0$
(c) $k_\phi = 1$, $c_\phi = \dfrac{0.3}{1 + \bar{I}}$, $\Delta I = 0$
(d) $k_\phi = 1$, $c_\phi = \dfrac{0.3}{1 + \bar{I}}$, $\Delta I = 0.24\dfrac{I}{\bar{I}}$

7.8. Two rigid two-dimensional basemats with mass m and width $2b$ rest at a distance d on the surface of a half-plane with a damping ratio $\zeta = 0.05$ (Fig. P7-8a). This system is excited by vertically incident seismic S-waves with an amplitude u_y^f ($= v_g$) of the out-of-plane free-field displacement. Determine the harmonic response of this coupled system, and compare it to that of one rigid base only (i.e., neglecting the through-soil coupling effect). Introduce a dimensionless mass ratio of the structure to the soil $\bar{m} = m/(\rho b^2)$. Plot the ratio of the absolute values of the total response $|v^t|$ (divided by $|v_g|$) versus $a_o = \omega b/c_s$ for $d/2b = 1$ and $\bar{m} = 5$. For this comparative study, model each basemat with only one substrip subjected to a constant distributed load having an amplitude q_o.

Solution:

The flexibility coefficient of the half-plane in the wave-number domain $F_{vv}(k)$ is equal to the inverse of the stiffness coefficient S_{SH}^R (Eq. 5.104)

$$F_{vv}(k) = (iktG^*)^{-1}$$

The out-of-plane displacement amplitudes $v(x)$ arising from a loaded substrip of width $2b$ with load amplitude q_o is equal to (Eq. 7.46)

$$v(x) = \frac{2}{\pi}\left(\int_0^\infty \frac{\sin kb}{ik^2tG^*}\cos kx\, dk\right)q_o$$

For $k \longrightarrow 0$, $kt \longrightarrow \omega/c_s^*$, so that $\sin kb/k^2t$ thus converges to bc_s^*/ω. For $k \longrightarrow \infty$, the integrand vanishes. The function $v(x)/q_o$ determines the flexibility-influence

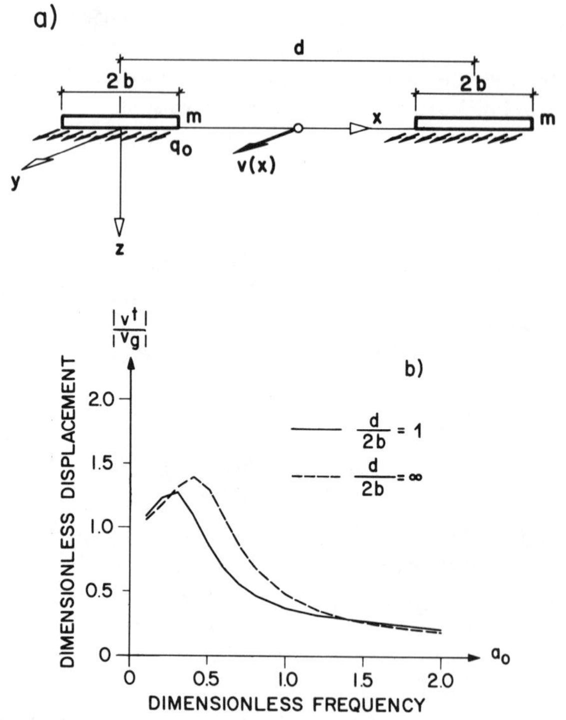

Figure P7-8 Through-soil coupling of two strip foundations. (a) Nomenclature; (b) amplification.

function $g(x)$, which expresses the displacement amplitude under the two basemats for the two unit-load intensities.

Although, because of the symmetry of the problem, the two load intensities will be equal, this property is not used when calculating the dynamic-stiffness matrix. The functions $L(x)$ and $N(x)$ are piecewise constant of unit value. Integrating $g(x)$ along the basemats numerically leads to the $[G]$ matrix of order 2×2 (Eq. 7.33). The elements of the diagonal matrix $[T]$ (Eq. 7.34) equal $2b$. The dynamic-stiffness matrix $[S_{bb}]$ of the coupled system follows from Eq. 7.40 and can be expressed as

$$[S_{bb}] = \pi G \left(\begin{bmatrix} k_{11} & k_{12} \\ k_{12}^T & k_{11} \end{bmatrix} + ia_o \begin{bmatrix} c_{11} & c_{12} \\ c_{12}^T & c_{11} \end{bmatrix} \right)$$

The spring coefficients k_{11} and k_{12} and damping coefficients c_{11} and c_{12} are functions of a_o, ζ, and d/b.

The equation of motion of the system with mass in the total displacement amplitude v^t for symmetric excitation equals

$$[-\omega^2 m + \pi G(k_{11} + k_{12} + ia_o(c_{11} + c_{12}))]v^t$$
$$= \pi G(k_{11} + k_{12} + ia_o(c_{11} + c_{12}))v_g$$

or introducing $\bar{m}$,

$$\frac{v^t}{v_g} = \frac{1}{1 - a_o^2\bar{m}/\pi[k_{11} + k_{12} + ia_o(c_{11} + c_{12})]}$$

The corresponding equation for the system with one rigid base follows from $d \longrightarrow \infty$. The amplification $|v^t|/|v_g|$ is plotted for $\bar{m} = 5$ in Fig. P7-8b. Although the two basemats are adjacent to each other, the through-soil coupling effects are small.

7.9. Calculate for relaxed contact the vertical dynamic-stiffness coefficient S_z of a circular basemat of radius a resting on the surface of an undamped elastic half-space having Poisson's ratio $= 0.33$. Use only one subdisk with a constant vertical load amplitude r_o, which will lead to a very crude approximation. Plot the real and imaginary parts of the influence function w as a function of r/a for various a_o. Plot the dimensionless spring and damping coefficients k_z and c_z versus $a_o = \omega a/c_s$ $[S_z = K_z(k_z + ia_oc_z)]$.

Solution:

For relaxed contact $F_{uw}(k)$ is disregarded in Eq. 7.57. The vertical-flexibility coefficient $F_{ww}(k)$ follows from the inversion of the stiffness matrix of order 2×2 specified in Eq. 5.135, and not as the inverse of the corresponding stiffness coefficient on the diagonal.

$$F_{ww}(k) = \frac{1}{kG}\,\frac{-is(1 + t^2)}{(1 - st)^2 + 4st}$$

The displacement amplitude $w(r)$ arising from the one subdisk equals (Eq. 7.57)

$$w(r) = a\left[\int_{k=0}^{\infty} J_1(ka)F_{ww}(k)J_o(kr)\,dk\right]r_o = g(r)r_o$$

This Green's function is plotted in Fig. P7-9a.

The two functions $L(r)$ and $N(r)$ are both constant and equal to 1. The flexibility coefficient G and T follow from Eqs. 7.33 and 7.34 as

$$G = \int_0^a 2\pi r g(r)\,dr$$

$$T = \pi a^2$$

The dynamic-stiffness matrix S_z (Eq. 7.40) is thus equal to

$$S_z = \frac{\pi a^4}{2\int_0^a rg(r)\,dr}$$

The static value $K_z = 1.82\pi Ga$ is used to nondimensionalize S_z. In the plot of k_z and c_z (Fig. P7-9b), the strong oscillations are visible that do not exist for a rigid basemat with the correct vertical load amplitudes (Fig. 7-19).

7.10. Calculate for relaxed contact the vertical dynamic-influence function $w(r)$ of an annular ring of radius a with the load amplitude R_o and which rests on the surface

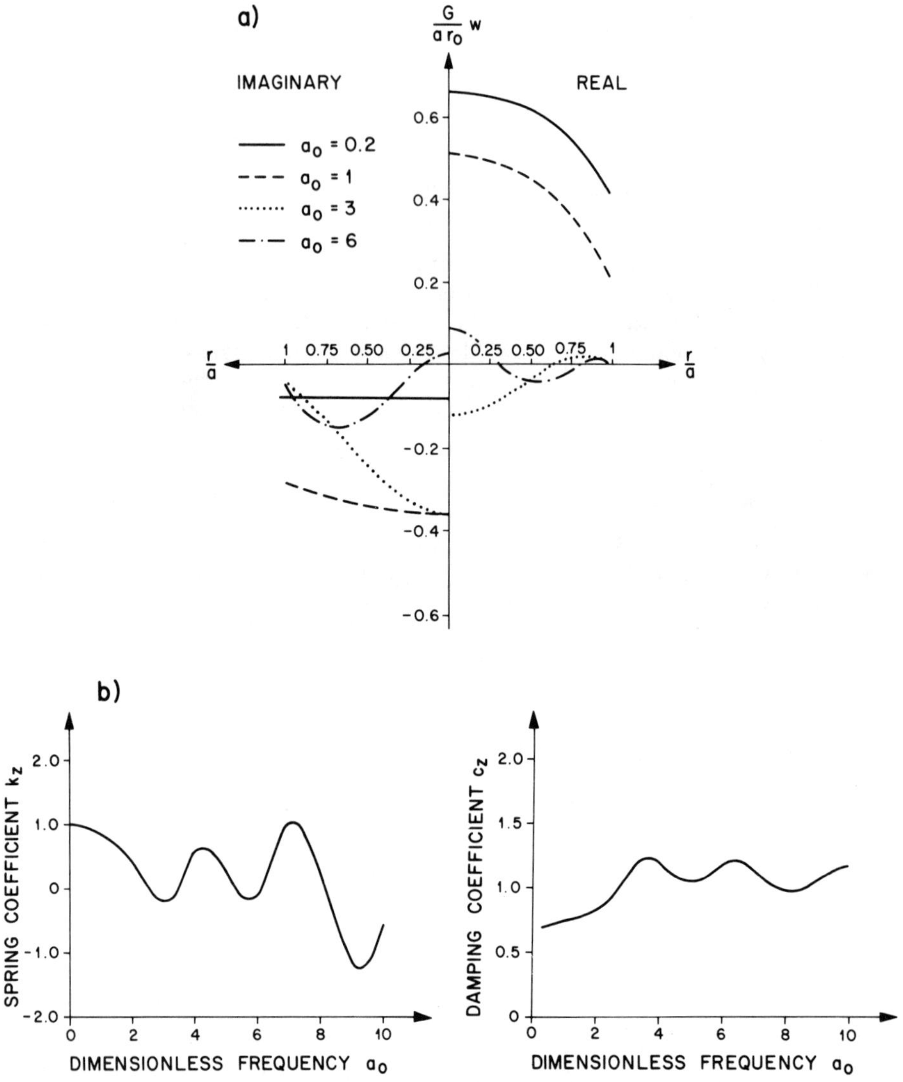

Figure P7-9 Rigid circular basemat with constant vertical load on half-space. (a) Influence function; (b) dynamic-stiffness coefficient.

of an undamped elastic half-space with Poisson's ratio $= 0.33$ (Fig. P7-10). Plot for the static case w versus r/a and note that the displacement amplitude is infinite under the ring load.

Solution:

The vertical load acting on the annular ring is assumed to have a constant amplitude r_o. This loading condition will occur when the loading on a circle with radius a is subtracted from that with $a + da$ (Fig. P7-10a). Performing the correspond-

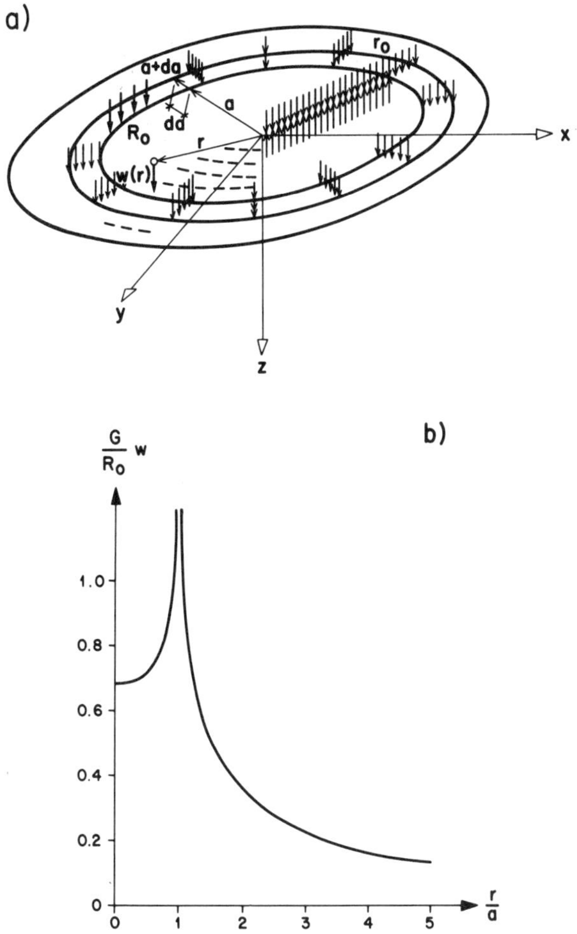

Figure P7-10 Annular ring load. (a) Nomenclature; (b) influence function.

ing operation on Eq. 7.57 results in the influence coefficient for a loaded annular ring:

$$w(r) = \left[(a + da) \int_{k=0}^{\infty} J_1(k(a + da)) F_{ww}(k) J_o(kr)\, dk \right.$$
$$\left. - a \int_{k=0}^{\infty} J_1(ka) F_{ww}(k) J_o(kr) dk \right] r_o$$

The total loading acting on the annular ring equals $2\pi a da r_o$, which is equal to $2\pi a R_o$, where R_o is the annular ring-load amplitude. This leads to

$$w(r) = \left[\int_{k=0}^{\infty} \frac{(a + da)J_1(k(a + da)) - aJ_1(ka)}{da} F_{ww}(k) J_o(kr)\, dk \right] R_o$$

or

$$w(r) = \left[\int_{k=0}^{\infty} [aJ_1(ka)]_{,a} F_{ww}(k) J_o(kr)\, dk \right] R_o$$

Comparing this expression with Eq. 7.57, it follows that the influence function for a ring load is equal to the derivative with respect to the radius of the influence function of a disk. (This applies to all degrees of freedom.) Using Eq. 7.52a results in

$$w(r) = \left[a \int_{k=0}^{\infty} k J_o(ka) F_{ww}(k) J_o(kr)\, dk \right] R_o$$

For $a_o = 0$, w is plotted versus r/a in Fig. P7-10b.

7.11. The procedure of Section 7.2, which discretizes the structure–soil interface into subdisks, allows the dynamic-stiffness coefficients of surface foundations of arbitrary geometry to be calculated. An example of a general case is discussed in Section 7.7. For special geometries, more efficient methods exist, which lead to identical results. For instance, for the circular basemat discussed in Section 7.4, annular rings with a finite width can be selected.

Calculate for relaxed contact the vertical dynamic-stiffness coefficient of

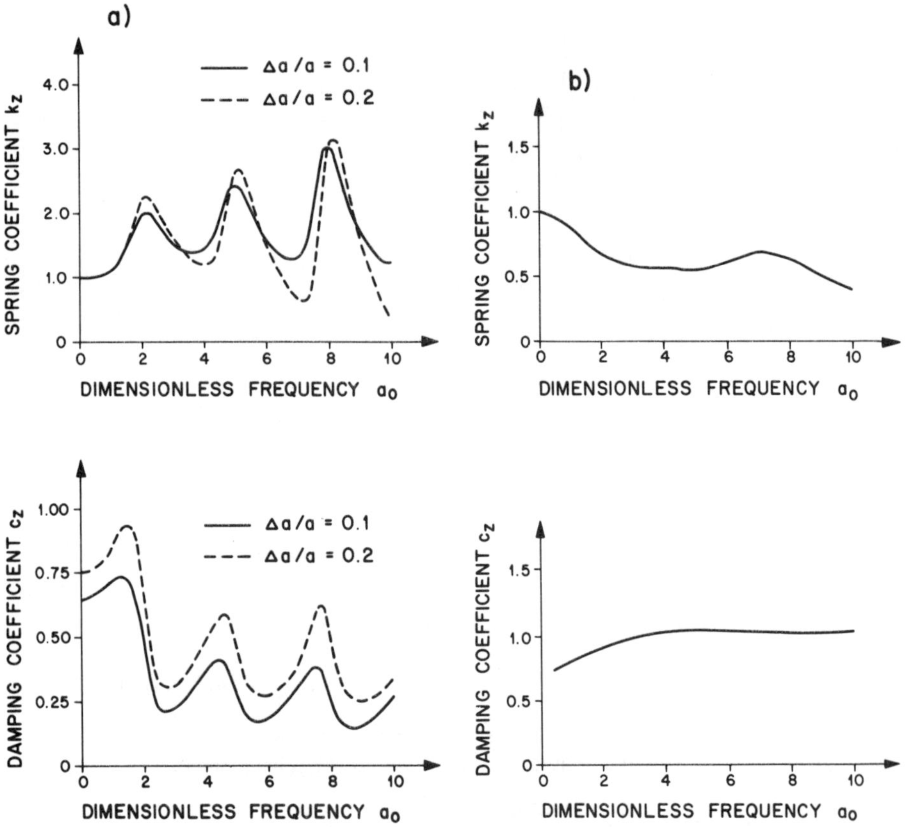

Figure P7-11 Discretization of circular basemat with annular rings. (a) Vertical dynamic-stiffness coefficient of annular ring with finite width; (b) vertical dynamic-stiffness coefficient of circular basemat.

a circular rigid basemat resting on the surface of a half-space using eight annular rings (Fig. P7-10a). Assume the load amplitude R_o to be constant over the (finite) width of each annular ring. Use the same parameters as chosen for the results shown in Fig. 7-19. Also plot the spring and damping coefficients of the non-dimensionalized dynamic-stiffness coefficient S_z for $\Delta a/a = 0.1$ and 0.2, whereby Δa denotes the finite width of the annular ring with the average radius a.

Results:

For the annular rings with $\Delta a/a = 0.1$ and $\Delta a/a = 0.2$, the static values K_z are equal to $1.618\pi Ga$ and $1.849\pi Ga$, respectively. The corresponding dynamic-stiffness coefficients are plotted in Fig. P7-11a. Note the strong dependency on frequency.

Using eight annular rings, the dynamic-stiffness coefficient of the circular basemat is calculated. The spring and damping coefficients are plotted in Fig. P7-11b as a function of a_o. The curves are quite smooth.

7.12. Calculate the flexibility-influence function $v(r)$ for the disk of radius Δa loaded torsionally as shown in Fig. P7-12. The load amplitude increases linearly from the center to the value q_o at Δa.

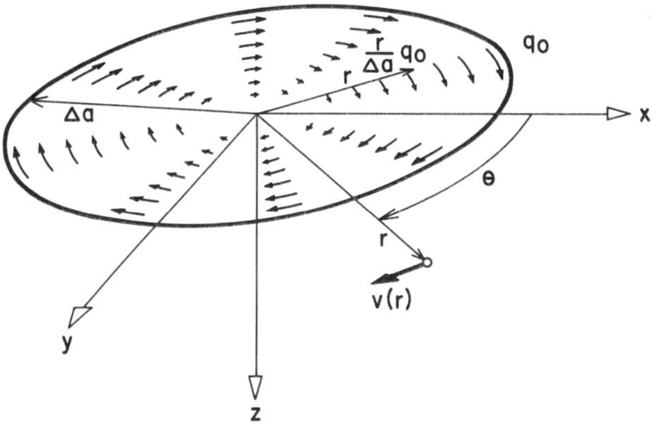

Figure P7-12 Subdisk loaded torsionally.

Solution:

Only the zeroth antimetric Fourier term is excited. With $q(r) = q_o r/\Delta a$, the amplitude of the load in the k-domain follows from Eq. 7.49a as

$$q(k) = \frac{q_o}{2\pi k\,\Delta a}\int_{r=0}^{\Delta a} rJ_o(kr)_{,r}r\int_{\theta=0}^{2\pi} d\theta\,dr = \frac{q_o}{k\,\Delta a}\int_{r=0}^{\Delta a} r^2 J_o(kr)_{,r}\,dr$$

Integrating by parts and using Eq. 7.52b lead to

$$q(k) = \frac{q_o\,\Delta a}{k}\left[J_o(k\,\Delta a) - \frac{2}{k\,\Delta a}J_1(k\,\Delta a)\right]$$

The (only) displacement amplitude in the k-domain results as

$$v(k) = F_{vv}(k)q(k)$$

The inverse transformation follows from Eq. 7.49b as

$$v(r) = \int_{k=0}^{\infty} J_o(kr),_r v(k)\, dk$$

Using Eq. 7.53,

$$v(r) = +\Delta a\left[\int_{k=0}^{\infty} \left(-J_o(k\,\Delta a) + \frac{2}{k\,\Delta a}J_1(k\,\Delta a)\right)F_{vv}(k)J_1(kr)\, dk\right]q_o$$

results.

7.13. Calculate the flexibility-influence function $w(r, \theta)$ for the disk of radius Δa loaded by a rocking moment for relaxed contact as shown in Fig. P7-13. The load amplitude increases linearly with x from the center to the value r_o at Δa.

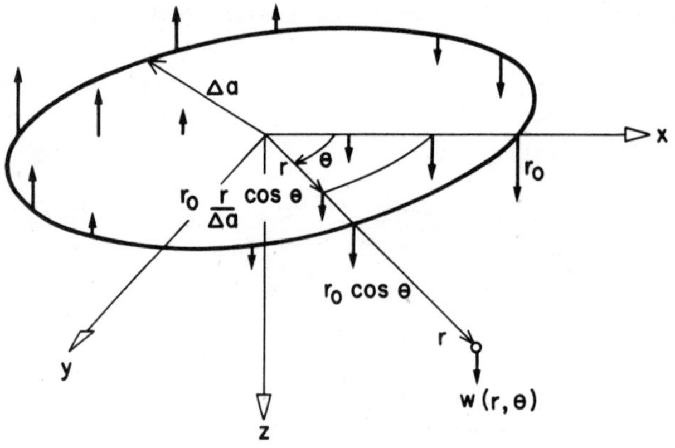

Figure P7-13 Subdisk loaded by a rocking moment.

Solution:

The first symmetric Fourier term is excited. With $r_o r \cos\theta/\Delta a$, the amplitude of the load in the k-domain results from Eq. 7.49a as

$$r(k) = -\frac{r_o}{\pi\,\Delta a}\int_{r=0}^{\Delta a} r^2 J_1(kr)\int_{\theta=0}^{2\pi}\cos^2\theta\, d\theta\, dr = -\frac{r_o}{\Delta a}\int_{r=0}^{\Delta a} r^2 J_1(kr)\, dr$$

$$= \frac{r_o\,\Delta a}{k}\left[J_o(k\,\Delta a) - \frac{2}{k\,\Delta a}J_1(k\,\Delta a)\right]$$

For relaxed contact, the displacement amplitude in the k-domain is equal to $w(k) = F_{ww}(k)r(k)$. The inverse transformation follows from Eq. 7.49b as

$$w(r, \theta) = \cos\theta\,\Delta a\left[\int_{k=0}^{\infty}\left(\frac{2}{k\,\Delta a}J_1(k\,\Delta a) - J_o(k\,\Delta a)\right)F_{ww}(k)J_1(kr)\, dk\right]r_o$$

7.14. In Problem 7.1 the dynamic-stiffness coefficient for the out-of-plane motion is derived, assuming a linear variation of the displacement with the depth of the homogeneous layer fixed at its base (Fig. P7-1). Assuming a source load that varies linearly from the free surface to zero at the base acting on a vertical line at an eccentricity e equal to the depth d, calculate the dynamic-stiffness coefficient S_y of the undamped case for $a_o = \pi/4$ and π, using the indirect boundary-element method. Compare with the analytical value specified in Problem 7.1. Also place the source line adjacent to the structure–soil interface and repeat the calculation.

Solution:

The procedure is analogous to that described in Section 7.5.2 for the static case. Analogous to Problem 7.1 the amplitude of the displacement $v(x, z)$ is specified as

$$v(x, z) = \sum_j a_j \cos k_j t z \exp(-ik_j x)$$

and the amplitude of the shear stress $\tau_{yx}(x, z)$,

$$\tau_{yx}(x, z) = -Gi \sum_j a_j k_j \cos k_j t z \exp(-ik_j x)$$

Formulating the boundary condition

$$\tau_{yx}(-e, z) = -\tfrac{1}{2} q(z) = -\tfrac{1}{2} L(z) q$$

with

$$L(z) = 1 - \frac{z}{d}$$

and expanding the load in a Fourier series leads to

$$a_j = \frac{\exp(-ik_j e)}{iGk_j(k_j td)^2} q$$

The coefficients $g_u(z)$ and $g_p(z)$ follow straightforwardly setting $x = 0$ in the equations for v and τ_{yx}. With

$$N(z) = 1 - \frac{z}{d}$$

The G- and T-coefficients of Eqs. 7.94 and 7.95 follow, which results in (Eq. 7.101)

$$S_y = 2G \frac{\left[\sum_j \dfrac{\exp(-ik_j e)}{(k_j td)^4}\right]^2}{\sum_j \dfrac{\exp(-2ik_j e)}{(k_j td)^5 \sqrt{1 - a_o^2/(k_j td)^2}}}$$

where

$$k_j td = \frac{2j-1}{2}\pi \qquad j = 1, 2, \ldots$$

and

$$k_j e = \frac{e}{id} \frac{k_j td}{\sqrt{1 - (a_o/(k_j td))^2}}$$

Decomposing S_y as in Problem 7.1, the spring and damping coefficients k_y and c_y follow. For $a_o = \pi/4$, $k_y = 0.824$ (= 95% of exact value) for $e/d = 1$ and $k_y = 0.844$ (= 97%) for $e/d = 0$ result. For $a_o = \pi$, $k_y = 0.583 \times 10^{-4}$ (< 1%), $c_y = 0.524$ (= 100%) for $e/d = 1$ and $k_y = 0.0175$ (= 45%), $c_y = 0.539$ (= 103%) for $e/d = 0$.

7.15. As described at the end of Section 7.1.3, a local boundary located on the surface of an embedded cyclinder with a vertical axis can be derived through the examination of an infinitesimally thin independent layer extending to infinity. The salient features can be discussed, calculating the dynamic-stiffness coefficient S_r, relating the amplitude of the radial displacement u_o at r = radius a of the cylinder to that of an applied load $2\pi ap$, where $p = -\sigma_r(r = a)$ is a constant in the circumferential direction θ (Fig. P7-15a). Plot the spring and damping coefficients k_r and c_r as a function of a suitably chosen dimensionless frequency a_o for $\nu = 0.3$.

Solution:

Only a radial displacement with an amplitude u can arise. The radial coordinate r is the only independent variable. The stress-displacement relationship follows from Eq. 5.152 a and b and Hooke's law for plain strain as

$$\sigma_r = \frac{(1-\nu)E}{(1+\nu)(1-2\nu)}\left(\frac{\nu}{1-\nu}\frac{u}{r} + u_{,r}\right)$$

$$\sigma_\theta = \frac{(1-\nu)E}{(1+\nu)(1-2\nu)}\left(\frac{u}{r} + \frac{\nu}{1-\nu}u_{,r}\right)$$

Substituting in Eq. 5.151a leads to

$$r^2 u_{,rr} + r u_{,r} + \left(\frac{a_o^2}{a^2}r^2 - 1\right)u = 0$$

with a_o defined as

$$a_o^2 = \frac{(1+\nu)(1-2\nu)}{(1-\nu)E}\rho\omega^2 a^2$$

This differential equation is the Bessel equation of order 1 with the parameter a_o/a (Eq. 5.172). The solution is specified analogously as in Eq. 5.173.

$$u(r) = c_1 J_1\left(\frac{a_o}{a}r\right) + c_2 Y_1\left(\frac{a_o}{a}r\right)$$

As the origin ($r = 0$) is not included in the problem, $c_2 \neq 0$ in contrast to the derivation in Section 5.5.2. Introducing so-called first-order Hankel functions of the first and second kind defined as

$$H_1^{(1)}\left(\frac{a_o}{a}r\right) = J_1\left(\frac{a_o}{a}r\right) + iY_1\left(\frac{a_o}{a}r\right)$$

$$H_1^{(2)}\left(\frac{a_o}{a}r\right) = J_1\left(\frac{a_o}{a}r\right) - iY_1\left(\frac{a_o}{a}r\right)$$

leads to

$$u(r) = d_1 H_1^{(1)}\left(\frac{a_o}{a}r\right) + d_2 H_1^{(2)}\left(\frac{a_o}{a}r\right)$$

The radiation condition demands that only outgoing waves arise, propagating in the positive r-direction. For large r

$$H_1^{(1)}\left(\frac{a_o}{a}r\right) \sim \sqrt{\frac{2a}{\pi a_o r}}\exp\left[i\left(\frac{a_o}{a}r - \frac{3\pi}{4}\right)\right]$$

$$H_1^{(2)}\left(\frac{a_o}{a}r\right) \sim \sqrt{\frac{2a}{\pi a_o r}}\exp\left[-i\left(\frac{a_o}{a}r - \frac{3\pi}{4}\right)\right]$$

Selecting the time factor as $\exp(+i\omega t)$, the wave propagating in the positive r-direction is characterized by a negative imaginary part of the exponent involving r. Only the term with $H_1^{(2)}$ is thus kept ($d_1 = 0$). This can also be verified

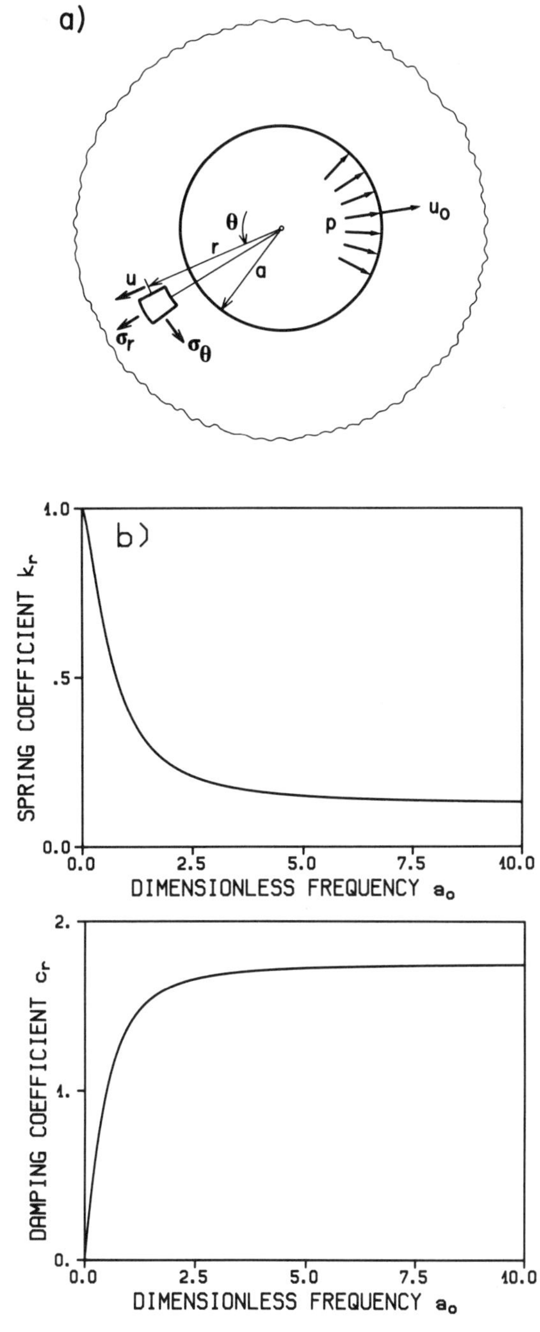

Figure P7-15 Infinitesimally thin layer extending to infinity with circular cavity. (a) Plan view and nomenclature; (b) dynamic-stiffness coefficient.

using Sommerfeld's radiation condition for two dimensions (see also Eq. 7.27)

$$\lim_{r\to\infty} \sqrt{r}\left(u_{,r} + i\frac{a_o}{a}u\right) = 0$$

The term with $H_1^{(1)}$ does not satisfy this equation. For a prescribed displacement amplitude $u_o = u(r = a)$

$$d_2 = \frac{u_o}{H_1^{(2)}(a_o)}$$

The solution thus equals

$$u(r) = \frac{H_1^{(2)}\left(\frac{a_o}{a}r\right)}{H_1^{(2)}(a_o)}$$

Substituting into the stress-displacement relationship yields $\sigma_r(r = a)$. Defining the dynamic-stiffness coefficient S_r

$$S_r = -\frac{2\pi a \sigma_r(r = a)}{u_o}$$

leads to

$$S_r = \frac{2\pi(1 - \nu)E}{(1 + \nu)(1 - 2\nu)}\left[\frac{1 - 2\nu}{1 - \nu} - a_o\frac{H_0^{(2)}(a_o)}{H_1^{(2)}(a_o)}\right]$$

$H_0^{(2)}$ is the zeroth-order Hankel function of the second kind. S_r is split up as

$$S_r = \frac{2\pi E}{1 + \nu}(k_r + i a_o c_r)$$

$2\pi E/(1 + \nu)$ represents the static-stiffness coefficient. k_r and c_r are plotted versus a_o in Fig. P7-15b.

This problem addresses the zeroth symmetric Fourier term in the circumferential direction. Other Fourier terms lead analogously to the stiffness coefficients of torsional, vertical, and horizontal degrees of freedom.

8

ALTERNATIVE FORMULATION OF EQUATION OF MOTION

8.1 DIRECT ANALYSIS OF TOTAL STRUCTURE–SOIL SYSTEM

The basic equation of motion is derived directly by the substructure method in Section 3.1. The same equation is again obtained below, starting from that of the direct solution of the total structure–soil system. This allows the direct method to be explained and its relationships to the procedure of the substructure analysis to be established. In particular, it will become evident that the two methods are equivalent, that is, that if they are implemented consistently, identical results will be obtained.

8.1.1 Equation of Motion in Time Domain

The total structure–soil system is shown in Fig. 8-1. The single structure with a flexible base (basemat and wall) is embedded in soil. In the direct method, the part of the soil included in the model is also discretized (e.g., with finite elements). The soil domain with some material damping is limited by a fictitious exterior boundary, which is placed so far away from the structure that during the total earthquake excitation, the waves generated along the structure–soil interface do not reach it. Subscripts are used to identify the nodes of the discretized model. The nodes along the structure–soil interface are denoted by b (for base), the remaining nodes of the structure by s. The subscripts i and r indicate the nodes of the soil in the interior region and on the exterior boundary, respectively. To differentiate between the various subsystems, superscripts are used when necessary. The letter s denotes the structure (when used with a property matrix), g the ground (soil with excavation), f the free field (continuous

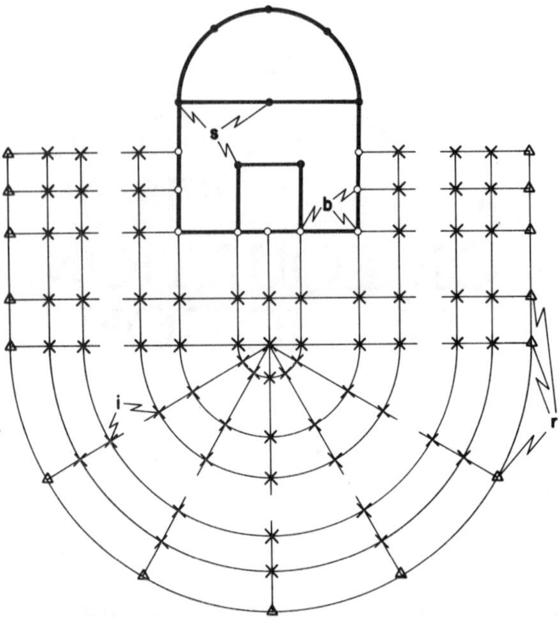

Figure 8-1 Total structure–soil system.

soil without excavation), and e the excavated soil. The reference soil systems, which are the same as those introduced in Section 3.1, are shown in Fig. 3-2.

Assembling the property matrices of the structure and of the soil, the equations of motion in the time domain of the total system for earthquake excitation can be formulated (analogous to Eq. 2.1).

$$
\begin{bmatrix}
[M_{ss}] & [M_{sb}] & & \\
[M_{bs}] & [M^s_{bb}] + [\bar{M}^g_{bb}] & [\bar{M}_{bi}] & \\
 & [\bar{M}_{ib}] & [\bar{M}_{ii}] & [\bar{M}_{ir}] \\
 & & [\bar{M}_{ri}] & [\bar{M}_{rr}]
\end{bmatrix}
\begin{Bmatrix} \{\ddot{r}^t_s\} \\ \{\ddot{r}^t_b\} \\ \{\ddot{r}^t_i\} \\ \{\ddot{r}^t_r\} \end{Bmatrix}
+
\begin{bmatrix}
[C_{ss}] & [C_{sb}] & & \\
[C_{bs}] & [C^s_{bb}] + [\bar{C}^g_{bb}] & [\bar{C}_{bi}] & \\
 & [\bar{C}_{ib}] & [\bar{C}_{ii}] & [\bar{C}_{ir}] \\
 & & [\bar{C}_{ri}] & [\bar{C}_{rr}]
\end{bmatrix}
\begin{Bmatrix} \{\dot{r}^t_s\} \\ \{\dot{r}^t_b\} \\ \{\dot{r}^t_i\} \\ \{\dot{r}^t_r\} \end{Bmatrix}
+
\begin{bmatrix}
[K_{ss}] & [K_{sb}] & & \\
[K_{bs}] & [K^s_{bb}] + [\bar{K}^g_{bb}] & [\bar{K}_{bi}] & \\
 & [\bar{K}_{ib}] & [\bar{K}_{ii}] & [\bar{K}_{ir}] \\
 & & [\bar{K}_{ri}] & [\bar{K}_{rr}]
\end{bmatrix}
\begin{Bmatrix} \{r^t_s\} \\ \{r^t_b\} \\ \{r^t_i\} \\ \{r^t_r\} \end{Bmatrix}
=
\begin{Bmatrix} \{0\} \\ \{0\} \\ \{0\} \\ \{R_r\} \end{Bmatrix}
\tag{8.1}
$$

The dynamic-equilibrium equations in the nodes of the exterior boundary of the soil (subscript r) are also specified. The matrices $[M]$, $[C]$, and $[K]$ represent the mass, the damping, and the static stiffness, respectively. A bar is used to indicate the property matrices of the soil corresponding to the discretization, including the internal nodes i. The property matrices corresponding to the degrees of freedom of the nodes b along the structure–soil interface are added, (e.g., $[K^s_{bb}] + [\bar{K}^g_{bb}]$). The vector $\{r\}$ represents the displacement; the superscript t indicates that the motion is total. The vector $\{R\}$ denotes the reaction forces.

For this direct-solution procedure to be valid, the exterior boundary of the soil must be placed sufficiently far away from the structure, where the free-field motion $\{r^f\}$ applies (it thus also has to be calculated in the direct-solution procedure, using, for example, the methods presented in Chapter 6).

$$\begin{aligned} \{\ddot{r}^t_r\} &= \{\ddot{r}^f_r\} \\ \{\dot{r}^t_r\} &= \{\dot{r}^f_r\} \\ \{r^t_r\} &= \{r^f_r\} \end{aligned} \tag{8.2}$$

Deleting the equilibrium equations in the exterior nodes r (which are used only to calculate the reaction forces $\{R_r\}$), substituting Eq. 8.2 and rearranging Eq. 8.1, the equations of motion in the time domain of the total system are modified as follows.

$$\begin{bmatrix} [M_{ss}] & [M_{sb}] & \\ [M_{bs}] & [M^s_{bb}] + [\bar{M}^g_{bb}] & [\bar{M}_{bi}] \\ & [\bar{M}_{ib}] & [\bar{M}_{ii}] \end{bmatrix} \begin{Bmatrix} \{\ddot{r}^t_s\} \\ \{\ddot{r}^t_b\} \\ \{\ddot{r}^t_i\} \end{Bmatrix} + \begin{bmatrix} [C_{ss}] & [C_{sb}] & \\ [C_{bs}] & [C^s_{bb}] + [\bar{C}^g_{bb}] & [\bar{C}_{bi}] \\ & [\bar{C}_{ib}] & [\bar{C}_{ii}] \end{bmatrix} \begin{Bmatrix} \{\dot{r}^t_s\} \\ \{\dot{r}^t_b\} \\ \{\dot{r}^t_i\} \end{Bmatrix}$$

$$+ \begin{bmatrix} [K_{ss}] & [K_{sb}] & \\ [K_{bs}] & [K^s_{bb}] + [\bar{K}^g_{bb}] & [\bar{K}_{bi}] \\ & [\bar{K}_{ib}] & [\bar{K}_{ii}] \end{bmatrix} \begin{Bmatrix} \{r^t_s\} \\ \{r^t_b\} \\ \{r^t_i\} \end{Bmatrix} = - \begin{Bmatrix} \{0\} \\ \{0\} \\ [\bar{M}_{ir}]\{\ddot{r}^f_r\} + [\bar{C}_{ir}]\{\dot{r}^f_r\} + [\bar{K}_{ir}]\{r^f_r\} \end{Bmatrix} \tag{8.3}$$

The effective load is described by the free-field motion of the external boundary and acts because of the band structure of the cross-coupling matrices $[\bar{K}_{ir}]$, and so on, only on those interior nodes which are adjacent to the exterior boundary.

8.1.2 Equation of Motion in Frequency Domain

Using the complex-response method (Section 2.6), the equations of motion (Eq. 8.3) in the frequency domain are as follows:

$$\begin{bmatrix} [S_{ss}] & [S_{sb}] & \\ [S_{bs}] & [S^s_{bb}] + [\bar{S}^g_{bb}] & [\bar{S}_{bi}] \\ & [\bar{S}_{ib}] & [\bar{S}_{ii}] \end{bmatrix} \begin{Bmatrix} \{u^t_s\} \\ \{u^t_b\} \\ \{u^t_i\} \end{Bmatrix} = - \begin{Bmatrix} \{0\} \\ \{0\} \\ [\bar{S}_{ir}]\{u^f_r\} \end{Bmatrix} \tag{8.4}$$

with the dynamic-stiffness matrix $[S]$ (Eq. 2.10)

$$[S] = [K] + i\omega[C] - \omega^2[M] \tag{8.5}$$

and with $\{u\}$, the vector of displacement amplitudes, and $\{r\}$ forming a Fourier transform pair (analogous to Eq. 2.17).

$$\{u\} = \int_{-\infty}^{+\infty} \{r\} \exp(-i\omega t)\, dt \tag{8.6a}$$

$$\{r\} = \frac{1}{2\pi} \int_{-\infty}^{+\infty} \{u\} \exp(i\omega t)\, d\omega \tag{8.6b}$$

Equations 8.5 and 8.6 apply to all submatrices and vectors, respectively, with the various subscripts and superscripts as specified in Eq. 8.4.

Eliminating all degrees of freedom of the interior nodes of the soil i (dynamic condensation) from Eq. 8.4 results in

$$\begin{bmatrix} [S_{ss}] & [S_{sb}] \\ [S_{bs}] & [S^s_{bb}] + [S^g_{bb}] \end{bmatrix} \begin{Bmatrix} \{u^t_s\} \\ \{u^t_b\} \end{Bmatrix} = -\begin{Bmatrix} \{0\} \\ [S_{br}]\{u^f_r\} \end{Bmatrix} \tag{8.7}$$

where

$$[S^g_{bb}] = [\bar{S}^g_{bb}] - [\bar{S}_{bi}][\bar{S}_{ii}]^{-1}[\bar{S}_{ib}] \tag{8.8a}$$

$$[S_{br}] = -[\bar{S}_{bi}][\bar{S}_{ii}]^{-1}[\bar{S}_{ir}] \tag{8.8b}$$

It should be remembered that eliminating variables in the frequency domain (and thus for harmonic excitation) is just as straightforward as in the static case.

Equation 8.7 represents a possible formulation of the soil–structure interaction analysis, expressed in the total motion. It is, however, not convenient to use, as the load vector (acting at the nodes b at the base) is expressed as the product of the cross-coupling dynamic-stiffness matrix $[S_{br}]$, and the vector of the free-field motion at the fictitious exterior boundary $\{u^f_r\}$. The matrix $[S_{br}]$ is difficult to establish. The coefficient matrix on the left-hand side is quite familiar. $[S^g_{bb}]$ represents the dynamic-stiffness matrix of the ground (soil with excavation, Fig. 3-3). Methods of calculating $[S^g_{bb}]$ are discussed in depth in Chapter 7. $[S_{ss}]$, $[S_{sb}]$, and $[S^s_{bb}]$ are the dynamic-stiffness submatrices of the structure (Chapter 4). For (frequency-independent) hysteretic damping in the structure, the following applies (Eq. 2.15):

$$[S_{ss}] = [K_{ss}](1 + 2\zeta i) - \omega^2[M_{ss}] \tag{8.9a}$$

$$[S_{sb}] = [K_{sb}](1 + 2\zeta i) - \omega^2[M_{sb}] \tag{8.9b}$$

$$[S^s_{bb}] = [K^s_{bb}](1 + 2\zeta i) - \omega^2[M^s_{bb}] \tag{8.9c}$$

The damping ratio ζ is assumed constant throughout the structure in the following. This is not imperative. It does, however, allow the formulas to be written more concisely.

The vector $\{u^t_s\}$ contains all dynamic degrees of freedom of the structure which are not associated with nodes located on the structure–soil interface. Besides displacement amplitudes, generalized displacements (modal amplitudes, generalized coordinates) will also arise in $\{u^t_s\}$ if introduced as described in Sections 4.3.6 and 4.4. In this case, Eq. 8.9 does not apply. It is worth noting that $[S^g_{bb}]$ cannot be formulated as in Eq. 8.9, as the radiation damping term cannot be written in that form ($[S^g_{bb}] = [K^g_{bb}] + i\omega\,[C^g_{bb}]$).

Equation 8.8a represents an intermediate formal result and should not be used actually to calculate $[S^g_{bb}]$. As discussed in Section 7.1, based on a discretization of a realistically bounded domain of soil with material damping using for example, finite elements, only a very crude approximation for the dynamic-stiffness matrix of the semi-infinite medium is obtained. This objection would disappear if the effect of the soil region not modeled were implicitly included in Eq. 8.1, using some kind of an absorbing boundary (dynamic-stiffness matrix). This would also enable a nonlinear system to be calculated (in the time domain), whereby that part of the soil not explicitly modeled (located on the exterior of the fictitious boundary) has to behave linearly. Strong nonlinearities could arise in the rest of the system, in particular in the structure. The Fourier transform of the dynamic-stiffness matrix (in the frequency domain) of the unbounded soil would result in the one applicable to the time domain. The convolution operator would then appear in Eq. 8.1, and this would make it necessary to store and process the complete time history of the displacements. The same also applies if a reduced equation after eliminating the internal nodes in the soil is used, for example, the equivalent in the time domain of the basic equation of motion (Eq. 8.13) for a nonlinear structure. This procedure is discussed briefly in Section 8.5.

8.2 SUBSTRUCTURE ANALYSIS WITH FLEXIBLE BASE

8.2.1 Basic Equation of Motion in Total Displacements

The load vector of the equation of motion of the structure–soil system (Eq. 8.7) can be reformulated as follows: In the dynamic system of the free field, the motion $\{u^f_r\}$ acts on the same exterior boundary (Fig. 8-2) that is introduced

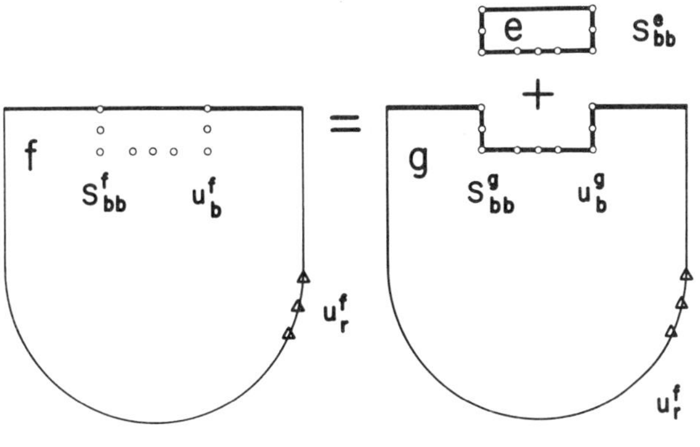

Figure 8-2 Dynamic-stiffness matrix and earthquake excitation referred to different reference systems of soil.

for the total structure–soil system (Fig. 8-1). The matrix and vector symbols are deleted in this figure. As explained in Section 3.1.1, the free-field system can be assembled from the ground and embedment systems, leading to

$$[S^f_{bb}] = [S^g_{bb}] + [S^e_{bb}] \tag{8.10}$$

By selecting the "structure" to consist of the excavated part of the soil only, Eq. 8.7 can be formulated as a special case. With $[S_{bs}] = [0]$, $[S^s_{bb}] = [S^e_{bb}]$, and $\{u^t_b\} = \{u^f_b\}$,

$$([S^e_{bb}] + [S^g_{bb}])\{u^f_b\} = -[S_{br}]\{u^f_r\} \tag{8.11}$$

results. With Eq. 8.10, Eq. 8.11 is formulated as

$$[S^f_{bb}]\{u^f_b\} = -[S_{br}]\{u^f_r\} \tag{8.12}$$

Substituting Eq. 8.12 in Eq. 8.7 leads to

$$\begin{bmatrix} [S_{ss}] & [S_{sb}] \\ [S_{bs}] & [S^s_{bb}] + [S^g_{bb}] \end{bmatrix} \begin{Bmatrix} \{u^t_s\} \\ \{u^t_b\} \end{Bmatrix} = \begin{Bmatrix} \{0\} \\ [S^f_{bb}]\{u^f_b\} \end{Bmatrix} \tag{8.13}$$

This represents the basic equation of motion formulated in total displacement amplitudes, also derived (and discussed in depth) in Section 3.1.1 (Eq. 3.9).

Instead of using the free field, the ground system (soil accounting for the excavation) could be used as a reference system for the earthquake excitation. The same free-field excitation $\{u^f_r\}$ acts at the exterior boundary shown in Fig. 8-2. The corresponding equations of motion can be established from Eq. 8.7, deleting all structural matrices. With $[S_{bs}] = [S^s_{bb}] = [0]$ and $\{u^t_b\} = \{u^g_b\}$,

$$[S^g_{bb}]\{u^g_b\} = -[S_{br}]\{u^f_r\} \tag{8.14}$$

results. This equation expresses that the interaction forces $[S^g_{bb}]\{u^g_b\} + [S_{br}]\{u^f_r\}$ of the substructure of the soil with excavation along the line having nodes b (where the motion is $\{u^g_b\}$) are zero. The term arising from the support motion $[S_{br}]\{u^f_r\}$ is clearly apparent and must be taken into account.

Substituting Eq. 8.12 in Eq. 8.14 leads to the following identity for the forces:

$$[S^g_{bb}]\{u^g_b\} = [S^f_{bb}]\{u^f_b\} \tag{8.15}$$

which is identical to Eq. 3.8. This equation could be used to determine the motion of the soil modified by the excavation $\{u^g_b\}$ (the so-called scattered-wave motion). Substituting Eq. 8.15 in Eq. 8.13 leads to the same equation as Eq. 3.5:

$$\begin{bmatrix} [S_{ss}] & [S_{sb}] \\ [S_{bs}] & [S^s_{bb}] + [S^g_{bb}] \end{bmatrix} \begin{Bmatrix} \{u^t_s\} \\ \{u^t_b\} \end{Bmatrix} = \begin{Bmatrix} \{0\} \\ [S^g_{bb}]\{u^g_b\} \end{Bmatrix} \tag{8.16}$$

There is, however, no advantage in using this equation, as $\{u^g_b\}$ would have to be calculated, which is unnecessary.

It is obvious that the direct method of analyzing soil–structure interaction leads to the same basic equation of motion (Eq. 8.13) as the derivation using the substructure method (Eq. 3.9). The two methods, the direct method, which in the frequency domain uses Eq. 8.4, and the substructure procedure based on Eq. 8.13 (or some analogous formulation, as Eq. 8.16), will thus lead to identical results. This, however, is the case only if both methods are applied consistently. For instance, $[S^g_{bb}]$ and the dynamic-stiffness matrices of the soil in Eq. 8.4 must be based on the same assumptions, the influence of the embedment has to be properly taken into account, and—very important—the spatial variation of the free-field motion $\{u^f_b\}$, $\{u^f_r\}$ must correspond to the same wave pattern and control point.

8.2.2 Base Response Motion Relative to Free Field

The free-field motion (superscript f of reference subsystem) should be used to characterize the earthquake excitation, especially for a system with a flexible base. It can be advantageous to define the response motion of the base of the total dynamic system relative to that of the corresponding free field. The motion of those parts of the structure not lying on the base (for which the free-field motion is not defined) is not modified. No superscript is used to denote relative motions. Substituting

$$\{u^t_b\} = \{u^f_b\} + \{u_b\} \tag{8.17}$$

in Eq. 8.13, and using Eq. 8.10 results in

$$\begin{bmatrix} [S_{ss}] & [S_{sb}] \\ [S_{bs}] & [S^s_{bb}] + [S^g_{bb}] \end{bmatrix} \begin{Bmatrix} \{u^t_s\} \\ \{u_b\} \end{Bmatrix} = - \begin{bmatrix} [S_{sb}] \\ [S^s_{bb}] - [S^e_{bb}] \end{bmatrix} \{u^f_b\} \tag{8.18}$$

In this formulation, where the additional motion of the base $\{u_b\}$ arising from placing the structure is determined, the load vector depends on the property matrices of the structure and of the excavated soil region and on the free-field motion along the base. Because of the band structure of $[S_{sb}]$, only those nodes of the structure are loaded which are directly coupled with the base (and of course all base nodes). The lower submatrix on the diagonal $[S^s_{bb}] + [S^g_{bb}]$ can also be written as $[S^s_{bb}] - [S^e_{bb}] + [S^f_{bb}]$. The first two submatrices are the same as those appearing in the corresponding subvector of the loads. This formulation describes the equations of motion of a discretized system (consisting of the structure and, in the embedded part, of the difference of the structure and the soil) supported on a generalized spring characterized by $[S^f_{bb}]$. The loads are equal to the negative reaction forces resulting when $\{u^f_b\}$ is enforced along the base of the reduced system without the generalized spring and when constraining the other nodes in the structure from moving. This is illustrated in Fig. 8-3, which should be compared with Fig. 3-4 for the equation of motion formulated with $\{u^t\}$ in all nodes. The subscripts as well as the vector and matrix symbols have been deleted in these figures.

If the relative motion $\{u_b\}$ were defined with respect to $\{u^g_b\}$ instead of $\{u^f_b\}$ as

$$\{u^t_b\} = \{u^g_b\} + \{u_b\} \tag{8.19}$$

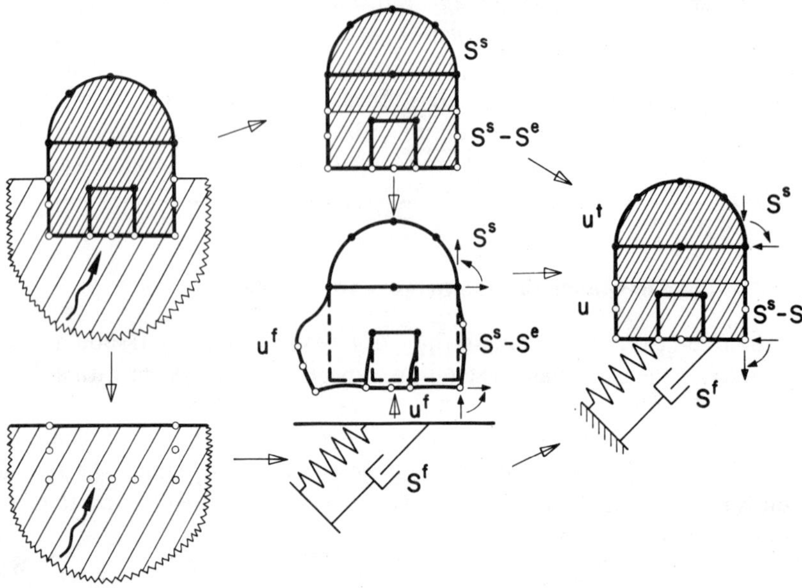

Figure 8-3 Physical interpretation of basic equation of motion with relative displacements of base.

the equivalent of Eq. 8.18 (which follows from Eq. 8.16) would read

$$\begin{bmatrix} [S_{ss}] & [S_{sb}] \\ [S_{bs}] & [S^s_{bb}] + [S^g_{bb}] \end{bmatrix} \begin{Bmatrix} \{u^t_s\} \\ \{u_b\} \end{Bmatrix} = -\begin{bmatrix} [S_{sb}] \\ [S^s_{bb}] \end{bmatrix} \{u^g_b\} \tag{8.20}$$

Although the load vector is formally simpler, this formulation should not be used, as it requires $\{u^g_b\}$ to be calculated. [In all formulations developed below, which are based on Eq. 8.18, $\{u^g_b\}$ instead of $\{u^f_b\}$ could be introduced. The corresponding formulas are, however, not specified in this section dealing with a flexible base (see Problem 8.3).]

Further formulations can be derived, starting with Eqs. 8.13 and 8.18. They are summarized in Fig. 8-5. One of the goals is to transform the load vector to the familiar form "mass of the structure times earthquake acceleration," which is well known from structural dynamics. This is also achieved in Section 3.2, where the total motion is split up into those of kinematic and inertial interactions. All formulations will lead to identical results (apart from numerical inaccuracies). Some of the more important equations are developed below.

8.2.3 Quasi-static Transmission of Free-Field Input Motion

In the first group, the so-called quasi-static motions are introduced. These are defined as the motions that arise when the known free-field motions at the base $\{u^f_b\}$ are applied statically either to the structure or to the complete structure–

soil system. The unknown (dynamic) motions are then defined relative to these quasi-static motions. Together they add up to the total motions.

In connection with Eq. 8.13 (total motion), the quasi-static displacement amplitudes $\{u_s^s\}$ can be expressed as a function of $\{u_b^f\}$, imposing $\{u_b^f\}$ statically on the structure only. The superscript s is used here to denote the static motion. The static part of the equilibrium equations of nodes s equals (Eq. 8.13)

$$[K_{ss}]\{u_s^s\} + [K_{sb}]\{u_b^f\} = \{0\} \tag{8.21}$$

or

$$\{u_s^s\} = -[K_{ss}]^{-1}[K_{sb}]\{u_b^f\} = [T_{sb}]\{u_b^f\} \tag{8.22}$$

The matrix $[T_{sb}]$ represents the quasi-static transformation. Each column of $[T_{sb}]$ can be visualized as the static displacements of the nodes s of the structure when a unit displacement is imposed at a specific node b. All other displacements at the nodes b are zero, and no loads are applied at the nodes s. Although the same transformation matrix $[T_{sb}]$ (Eq. 3.24) as in the kinematic interaction part of the analysis (Section 3.2.1) appears, the two methods are not identical for the general case, as $\{u_b^k\} \neq \{u_b^f\}$.

Defining the (dynamic-)displacement amplitudes $\{u_s^d\}$ and $\{u_b^d\}$ relative to these quasi-static displacements (superscript d indicates the dynamic motion) as

$$\{u_s^t\} = [T_{sb}]\{u_b^f\} + \{u_s^d\} \tag{8.23a}$$

$$\{u_b^t\} = \{u_b^f\} + \{u_b^d\} \tag{8.23b}$$

substituting in Eq. 8.13, using Eqs. 8.9 and 8.22, and transferring all terms associated with the free-field motions to the right-hand side results in

$$\begin{bmatrix} [S_{ss}] & [S_{sb}] \\ [S_{bs}] & [S_{bb}^s] + [S_{bb}^g] \end{bmatrix} \begin{Bmatrix} \{u_s^d\} \\ \{u_b^d\} \end{Bmatrix} = \omega^2 \begin{bmatrix} [M_{sb}] + [M_{ss}][T_{sb}] \\ [M_{bb}^s] + [M_{bs}][T_{sb}] \end{bmatrix} \{u_b^f\} + \begin{bmatrix} [0] \\ [S_{bb}^e] - (1 + 2\zeta i)([K_{bb}^s] + [K_{bs}][T_{sb}]) \end{bmatrix} \{u_b^f\} \tag{8.24}$$

The term $\omega^2\{u_b^f\}$ represents the negative acceleration amplitude $\{\ddot{u}_b^f\}$ of the free-field motion. The first term on the right-hand side can thus be interpreted as the negative inertia loads calculated as the product of the mass of the structure and the acceleration amplitude of the free-field motion imposed statically on the structure. The term $[K_{bb}^s] + [K_{bs}][T_{sb}]$ represents self-equilibrating loads which act on the base and does not in general vanish.

The resulting total response is equal to the sum of those of the quasi-static transmission and of the actual dynamic analysis. For the displacement amplitudes (and the acceleration amplitudes), this superposition is expressed in Eq. 8.23.

An analogous relationship applies to the stress resultants. In particular, the quasi-static transmission of the free-field motion will, in general, lead to nonvanishing stress resultants in the structure, including the base.

For the important case of a surface structure excited by vertically incident body waves, this analysis, which introduces the dynamic motion relative to the quasi-static transmission of the free-field excitation, is identical to the two-step procedure of kinematic and inertial interactions ($\{u_b^k\} = \{u_b^f\}$, $\{u_b^i\} = \{u_b^d\}$). The second term of the load vector vanishes for this special case.

This procedure described by Eqs. 8.23 and 8.24 is illustrated for the structure shown in Fig. 8-1 in Fig. 8-4a. The inner part of the structure is omitted for clarity. The undeformed structure is shown as a thin line, and the position after the quasi-static transmission of the free-field base motion $\{u_b^f\}$ into the structure as a dashed line. This position should be compared to that shown in Fig. 8-3, where the nodes s of the structure are held fixed. The final position is indicated as a solid line. The vector and matrix symbols are deleted in this figure.

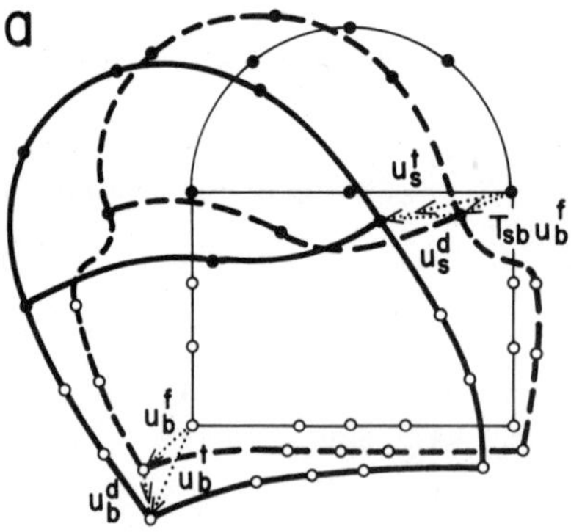

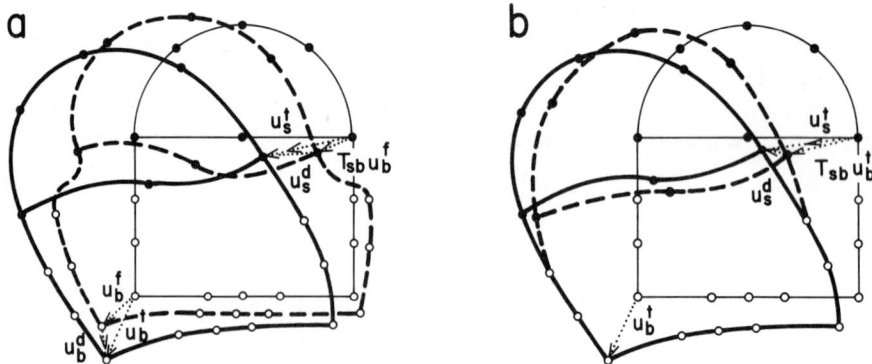

Figure 8-4 Quasi-static transmission. (a) Free-field base motion; (b) base response motion.

In connection with Eq. 8.18 (base motion relative to free field), the quasi-static displacement amplitudes $\{u_s^s\}$ and $\{u_b^s\}$ of the complete system can be related to $\{u_b^f\}$, which is imposed on the total system and not only on the structure. Omitting all dynamic terms in Eq. 8.18 leads to

$$\begin{bmatrix} [K_{ss}] & [K_{sb}] \\ [K_{bs}] & [K_{bb}^s] + [K_{bb}^g] \end{bmatrix} \begin{Bmatrix} \{u_s^s\} \\ \{u_b^s\} \end{Bmatrix} = -\begin{bmatrix} [K_{sb}] \\ [K_{bb}^s] - [K_{bb}^e] \end{bmatrix} \{u_b^f\} \tag{8.25}$$

which after solution can also be written as

$$\begin{Bmatrix} \{u_s^s\} \\ \{u_b^s\} \end{Bmatrix} = [T]\{u_b^f\} = \begin{bmatrix} [T_s] \\ [T_b] \end{bmatrix} \{u_b^f\} \tag{8.26}$$

Defining the (dynamic-)displacement amplitudes $\{u_s^d\}$ and $\{u_b^d\}$ relative to the quasi-static displacements as

$$\{u_s^t\} = [T_s]\{u_b^f\} + \{u_s^d\} \tag{8.27a}$$

$$\{u_b\} = [T_b]\{u_b^f\} + \{u_b^d\} \tag{8.27b}$$

and substituting in Eq. 8.18 leads to

$$\begin{bmatrix} [S_{ss}] & [S_{sb}] \\ [S_{bs}] & [S^s_{bb}] + [S^g_{bb}] \end{bmatrix} \begin{Bmatrix} \{u^d_s\} \\ \{u^d_b\} \end{Bmatrix}$$

$$= \omega^2 \begin{bmatrix} [M_{ss}][T_s] + [M_{sb}][T_b] + [M_{sb}] \\ [M_{bs}][T_s] + [M^s_{bb}][T_b] + [M^s_{bb}] - [M^e_{bb}] \end{bmatrix} \{u^f_b\} \tag{8.28}$$

$$- i \begin{bmatrix} [O] \\ 2\zeta(-[K^g_{bb}][T_b] + [K^e_{bb}]) + \omega([C^g_{bb}][T_b] - [C^e_{bb}]) \end{bmatrix} \{u^f_b\}$$

The stiffness term does not appear in the load vector, as (Eqs. 8.25 and 8.26)

$$\begin{bmatrix} [K_{ss}] & [K_{sb}] \\ [K_{bs}] & [K^s_{bb}] + [K^g_{bb}] \end{bmatrix} [T] = \begin{bmatrix} -[K_{sb}] \\ -[K^s_{bb}] + [K^e_{bb}] \end{bmatrix} \tag{8.29}$$

This is an advantage. As an approximation, the damping term can be neglected on the right-hand side (which is equivalent to assuming constant hysteretic damping also for the soil), resulting in the familiar inertia-load vector "mass times earthquake excitation." In this formulation, the analysis of the soil–structure interaction is actually performed in three steps. In the first, a dynamic analysis of the free field is performed (this step is necessary for all procedures). The resulting free-field motions are then applied statically at the base to the total structure–soil system in the second step. In the third, a dynamic analysis of the complete structure–soil system is performed to determine the relative motions with respect to those of the second step.

8.2.4 Quasi-static Transmission of Base Response Motion

In the second group, quasi-static displacements are also introduced, but they are defined differently, namely as the motions that arise when the unknown response base motions (either total or relative to $\{u^f_b\}$) are applied statically to the structure. As the response base motions are unknown a priori, the quasi-static displacements cannot be determined before the actual calculation as they could in the first group. It will become apparent that this definition results in a transformation of the dynamic-stiffness matrix of the total system.

In connection with Eq. 8.13 (total motion), the (dynamic-)displacement amplitudes $\{u^d_s\}$ are defined as follows:

$$\{u^t_s\} = [T_{sb}]\{u^t_b\} + \{u^d_s\} \tag{8.30}$$

where $[T_{sb}]$ is specified in Eq. 8.22. Comparing this equation with Eq. 8.23, the difference between the methods of the two groups becomes apparent. The following transformation matrix can be established:

$$\begin{Bmatrix} \{u^t_s\} \\ \{u^t_b\} \end{Bmatrix} = \begin{bmatrix} [I] & [T_{sb}] \\ & [I] \end{bmatrix} \begin{Bmatrix} \{u^d_s\} \\ \{u^t_b\} \end{Bmatrix} \tag{8.31}$$

Substituting Eq. 8.31 in Eq. 8.13 and premultiplying with the transposed transformation matrix leads to

$$
\begin{bmatrix} [S_{ss}] & -\omega^2([M_{sb}] + [M_{ss}][T_{sb}]) \\ -\omega^2([M_{sb}]^T + [T_{sb}]^T[M_{ss}]) & [S^s_{bb}] + [S_{bs}][T_{sb}] - \omega^2[T_{sb}]^T([M_{sb}] + [M_{ss}][T_{sb}]) + [S^g_{bb}] \end{bmatrix} \begin{Bmatrix} \{u^d_s\} \\ \{u^t_b\} \end{Bmatrix} = \begin{Bmatrix} \{0\} \\ [S^f_{bb}]\{u^f_b\} \end{Bmatrix} \tag{8.32}
$$

The load vector is unchanged.

This formulation, defined by Eqs. 8.31 and 8.32, is illustrated in Fig. 8-4b. The dashed line indicates the deformed position of the structure after transmitting quasi-statically the total base motion. The difference between this method and that of the first group shown in Fig. 8-4a is apparent.

In connection with Eq. 8.18 (base motion relative to free field), the following transformation is introduced:

$$
\{u^t_s\} = [T_{sb}]\{u_b\} + \{u^d_s\} \tag{8.33}
$$

The relative motion $\{u_b\}$ is defined by Eq. 8.17.

Proceeding as above, the coefficient matrix of the equations of motion is the same as in Eq. 8.32, but with the unknown vectors $\{u^d_s\}$ and $\{u_b\}$. The load vector equals

$$
\omega^2 \begin{bmatrix} [M_{sb}] + [M_{ss}][T_{sb}] \\ [T_{sb}]^T([M_{sb}] + [M_{ss}][T_{sb}]) \end{bmatrix} \{u^f_b\} + \begin{bmatrix} [0] \\ -[S_{bs}][T_{sb}] - [S^s_{bb}] + [S^e_{bb}] \end{bmatrix} \{u^f_b\} \tag{8.34}
$$

8.2.5 Transformation to Modal Amplitudes of Fixed-Base Structure

The formulation of the second group, where the unknown base motion is transmitted into the structure (Eqs. 8.32 and 8.34), can be expanded to reduce the number of degrees of freedom of the structure. It is well known that even for quite complicated structures, the seismic response is governed by only a few vibrational modes (see Chapter 4). As the dynamic-stiffness matrix of the soil $[S^g_{bb}]$ is frequency dependent and the orthogonality condition does not apply to the damping matrix, the structure–soil system does not possess vibrational modes in the classical sense, as described in Chapter 2. A dynamic subsystem (i.e., the structure on a fixed base) which thus does not involve $[S^g_{bb}]$, can, however, be used to define vibrational modes. This allows the concept of the substructure-mode synthesis discussed in Section 4.4.3 to be applied. Setting $\{u^t_b\} = \{0\}$ (fixed-base structure) in Eq. 8.30, the only nonzero displacement amplitudes of the structure are $\{u^d_s\}$. It is thus perfectly acceptable to formulate $\{u^d_s\}$ as a linear combination of the first few vibrational modes of the structure on a fixed base. The number of modes is selected smaller than the number of degrees of freedom of the built-in structure (order of vector $\{u_s\}$).

The eigenvalue problem equals (analogous to Eq. 2.2)

$$
[K_{ss}]\{\phi_j\} = \omega_j^2[M_{ss}]\{\phi_j\} \tag{8.35}
$$

where $\{\phi_j\}$ represents the jth eigenvector (mode shape) and ω_j^2 the corresponding eigenvalue (square of the circular frequency). Assembling all suitably scaled $\{\phi_j\}$ in $[\Phi]$ and all ω_j^2 as the elements in the diagonal matrix $[\Omega]$, the orthogonality properties are as follows (Eq. 2.3):

$$[\Phi]^T[M_{ss}][\Phi] = [I] \tag{8.36a}$$

$$[\Phi]^T[K_{ss}][\Phi] = [\Omega] \tag{8.36b}$$

The following relationship is defined

$$\{u_s^d\} = [\Phi]\{z\} \tag{8.37}$$

where $\{z\}$ represents the vector of the modal amplitudes.

The transformation matrix relating the old to the new variables for the case where the total base motion is transmitted quasi-statically into the structure equals

$$\begin{Bmatrix} \{u_s^d\} \\ \{u_b^t\} \end{Bmatrix} = \begin{bmatrix} [\Phi] & \\ & [I] \end{bmatrix} \begin{Bmatrix} \{z\} \\ \{u_b^t\} \end{Bmatrix} \tag{8.38}$$

Substituting Eq. 8.38 in Eq. 8.32, premultiplying by the transposed transformation matrix and using Eq. 8.36 results in

$$\begin{bmatrix} [\Omega](1 + 2\zeta i) - \omega^2[I] & -\omega^2[\Phi]^T([M_{sb}] + [M_{ss}][T_{sb}]) \\ -\omega^2([M_{sb}]^T + [T_{sb}]^T[M_{ss}])[\Phi] & [S_{bb}^s] + [S_{bs}][T_{sb}] - \omega^2[T_{sb}]^T([M_{sb}] + [M_{ss}][T_{sb}]) + [S_{bb}^g] \end{bmatrix} \begin{Bmatrix} \{z\} \\ \{u_b^t\} \end{Bmatrix} = \begin{Bmatrix} \{0\} \\ [S_{bb}^f]\{u_b^f\} \end{Bmatrix} \tag{8.39}$$

The contribution of the structure to the coefficient matrix of this equation is analogous to the coefficient matrix of the reduced equation, applying substructure-mode synthesis (Eq. 4.27). In the derivation of the latter equation, a lumped-mass matrix is assumed ($[M_{sb}] = [0]$).

For the case in which the base motion relative to that of the free field is transmitted quasi-statically into the structure, the same transformation matrix Eq. 8.38 applies, but with $\{u_b\}$ instead of $\{u_b^t\}$ being one of the variables. Proceeding as before, Eq. 8.34 is transformed. The coefficient matrix with the unknowns $\{z\}$ and $\{u_b\}$ of the equations of motion is identical to that in Eq. 8.39. The load vector equals

$$\omega^2 \begin{bmatrix} [\Phi]^T([M_{sb}] + [M_{ss}][T_{sb}]) \\ [T_{sb}]^T([M_{sb}] + [M_{ss}][T_{sb}]) \end{bmatrix} \{u_b^f\} + \begin{bmatrix} [0] \\ -[S_{bs}][T_{sb}] - [S_{bb}^s] + [S_{bb}^e] \end{bmatrix} \{u_b^f\} \tag{8.40}$$

The matrix $[\Phi]^T([M_{sb}] + [M_{ss}][T_{sb}])$, which also appears on the left-hand side, contains the generalized modal loads (participation factors) for the structure built in at its base.

The order of the equation of motion in Eq. 8.39, which has to be solved for each frequency ω, is generally considerably smaller than in the other formulations. In addition, valuable insight is gained in determining the frequencies and mode shapes of the structure fixed at its base. If the number of mode shapes included in Eq. 8.37 is the same as the number of degrees of freedom of the built-in structure, the method, of course, leads to identical results (apart from numerical inaccuracies).

All the different formulations are schematically represented in Fig. 8-5. The unknowns at the base and in the other nodes of the structure which appear in the specified equations are indicated with a solid line. The vector and matrix symbols are deleted. The part of the motion that is transmitted quasi-statically is indicated with a dashed line.

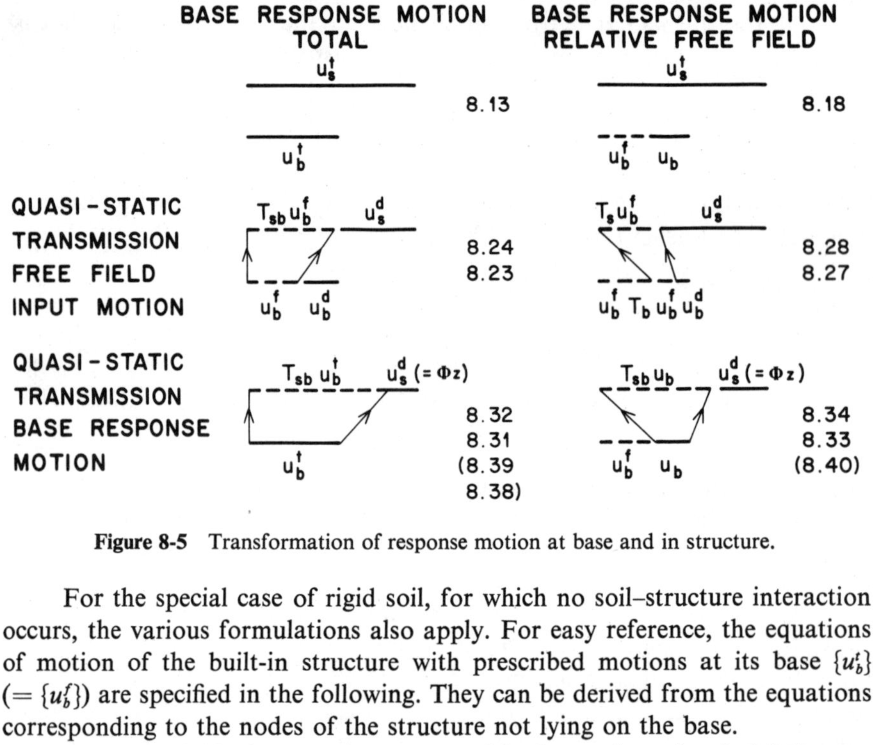

Figure 8-5 Transformation of response motion at base and in structure.

For the special case of rigid soil, for which no soil–structure interaction occurs, the various formulations also apply. For easy reference, the equations of motion of the built-in structure with prescribed motions at its base $\{u_b^t\}$ $(= \{u_b^f\})$ are specified in the following. They can be derived from the equations corresponding to the nodes of the structure not lying on the base.

From Eq. 8.13, the equation expressed in the total motion $\{u_s^t\}$ follows.

$$[S_{ss}]\{u_s^t\} = -[S_{sb}]\{u_b^t\} \tag{8.41}$$

Enforcing $\{u_b^t\}$ statically on the structure leads to (Eq. 8.24)

$$[S_{ss}]\{u_s^d\} = \omega^2([M_{sb}] + [M_{ss}][T_{sb}])\{u_b^t\} \tag{8.42}$$

with

$$\{u_s^t\} = [T_{sb}]\{u_b^t\} + \{u_s^d\} \tag{8.43}$$

The same formula is also derived from Eq. 8.28, 8.32, or 8.34.

In modal amplitudes (Eq. 8.37), Eq. 8.39 results in

$$((1 + 2\zeta i)[\Omega] - \omega^2[I])\{z\} = \omega^2[\Phi]^T([M_{sb}] + [M_{ss}][T_{sb}])\{u_b^t\} \tag{8.44}$$

Valuable physical insight can be gained from the formulations summarized in Fig. 8-5. However, they are just alternative algebraic formulations of the basic equation of motion and do not contain any additional structural mechanics. All that is actually needed to solve the most general case of soil–structure interaction is this basic equation expressed in total displacements, which, for a flexible base, is specified in Eq. 8.13.

8.3 SUBSTRUCTURE ANALYSIS WITH RIGID BASE

As discussed in Chapter 4, the base consisting of the basemat and the adjacent walls can be assumed to be rigid in many cases. This compatibility constraint on the structure–soil interface leads to a slight modification of the various formulations developed in Section 8.2. As the number of degrees of freedom is reduced compared to a flexible base, the various formulations can more easily be interpreted physically.

8.3.1 Basic Equation of Motion in Total Displacements

The same structure–soil system of Fig. 8-1 is shown with a rigid base in Fig. 8-6. The total motion at the base $\{u_b^t\}$ can be expressed as a function of the rigid-body total motions of a point O $\{u_o^t\}$ as

$$\{u_b^t\} = [A]\{u_o^t\} \tag{8.45}$$

The matrix $[A]$ represents the kinematic transformation with geometric quantities only.

Starting from Eq. 8.13 and proceeding analogously as in Section 3.1.2, the basic equation expressed in total motion is formulated as

$$\begin{bmatrix} [S_{ss}] & [S_{so}] \\ [S_{os}] & [S_{oo}^s] + [S_{oo}^g] \end{bmatrix} \begin{Bmatrix} \{u_s^t\} \\ \{u_o^t\} \end{Bmatrix} = \begin{Bmatrix} \{0\} \\ [A]^T[S_{bb}^f]\{u_b^f\} \end{Bmatrix} \tag{8.46}$$

This equation is identical to Eq. 3.15, in connection with which the nomenclature is defined. In particular, $[S_{oo}^g]$ represents the dynamic-stiffness matrix of the soil (with excavation) for a rigid structure–soil interface (Fig. 3-6), as specified

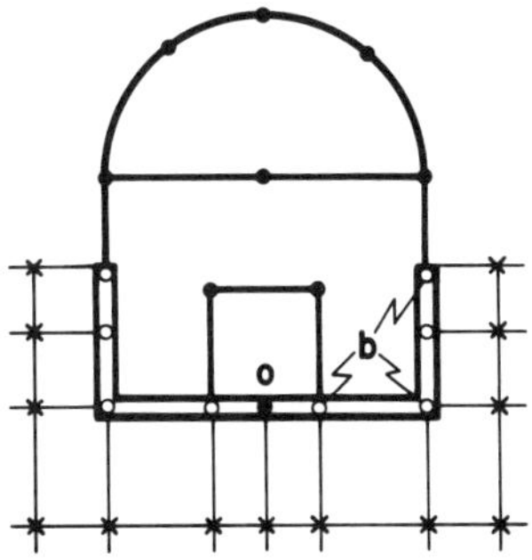

Figure 8-6 Structure–soil system with rigid base.

in Eq. 3.16c:

$$[S^g_{oo}] = [A]^T[S^g_{bb}][A] \tag{8.47}$$

The corresponding scattered-wave motion $\{u^g_o\}$ is given in Eq. 3.19 as

$$\{u^g_o\} = [S^g_{oo}]^{-1}[A]^T[S^f_{bb}]\{u^f_b\} \tag{8.48}$$

Using this equation, Eq. 8.46 is reformulated as (Eq. 3.20)

$$\begin{bmatrix} [S_{ss}] & [S_{so}] \\ [S_{os}] & [S^s_{oo}] + [S^g_{oo}] \end{bmatrix} \begin{Bmatrix} \{u^t_s\} \\ \{u^t_o\} \end{Bmatrix} = \begin{Bmatrix} \{0\} \\ [S^g_{oo}]\{u^g_o\} \end{Bmatrix} \tag{8.49}$$

As this equation is often referenced below, it is restated here.

8.3.2 Base Response Motion Relative to Scattered Motion

When trying to proceed as in Section 8.2, it is apparent that it is not possible to define a motion of the base relative to the free-field motion as in Eq. 8.17. This follows from the fact that $\{u^f_b\}$ is not compatible with $\{u^t_o\}$. Only for surface structures being excited by vertically propagating body waves are the two motions compatible. This special case, however, is contained in the formulation based on the following reference soil system.

The response motion of point O of the base can be defined relative to $\{u^g_o\}$, as the latter incorporates the constraints of the rigid-body motion. With

$$\{u^t_o\} = \{u^g_o\} + \{u_o\} \tag{8.50}$$

Eq. 8.49 is reformulated as

$$\begin{bmatrix} [S_{ss}] & [S_{so}] \\ [S_{os}] & [S^s_{oo}] + [S^g_{oo}] \end{bmatrix} \begin{Bmatrix} \{u^t_s\} \\ \{u_o\} \end{Bmatrix} = -\begin{bmatrix} [S_{so}] \\ [S^s_{oo}] \end{bmatrix} \{u^g_o\} \tag{8.51}$$

This equation corresponds to Eq. 8.20 for a flexible base.

8.3.3 Quasi-static Transmission of Scattered Motion

As in Section 8.2, further formulations can be developed starting from Eq. 8.49 (or Eq. 8.46) and from Eq. 8.51. In this first group, the dynamic motion $\{u^d\}$ is defined relative to that resulting when the known motion of the ground $\{u^g_o\}$ is applied statically. In connection with the formulation in total motion (Eq. 8.49), the following equations apply (Eq. 8.23):

$$\{u^t_s\} = [T_{so}]\{u^g_o\} + \{u^d_s\} \tag{8.52a}$$

$$\{u^t_o\} = \{u^g_o\} + \{u^d_o\} \tag{8.52b}$$

where

$$[T_{so}] = -[K_{ss}]^{-1}[K_{so}] \tag{8.53}$$

The quasi-static transformation matrix $[T_{so}]$ follows directly from rigid-body kinematics, analogously to the matrix $[A]$. As discussed in connection with Eq. 3.30, $[T_{so}]$ depends on geometric quantities only and not on the stiffness properties of the structure. This is shown in Fig. 8-7a. For the sake of clarity,

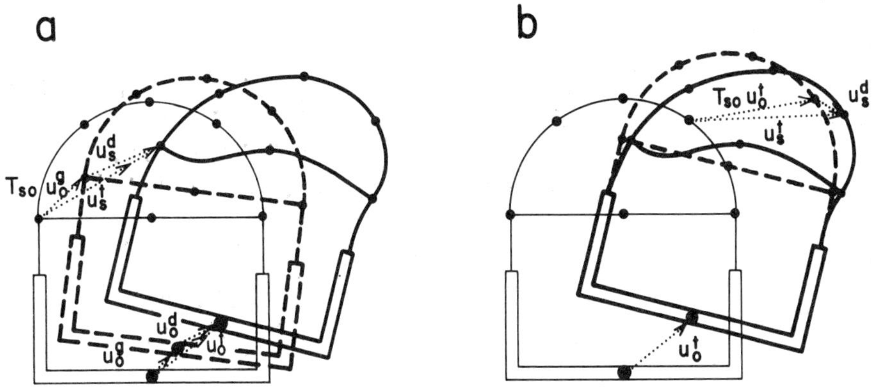

Figure 8-7 Quasi-static transmission. (a) Base input motion; (b) base response motion.

the inner part of the structure shown in Fig. 8-6 is deleted. The undeformed structure is shown as a thin line, the structure after the quasi-static transmission of $\{u_o^g\}$ as a dashed line, and the deformed structure exhibiting the total motion as a solid line. The symbols for vectors and matrices are omitted.

Substituting Eq. 8.52 in Eq. 8.49 leads to

$$\begin{bmatrix} [S_{ss}] & [S_{so}] \\ [S_{os}] & [S_{oo}^s] + [S_{oo}^g] \end{bmatrix} \begin{Bmatrix} \{u_s^d\} \\ \{u_o^d\} \end{Bmatrix} = \omega^2 \begin{bmatrix} [M_{so}] + [M_{ss}][T_{so}] \\ [M_{oo}^s] + [M_{os}][T_{so}] \end{bmatrix} \{u_o^g\} \tag{8.54}$$

The load vector does not contain any static-stiffness terms, inasmuch as

$$[K_{oo}^s] - [K_{os}][K_{ss}]^{-1}[K_{so}] = [0] \tag{8.55}$$

and this expresses static equilibrium of the structure. Equation 8.54 can, of course, also be derived directly from Eq. 8.24. The load vector is equal to the negative product of the mass of the structure and of the seismic-acceleration amplitude, the latter being determined by applying the acceleration amplitude $\{\ddot{u}_o^g\}$ $(= -\omega^2\{u_o^g\})$ in point O at the base and applying rigid-body kinematics. See Problem 8.1 for a numerical example.

In connection with the formulation of Eq. 8.51 (base motion relative to $\{u_o^g\}$), it is straightforward to demonstrate that transmitting quasi-statically $\{u_o^g\}$ into the complete structure–soil system, the same equation, Eq. 8.54, results $([T_b] = 0)$.

This procedure, which defines the dynamic motion $\{u^d\}$ relative to that resulting when the motion of the ground $\{u_o^g\}$ is applied statically, should be compared to the two-step approach which splits the total motion into those caused by kinematic and by inertial interactions. Comparing Eq. 8.52 with Eqs. 3.27 and 3.29 as well as Eq. 8.54 with Eq. 3.32, it is obvious that for a rigid base, the two formulations are the same: the quasi-static transmission of the ground motion corresponds to the kinematic motion $(\{u_o^g\} = \{u_o^k\}, [T_{so}]\{u_o^g\}$

$= \{u_s^k\}$) and the dynamic part forms the inertial motion ($\{u^d\} = \{u^i\}$). This is also apparent comparing Figs. 8-7a and 3-12.

8.3.4 Quasi-static Transmission of Base Response Motion

For the sake of completeness, the formulas are also specified for the second group, where the unknown response base motion is transmitted quasi-statically. In connection with Eq. 8.49, the following transformation, which is illustrated in Fig. 8-7b, is introduced.

$$\begin{Bmatrix} \{u_s^t\} \\ \{u_o^t\} \end{Bmatrix} = \begin{bmatrix} [I] & [T_{so}] \\ & [I] \end{bmatrix} \begin{Bmatrix} \{u_s^d\} \\ \{u_o^t\} \end{Bmatrix} \tag{8.56}$$

Substituting Eq. 8.56 in Eq. 8.49, premultiplying with the transposed transformation matrix and using Eq. 8.55 results in (see also Eq. 8.32)

$$\begin{bmatrix} [S_{ss}] & -\omega^2([M_{so}] + [M_{ss}][T_{so}]) \\ -\omega^2([M_{so}]^T + [T_{so}]^T[M_{ss}]) & -\omega^2([M_{oo}^s] + [M_{os}][T_{so}] + [T_{so}]^T[M_{so}] + [T_{so}]^T[M_{ss}][T_{so}]) + [S_{oo}^g] \end{bmatrix} \begin{Bmatrix} \{u_s^d\} \\ \{u_o^t\} \end{Bmatrix} = \begin{Bmatrix} \{0\} \\ [S_{oo}^g]\{u_o^g\} \end{Bmatrix} \tag{8.57}$$

The load vector can be transformed using Eq. 8.48, thus introducing $\{u_b^f\}$.

In connection with Eq. 8.51, the following transformation applies:

$$\{u_s^t\} = \{u_s^d\} + [T_{so}]\{u_o\} \tag{8.58}$$

The relative motion $\{u_o\}$ is defined by Eq. 8.50.

Proceeding as above, the coefficient matrix of the equations is the same as in Eq. 8.57, but with the unknowns $\{u_s^d\}$ and $\{u_o\}$. The load vector equals (see also Eq. 8.34)

$$\omega^2 \begin{bmatrix} [M_{so}] + [M_{ss}][T_{so}] \\ [M_{oo}^s] + [M_{os}][T_{so}] + [T_{so}]^T[M_{so}] + [T_{so}]^T[M_{ss}][T_{so}] \end{bmatrix} \{u_o^g\} \tag{8.59}$$

The matrix $[M_{oo}^s] + [M_{os}][T_{so}] + [T_{so}]^T[M_{so}] + [T_{so}]^T[M_{ss}][T_{so}]$, which also appears on the left-hand side, contains the generalized mass of the structure with respect to the base, that is, the rigid-structure properties: the mass, the static mass moment, and the mass moment of inertia.

8.3.5 Transformation to Modal Amplitudes of Fixed-Base Structure

Finally, $\{u_s^d\}$ can also be expressed as a linear combination with the amplitudes $\{z\}$ of the first few vibrational modes $[\Phi]$ of the structure on a fixed base. This transformation is specified in Eq. 8.37. Equation 8.57 is then transformed as follows. The first three submatrices of the coefficient matrix with the unknowns

$\{z\}$ and $\{u_o^t\}$ are specified in Eq. 8.39, with the subscript o replacing b. The fourth submatrix (lower diagonal matrix) and the load vector are given in Eq. 8.57.

In connection with Eq. 8.58, the coefficient matrix of the resulting equations with the unknowns $\{z\}$ and $\{u_o\}$ is the same as the one described in the preceding paragraph. The upper subvector of the load is specified in Eq. 8.40, the subscript o replacing b and the superscript g the letter f. The lower subvector is given by the corresponding quantity in Eq. 8.59.

The formulations developed in this chapter are applicable to much more general structural configurations than the single structure embedded in soil (Figs. 8-1 and 8-6). They can be used to analyze soil–structure interaction of complicated systems as shown in Figs. 1-3 and 3-9.

8.4 APPROXIMATE FORMULATION IN TIME DOMAIN

It is possible to analyze soil–structure interaction remaining in the time domain by introducing approximations. The most common one neglects the frequency dependence of the dynamic stiffness of the soil. The latter's force–displacement relationship is represented by means of a frequency-independent spring and dashpot. This procedure is adapted in the introductory example in Section 3.4.4. In addition, in certain cases, the total structure–soil system (and not only the structure, built in at its base) is assumed to have classical modes. Although a comprehensive treatment of these methods lies outside the scope of this text, certain aspects are discussed in the following, using the equations for a rigid base of Section 8.3 for illustration. Analogous formulations also apply for a flexible base.

8.4.1 Basic Equation of Motion in Total Displacements

In principle, all dynamic equations in the frequency domain can be approximated as follows. For instance, Eq. 8.49 is written as

$$\begin{bmatrix}[M_{ss}] & [M_{so}]\\ [M_{os}] & [M_{oo}^s]\end{bmatrix}\begin{Bmatrix}\{\ddot{r}_s^t\}\\ \{\ddot{r}_o^t\}\end{Bmatrix} + \begin{bmatrix}[C_{ss}] & [C_{so}]\\ [C_{os}] & [C_{oo}^s]+[C_{oo}^g]\end{bmatrix}\begin{Bmatrix}\{\dot{r}_s^t\}\\ \{\dot{r}_o^t\}\end{Bmatrix}$$
$$+ \begin{bmatrix}[K_{ss}] & [K_{so}]\\ [K_{os}] & [K_{oo}^s]+[K_{oo}^g]\end{bmatrix}\begin{Bmatrix}\{r_s^t\}\\ \{r_o^t\}\end{Bmatrix} = \begin{Bmatrix}\{0\}\\ [K_{oo}^g]\{r_o^g\}+[C_{oo}^g]\{\dot{r}_o^g\}\end{Bmatrix} \tag{8.60}$$

The letter r denotes a displacement as a function of time. The vectors $\{r_o^g\}$ and $\{\dot{r}_o^g\}$ represent the displacement and velocity, respectively, of the scattered motion. The dynamic-stiffness matrix of the soil $[S_{oo}^g]$ is approximated as

$$[S_{oo}^g(\omega)] = [K_{oo}^g] + i\omega[C_{oo}^g] \tag{8.61}$$

where the static-stiffness matrix $[K_{oo}^g]$ containing the spring coefficients and the damping matrix $[C_{oo}^g]$ are both independent of the frequency. They are calculated for the fundamental frequency of the structure–soil system, for example. Equation 8.60 can be solved, for example, by direct integration. This even allows a nonlinear structure to be investigated.

8.4.2 Quasi-static Transmission of Base Response Motion

Various transformations in the time domain are also possible. To be able to simplify the equations, the mass matrix is assumed to be diagonal ($[M_{so}] = [0]$). For instance, in analogy to Eqs. 8.50 and 8.58, introducing $\{r_o\}$, the displacement of the base relative to the scattered motion, and $\{r_s^d\}$ defined by

$$\{r_o^t\} = \{r_o^g\} + \{r_o\} \tag{8.62a}$$

$$\{r_s^t\} = [T_{so}]\{r_o\} + \{r_s^d\} \tag{8.62b}$$

Eq. 8.60 is transformed to

$$\begin{bmatrix} [M_{ss}] & [M_{ss}][T_{so}] \\ [T_{so}]^T[M_{ss}] & [M_{oo}^s] + [T_{so}]^T[M_{ss}][T_{so}] \end{bmatrix} \begin{Bmatrix} \{\ddot{r}_s^d\} \\ \{\ddot{r}_o\} \end{Bmatrix} + \begin{bmatrix} [C_{ss}] & \\ & [C_{oo}^g] \end{bmatrix} \begin{Bmatrix} \{\dot{r}_s^d\} \\ \{\dot{r}_o\} \end{Bmatrix}$$
$$+ \begin{bmatrix} [K_{ss}] & \\ & [K_{oo}^g] \end{bmatrix} \begin{Bmatrix} \{r_s^d\} \\ \{r_o\} \end{Bmatrix} = - \begin{bmatrix} [M_{ss}][T_{so}] \\ [M_{oo}^s] + [T_{so}]^T[M_{ss}][T_{so}] \end{bmatrix} \{\ddot{r}_o^g\} \tag{8.63}$$

In the frequency domain, the corresponding left- and right-hand sides are specified in Eqs. 8.57 and 8.59, respectively.

8.4.3 Transformation to Modal Amplitudes of Total System

In general, the transformation to normal modes of the total soil–structure system does not uncouple Eq. 8.63, that is, classical normal modes do not exist (Chapter 2). Disregarding as a further approximation the off-diagonal terms of the transformed damping matrix, a simple solution procedure results. This modal analysis in the time domain proceeds as follows: A tilde (˜) denotes the values corresponding to the uncoupled system of equations of the total system. The eigenvalue problem equals (Eq. 2.2)

$$\begin{bmatrix} [K_{ss}] & \\ & [K_{oo}^g] \end{bmatrix} \{\tilde{\phi}_j\} = \tilde{\omega}_j^2 \begin{bmatrix} [M_{ss}] & [M_{ss}][T_{so}] \\ [T_{so}]^T[M_{ss}] & [M_{oo}^s] + [T_{so}]^T[M_{ss}][T_{so}] \end{bmatrix} \{\tilde{\phi}_j\} \tag{8.64}$$

The following orthogonality equations (Eq. 2.3) apply for the suitably scaled mode shape $\{\tilde{\phi}_j\}$ and circular frequency $\tilde{\omega}_j$:

$$\{\tilde{\phi}_j\}^T \begin{bmatrix} [M_{ss}] & [M_{ss}][T_{so}] \\ [T_{so}]^T[M_{ss}] & [M_{oo}^s] + [T_{so}]^T[M_{ss}][T_{so}] \end{bmatrix} \{\tilde{\phi}_j\} = 1 \tag{8.65a}$$

$$\{\tilde{\phi}_j\}^T \begin{bmatrix} [K_{ss}] & \\ & [K_{oo}^g] \end{bmatrix} \{\tilde{\phi}_j\} = \tilde{\omega}_j^2 \tag{8.65b}$$

Assuming all $\{\tilde{\phi}_j\}$ as columns in $[\tilde{\Phi}]$, the following transformation is defined (Eq. 2.4):

$$\begin{Bmatrix} \{r_s^d\} \\ \{r_o\} \end{Bmatrix} = [\tilde{\Phi}]\{y\} \tag{8.66}$$

where $\{y\}$ represents the vector of the (first few) modal amplitudes. With this transformation, Eq. 8.63 is approximated as

$$\ddot{y}_j + 2\tilde{\zeta}_j\tilde{\omega}_j\dot{y}_j + \tilde{\omega}_j^2 y_j = -\{\tilde{\phi}_j\}^T \begin{bmatrix} [M_{ss}][T_{so}] \\ [M^s_{oo}] + [T_{so}]^T[M_{ss}][T_{so}] \end{bmatrix} \{\ddot{r}^g_o\} \tag{8.67}$$

whereby such an equation can be formulated for every amplitude y_j. The equivalent modal-damping ratio $\tilde{\zeta}_j$ is given by

$$\tilde{\zeta}_j = \frac{1}{2\tilde{\omega}_j}\{\tilde{\phi}_j\}^T \begin{bmatrix} [C_{ss}] & \\ & [C^g_{oo}] \end{bmatrix} \{\tilde{\phi}_j\} \tag{8.68}$$

The off-diagonal terms in the transformation $[\tilde{\Phi}]^T \begin{bmatrix} [C_{ss}] & \\ & [C^g_{oo}] \end{bmatrix} [\tilde{\Phi}]$ are thus neglected. When solving Eq. 8.64, it should be remembered that $[K^g_{oo}]$ is frequency dependent and thus is a function of $\tilde{\omega}_j$. An iterative procedure should be applied when determining $\{\tilde{\phi}_j\}$ and $\tilde{\omega}_j$. The dependence of $[C^g_{oo}]$ on frequency should also be taken into account when calculating $\tilde{\zeta}_j$ (Eq. 8.68). This is especially important when the dynamic-stiffness matrix of the soil varies strongly with frequency (layered sites; see Sections 7.3 and 7.4.1). Application of Eq. 8.68 (or of similar approximations) is, however, not without danger. Experience has shown that such a weighting process with the mode shapes can overestimate the damping ratio of a specific mode which is in reality hardly damped. This can lead to an underestimation of the final result if this mode has a large generalized modal load.

8.4.4 Transformation to Modal Amplitudes of Fixed-Base Structure

Finally, another formulation for which $\{r^d_s\}$ is expressed as a linear combination of the first few vibrational mode shapes $[\Phi]$ with frequencies $[\Omega]$ of the structure on a fixed base with the amplitudes $\{z\}$ is often applied (Eq. 8.35). In analogy to Eq. 8.37, this leads to

$$\{r^d_s\} = [\Phi]\{z\} \tag{8.69}$$

where $\{z\}$ is a function of time. This formulation allows the structure's fixed-base ratios of damping, which is, in addition, assumed to be viscous, to be introduced directly. The latter form the diagonal elements of the matrix $[\zeta]$. Eq. 8.63 is then transformed, using the same procedure as in the frequency domain (Eqs. 8.39, 8.58, 8.40, and 8.59), to

$$\begin{bmatrix} [I] & [\Phi]^T[M_{ss}][T_{so}] \\ [T_{so}]^T[M_{ss}][\Phi] & [M^s_{oo}] + [T_{so}]^T[M_{ss}][T_{so}] \end{bmatrix} \begin{Bmatrix} \{\ddot{z}\} \\ \{\ddot{r}_o\} \end{Bmatrix} + \begin{bmatrix} 2[\zeta][\Omega]^{1/2} & \\ & [C^g_{oo}] \end{bmatrix} \begin{Bmatrix} \{\dot{z}\} \\ \{\dot{r}_o\} \end{Bmatrix}$$

$$+ \begin{bmatrix} [\Omega] & \\ & [K^g_{oo}] \end{bmatrix} \begin{Bmatrix} \{z\} \\ \{r_o\} \end{Bmatrix} = -\begin{bmatrix} [\Phi]^T[M_{ss}][T_{so}] \\ [M^s_{oo}] + [T_{so}]^T[M_{ss}][T_{so}] \end{bmatrix} \{\ddot{r}^g_o\} \tag{8.70}$$

It is worth mentioning that in the formulation in the frequency domain the damping of the structure is assumed to be hysteretic, while that in the time domain (Eq. 8.70) is viscous.

Again, Eq. 8.70 can be solved approximately, assuming normal modes to exist. The transformation of Eq. 8.66 applies with $\{r_s^d\}$ replaced by $\{z\}$. The decoupled equations of motion are equal to

$$\ddot{y}_j + 2\tilde{\zeta}_j\tilde{\omega}_j\dot{y}_j + \tilde{\omega}_j^2 y_j = -\{\tilde{\phi}_j\}^T \begin{bmatrix} [\Phi]^T[M_{ss}][T_{so}] \\ [M_{oo}^s] + [T_{so}]^T[M_{ss}][T_{so}] \end{bmatrix} \{\ddot{r}_o^g\} \qquad (8.71)$$

where $\{\tilde{\phi}_j\}$ (appropriately scaled) and $\tilde{\omega}_j$ follow from

$$\begin{bmatrix} [\Omega] & \\ & [K_{oo}^g] \end{bmatrix} \{\tilde{\phi}_j\} = \tilde{\omega}_j^2 \left[\begin{array}{c|c} [I] & [\Phi]^T[M_{ss}][T_{so}] \\ [T_{so}]^T[M_{ss}][\Phi] & [M_{oo}^s] + [T_{so}]^T[M_{ss}][T_{so}] \end{array}\right] \{\tilde{\phi}_j\} \qquad (8.72)$$

The equivalent modal-damping ratio $\tilde{\zeta}_j$ equals

$$\tilde{\zeta}_j = \frac{1}{2\tilde{\omega}_j}\{\tilde{\phi}_j\}^T \begin{bmatrix} 2[\zeta][\Omega]^{1/2} & \\ & [C_{oo}^g] \end{bmatrix} \{\tilde{\phi}_j\} \qquad (8.73)$$

Alternatively, if the structural damping is assumed to be hysteretic with a ratio ζ (Eq. 8.39),

$$\tilde{\zeta}_j = \frac{1}{2\tilde{\omega}_j}\{\tilde{\phi}_j\}^T \begin{bmatrix} 2\dfrac{\zeta}{\tilde{\omega}_j}[\Omega] & \\ & [C_{oo}^g] \end{bmatrix} \{\tilde{\phi}_j\} \qquad (8.74)$$

results.

Valuable physical insight can be gained by comparing the natural frequencies ω_j, the mode shapes $\{\phi_j\}$, and the damping ratios ζ of the structure built in at its base with the corresponding values $\tilde{\omega}_j$, $\{\tilde{\phi}_j\}$, and $\tilde{\zeta}_j$ of the total structure–soil system. The procedure is, however, as already mentioned, approximate, as, in general, classical normal modes do not exist, and the dynamic stiffness of the soil depends on the frequency of excitation. The latter can only crudely be taken into account by considering the values at the natural frequencies of the total system.

8.5 ANALYSIS OF NONLINEAR STRUCTURE WITH LINEAR SOIL (FAR FIELD)

8.5.1 Types of Nonlinearities

The procedures developed in this text assume linear or at least quasi-linear behavior of the structure and of the unbounded soil. It is, however, well known that structures are designed by providing sufficient ductility to perform in the nonlinear range for high seismic excitation. Base-isolation systems with friction plates which exhibit strong nonlinear characteristics for the design basis earthquake are routinely being used, even for nuclear-power plants (see Section

9.3.4). Other local nonlinear effects include the partial uplift of the basemat, the separation occurring between the walls of the base and the neighboring soil in the case of embedded structures, and the highly nonlinear soil behavior arising adjacent to the basemat. In all these cases, the nonlinear behavior is restricted to the structure and possibly an irregular soil region adjacent to the structure (the near field), while the far field of the unbounded soil is assumed to remain linearly elastic with material damping. Referring to Fig. 3-8, the line joining the nodes with subscript b separates these two regions.

8.5.2 Equation of Motion in Time Domain Using Convolution Integrals

To analyze such cases, procedures that work directly in the time domain are presently being developed. Of the many approaches being proposed (e.g., the superposition of solutions which cancel all reflections in an explicit time-domain solution), the following one is a direct generalization of the method used for the linear case. A brief discussion to demonstrate certain features is appropriate; a thorough treatment, however, lies beyond the scope of this text.

Transforming the basic equation of motion (Eq. 8.16) to the time domain leads to

$$\begin{bmatrix}[M_{ss}][M_{sb}]\\ [M_{bs}][M^s_{bb}]\end{bmatrix}\begin{Bmatrix}\{\ddot{r}^t_s\}\\ \{\ddot{r}^t_b\}\end{Bmatrix}+\begin{bmatrix}[C_{ss}][C_{sb}]\\ [C_{bs}][C^s_{bb}]\end{bmatrix}\begin{Bmatrix}\{\dot{r}^t_s\}\\ \{\dot{r}^t_b\}\end{Bmatrix}+\begin{bmatrix}[K_{ss}][K_{sb}]\\ [K_{bs}][K^s_{bb}]\end{bmatrix}\begin{Bmatrix}\{r^t_s\}\\ \{r^t_b\}\end{Bmatrix}$$
$$+\begin{Bmatrix}\{0\}\\ \int_0^t [S^g_{bb}(t-\tau)]\{r^t_b(\tau)\}\,d\tau\end{Bmatrix}=\begin{Bmatrix}\{0\}\\ \int_0^t [S^g_{bb}(t-\tau)]\{r^g_b(\tau)\}\,d\tau\end{Bmatrix} \tag{8.75}$$

If the structure and the irregular soil region exhibit a nonlinear behavior, the second and third terms on the left-hand side are replaced by a vector with $\{R_s\}$ and $\{R^s_b\}$, where $\{R\}$ denotes the (nonlinear) internal forces. In this convolution-integral approach for the contribution of the soil to the equations of motion, the dynamic-stiffness matrix in the time domain $[S^g_{bb}(t)]$ contains the forces required to produce unit-impulse displacements. Denoting the interaction forces of the soil in the time domain as $\{R^g_b(t)\}$, in analogy to the contribution of the soil to the terms in Eq. 3.4,

$$\{R^g_b(t)\} = \int_0^t [S^g_{bb}(t-\tau)](\{r^t_b(\tau)\} - \{r^g_b(\tau)\})\,d\tau \tag{8.76}$$

applies. Analogously,

$$\{r^t_b(t)\} - \{r^g_b(t)\} = \int_0^t [F^g_{bb}(t-\tau)]\{R^g_b(\tau)\}\,d\tau \tag{8.77}$$

holds, whereby the displacements for unit-impulse forces form the elements of the dynamic-flexibility matrix in the time domain $[F^g_{bb}(t)]$. The vector $\{r^g_b(t)\}$ describes the scattered input motion in the time domain, which can be calculated from the free-field response $\{r_b^f(t)\}$ based on Eq. 8.15. (Alternatively, Eq. 8.13 could be transformed to the time domain, involving the convolution integral $\int_0^t [S_{bb}^f(t-\tau)]\{r_b^f(\tau)\}\, d\tau$ on the right-hand side of the equation.)

Equation 8.75 is solved directly in the time domain, taking the nonlinearities into account, using, for example, an explicit time-integration scheme. Such a procedure is also possible using a flexibility formulation (Eq. 8.77) for the contribution of the soil (far field) together with the direct stiffness method for the structure and the irregular soil region (near field).

8.5.3 Transformation of Stiffness Matrix

By definition, the dynamic-stiffness coefficient in the time domain $S(t)$ is equal to the force that produces a unit impulse displacement in the time domain (Dirac delta function). Transforming the unit impulse displacement into the frequency domain (Eq. 2.17a),

$$u(\omega) = \int_{-\infty}^{+\infty} r(t) \exp(-i\omega t)\, dt = \int_{-\infty}^{+\infty} \delta(t) \exp(-i\omega t)\, dt = 1 \tag{8.78}$$

formulating the force–displacement relationship in the frequency domain,

$$P(\omega) = S(\omega)u(\omega) = S(\omega) \tag{8.79}$$

and applying the inverse transformation (Eq. 2.17b) leads to

$$R(t) = \frac{1}{2\pi}\int_{-\infty}^{+\infty} P(\omega) \exp(i\omega t)\, d\omega = \frac{1}{2\pi}\int_{-\infty}^{+\infty} S(\omega) \exp(i\omega t)\, d\omega \tag{8.80}$$

which means, as $S(t) = R(t)$,

$$S(t) = \frac{1}{2\pi}\int_{-\infty}^{+\infty} S(\omega) \exp(i\omega t)\, d\omega \tag{8.81}$$

that is, $S(t)$ and $S(\omega)$ form a Fourier transform pair. Analogously for the dynamic-flexibility term

$$F(t) = \frac{1}{2\pi}\int_{-\infty}^{+\infty} F(\omega) \exp(i\omega t)\, d\omega \tag{8.82}$$

applies, with

$$F(\omega) = S(\omega)^{-1} \tag{8.83}$$

In analogy to $S(\omega)$, which is formulated as a function of the dimensionless frequency $a_o = \omega l/c_s$ (Eq. 7.64), it is convenient to introduce the dimensionless time $\bar{t}$ in $S(t)$:

$$\bar{t} = \frac{c_s}{l} t \tag{8.84}$$

Using dimensionless parameters, the Fourier transform pair for the dynamic-stiffness coefficient is equal to

$$S(\bar{t}) = \frac{c_s}{2\pi l}\int_{-\infty}^{+\infty} S(a_o) \exp(i a_o \bar{t})\, da_o \tag{8.85}$$

and for the dynamic-flexibility coefficient

$$F(\bar{t}) = \frac{c_s}{2\pi l}\int_{-\infty}^{+\infty} F(a_o)\exp(ia_o\bar{t})\,da_o \tag{8.86}$$

As the frequency tends to infinity, $S(\omega)$ will be infinite. It is thus necessary to decompose $S(\omega)$ into a regular part and a singular part whose transformation is valid only in the sense of a distribution. As $F(\omega)$ tends to zero for $\omega \longrightarrow \infty$, working with dynamic flexibilities may turn out to be computationally simpler.

8.5.4 Rod with Exponentially Increasing Area

As an example, the rod with the exponentially increasing area introduced in Section 5.1 is examined. Dropping the subscript 1 in Eq. 5.31, the dimensionless stiffness coefficient $\bar{S}(a_o)$ of the undamped rod with $a_o = \omega f/c_l$ is equal to

$$\bar{S}(a_o) = \tfrac{1}{2}(1 + \sqrt{1 - 4a_o^2}) \tag{8.87}$$

with the static value K specified by

$$K = \frac{EA_o}{f} \tag{8.88}$$

For $a_o \longrightarrow \infty$, $\bar{S}(a_o)$ converges to $\frac{1}{2} + ia_o$.

Substituting Eq. 8.87 in Eq. 8.85 and replacing l by f and c_s by c_l (Eq. 5.7) leads to

$$\bar{S}(\bar{t}) = \frac{c_l}{f}\left[\frac{1}{2\pi}\int_{-\infty}^{+\infty}\frac{1}{2}\exp(ia_o\bar{t})\,da_o + \frac{1}{2\pi}\int_{-\infty}^{+\infty}\frac{\sqrt{1-4a_o^2}}{2}\exp(ia_o\bar{t})\,da_o\right] \tag{8.89}$$

The first term is equal to $\delta(\bar{t})/2$ (see Eq. 8.78). For $\bar{t} > 0$, the second term can be shown to result in $J_1(\bar{t}/2)/(2\bar{t})$, where J_1 is the Bessel function of the first kind and of the first order. The remaining contribution of the second term (for $\bar{t} = 0$) leads to $d\delta(\bar{t})/d\bar{t}$. Equation 8.89 is thus transformed to (for $\bar{t} \geq 0$)

$$\bar{S}(\bar{t}) = \frac{c_l}{f}\left[\frac{1}{2}\,\delta(\bar{t}) + \frac{d\delta(\bar{t})}{d\bar{t}} + \frac{1}{2\bar{t}}J_1\left(\frac{\bar{t}}{2}\right)\right] \tag{8.90}$$

and for $\bar{t} < 0$, $\bar{S}(\bar{t}) = 0$. The regular part $J_1(\bar{t}/2)/(2\bar{t})$ is plotted as a function of $\bar{t}$ in Fig. 8-8. For $\bar{t} = +0$, the value equals 0.125.

Eliminating the dimensionless time $\bar{t}$ using

$$\bar{t} = \frac{c_l}{f}t \tag{8.91a}$$

$$\delta(\bar{t}) = \frac{f}{c_l}\delta(t) \tag{8.91b}$$

$$\frac{d\delta(\bar{t})}{d\bar{t}} = \frac{f^2}{c_l^2}\dot{\delta}(t) \tag{8.91c}$$

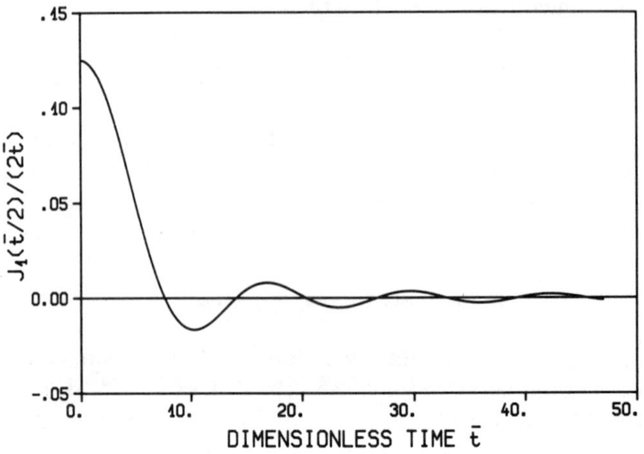

Figure 8-8 Regular part of dynamic-stiffness coefficient (in time domain) of rod with exponentially increasing area.

Eq. 8.90 is transformed to

$$\bar{S}(t) = \frac{1}{2}\delta(t) + \frac{f}{c_l}\dot{\delta}(t) + \frac{1}{2t}J_1\left(\frac{c_l}{2f}t\right) \tag{8.92}$$

The force $R(t)$ follows from the convolution integral

$$R(t) = K\int_0^t \bar{S}(t-\tau)r(\tau)\,d\tau \tag{8.93}$$

Substituting Eq. 8.92 in Eq. 8.93 leads to

$$R(t) = K\left[\frac{1}{2}r(t) + \frac{f}{c_l}\dot{r}(t) + \frac{1}{2}\int_0^t \frac{1}{t-\tau}J_1\left(\frac{c_l}{2f}(t-\tau)\right)r(\tau)\,d\tau\right] \tag{8.94}$$

Alternatively, the "dimensionless"-flexibility coefficient in the time domain $\bar{F}(\bar{t})$ can be calculated. The latter multiplied by f/c_l corresponds to the integral $(1/2\pi)\int_{-\infty}^{+\infty} [1/\bar{S}(a_o)] \exp(ia_o\bar{t})\,da_o$ and is plotted as a solid line in Fig. 8-9. No singular part exists. The displacement $r(t)$ follows, using this flexibility formulation after transforming back to time t, as

$$r(t) = \frac{1}{K}\int_0^t \bar{F}(t-\tau)R(\tau)\,d\tau \tag{8.95}$$

Viscous material damping can be defined, again using the correspondence principle, as

$$E^* = E(1 + 2\zeta_v\omega i) \tag{8.96a}$$

or introducing a nondimensionalized damping ratio $\bar{\zeta}_v$ as

$$E^* = E(1 + 2\bar{\zeta}_v a_o i) \tag{8.96b}$$

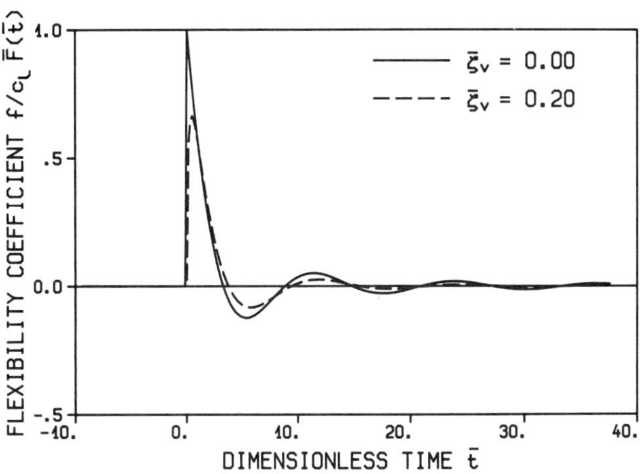

Figure 8-9 Dynamic-flexibility coefficient (in time domain) of rod with exponentially increasing area.

whereby

$$\bar{\zeta}_v = \frac{c_l}{f}\zeta_v \tag{8.96c}$$

applies. The subscript v stands for viscous. The corresponding flexibility coefficient $\bar{F}(\bar{t})$ is also shown as a dashed line for $\bar{\zeta}_v = 0.2$ in Fig. 8-9. In contrast to the undamped case, $\bar{F}(\bar{t} = +0) = 0$ for the viscously damped one. For the flexibility coefficients, the calculations are performed numerically.

8.5.5 Disk on Half-Space and on Layer

Finally, the "dimensionless"-flexibility coefficient in the time domain for the rocking motion $\bar{F}_{\phi v}(\bar{t})$ of a rigid disk of radius a resting on a half-space with Poisson's ratio $\nu = 0.33$ is determined. Multiplied by a/c_s, this value represents the integral $(1/2\pi)\int_{-\infty}^{+\infty} [1/(k_{\phi_v}(a_o) + ia_o c_{\phi_v}(a_o))] \exp(ia_o\bar{t})\, da_o$. The dynamic-stiffness coefficient in the frequency domain $k_{\phi_v}(a_o) + ia_o c_{\phi_v}(a_o)$ is taken for the undamped case from Fig. 7-19. The results determined numerically are shown in Fig. 8-10. Uniform viscous material damping is also introduced, defined by

$$G^* = G(1 + 2\bar{\zeta}_v a_o i) \tag{8.97a}$$

$$\lambda^* + 2G^* = (\lambda + 2G)(1 + 2\bar{\zeta}_v a_o i) \tag{8.97b}$$

The results for the damped case with $\bar{\zeta}_v = 0.10$ are also shown in Fig. 8-10. It should be noted that $\bar{F}_{\phi v}(\bar{t})$ determined numerically, is, from a practical point of view, equal to zero for $\bar{t} < 0$.

The corresponding results for two damping values for a disk resting on a

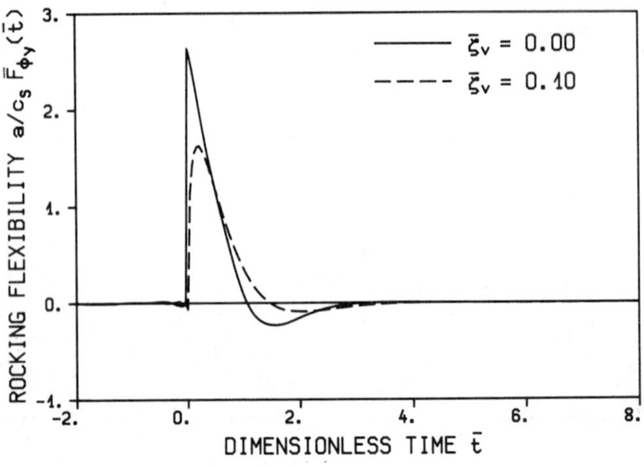

Figure 8-10 Dynamic-flexibility coefficient for rocking (in time domain) of disk on half-space.

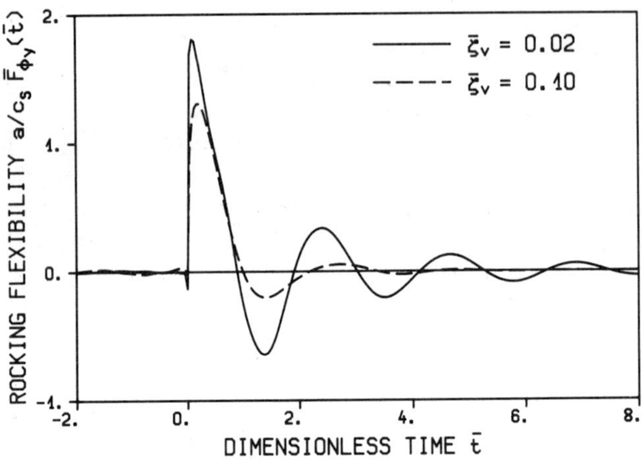

Figure 8-11 Dynamic-flexibility coefficient for rocking (in time domain) of disk on layer built in at its base.

layer built in at its base with the radius equal to the depth of the layer ($\nu = 0.33$) are presented in Fig. 8-11. As expected, the oscillations are more pronounced in this case.

SUMMARY

1. The basic equation of motion expressed in total-displacement amplitudes of the nodes in the structure and on the structure–soil interface (base) can also be derived starting from the equation of motion of the total structure–soil

system. The direct method and the substructure procedure will thus lead to identical results, if implemented consistently.

2. The motion of the base can be defined relative to that of the free field, keeping the total motion of the other nodes in the structure as unknowns. The coefficients of the basic equation of motion are unaffected. The load vector equals the negative reaction forces resulting when the free-field motion is enforced along the base of the system consisting of the structure and, in the embedded part, of the difference of the structure and the soil, with the other nodes fixed. Alternatively, the motion of the base can be defined relative to that of the scattered seismic-input motion of the soil with excavation. Enforcing this scattered motion along the base of the structure with its other nodes fixed results in the negative load vector. This alternative formulation can also be used for a rigid base.
3. Transmitting the (known) free-field motion along the base quasi-statically into the structure does not, in general, result in the kinematic motion. Defining the dynamic motion relative to this quasi-static reference system does not change the coefficient matrix and leads to a load vector consisting of inertial loads and contributions involving the stiffness matrix of the structure and of the embedded soil.
4. Transmitting the scattered motion of the soil with excavation (with the compatibility constraints enforced) quasi-statically into the structure with a rigid base leads to the kinematic motion. The remaining dynamic motion is equal to that of the inertial-interaction part with a load vector consisting of inertial loads based on the acceleration of the kinematic interaction.
5. Transmitting the (unknown) total base motion quasi-statically into the structure and defining the dynamic motion relative to this reference system result in a transformation of the coefficient matrix. The load vector is unchanged. This formulation can also be used to reduce the number of degrees of freedom of the structure by introducing the first few amplitudes of the vibrational modes of the structure fixed at its base.
6. Representing the soil by a frequency-independent spring and dashpot (calculated, for example, at the fundamental frequency of the system), it is possible to analyze soil–structure interaction approximately, remaining in the time domain. In addition, classical normal modes can be assumed to exist. When calculating the natural frequencies and the mode shapes, the frequency dependency can be crudely taken into account by iteration. The damping terms coupling the different modes are disregarded.
7. Generalizing the calculational procedure developed in this text, it is possible in principle, to analyze a nonlinear system, if the nonlinearities are restricted to the structure and the adjacent soil. The unbounded soil (the far field) must remain linear with material damping. The equation of motion in the time domain contains convolution integrals involving the dynamic-stiffness coefficients of the soil in the time domain (i.e., the force as a function of time

which produces a unit impulse displacement). The dynamic-stiffness coefficients in the time and frequency domains form a Fourier transform pair. As the dynamic-stiffness coefficients are infinite for the frequency tending toward infinity, the transformation of the singular part is valid only in the sense of a distribution. It may be computationally simpler to work with dynamic-flexibility coefficients and thus to formulate the contribution of the unbounded soil to the equations of motion by using the flexibility approach.

PROBLEMS

8.1. For the system shown in Fig. P3-1, derive the harmonic equations of motion for which the (unknown) response-base motion is transmitted quasi-statically (Eqs. 8.57 and 8.59). Also establish the equations of motion using the amplitude of the fundamental fixed-base mode as generalized coordinate. Certain property matrices of this system are specified in Problem 3.1.

Solution:

At first, the displacement amplitudes of the structure are defined relative to those of the total rigid-body motion of the base as (Eq. 8.56)

$$\begin{Bmatrix} u^t \\ w^t \end{Bmatrix} = \begin{bmatrix} 1 & & 1 & & \frac{l}{2}\sqrt{3} \\ & 1 & & 1 & \end{bmatrix} \begin{Bmatrix} u^d \\ w^d \\ u^t_o \\ w^t_o \\ \phi^t \end{Bmatrix}$$

Using $[S_{ss}]$ derived in Problem 3.1, and with $[M_{so}] = 0$, Eq. 8.57 is equal to

$$\left(-m\omega^2 \begin{bmatrix} 1 & & 1 & & \frac{l}{2}\sqrt{3} \\ & 1 & & 1 & \\ 1 & & 1 & & \frac{l}{2}\sqrt{3} \\ & 1 & & 1 & \\ \frac{l}{2}\sqrt{3} & & \frac{l}{2}\sqrt{3} & & \frac{3l^2}{4} \end{bmatrix} + \frac{EA}{l}(1 + 2\zeta i) \begin{bmatrix} \frac{1}{2} & & & & \\ & \frac{3}{2} & & & \\ & & & & \\ & & & & \\ & & & & \end{bmatrix} \right.$$

$$\left. + \begin{bmatrix} & & & & \\ & & & & \\ & & S^g_h & & \\ & & & S^g_v & \\ & & & & S^g_r \end{bmatrix} \right) \begin{Bmatrix} u^d \\ w^d \\ u^t_o \\ w^t_o \\ \phi^t \end{Bmatrix} = \begin{Bmatrix} & \\ & \\ \begin{bmatrix} S^g_h & & \\ & S^g_v & \\ & & S^g_r \end{bmatrix} \begin{Bmatrix} u^g_o \\ w^g_o \\ \beta^g_o \end{Bmatrix} \end{Bmatrix}$$

Expressing the amplitudes of the structure relative to those of the rigid-body motion of the base relative to the scattered motion as (Eqs. 8.50 and 8.58)

$$\begin{Bmatrix} u^t \\ w^t \\ u_o^t \\ w_o^t \\ \phi^t \end{Bmatrix} = \begin{bmatrix} 1 & & 1 & & \frac{l}{2}\sqrt{3} \\ & 1 & & 1 & \\ & & 1 & & \\ & & & 1 & \\ & & & & 1 \end{bmatrix} \begin{Bmatrix} u^d \\ w^d \\ u_o \\ w_o \\ \phi \end{Bmatrix} + \begin{Bmatrix} \\ \\ u_o^g \\ w_o^g \\ \beta_o^g \end{Bmatrix}$$

the coefficient matrix of the equations is the same as above, but with the unknowns u^d, w^d, u_o, w_o, and ϕ. The right-hand side equals (Eq. 8.59)

$$\omega^2 m \begin{bmatrix} 1 & & \frac{l}{2}\sqrt{3} \\ & 1 & \\ 1 & & \frac{l}{2}\sqrt{3} \\ & 1 & \\ \frac{l}{2}\sqrt{3} & & \frac{3l^2}{4} \end{bmatrix} \begin{Bmatrix} u_o^g \\ w_o^g \\ \beta_o^g \end{Bmatrix}$$

The eigenvalue problem of the fixed-base structure equals (Eq. 8.35)

$$\left(\frac{EA}{l}\begin{bmatrix} \frac{1}{2} & \\ & \frac{3}{2} \end{bmatrix} - \omega_j^2 m \begin{bmatrix} 1 & \\ & 1 \end{bmatrix}\right)\{\phi_j\} = 0$$

Setting the determinant equal to zero,

$$\left(\frac{EA}{2l} - \omega_j^2 m\right)\left(\frac{3EA}{2l} - \omega_j^2 m\right) = 0$$

leads to

$$\omega_1^2 = \frac{EA}{2lm}$$

$$\omega_2^2 = \frac{3EA}{2lm}$$

Suppressing the higher mode [i.e., selecting $\omega_1^2 = EA/(2lm)$], the mode shape follows as

$$\{\phi_1\} = \begin{Bmatrix} \frac{1}{\sqrt{m}} \\ 0 \end{Bmatrix}$$

whereby the scaling factor has been selected such that Eq. 8.36a is satisfied.

$$\begin{bmatrix} \frac{1}{\sqrt{m}} & 0 \end{bmatrix} \begin{bmatrix} m & \\ & m \end{bmatrix} \begin{Bmatrix} \frac{1}{\sqrt{m}} \\ 0 \end{Bmatrix} = 1$$

The following transformation applies (Eq. 8.37):

$$\begin{Bmatrix} u^d \\ w^d \end{Bmatrix} = \begin{Bmatrix} \dfrac{1}{\sqrt{m}} \\ 0 \end{Bmatrix} z_1$$

The vertical fixed-base mode is disregarded when using this relation, and doing so has the effect that, in the vertical direction, the structure (i.e., the two truss bars) is rigid. The corresponding equation of motion is equal to (Eqs. 8.39 and 8.57).

$$\left(-\omega^2 \begin{bmatrix} 1 & \sqrt{m} & & \frac{l}{2}\sqrt{3}\sqrt{m} \\ \sqrt{m} & m & & \frac{l}{2}\sqrt{3}\,m \\ & & m & \\ \frac{l}{2}\sqrt{3}\sqrt{m} & \frac{l}{2}\sqrt{3}\,m & & \frac{3l^2}{4}m \end{bmatrix} + \begin{bmatrix} \frac{EA}{2lm}(1+2\zeta i) & & & \\ & S_h^g & & \\ & & S_v^g & \\ & & & S_r^g \end{bmatrix} \right) \begin{Bmatrix} z_1 \\ u_o^t \\ w_o^t \\ \phi^t \end{Bmatrix} = \left\{ \begin{bmatrix} S_h^g & & \\ & S_v^g & \\ & & S_r^g \end{bmatrix} \begin{Bmatrix} u_o^g \\ w_o^g \\ \beta_o^g \end{Bmatrix} \right\}$$

Selecting z_1, u_o, w_o, and ϕ as the unknowns does not change the coefficient matrix. The load vector equals (Eqs. 8.40 and 8.59)

$$\omega^2 \begin{bmatrix} \sqrt{m} & & \frac{l}{2}\sqrt{3}\sqrt{m} \\ m & & \frac{l}{2}\sqrt{3}\sqrt{m} \\ & m & \\ \frac{l}{2}\sqrt{3}\,m & & \frac{3l^2}{4}m \end{bmatrix} \begin{Bmatrix} u_o^g \\ w_o^g \\ \beta_o^g \end{Bmatrix}$$

8.2. Verify that for a surface structure excited by vertically incident waves, the second term of the load vector (with the static stiffnesses of the structure) of the equations of motion based on the quasi-static transmission of the free-field input motion (Eq. 8.24) vanishes. In this case the equation of motion is identical to that of inertial interaction (Eq. 3.26).

8.3. Derive the equation of motion based on the quasi-static transmission of the scattered input motion $\{u_b^g\}$.

$$\begin{bmatrix} [S_{ss}] & [S_{sb}] \\ [S_{bs}] & [S_{bb}^s] + [S_{bb}^g] \end{bmatrix} \begin{Bmatrix} \{u_s^d\} \\ \{u_b^d\} \end{Bmatrix} = \omega^2 \begin{bmatrix} [M_{sb}] + [M_{ss}][T_{sb}] \\ [M_{bb}^s] + [M_{bs}][T_{sb}] \end{bmatrix} \{u_b^g\} + \begin{bmatrix} [0] \\ -(1+2\zeta i)([K_{bb}^s] + [K_{bs}][T_{sb}]) \end{bmatrix} \{u_b^g\}$$

Hints:

Substitute the transformation

$$\{u_s^t\} = [T_{sb}]\{u_b^g\} + \{u_s^d\}$$

$$\{u_b^t\} = \{u_b^g\} + \{u_b^d\}$$

in Eq. 8.16, where

$$[T_{sb}] = -[K_{ss}]^{-1}[K_{sb}]$$

8.4. Derive the following equation of motion:

$$\begin{bmatrix} [S_{ss}] & [S_{sb}] & \\ [S_{bs}] & [S_{bb}^s] + [\bar{S}_{bb}^g] & [\bar{S}_{bi}] \\ & [\bar{S}_{ib}] & [\bar{S}_{ii}] \end{bmatrix} \begin{Bmatrix} \{u_s^t\} \\ \{u_b^t\} \\ \{u_i\} \end{Bmatrix} = \begin{bmatrix} [0] \\ [\bar{S}_{bb}^f] \\ [\bar{S}_{ib}] \end{bmatrix} \{u_b^f\}$$

where

$$[\bar{S}_{bb}^f] = [S_{bb}^e] + [\bar{S}_{bb}^g]$$

$$\{u_i\} = \{u_i^t\} - \{u_i^f\}$$

The nomenclature is defined in Section 8.1.

This equation of motion corresponds to a direct method, in which the motions of the structure and of the soil are calculated simultaneously. Note that the free-field motion $\{u_b^f\}$ is required in those nodes only which subsequently will lie on the structure–soil interface.

Hints:

Formulate Eq. 8.4 first for the structure–soil system and then for the free-field system (with the "structure" representing the excavated part of the soil). The right-hand sides are the same.

8.5. Derive the equation for the response of the base of a structure with a rigid base, and identify the contributions of the modes of the fixed-base structure. Verify that for low- and high-frequency excitations, the total mass of the structure and the mass of the base, respectively, are excited.

Solution:

For the sake of a concise notation, the mass matrix of the structure is assumed to be diagonal ($[M_{so}] = 0$). Using the amplitudes $\{z\}$ of the modes of the fixed-base structure and the total displacement amplitudes $\{u_o^t\}$ of the base, the equations of motion are equal to (Eqs. 8.39 and 8.57)

$$\begin{bmatrix} [\Omega](1 + 2\zeta i) - \omega^2[I] & -\omega^2[\Phi]^T[M_{ss}][T_{so}] \\ -\omega^2[T_{so}]^T[M_{ss}][\Phi] & -\omega^2([M_{oo}^s] + [T_{so}]^T[M_{ss}][T_{so}]) + [S_{oo}^g] \end{bmatrix} \begin{Bmatrix} \{z\} \\ \{u_o^t\} \end{Bmatrix} = \begin{Bmatrix} \{0\} \\ [S_{oo}^g]\{u_o^g\} \end{Bmatrix}$$

Introducing the following convenient nomenclature for the diagonal matrix $[D]$, the generalized modal load matrix (participation factor matrix) $[\Gamma]$, and the generalized mass matrix $[M_o]$ (referred to the base)

$$[D] = [\Omega](1 + 2\zeta i) - \omega^2[I]$$

$$[\Gamma] = [\Phi]^T[M_{ss}][T_{so}]$$

$$[M_o] = [M_{oo}^s] + [T_{so}]^T[M_{ss}][T_{so}]$$

and eliminating $\{z\}$ results in

$$\{u_o^t\} = (-\omega^2[M_o] - \omega^4[\Gamma]^T[D]^{-1}[\Gamma] + [S_{oo}^g])^{-1}[S_{oo}^g]\{u_o^g\}$$

For a specified scattered motion $\{u_o^g\}$, the amplitudes of the base $\{u_o^t\}$ are a function of the generalized mass $[M_o]$ (i.e., a property of the rigid structure), of the participation factors $[\Gamma]$, of the natural frequencies $[\Omega]$, and of the damping ratio ζ of the fixed-base structure and of the dynamic-stiffness matrix of the soil $[S_{oo}^g]$. The second term of this equation, rewritten slightly,

$$\{u_o^t\} = (-\omega^2([M_o] + \omega^2[\Gamma]^T[D]^{-1}[\Gamma]) + [S_{oo}^g])^{-1}[S_{oo}^g]\{u_o^g\}$$

can be interpreted as a frequency-dependent mass. The diagonal matrix $\omega^2[D]^{-1}$ has as its jth element $\omega^2/(\omega_j^2 + \omega_j^2 2\zeta i - \omega^2) = 1/(\omega_j^2/\omega^2 + \omega_j^2 2\zeta i/\omega^2 - 1)$. For low frequencies (compared to the fundamental built-in base frequency ω_1), that is, for a large value of ω_1/ω, these elements tend to vanish, leading to

$$\{u_o^t\} = (-\omega^2[M_o] + [S_{oo}^g])^{-1}[S_{oo}^g]\{u_o^g\}$$

This equation states that, for low frequencies, the system behaves essentially as a rigid body with the total mass of the structure resting on flexible soil. For high frequencies (compared to the highest frequency ω_n), that is, for a small value of ω_n/ω, these elements will approach the value -1. As

$$[\Gamma]^T[\Gamma] = [T_{so}]^T[M_{ss}][\Phi][\Phi]^T[M_{ss}][T_{so}] = [T_{so}]^T[M_{ss}][T_{so}]$$

holds,

$$\begin{aligned}\{u_o^t\} &= (-\omega^2([M_o] - [T_{so}]^T[M_{ss}][T_{so}]) + [S_{oo}^g])^{-1}[S_{oo}^g]\{u_o^g\}\\ &= (-\omega^2[M_{oo}^s] + [S_{oo}^g])^{-1}[S_{oo}^g]\{u_o^g\}\end{aligned}$$

results. (Premultiplying and postmultiplying $[\Phi]^T[M_{ss}][\Phi] = [I]$ by $[\Phi]$ and by $[\Phi]^{-1}$, respectively, result in $[\Phi][\Phi]^T[M_{ss}] = [I]$.) This equation means that, for high frequencies, the effective system consists only of a rigid base with its corresponding mass resting on flexible soil. The actual structure is of no importance in this case.

To derive a clearer physical interpretation of the contributions of the modes of the fixed-base structure, the following transformations are performed. Substituting

$$[M_o] = [M_{oo}^s] + [T_{so}]^T[M_{ss}][T_{so}] = [M_{oo}^s] + [\Gamma]^T[\Gamma]$$

in

$$[M_o] + \omega^2[\Gamma]^T[D]^{-1}[\Gamma] = [M_{oo}^s] + [\Gamma]^T(\omega^2[D]^{-1} + [I])[\Gamma]$$

and defining

$$[H] = \omega^2[D]^{-1} + [I]$$

leads to

$$\{u_o^t\} = (-\omega^2([M_{oo}^s] + [\Gamma]^T[H][\Gamma]) + [S_{oo}^g])^{-1}[S_{oo}^g]\{u_o^g\}$$

The diagonal matrix $[H]$ has as its jth element (corresponding to the jth mode of the structure built in at its base) the value

$$h_j = \frac{\omega_j^2 + \omega_j^2 2\zeta i}{\omega_j^2 + \omega_j^2 2\zeta i - \omega^2}$$

This is equal to the amplitude of the total displacement u_s of a one-degree-of-freedom system with frequency ω_j and hysteretic-damping ratio ζ excited by a support displacement of unit amplitude u_o of frequency ω. The corresponding equation equals

$$[-m\omega^2 + k(1 + 2\zeta i)]u_s = k(1 + 2\zeta i)u_o$$

which, after introducing $\omega_j^2 = k/m$, results in

$$\frac{u_s}{u_o} = \frac{\omega_j^2(1 + 2\zeta i)}{\omega_j^2(1 + 2\zeta i) - \omega^2} = h_j$$

The matrix $[H]$ can thus be interpreted as a transfer matrix. The matrix $[\Gamma]^T[H][\Gamma]$ contains the (frequency-dependent) effective mass of the modes of the structure built in at its base $[M_{\text{eff}}]$. This matrix can be decomposed into frequency-independent terms $[M_j]$ (fictitious mass) and frequency-dependent terms h_j. In general, each mode j will contribute to the six components of the base as

$$[M_j] = [\Gamma_j]^T[\Gamma_j]$$

where $[\Gamma_j]$ contains the participation factors of mode j. Summing all these $[M_j]$ over all modes results in

$$\sum_j [M_j] = \sum_j [\Gamma_j]^T[\Gamma_j] = [\Gamma]^T[\Gamma] = [T_{so}]^T[M_{ss}][T_{so}]$$

The effective mass of the built-in modes of the structure is equal to

$$[M_{\text{eff}}] = \sum_j [M_j]h_j$$

When using the equation $[\Gamma]^T[\Gamma] = [T_{so}]^T[M_{ss}][T_{so}]$, it is assumed that the number of modes is equal to that of the degrees of freedom of the structure built in at its base. If this is not the case, that is, if only the first few modes are considered, the difference of $[T_{so}]^T[M_{ss}][T_{so}]$ (rigid-structure property) and of $[\Gamma]^T[\Gamma]$ can be regarded as the so-called lost mass. This expression arises because for these additional modes which were disregarded, ω/ω_j is a small value, resulting in h_j approaching 1. This lost mass can thus directly be added to M^s_{oo}. [Actually, if the first equation for $\{u^t_o\}$ is used, which involves $[M_o]$, the term involving the frequency-dependent mass $\omega^2[\Gamma]^T[D]^{-1}[\Gamma]$ vanishes for those modes not included in the analysis.] In general, M_j of the higher modes is much smaller than those of the lower ones, and assuming a reasonable structural damping which leads to a limited transfer coefficient h_j, the motion of the base is hardly affected by the higher modes.

As an example, the vertical response w^t_o at the base (i.e., at $z = 0$) of a prismatic rod of length l and total mass m for vertical excitation w_g is calculated (Fig. P8-5). The nomenclature is specified in the figure; the vertical dynamic-stiffness coefficient of the soil is denoted as S^g_v. The straightforward model of the rod consists of (equally spaced) mass points connected by springs. As an alternative, an analytical solution can be used. The exact dynamic-stiffness matrix of the prismatic rod is specified in Eq. 5.57. Identifying the nodes 1 and 2 with the base and the top, respectively, of the rod enforcing the boundary conditions $w_1 = 0, P_2 = 0$, and setting the resulting stiffness coefficient for the undamped case equal to zero

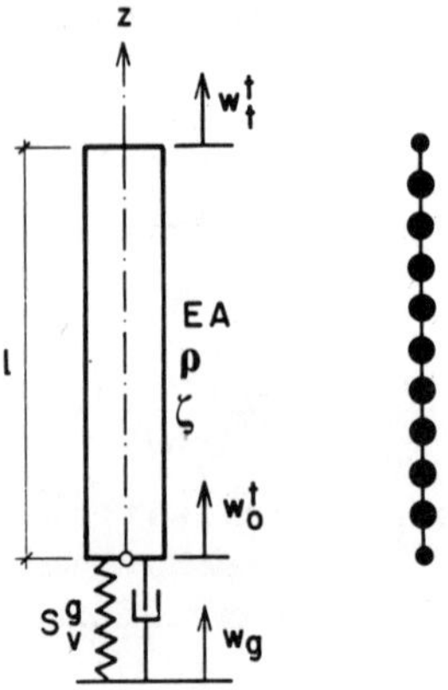

Figure P8-5 Vertical rod loaded by vertical seismic excitation.

results in the natural frequencies of the rod fixed at the base:

$$\omega_j = \frac{2j-1}{2}\pi\frac{c_l}{l} \qquad j = 1, 2, 3, \ldots$$

The corresponding mode shapes are equal to

$$\phi_j(z) = \sqrt{\frac{2}{m}} \sin\frac{2j-1}{2}\pi\frac{z}{l}$$

whereby they are normalized to result in

$$\int_0^l \phi_j^2(z)\rho A\, dz = 1$$

The participation factor of the jth mode is calculated as

$$\Gamma_j = \int_0^l \phi_j(z) 1 \rho A\, dz = \frac{2\sqrt{2}\sqrt{m}}{(2j-1)\pi}$$

the fictitious mass as

$$M_j = \frac{8}{(2j-1)^2}\frac{m}{\pi^2}$$

In particular, for $j = 1$,

$$M_1 = \frac{8}{\pi^2}m$$

which is already a dominant contribution to $M_o = m$. With $M_{oo}^s = 0$, the equations of motion are formulated as

$$w_o^t = \left(-\omega^2 \sum_j M_j h_j + S_v^g\right)^{-1} S_v^g w_g$$

9

ENGINEERING APPLICATIONS

9.1 EVALUATION OF INTERACTION EFFECTS

In Section 3.4 the system with one dynamic degree of freedom shown in Fig. 3-18 is examined to identify the key parameters affecting soil–structure interaction. In addition to the parameters introduced earlier, which in this section are varied in different ranges corresponding to other types of structures, a mass m_o and a mass moment of inertia I_o are now associated with the base (Fig. 9-1). The frequency dependency of the dynamic stiffness of the soil is also taken into account. Besides the half-space, a layer of varying depth d built in at its base is addressed.

9.1.1 Dimensionless Parameters

The same dimensionless parameters as introduced in Section 3.4.4 are again used to describe the response of this coupled system, which has three dynamic degrees of freedom: the stiffness ratio of the structure and of the soil $\bar{s} = \omega_s h/c_s$ (where ω_s denotes the fixed-base frequency of the structure), the slenderness ratio $\bar{h} = h/a$, the mass ratio $\bar{m} = m/(\rho a^3)$, Poisson's ratio of the soil ν, and the hysteretic-damping ratios of the structure ζ and of the soil ζ_g. In addition, the ratios m_o/m and $I_o/(h^2 m)$ and d/a are introduced. The dynamic-stiffness matrix of the soil is formulated analogously as in Section 7.2.5 as $[K_{oo}]([k] + ia_o[c])$, where the diagonal matrix $[K_{oo}]$ contains the static-stiffness coefficients, and $[k]$ and $[c]$ are the matrices of the (nondimensionalized) spring and damping coefficients, respectively. The latter two matrices depend on d/a,

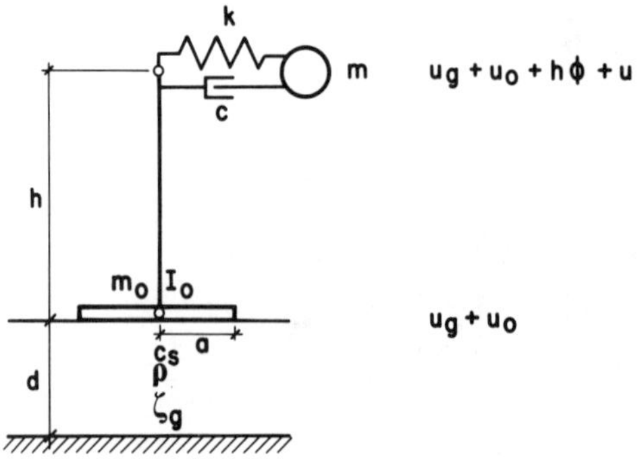

Figure 9-1 Structural model with soil.

ν, ζ_g, and on the dimensionless frequency $a_o = \omega a/c_s$. The elements of $[k]$ and $[c]$ shown in Section 7.4.1 (or the corresponding values for the other ratio of d/a) are used in the following. For instance, for the half-space, $[K_{oo}]$ contains the elements $8Ga/(2-\nu)$ and $8Ga^3/[3(1-\nu)]$ for the horizontal and rocking directions, respectively (Table 7-1).

9.1.2 Equivalent One-Degree-of-Freedom System

Denoting the corresponding terms in $[k]$ as k_x and k_ϕ and in $[c]$ as c_x and c_ϕ (these coefficients are thus used differently from in Section 3.4.1, where they have a dimension, where a_o is not used in the definition of the dynamic-stiffness coefficients, and where they do not include the influence of ζ_g), the properties of the equivalent dynamic one-degree-of-freedom system can be calculated (neglecting the influence of m_o and I_o). It follows from Eq. 3.56 that the frequency $\tilde{\omega}$ of the soil–structure system equals

$$\frac{\tilde{\omega}^2}{\omega_s^2} = \frac{1}{1 + \dfrac{\bar{m}\bar{s}^2}{8}\left[\dfrac{2-\nu}{\bar{h}^2 k_x} + \dfrac{3(1-\nu)}{k_\phi}\right]} \tag{9.1}$$

The equivalent damping ratio $\tilde{\zeta}$ (Eq. 3.61) is specified by

$$\tilde{\zeta} = \frac{\tilde{\omega}^2}{\omega_s^2}\zeta + \left(\frac{\tilde{\omega}}{\omega_s}\right)^3 \frac{\bar{s}^3\bar{m}}{16\bar{h}}\left[\frac{2-\nu}{\bar{h}^2}\frac{c_x}{k_x^2} + 3(1-\nu)\frac{c_\phi}{k_\phi^2}\right] \tag{9.2}$$

The damping ratio ζ_g does not occur explicitly in this equation, as the material damping of the soil directly affects c_x, k_x, c_ϕ, and k_ϕ. The coefficients k_x, k_ϕ, c_x, and c_ϕ are functions of a_o evaluated at the frequency $\tilde{\omega}$. Equation 9.1 can thus be solved only by iteration.

In the following, the equivalent properties are not determined from Eqs. 9.1 and 9.2, but from solving the system of the three dynamic equations of motion, varying the frequency of excitation ω. At the natural frequency $\tilde{\omega}$ of the soil–structure system, the structural distortion u will be a maximum; the equivalent damping ratio $\tilde{\zeta}$ follows from the product of $|u_g|/|2u|$ and $(\tilde{\omega}/\omega_s)^2$, evaluated at $\omega = \tilde{\omega}$. The amplitude of excitation is denoted as u_g. The second factor reflects the effective seismic input to be applied to the equivalent oscillator (see Section 3.4.3). The values $\tilde{\omega}/\omega_s$ and $\tilde{\zeta}/\zeta$ are regarded only as a convenient measure to characterize the effect of soil–structure interaction. All actual response calculations are based on the three dynamic equations of motion.

9.1.3 Depth of Layer

At first, the influence of the depth of the layer of soil on the response of a structure is investigated, in particular the effect of the vanishing radiation damping below the fundamental frequency of the site. In the horizontal direction, the latter equals $\pi c_s/(2d)$ (in radians) for a homogeneous layer. The structure with a circular basemat of radius a is founded on the surface of the homogeneous layer with depth d. The parameters are specified in the caption of Fig. 9-2. Layers with the ratios $d/a = 1$ and $= 2$ as well as the half-space are investigated. For this squat structure, the shape of the fundamental mode of the soil–structure system will consist of a predominantly translational motion. The cutoff frequency

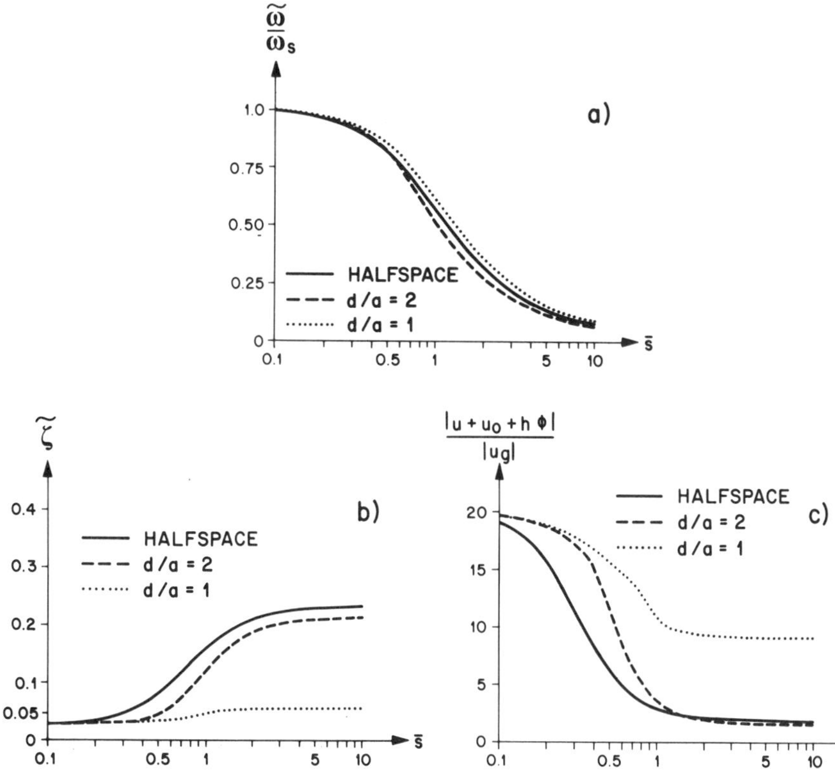

Figure 9-2 Equivalent properties and harmonic response ($\bar{h} = 0.67$, $\bar{m} = 3$, $m_o = I_o = 0$, $\nu = 0.33$, $\zeta = 0.025$, $\zeta_g = 0.05$) varying depth of layer. (a) Natural frequency; (b) damping; (c) displacement of mass relative to free field.

above which radiation damping occurs will thus be determined essentially by the fundamental frequency of the site in the horizontal and not in the vertical direction, as would be the case if rocking were to dominate. The properties of the equivalent one-degree-of-freedom system $\tilde{\omega}/\omega_s$ and $\tilde{\zeta}$ are plotted as a function of $\bar{s}$ for the three sites in Fig. 9-2a and b. The ratio $\tilde{\omega}/\omega_s$ depends only weakly on the site. For a negligible effect of the soil–structure interaction ($\bar{s} = 0.1$), all curves for $\tilde{\zeta}$ start at the structural-damping ratio $\zeta = 0.025$. For the half-space (solid line), $\tilde{\zeta}$ increases significantly for increasing $\bar{s}$, as the effect of the large amount of radiation damping (predominantly in the horizontal direction) applies throughout the range. For the other extreme site, $d/a = 1$ (dotted line), radiation damping is never activated, as $\tilde{\omega}$ is always smaller than the horizontal fixed-base frequency of the site. For a significant soil–structure interaction effect, the ratio $\tilde{\zeta}$ converges essentially to the material-damping ratio of the soil $\zeta_g = 0.05$. For the intermediate site $d/a = 2$ (dashed line), a transmission occurs. For a large c_s ($\bar{s}$ small), $\tilde{\omega}$ is smaller than the fundamental frequency of the site, and thus no radiation damping occurs. Decreasing c_s (and keeping all other parameters constant), which corresponds to increasing $\bar{s}$, will lead to a smaller decrease in $\tilde{\omega}$ (Fig. 9-2a) than in the fundamental frequency of the site (which is proportional to c_s). In the range of $\bar{s}$ from 0.5 to 1, these two frequencies will be very similar. For smaller c_s (and thus larger $\bar{s}$), radiation damping will arise with a tendency to reach a value approaching that of the half-space. The amplitude of the peak-structural distortion $|u|$ (Fig. 3-18) occurring at $\omega = \tilde{\omega}$ is inversely proportional to $\tilde{\zeta}$, the factor being equal to $u_g\tilde{\omega}^2/(2\omega_s^2)$. In Fig. 9-2c, the amplitude of the maximum displacement of the mass relative to the free field $|u + u_o + h\phi|/|u_g|$ for harmonic excitation is plotted versus $\bar{s}$ for the three sites. The conclusions reached discussing the curves $\tilde{\zeta}$ in Fig. 9-2b are confirmed. The influence of the depth d of the layer on the equivalent damping ratio $\tilde{\zeta}$ (assuming all other properties of the soil and of the structure are unchanged) is thus extremely important. This represents probably the most dangerous pitfall encountered in actual practice, where a layered half-space is often present, which makes it very difficult to calculate a reliable, but not overly conservative, equivalent damping ratio. The strong influence of d/a on the structural response is also visible for a transient excitation, although to a somewhat lesser extent. For the artificial time history (Fig. 3-17a), normalized to $0.1g$, the maximum values of the structural distortion u_{max} and of the displacement of the mass relative to the free field $(u + u_o + h\phi)_{max}$ are plotted for the same three sites and for a frequency $\omega_s/2\pi = 4$ of the fixed-base structure in Fig. 9-3. As expected, the responses for the half-space and for the layer with $d/a = 2$ are similar for $\bar{s} > 1$ and significantly smaller than that of the layer with $d/a = 1$.

9.1.4 Mass of Base

The influence of a mass m_o and of a mass moment of inertia I_o (associated with the base) on the equivalent parameters and on the structural response is examined next. For reasonable values of these additional parameters, the results

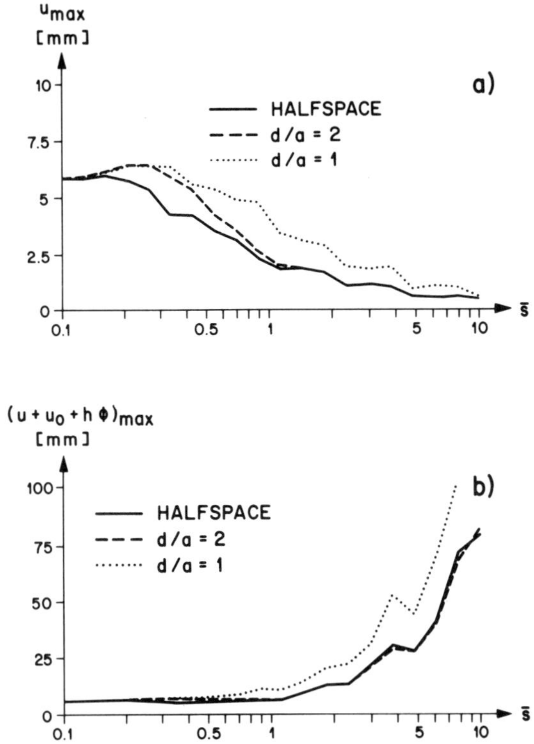

Figure 9-3 Maximum response, artificial time history ($\bar{h} = 0.67$, $\bar{m} = 3$, $m_o = I_o = 0$, $\omega_s/2\pi = 4$, $\nu = 0.33$, $\zeta = 0.025$, $\zeta_g = 0.05$) varying depth of layer. (a) Structural distortion; (b) displacement of mass relative to free field.

are hardly changed. For instance, for $m_o/m = 0.2$ and $I_o/(h^2 m) = 0.05$, the corresponding curves coincide completely from a practical point of view with those shown in Fig. 9-2 for $m_o = I_o = 0$ (comparison not shown).

9.1.5 Bridge Structure

The range of parameters selected up to now in this section (and also in Section 3.4.5) corresponds to that encountered in nuclear-power plants, where stiff massive structures occur. Other types of structures also show significant soil–structure interaction effects. There are, however, also many types of structures where certain aspects of soil–structure interaction can be neglected, resulting in much simpler structural behavior. As an example, the continuous bridge shown in Fig. 3-19 is investigated. For this type of structure, the slenderness ratio $\bar{h}$ is large (> 3). In this case, the properties of the equivalent one-degree-of-freedom system depend only weakly on $\bar{h}$, as is visible in Fig. 3-20. The response will depend strongly on the material-damping ratios ζ and ζ_g, as well as on $\bar{s}$ and $\bar{m}$. As the characteristic length a of the basemat is small and

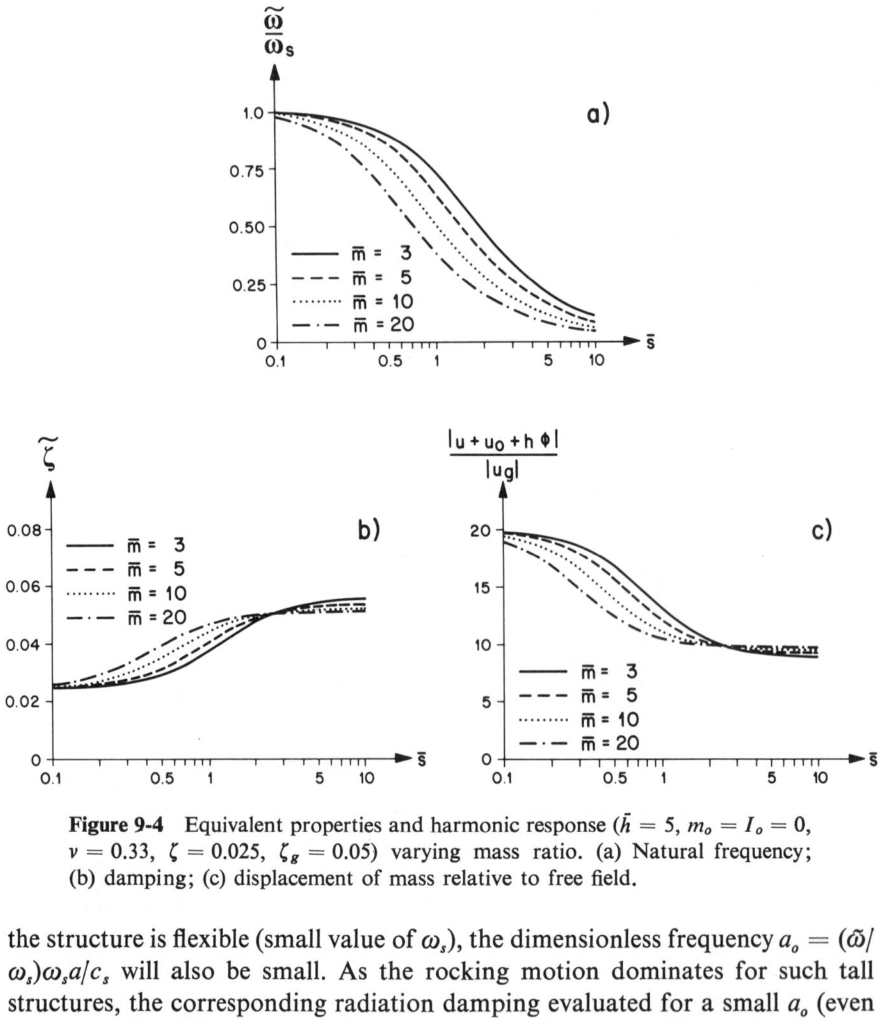

Figure 9-4 Equivalent properties and harmonic response ($\bar{h} = 5$, $m_o = I_o = 0$, $\nu = 0.33$, $\zeta = 0.025$, $\zeta_g = 0.05$) varying mass ratio. (a) Natural frequency; (b) damping; (c) displacement of mass relative to free field.

the structure is flexible (small value of ω_s), the dimensionless frequency $a_o = (\tilde{\omega}/\omega_s)\omega_s a/c_s$ will also be small. As the rocking motion dominates for such tall structures, the corresponding radiation damping evaluated for a small a_o (even for a half-space) will be negligible. This is confirmed in Fig. 9-4, where the equivalent properties and the amplitude of the displacement of the mass relative to the free field of typical bridge structures founded on a half-space are plotted for the specified parameters. The importance of accurately determining ζ_g is accentuated. For the artificial time history normalized to 0.1g, the maximum structural distortion and displacement of the mass relative to the free field are plotted in Fig. 9-5, varying the fixed-base frequency in the range applicable to bridge structures. The structural distortion $u_{\max}$ again decreases for increasing $\bar{s}$, the decrease being more pronounced for the more flexible structure. The value $(u + u_o + h\phi)_{\max}$ also increases, but only for the stiffest structure examined ($\omega_s/2\pi = 2$ Hz) as is the case for the stiffer structures examined in Figs. 9-3 and 3-23. For the more flexible structures, however, the displacement of the mass

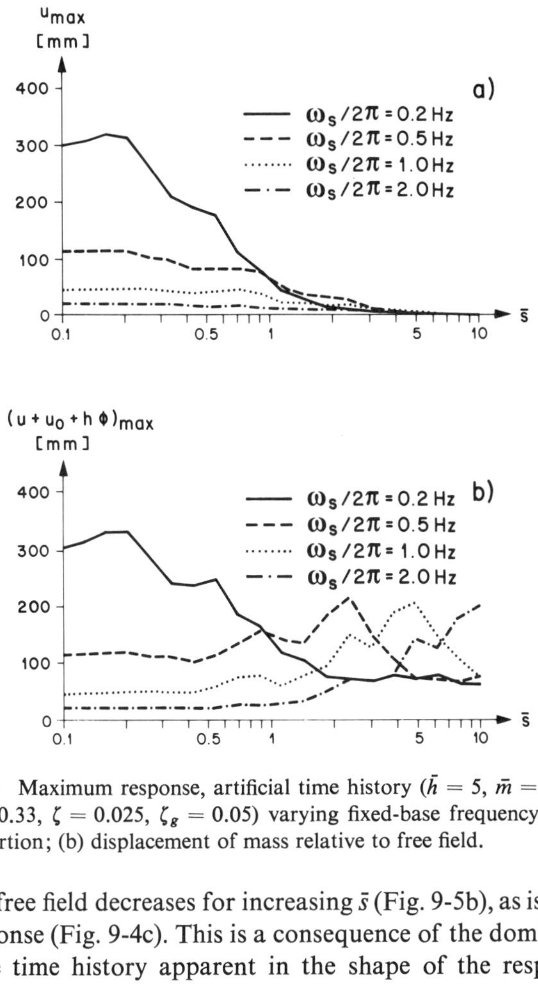

Figure 9-5 Maximum response, artificial time history ($\bar{h} = 5$, $\bar{m} = 5$, $m_o = I_o = 0$, $\nu = 0.33$, $\zeta = 0.025$, $\zeta_g = 0.05$) varying fixed-base frequency. (a) Structural distortion; (b) displacement of mass relative to free field.

relative to the free field decreases for increasing $\bar{s}$ (Fig. 9-5b), as is the case for the harmonic response (Fig. 9-4c). This is a consequence of the dominant-frequency content of the time history apparent in the shape of the response spectrum compared to the natural frequency of the structure–soil system.

9.1.6 Second Mode

Finally, the influence of soil–structure interaction on the response of the second mode is examined. The corresponding model is shown in Fig. 9-6. The subscript 1 is added to all variables associated with the fundamental mode, for which the dimensionless parameters introduced earlier are defined. In addition, ω_{s2}/ω_{s1}, m_2/m_1, and h_2/h_1 have to be specified, assuming the same structural-material damping for the first and second modes. The base shears corresponding to the first and second modes are denoted as H_1 and H_2, respectively. The ratio $|H_{max} - H_{1\,max}|/|H_{max}|$ represents the relative contribution of the second mode to the total base shear, where H_{max} represents the maximum value of $H_1 + H_2$. This ratio is plotted for the artificial time history in Fig. 9-7 as a function of the

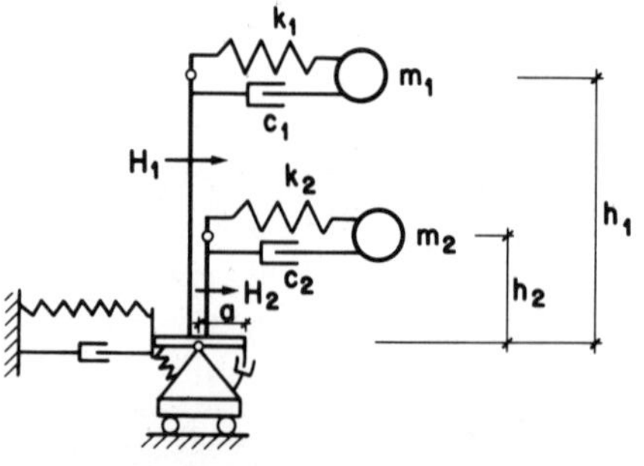

Figure 9-6 Model with two modes.

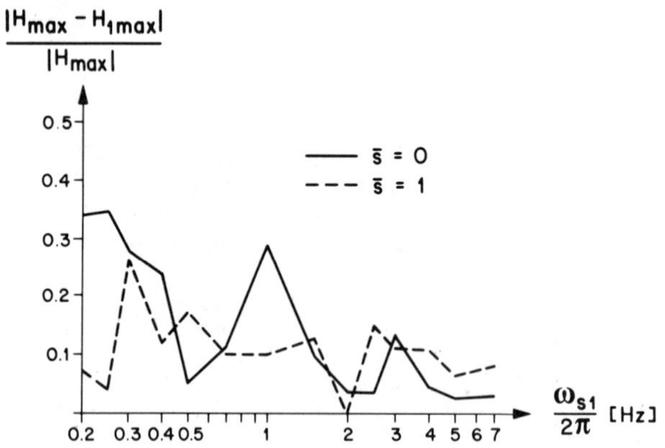

Figure 9-7 Relative contribution of second mode to total base shear ($\bar{h} = 1$, $\bar{m} = 3$, $\nu = 0.33$, $\zeta = 0.025$, $\zeta_g = 0.05$, $\omega_{s2}/\omega_{s1} = 3$, $m_2/m_1 = 0.25$, $h_2/h_1 = 0.3$) with and without soil–structure interaction.

fixed-base frequency of the fundamental mode $\omega_{s1}/2\pi$ for $\bar{s} = 0$ (no soil–structure interaction) and $\bar{s} = 1$. The other parameters are specified in the caption. As expected, the contribution of the second mode in the lower-frequency range is significant. (For the overturning moment, the contribution of the second mode to the total response is smaller than for the base shear.) Comparing the results for $\bar{s} = 1$ to those for $\bar{s} = 0$ leads to a nonuniform tendency, which is so often the case when discussing the effects of soil–structure interaction. Sweeping generalizations regarding its influence are dangerous!

9.2 EFFECTS OF HORIZONTALLY PROPAGATING WAVES

Inclined body waves and surface waves propagate horizontally across the site. The corresponding apparent velocity c_a is a function of the phase velocity c (Eq. 6.6). The latter follows from the wave velocity (c_s^* or c_p^*) and from the angle of incidence (ψ_{SH}, ψ_{SV}, or ψ_P) for body waves (Eqs. 5.93 and 5.126) and, for surface waves, from the eigenvalue problem (Section 6.2.3). In the latter case, c

is, in general, a function of ω. The apparent velocity can be used to classify the nature of the motion, as discussed in point 9 of the Summary of Chapter 6. The amplitudes of the horizontal and vertical components of the motion and the phase angle between them are thereby determined.

The effect of the horizontally propagating waves on the seismic loading is addressed qualitatively for the three components of the free-field motion in Section 4.2 (Figs. 4-3 and 4-4) and in Chapter 6 (Figs. 6-28, 6-52, 6-69, 6-70, 6-78, and 6-79). For a structure with a rigid base, it is meaningful to describe the seismic-input motion acting on the structure using the so-called scattered motion $\{u_o^g\}$, which is equal to that of kinematic interaction $\{u_o^k\}$ (Eq. 3.29b). For a surface structure with a circular rigid basemat, the wave effects can essentially be characterized by the ratio $\omega a/c_a$, where a is the radius of the basemat. The out-of-plane component of the free-field motion leads, for relaxed contact, to a translational component with an amplitude v_o^g and to a torsional component with an amplitude γ_o^g. The in-plane motion results in two translational components with amplitudes u_o^g and w_o^g and in a rocking component with an amplitude β_o^g. The amplitudes as a function of $\omega a/c_s$ are examined in Section 7.4.3 (Figs. 7-30 and 7-31).

The free-field motion is commonly postulated to consist only of vertically incident body waves. To determine the influence of this assumption, the response of simple structures of increasing complexity to different types of horizontally propagating waves is systematically investigated. Extreme cases are also examined to overemphasize the effects studied. Comparisons with the results of the same structures, calculated for the standard vertically incident body waves of the same amplitudes, are performed.

The structures with a rigid basemat of radius a are founded on the surface of a homogeneous half-space (Fig. 9-8). Its Poisson's ratio ν is selected as 0.33.

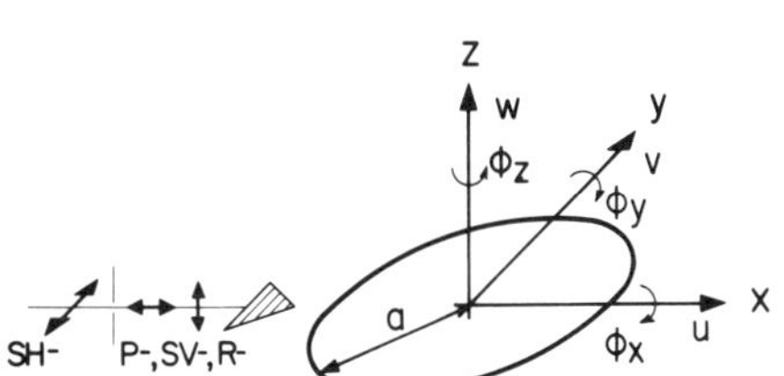

Figure 9-8 Nomenclature of circular rigid basemat excited by horizontally propagating waves (Ref. [4]).

The hysteretic damping ratio of the soil ζ_g equals 0.05. The corresponding dynamic-stiffness coefficients of the soil are plotted in Fig. 7-20. The waves propagate horizontally in the x-direction. The nomenclature of the displacements is defined in Fig. 9-8. The amplitudes of the displacement in the horizontal direction and of the rotation around the x-axis are denoted as u and ϕ_x, respectively. Harmonic excitation and an earthquake acceleration–time history as a transient are examined. The artificial 30-s time history that follows the U.S. NRC response spectrum (Fig. 3-17a) is used.

9.2.1 Investigated Structures

For the structures investigated, the dimensionless parameters (for harmonic excitation), which are varied in this study, and the symbols used later for the structures are listed in Fig. 9-9. The simplest structure analyzed is the circular massless basemat of radius a. It represents the model used for kinematic interaction. Its response is thus equal to the seismic input used for more complex structures in the inertial interaction step. The response of this basemat depends, for selected ν and ζ_g, on the dimensionless frequency $a_o = \omega a/c_s$ for harmonic excitation. To be able to determine the response at a specific level, the massless structure of height h is also examined. The slenderness ratio $\bar{h} = h/a$ is introduced as an additional parameter. For the basemat with mass m, the mass ratio $\bar{m} = m/(\rho a^3)$ is specified together with a_o (ρ = mass density of the elastic half-space). A translational mass–spring system modeling the simple structure (shown in Fig. 9-1) connected to a basemat with mass m_o is also analyzed. The ratio of this mass to that of the mass point m_o/m equals 0.33. To calculate the mass moment of inertia I_o, the mass is assumed to be evenly distributed over the circular basemat. The stiffness ratio of the structure and of the soil $\bar{s} = \omega_s h/c_s$ and the hysteretic damping ratio ζ of the translational mass–spring system are

TYPE OF STRUCTURE	SYMBOL	DIMENSIONLESS PARAMETERS
MASSLESS BASEMAT	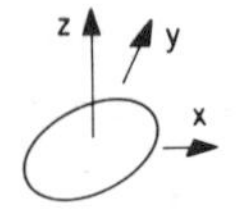	a_o
MASSLESS STRUCTURE	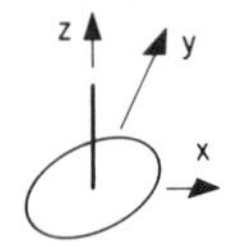	$a_o, \bar{h}$
BASEMAT WITH MASS	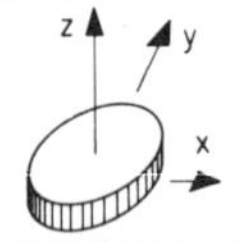	$a_o, \bar{m}$
TRANSLATIONAL MASS-SPRING SYSTEM CONNECTED TO BASEMAT WITH MASS	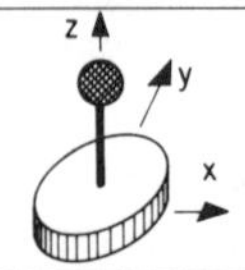	$a_o, \bar{h}, \bar{m}, \bar{s}, \zeta$
TRANSLATIONAL AND TORSIONAL MASS-SPRING SYSTEMS CONNECTED TO BASEMAT WITH MASS	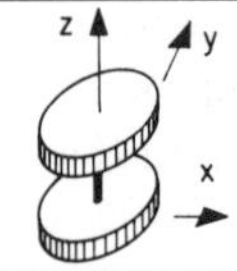	$a_o, \bar{h}, \bar{m}, \bar{s}, \zeta, \bar{s}_\phi$

Figure 9-9 Investigated structures (Ref. [4]).

introduced as additional parameters, where ω_s denotes the translational fixed-base frequency. To be able to account for the torsional motion of the structure for SH-waves, the translational and torsional mass–spring systems connected to a basemat with mass are finally introduced. In addition to the structural configuration present in the translational mass–spring system connected to a basemat with mass, a torsional degree of freedom of the structure consistent with the same mass distribution is included. The same damping ratio ζ is assumed for the torsional motion. The only additional dimensionless parameter consists of $\bar{s}_\phi = \omega_\phi h/c_s$, where ω_ϕ denotes the fixed-base torsional frequency. For a transient excitation, a_o is replaced by a/c_s for all structures.

The (total) response for the standard assumption of vertically incident waves is denoted by a superscript 90°. The superscript ∞ indicates that the apparent velocity c_a is infinite. The superscript f denotes the free-field motion. The amplitudes of the corresponding displacements in the three directions are denoted as u^f $(= u_x^f)$, v^f $(= u_y^f)$, and w^f $(= u_z^f)$.

9.2.2 Inclined SH-Waves

The effect of the out-of-plane motion caused by SH-waves is addressed first. The total displacement amplitude $|v^t|$ for harmonic excitation is plotted in Fig. 9-10 versus a_o in three points of the massless basemat varying ψ_{SH}. For increasing a_o and decreasing ψ_{SH}, the response is diminished and is always less than that for vertical incidence. For $a_o \longrightarrow 0$ the phase angle φ (by which the torsional rotation $\phi_z = \gamma_o^g$ lags behind the translational $v^t = v_o^g$) equals 90°. For larger a_o, φ increases, always causing a larger response in point 2 (facing the arriving wave) than in point 3. For $a_o > 4$, φ is approximately 180°. The response of this massless basemat for horizontally propagating SH-waves is always smaller than that based on the standard assumption of vertical incidence.

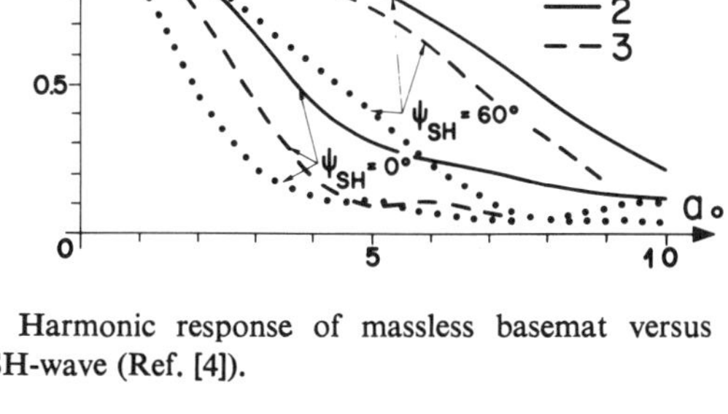

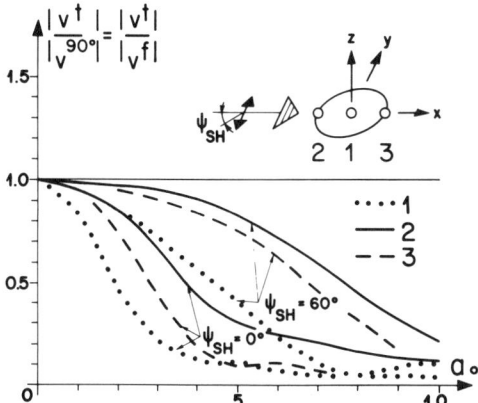

Figure 9-10 Harmonic response of massless basemat versus dimensionless frequency, SH-wave (Ref. [4]).

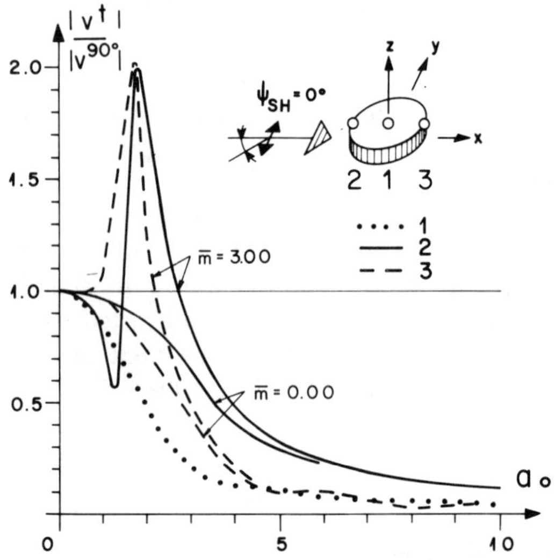

Figure 9-11 Harmonic response of basemat with mass versus dimensionless frequency, SH-wave (Ref. [4]).

For a basemat with a mass ratio $\bar{m} = 3$, the response ratio $|v^t|/|v^{90°}|$ is plotted for a horizontally propagating SH-wave versus a_0 in Fig. 9-11. In contrast to the massless basemat ($\bar{m} = 0$), the response ratio in points 2 and 3 exceeds the value 1 quite drastically. As the radiation damping of the torsional mode is less than that of the translational, a large amplification arises for intermediate a_o. It is worthwhile to examine the response characteristics for increasing a_o. For small a_o, the resulting torsional motion ϕ_z^t lags behind the center's translational v^t by approximately 90°, as discussed above for the massless basemat. For increasing a_o, the fundamental frequency of the translational motion is reached first ($a_o \sim 1.2$), resulting in a change of phase of $\sim$90° for v^t. Combined with that of the input, v^t and ϕ_z^t, are nearly in phase, so that their motions are added in point 3 (dashed line) and subtracted in point 2 (solid line). The dip in the solid line is clearly visible. The response in point 2 is even smaller than that in the center (point 1) in this range of a_o. For larger a_o, the fundamental frequency of the torsional motion occurs ($a_o \sim 1.6$), resulting in a phase change of $\sim$90° for ϕ_z^t. For even larger a_o, the phase difference of v^t and ϕ_z^t is the same as that for the massless basemat. The ratio of the maximum acceleration ($\ddot{v}^t_{\max}/\ddot{v}^{90°}_{\max}$) for the earthquake excitation is shown in Fig. 9-12a versus the angle of incidence ψ_{SH} of the SH-wave and in Fig. 9-12b as a function of the mass ratio $\bar{m}$ for the specified parameters. As most of the earthquake transient corresponds to small and intermediate a_o, for which the response ratio for harmonic excitation in point 3 is larger than in point 2 (Fig. 9-11), the response in the basemat's point facing away from the earthquake thus exceeds that in the point facing toward the arriving wave for all parameters examined (Fig. 9-12). It is, in general, even

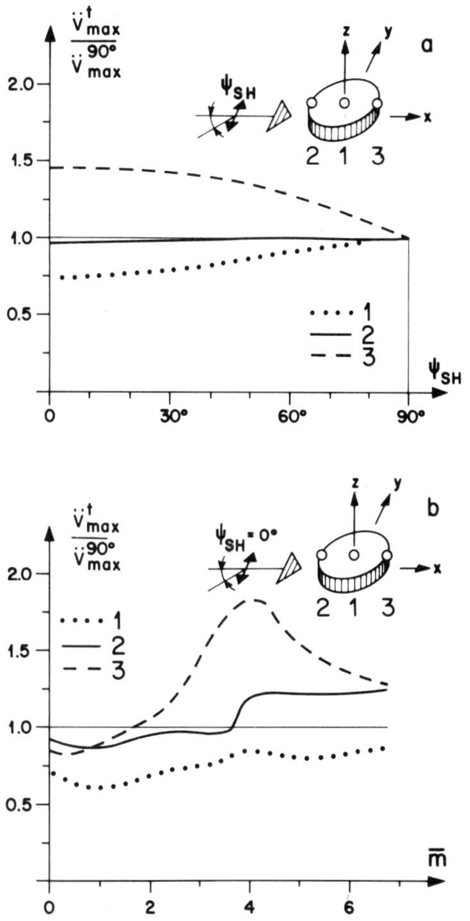

Figure 9-12 Maximum acceleration of basemat with mass, SH-wave as transient (Ref. [4]). (a) Versus angle of incidence $a/c_s = 0.10$ s, $\bar{m} = 3.0$; (b) versus mass ratio $a/c_s = 0.10$ s.

larger than that based on vertically incident waves. It should be remembered, however, that for most structures, $\bar{m}$ is < 3, except for tall structures, which cannot be modeled as a basemat with mass anyway. For many sites, the waves can travel in all directions. Thus the largest values (point 3) have to be used for design.

Finally, the translational and torsional mass–spring system connected to a basemat with mass (Fig. 9-9) is examined for a horizontally propagating SH-wave selecting the following parameters: $\bar{h} = 1$, $\bar{m} = 2.25$, $\bar{s} = \bar{s}_\phi = 2.26$, and $\zeta = 0.07$. The steady-state torsional rotation $|a\phi_z^t|$ (divided by $|v^f|$) is plotted versus a_o in Fig. 9-13. A large amplification arises in both points at the fundamental torsional frequency of the structure–soil system. For point 1, located on the basemat, the total resulting response shows a significant dip at the fixed-base torsional frequency. It would be zero for $\zeta = 0$. This effect is well known from the steady-state response of a system consisting of two spring–masses in series.

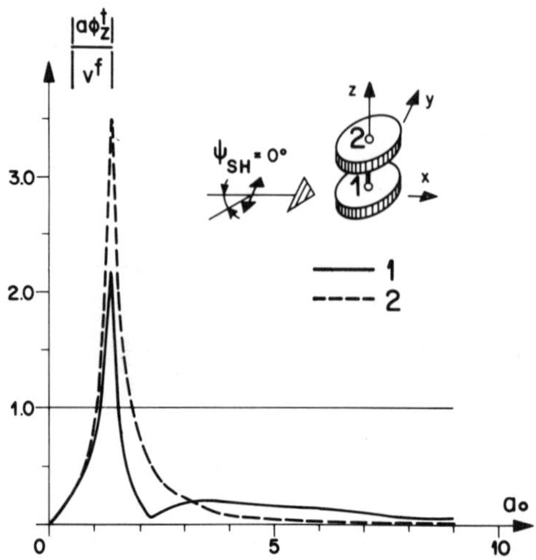

Figure 9-13 Harmonic response of translational and torsional mass-spring systems connected to basemat with mass versus dimensionless frequency, SH-wave ($\bar{h} = 1.0$, $\bar{m} = 2.25$, $\bar{s} = \bar{s}_\phi = 2.26$, $\zeta = 0.07$) (Ref. [4]).

The same structure is analyzed for the earthquake motion normalized to $0.2g$. Selecting $a/c_s = 0.06s$, the fixed-base frequencies of the translational and of the torsional motion both equal 6 Hz. Total acceleration-response spectra are calculated for the SH-wave propagating horizontally and for vertical incidence. This in-structure horizontal-response spectrum at the elevation h and radius a (Fig. 9-14) is typical. Because of the small filtering effect at the translational-rocking fundamental frequency of the structure–soil system, the corresponding peak is only somewhat reduced for $\psi_{SH} = 0$, while an additional strong peak at the torsional fundamental frequency (Fig. 9-13) appears.

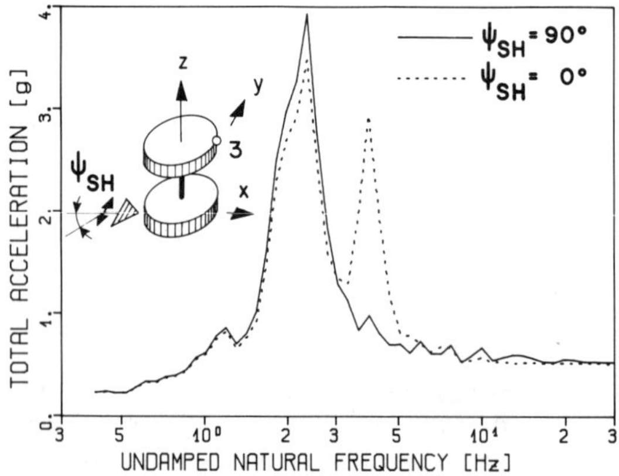

Figure 9-14 In-structure response spectra (1% damping) translational and torsional mass–spring systems connected to basemat with mass, SH-wave, point 3 ($a/c_s = 0.06$ s, $\bar{h} = 1.0$, $\bar{m} = 2.25$, $\omega_s/2\pi = \omega_\phi/2\pi = 6.0$ Hz, $\zeta = 0.07$) (Ref. [4]).

Love waves are not investigated, as, first, they do not exist for a half-space, and second, if a layered half-space were introduced, no new aspects would occur.

9.2.3 Inclined P- and SV-Waves

The in-plane body waves are examined next. Nonvertically propagating P- and SV-waves both generally lead to a horizontal and vertical free-field motion u^f, w^f. The resulting horizontal and vertical displacements of the center of the massless basemat, $u^t = u_o^g$ and $w^t = w_o^g$, for harmonic excitation are reduced similarly as for $v^t = v_o^g$ in the case of the SH-wave (see Fig. 7-30). The induced rotational input motion, that is, the rocking rotation $\phi_y = \beta_o^g$, caused by the propagating component w^f (which could arise from an inclined P- or SV-wave), is, however, approximately twice as large as $\phi_z = \gamma_o^g$ for the SH-wave. The shape of the curves of β_o^g and γ_o^g as a function of a_o hardly differs (Fig. 7-31). Only the results of kinematic interaction are presented in the following. The displacement amplitude $|w^t|$ for a propagating free-field harmonic vertical component w^f is plotted in three points of the massless basemat in Fig. 9-15. The selected parameters $c_s/c_a = 0.5$ and $= 1.0$ correspond to the limiting cases of a P- and an SV-wave, respectively, for the corresponding angle of incidence approaching zero. The value $|w^t|$ in the center is very similar to $|v^t|$ for the SH-wave. Actually, the two dotted curves of Fig. 9-15 coincide, from a practical point of view, with those of Fig. 9-10. The vertical-response ratio $|w^t|/|w^f|$ in points 2 and 3 exceeds 1, because of the large rocking component β_o^g. This is in contrast to the response of the massless basemat for SH-waves, where horizontally propagating-wave effects for harmonic excitation never dominate. As discussed for SH-waves, the response in point 2, facing the arriving wave again exceeds that in point 3.

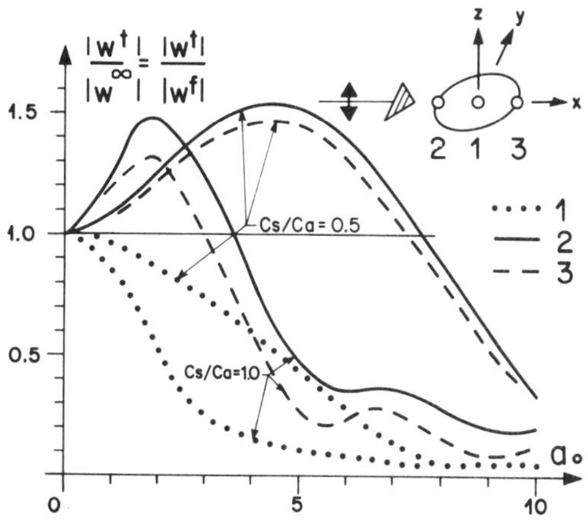

Figure 9-15 Harmonic response of massless basemat versus dimensionless frequency, vertical excitation (Ref. [4]).

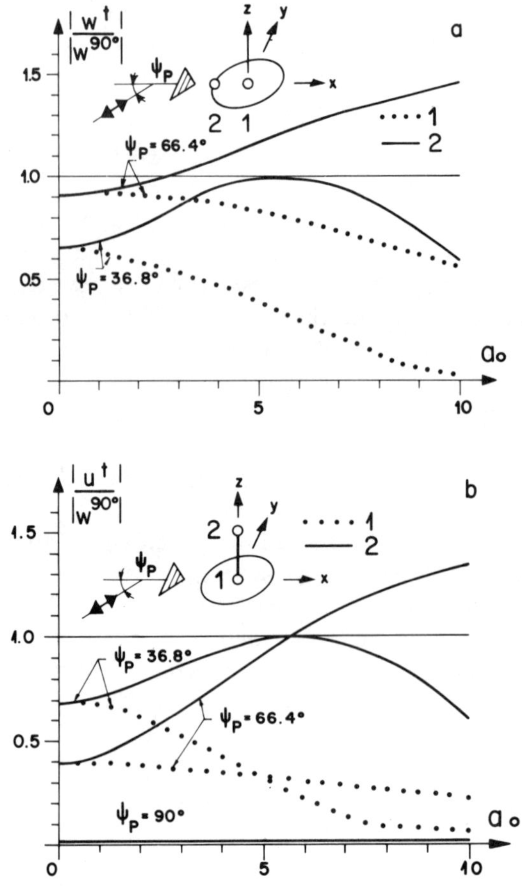

Figure 9-16 Harmonic response, P-wave (Ref. [4]). (a) Massless basemat, vertical versus dimensionless frequency; (b) massless structure, horizontal versus dimensionless frequency ($\bar{h} = 1.0$).

The vertical response ratio $|w^t|/|w^{90°}|$ of the massless basemat for propagating P-waves for harmonic excitation is specified in Fig. 9-16a. Of the many angles of incidence that were investigated, $\psi_P = 66.40°$ results in the largest response ratio for $a_o > 3$. As expected from the free-field response (Fig. 6-5), changing ψ_P from 90° does not lead to an excessive increase in the vertical response. The horizontal response for P-waves is presented for the massless structure ($\bar{h} = 1$) in Fig. 9-16b. The two selected angles of incidence lead to the largest ratios. Significant horizontal response does arise.

In Fig. 9-17a, the horizontal response ratio $|u^t|/|u^{90°}|$ of the massless structure ($\bar{h} = 1$) for SV-waves is examined. Decreasing ψ_{SV} from 90° at first leads to a ratio < 1, as w^f is quite small (Fig. 6-7). The large ratio that results, even in the center, in point 1, when $\psi_{cr} = 60°$ is approached, is caused by the

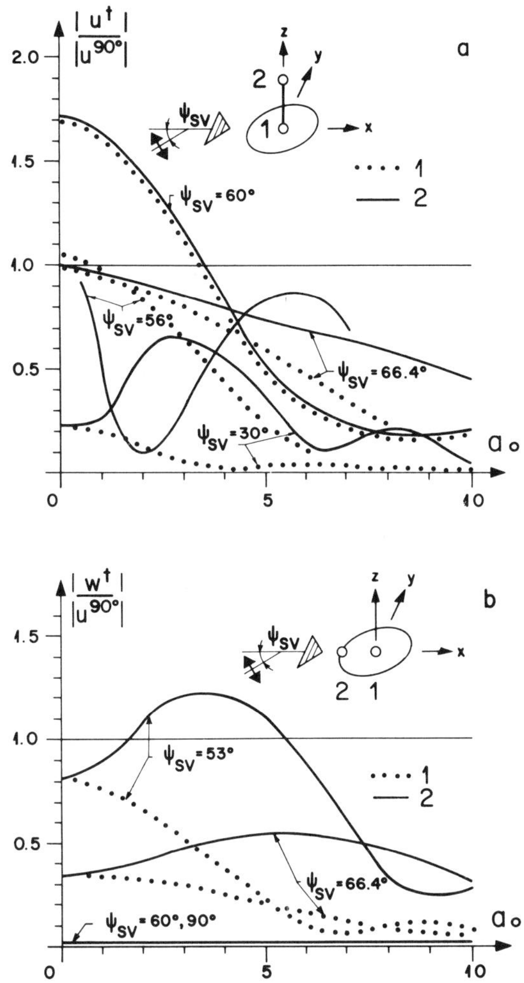

Figure 9-17 Harmonic response, SV-wave (Ref. [4]). (a) Massless structure, horizontal versus dimensionless frequency ($\bar{h} = 1.0$); (b) massless basemat, vertical versus dimensionless frequency.

strong increase in u^f. For $\psi_{SV} = 56°$, u^f and w^f are significant, but as this motion is prograde, the horizontal component from the rocking in point 2 is approximately out of phase (for small a_o) with that arising from the horizontal component, resulting in a diminished response. For $\psi_{SV} < 45°$, where the free-field motion is retrograde, the two horizontal components are added, as is discussed in detail later in connection with R-waves (Section 9.2.4). But as u^f is small, this effect is not clearly visible. The vertical response for SV-waves is represented for the massless basemat in Fig. 9-17b. For $20° < \psi_{SV} < 60°$, w^f is considerable, resulting in a substantial w^t (i.e., $\psi_{SV} = 53°$).

Many possibilities exist of combining P- and SV-waves to create specified, statistically independent horizontal and vertical free-field motions. Even selecting the same apparent velocity for the P- and SV-waves, which does not distort the earthquake motion as it propagates, many quite arbitrary assumptions have to be made. Two possibilities are discussed, and the corresponding responses of a massless structure are compared to that based on the standard assumption of vertically incident waves. In the first procedure, a horizontal and a vertical earthquake motion are selected, where $\ddot{w}^f_{max} = 0.67\ddot{u}^f_{max}$. For the horizontal-acceleration time history, the record of Fig. 3-17a is chosen. The vertical earthquake motion is generated from the latter by assigning a random phase angle to each Fourier term and by scaling the resulting motion appropriately. The motions of the SV- and P-waves, both vertically incident, are associated with the horizontal and vertical earthquake records, respectively. This determines the (complex) amplitudes of the two body waves. With these fixed amplitudes, different angles of incidence for the P-wave are selected. As the apparent velocity of both body waves is the same, ψ_{SV} of the SV-wave follows. This method will result in different horizontal and vertical free-field motions for each ψ_P. The horizontal response ratios ($=\ddot{u}^t_{max}/\ddot{u}^{\infty}_{max}$) for the massless structure ($\bar{h} = 1$) versus a/c_s, selecting $\psi_P = 30°$, are plotted in Fig. 9-18. The corresponding curves for $\psi_P = 60°$ are very similar. In the second procedure, the horizontal motion, for which again the record of Fig. 3-17a is chosen, remains unchanged when varying the angles of incidence. For each selected ψ_P, the (complex) amplitudes of the SV- and P-waves are determined from the horizontal motion by postulating that the ratio of the (real) magnitudes of the amplitudes of the P- and SV-waves for each frequency equals 0.67 and by choosing a random phase angle between the amplitudes of the two body waves. The vertical motion thus depends on ψ_P. In Fig. 9-19, the horizontal ratios $\ddot{u}^t_{max}/\ddot{u}^{90°}_{max}$ of the same massless structure versus a/c_s for $\psi_P = 30°$ are shown as solid and dotted lines. The most straightforward procedure, which does not determine the individual contributions of the two waves, is to assume that the prescribed horizontal and vertical motions propagate horizontally with the same apparent velocity, consistent with a selected angle

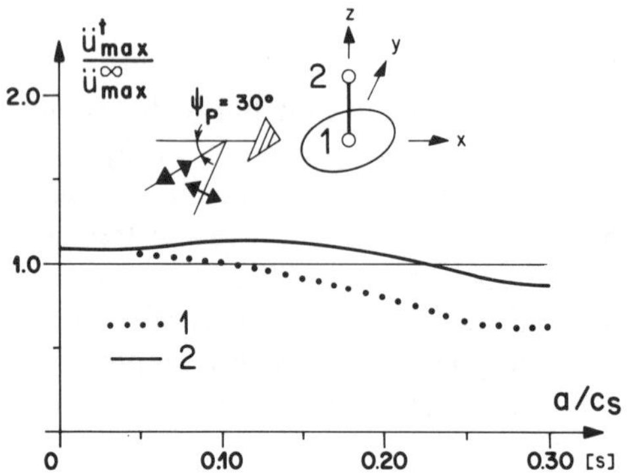

Figure 9-18 Maximum acceleration of massless structure versus radius/shear-wave velocity, vertically incident SV- and P-waves associated with horizontal and vertical earthquakes, respectively ($\bar{h} = 1.0$) (Ref. [4]).

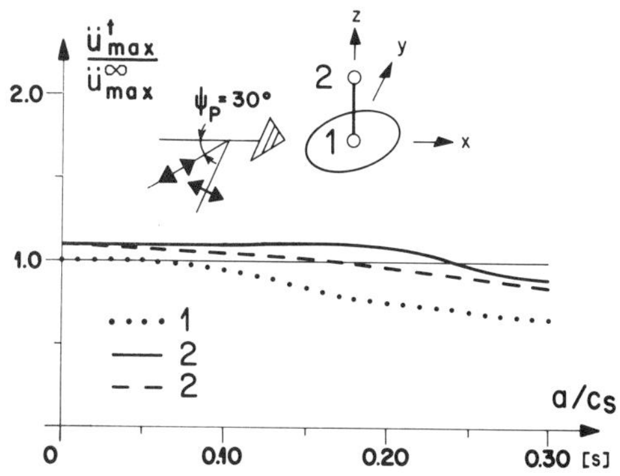

Figure 9-19 Maximum acceleration of massless structure versus radius/shear-wave velocity, SV- and P-waves determined from horizontal earthquake ($\bar{h} = 1.0$) (Ref. [4]).

of incidence. For comparison, the same response ratio, with c_a corresponding to $\psi_P = 30°$, is also plotted as a dashed line in Fig. 9-19. The horizontal and vertical earthquake motions are used, as determined in the first procedure for vertical incidence. It can be concluded from Figs. 9-18 and 9-19 that the horizontally propagating wave effects are quite small, as the apparent velocities for inclined body waves are large. It should be remembered, however, that the angle of incidence of the SV-wave is selected to be larger than ψ_{cr}, in order to be able to generate a P-wave with the same apparent velocity.

9.2.4 Rayleigh Waves

Finally, Rayleigh waves are addressed. The characteristics of these surface waves of a half-space are discussed in Section 6.3. The retrograde motion in a specific point as a function of time is shown in Fig. 6-6. Another representation showing the wave at a specific time along the horizontal axis is contained in Fig. 9-20. The horizontal and vertical components are denoted as u^f and w^f. For the uniformly damped half-space, w^f lags behind u^f by a phase angle of 90° for the coordinate system selected. This retrograde motion causes the horizontal displacement from the horizontal component of the wave u^h to be in phase with

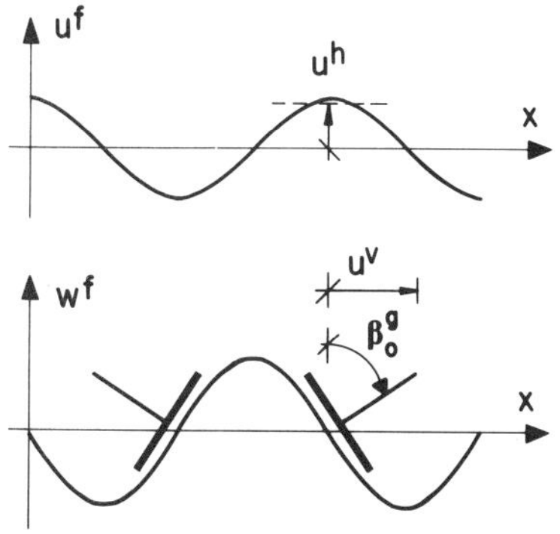

Figure 9-20 Addition of horizontal response, R-wave (Ref. [4]).

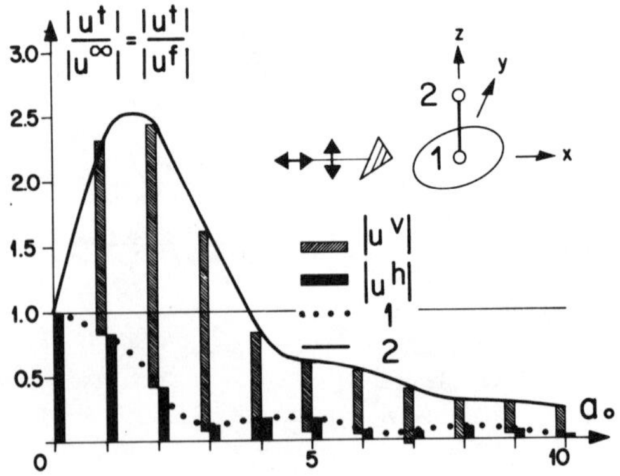

Figure 9-21 Harmonic response of massless structure versus dimensionless frequency, R-wave ($\bar{h} = 1.0$) (Ref. [4]).

the horizontal displacement u^v from the rocking-seismic input β_o^g, which arises from the vertical component. This effect of kinematic interaction is also visible in Fig. 9-21, where the horizontal displacement amplitude $|u^t|$ is plotted for points located on the massless structure ($\bar{h} = 1$) for harmonic excitation. For point 1 in the center of the basemat, the response ratio is strongly reduced for increasing a_o. This dotted curve is similar to that corresponding to a horizontally propagating SH-wave (Fig. 9-10, dotted line for $\psi_{SH} = 0$). For point 2, the contribution to u^t arising from u^f, u^h, and that from w^f, u^v, are shown separately. As $|w^f| \sim 1.5|u^f|$, $|u^v|$ is considerable larger than $|u^h|$. For $a_o < 3$, u^h and u^v are in phase, resulting in a response ratio that reaches 2.5. For larger a_o, β_o^g and u_o^g are no longer 90° out of phase, as is observed in the case of the discussed SH-wave. This results in u^h and u^v no longer being in phase. The response ratio is < 1 for $a_o > 4$, in contrast to the P-wave (Fig. 9-16b, $\psi_P = 66.40°$). This follows from the much larger apparent velocity of the inclined P-wave compared to the corresponding value of the R-wave. In Fig. 9-22, the maximum horizontal acceleration $\ddot{u}^t_{max}$ versus a/c_s in two points of the same massless structure is plotted for the earthquake excitation of Fig. 3-17a as solid and dotted lines for an R-wave. Over a wide range of a/c_s, the response ratio exceeds 2 for this transient excitation, which confirms the steady-state results of Fig. 9-21. For comparison, a retrograde wave, but with $|w^f| = |u^f|$, and a prograde wave with the same free-field amplitude ratio, both propagating at the same apparent velocity, are also examined. These two additional waves violate the equations of an elastic half-space. However, for a layered medium, surface waves can bear a certain resemblance with these two waves, although the wave pattern is much more complicated. Whereas for the retrograde wave (dashed line) horizontally propagating wave effects, although smaller, are still important, this is no longer the case for the prograde wave (dashed-dotted line).

The response ratios $\ddot{u}^t_{max}/\ddot{u}^\infty_{max}$ and $\ddot{w}^t_{max}/\ddot{w}^\infty_{max}$ for the translational mass–

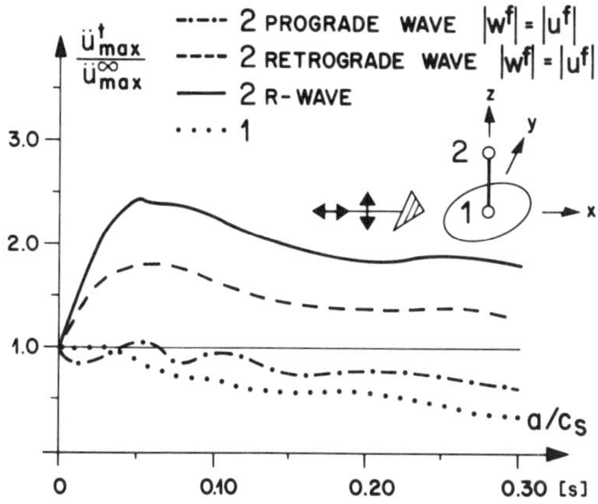

Figure 9-22 Maximum acceleration of massless structure versus radius/shear-wave velocity, surface waves as transient ($\bar{h} = 1.0$) (Ref. [4]).

spring system connected to a basemat with mass are plotted versus a/c_s (Fig. 9-23a), versus $\bar{m}$ (Fig. 9-23b), and versus the fixed-base frequency $\omega_s/2\pi$ (Fig. 9-23c) for the specified parameters. The damping ratio $\zeta = 0.07$ is selected. The values $\ddot{u}^v_{\max}$ and $\ddot{u}^h_{\max}$ represent the maximum horizontal acceleration (taking kinematic and inertial interactions into account) caused by $\ddot{w}^f$ and $\ddot{u}^f$, respectively. The ratios of the horizontal and vertical accelerations in point 1 at the base are, in general, < 1. In this point, $\ddot{u}^v_{\max}$ and $\ddot{u}^h_{\max}$ appear to be combined in $\ddot{u}^t_{\max}$ as if they were arising from statistically independent motions. The ratio of the horizontal acceleration in point 2 averages 2.5 for a substantial parameter variation. The maxima $\ddot{u}^v_{\max}$ and $\ddot{u}^h_{\max}$ arise at the same time. R-waves dominate at higher elevations of the structure.

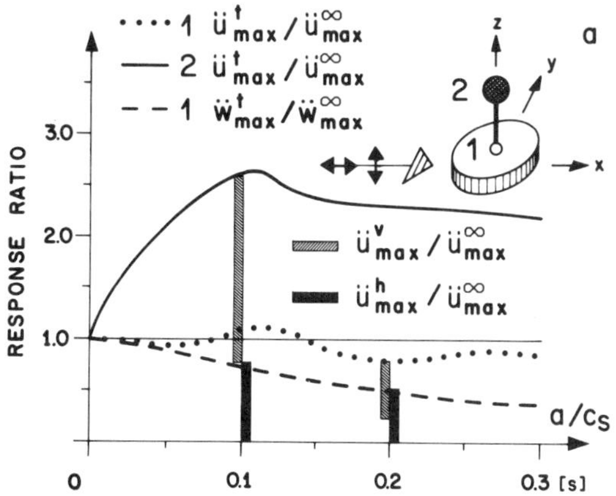

Figure 9-23 Maximum acceleration, translational mass–spring system connected to basemat with mass, R-wave as transient (Ref. [4]). (a) Versus radius/shear-wave velocity ($\bar{h} = 1.0$, $\bar{m} = 0.75$, $\omega_s/2\pi = 4.0$ Hz, $\zeta = 0.07$);

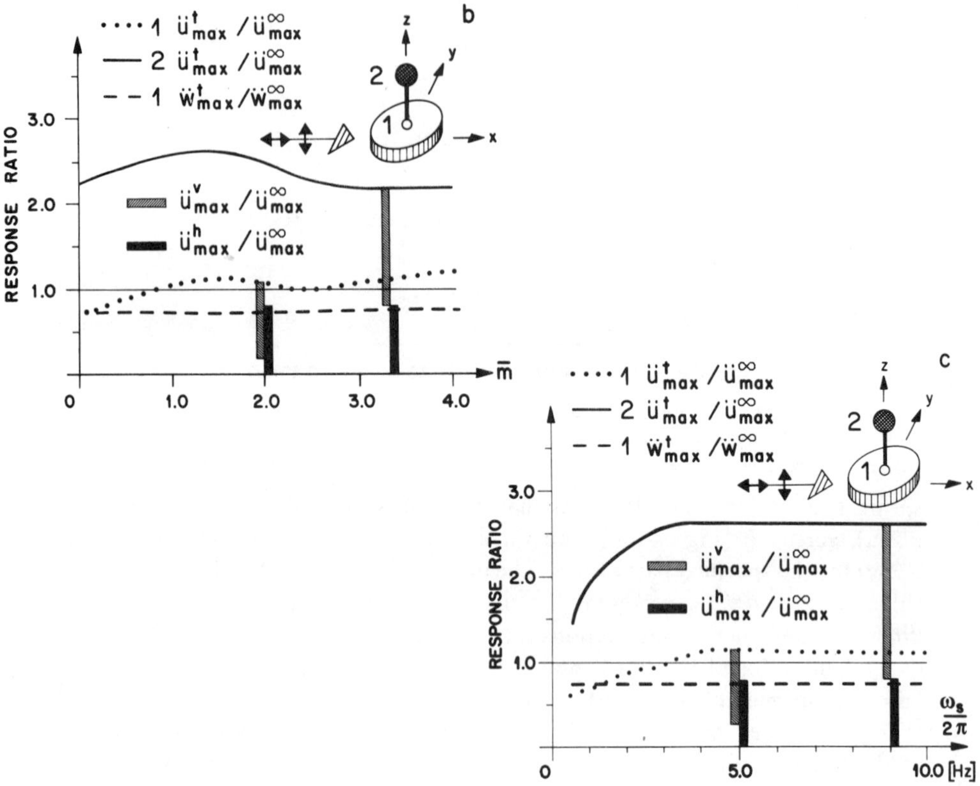

Figure 9-23 (Continued) (b) versus mass ratio ($a/c_s = 0.10$ s, $\bar{h} = 1.0$, $\omega_s/2\pi = 4.0$ Hz, $\zeta = 0.07$); (c) versus fixed-base frequency ($a/c_s = 0.10$ s, $\bar{h} = 1.0$, $\bar{m} = 0.75$, $\zeta = 0.07$).

9.3 EXAMPLES FROM ACTUAL PRACTICE

9.3.1 Through-Soil Coupling of Reactor Building and of Reactor-Auxiliary and Fuel-Handling Building

As the first practical application, the interaction of the reactor building through the soil with the surrounding reactor-auxiliary and fuel-handling building is analyzed. The three-dimensional dynamic models of the two structures for seismic loads are shown in Figs. 4-15 and 4-16. The seismic excitation consists of two statistically independent horizontal-acceleration time histories, normalized to $0.22g$, and of a vertical one with a maximum value of 0.67 times that in the horizontal direction. The layered site is discretized in Section 7.7 (Fig. 7-55). The method of substructure-mode synthesis is used to reduce the number of dynamic degrees of freedom. Besides the 12 unknowns of the two rigid basemats, 32 amplitudes of the mode shapes of the structures built in at their bases are used. The fundamental frequency of the total system equals 1.9 Hz.

TABLE 9-1 Maximum Response of Reactor Building in x-Direction

		Coupled System	Reactor Building Alone
Horizontal displacement	(mm)		
Basemat		17.8	12.0
Top		127.6	113.0
Horizontal acceleration	(g)		
Basemat		0.34	0.33
Top		1.48	1.30
Base shear	(GN)	0.31	0.28
Overturning moment	(GNm)	12.51	10.74

The corresponding mode shape shows a strong rocking motion of the reactor building. As discussed in connection with Fig. 7-58, the radiation damping will not contribute to the modal damping. In Table 9-1, the reactor building's maximum response determined from the coupled analysis is compared to that determined from examining the structure alone (i.e., without the surrounding building). The exact location of the nodes is specified in Fig. 4-15. Taking the through-soil coupling effect into account increases the response, especially at higher levels of the reactor building.

The influence of the flexibility of the basemat is addressed next. The corresponding dynamic model is shown in Fig. 4-17. Comparing the results for the rigid basemat with those for a thickness of 3.0 m and (as an extreme case) of 1.5 m, a nonuniform tendency is observed (not shown), the effect not being overwhelming. In a specific location, the effect can, however, be quite large. This is, for example, visible in the in-structure response spectrum at the top of the pressure vessel (Fig. 9-24). For a rigid basemat (solid line), the highest peak

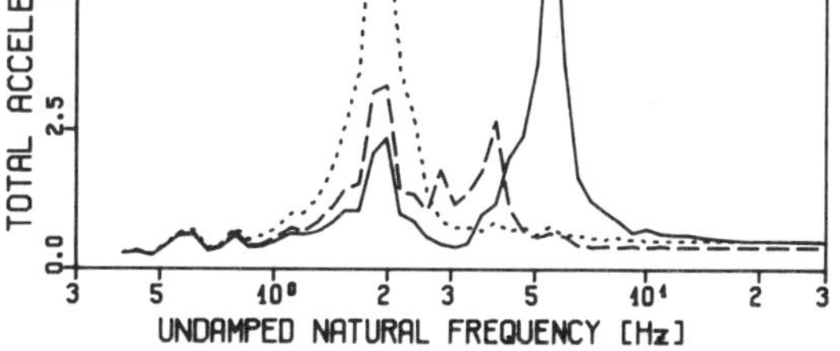

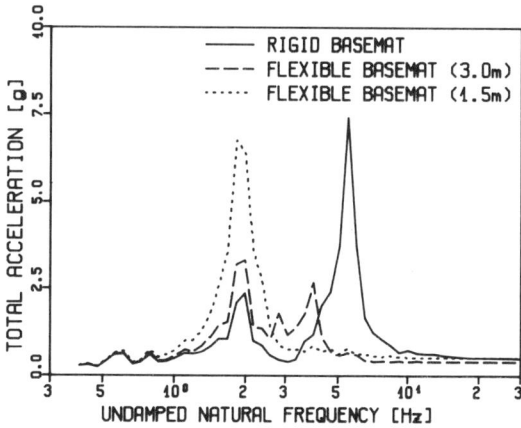

Figure 9-24 Horizontal in-structure response spectra (1% damping), reactor building with flexible basemat, top pressure vessel (Ref. [4]).

arises at 5.4 Hz, which is approximately equal to the fixed-base fundamental frequency of the pressure vessel. The frequency at which this (second) peak occurs is reduced to 3.8 Hz for a thickness of 3 m (dashed line). The peak at the structure–soil system's fundamental frequency of 1.9 Hz, which is hardly modified by the flexible basemat, becomes dominant. Introducing a thickness of 1.5 m (dotted line) results in only one peak at 1.9 Hz. Turning to horizontally propagating waves, while the maximum response in the horizontal direction remains nonuniform, the vertical response along the pressure vessel increases significantly for decreasing thickness of the basemat. For a flexible basemat, the "effective" radius and thus the self-canceling effect are reduced, resulting in increased seismic input in the vertical direction for the pressure vessel (results not shown).

9.3.2 Pile–Soil–Pile Interaction

As the next practical application, the seismic analysis of a reactor building founded on piles is discussed (Fig. 9-25). Below the basemat, a strongly horizontally layered compressive soil of 36 m thickness rests on bedrock (Fig. 9-26). The pile foundation consists of 202 end-bearing piles (146 with $\varnothing = 1.30$ m, 56 with $\varnothing = 1.10$ m) and of 88 floating piles of 15 m length (80 with $\varnothing = 1.80$ m, 8 with $\varnothing = 1.30$ m). Broad-banded design-response spectra specify the seismic excitation as outcropping on the level of bedrock. The maximum accelerations of the corresponding artificial time histories of duration 15 s equal $0.1g$ and $0.067g$ in the horizontal and vertical directions, respectively. The free-field properties of the six soil layers at the site are shown in Fig. 9-26. The shear moduli (and the shear-wave velocities) are compatible with the strains derived from the horizontal earthquake record, which is assumed to consist of vertically propagating waves. The top two layers are made up of extremely soft soil. It should also be noted that the properties of the second layer are even worse than those of the first one. Figure 9-26 also shows the section $y = 0$ through the piles. The three end-bearing piles of $\varnothing = 1.30$ m are denoted later as the "center pile," the "intermediate pile," and the "boundary pile." The fourth marked pile of $\varnothing = 1.80$ m will be referred to as the "floating pile."

The reactor building with a rigid basemat is modeled with three vertical beams introducing, in a three-dimensional analysis, a total of 198 degrees of freedom. All piles are built in to the rigid basemat. The end-bearing piles are assumed to be hinged at their tips. Interaction forces arise (and thus compatibility of the pile and soil displacements is formulated) along the axes of the piles at the nodes located at the top of each layer. In addition, compatibility is formulated for all piles at the level of the tips of the floating piles ($z = -15.0$ m). Decomposing the earthquake motion into a symmetrical and an antisymmetrical part, the dynamic model of the pile–soil system can (approximately) be reduced to a quarter of the foundation. This results in 86 piles, for which, in a total of 556 nodes, compatibility with the neighboring soil is enforced. The resulting

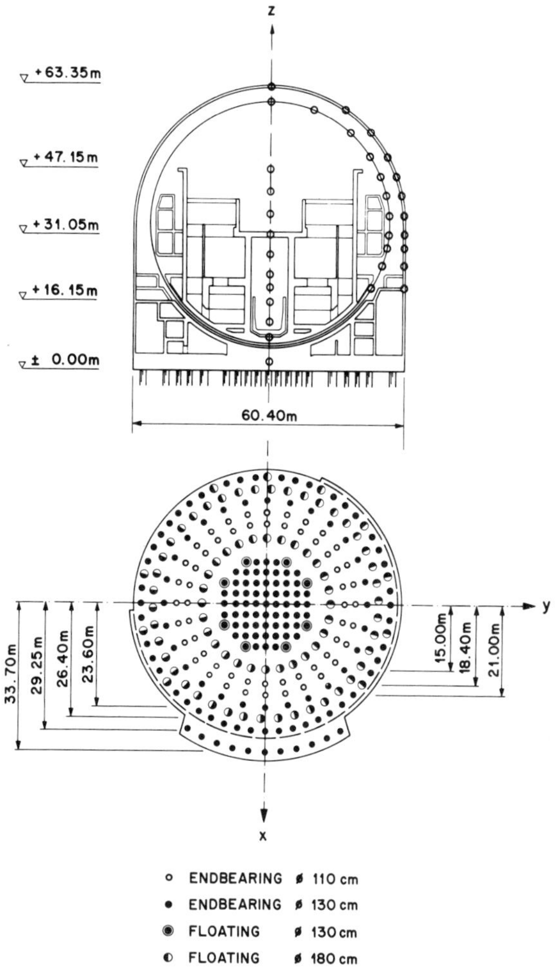

Figure 9-25 Elevation and plan view of reactor building founded on piles (Ref. [14]).

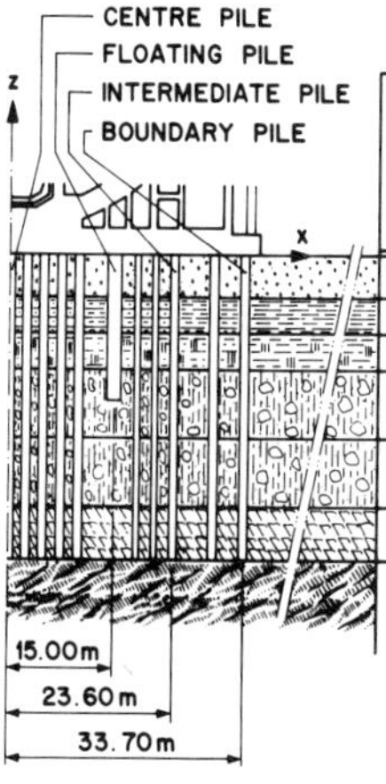

SOIL - TYPE	SHEAR MODULUS [GN/m²]	MASS DENSITY [Mg/m³]	SHEAR WAVE VELOCITY [m/s]	POISSON'S RATIO	HYSTERETIC DAMPING
SAND	0.021	1.95	104	0.42	0.110
CLAY	0.016	1.70	97	0.49	0.100
TRANSITION	0.670	2.10	565	0.30	0.020
RESIDUAL 1	0.580	2.05	532	0.35	0.025
RESIDUAL 2	1.350	2.05	812	0.35	0.020
WEATHERED ROCK	2.700	2.40	1061	0.30	0.020
BEDROCK	12.260	2.65	2151	0.30	0.000

± 0.00m
\- 5.50m
\- 8.50m
\- 11.00m
\- 20.00m
\- 29.00m
\- 36.00m

Figure 9-26 Free-field properties of soil layers (Ref. [14]).

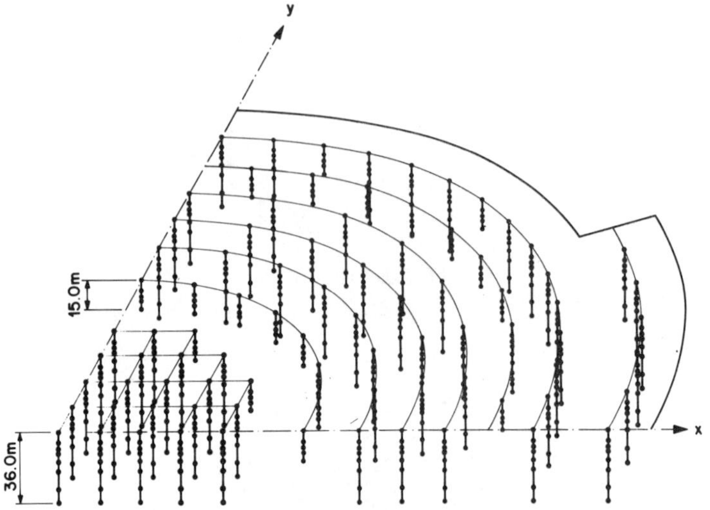

Figure 9-27 Discretized pile–soil system (Ref. [14]).

discretized pile–soil system is shown in Fig. 9-27. As the dynamic stiffness of the pile foundation is frequency dependent, the natural frequencies of the pile–soil–structure system have to be determined by iteration. Its fundamental frequencies for the horizontal and vertical vibrations equal, for rigid bedrock, 2.10 Hz and 9.20 Hz, respectively. As the fundamental frequencies of the layered site (free field) are higher (2.63 Hz in the horizontal and 12.2 Hz in the vertical direction) than those of the pile–soil–structure system, radiation damping for these vibrational modes with a large generalized modal load (participation factor) will be extremely small (zero for no hysteretic damping).

For the horizontal earthquake in the x-direction, the maximum forces at the pile heads are shown in Fig. 9-28. As can be seen from the shear forces in the x-direction (Fig. 9-28a), the boundary pile is loaded much more heavily (1.26 times the average) than that in the center (0.60 times the average). As expected, all piles located on the same circle carry approximately the same load. The floating piles of large diameter attract more shear force than would follow from the ratio of the cross-sectional areas. The bending moments acting around the y-axis (Fig. 9-28b) are spatially much more evenly distributed than the shear forces. Piles of different diameters are loaded approximately proportionally to their moments of inertia. The axial forces caused by the global overturning moment around the y-axis are shown in Fig. 9-28c.

The following figures show the distribution of the forces and of the displacements along the axes of selected piles at the time when the corresponding value at the head reaches its maximum. For the horizontal earthquake, the distribution of the horizontal displacements and of the bending moments hardly depends on the location of the pile. In addition, end-bearing and floating piles

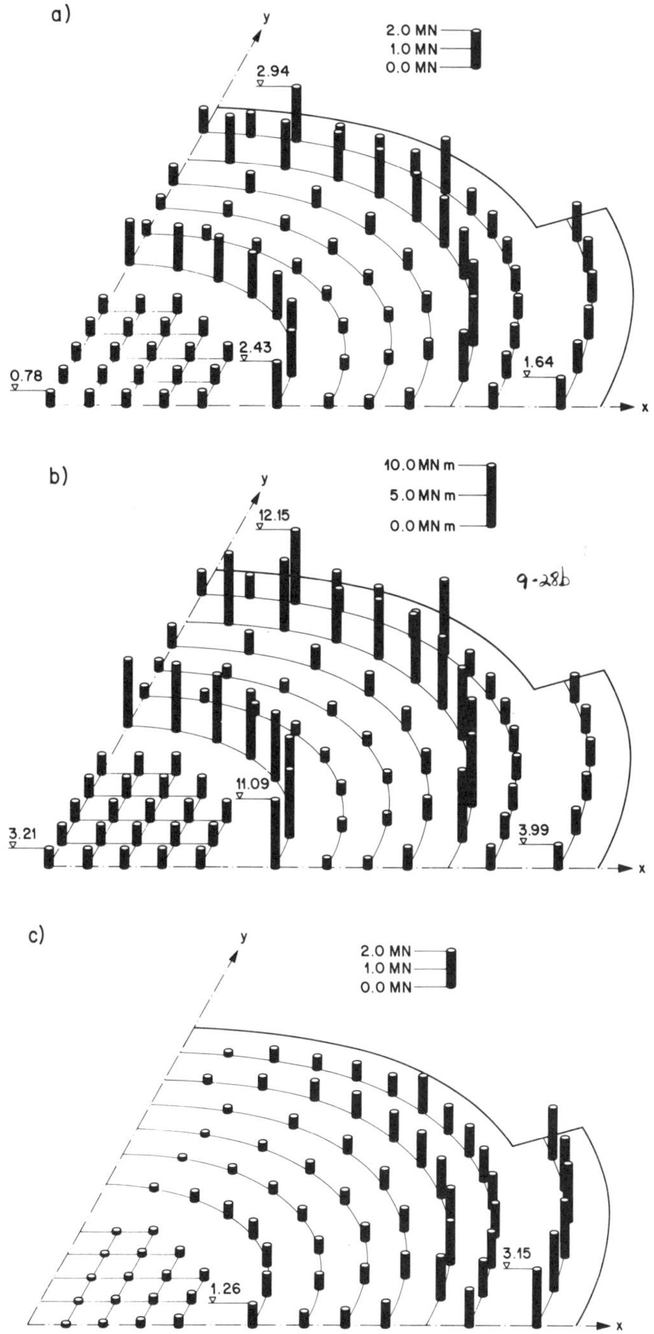

Figure 9-28 Maximum forces at pile heads, horizontal earthquake (Ref. [14]). (a) Shear forces; (b) bending moments; (c) axial forces.

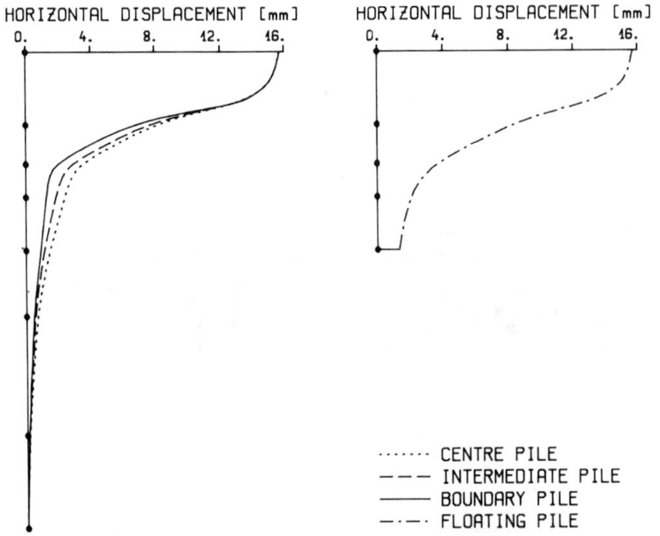

Figure 9-29 Maximum horizontal displacements along piles, horizontal earthquake (Ref. [14]).

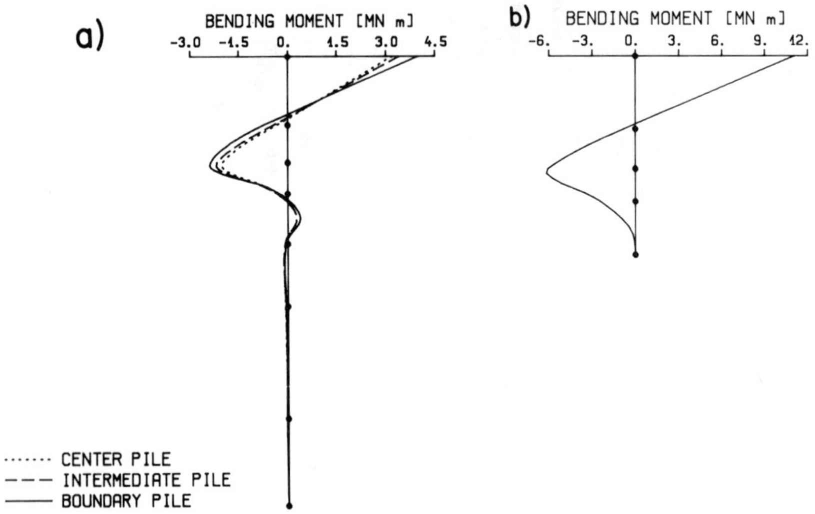

Figure 9-30 Maximum bending moments, horizontal earthquake (Ref. [14]). (a) Along end-bearing piles; (b) along floating pile.

exhibit almost the same shape (Figs. 9-29 and 9-30). For the vertical earthquake, the vertical displacement and the axial force of the boundary pile decrease more rapidly than the corresponding values of the center pile (Fig. 9-31). Even for a larger value of the axial force at the head, a smaller value at the tip results.

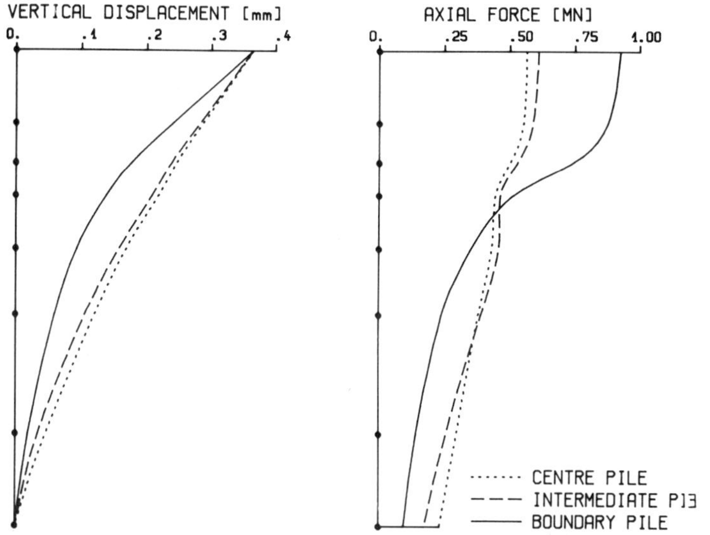

Figure 9-31 Maximum vertical displacements and axial forces along end-bearing piles, vertical earthquake (Ref. [14]).

9.3.3 Horizontally Propagating Waves on Hyperbolic Cooling Tower

The seismic response of a hyperbolic cooling tower to horizontally propagating waves represents an important practical problem. The dimensions of this structure at its base are so large that even for quite large apparent velocities, traveling-wave effects are significant. Often for a structure of this type, which is not a high-safety class structure, only approximate dynamic-soil investigations are performed for economical reasons. This does not allow the wave pattern, which can occur at the site, to be calculated rigorously, as is discussed in Sections 6.7 and 6.8. As presented in Section 9.2, certain "reasonable" wave patterns are assumed to exist.

The dimensions of the hyperbolic cooling tower with a height of 144 m and a diameter at the base of 117.08 m are shown in Fig. 4-13. The dynamic-shear modulus (compatible with the strains developed for the design basis earthquake), the density, and the Poisson's ratio of the soil are estimated as 0.12 GPa, 2.4 Mg/m^3, and 0.4, respectively. The damping ratio of the soil and of the structure is assumed to be equal to 0.04. The design of the supporting columns, and thus also of the lower part of the shell (edge beam) and of the foundations, is governed even for moderate earthquake excitation by the seismic loads instead of the wind. The tower is designed for a design-basis earthquake with a peak horizontal ground acceleration $= 0.12g$. The response spectrum of the 10-s artificial time history for the horizontal direction closely follows that of the U.S. NRC Regulatory Guide 1.60 for a damping ratio = 0.04 (Fig. 3-17a). Two horizontally propagating wave patterns are examined. The vertical component, w^f ($= u_z^f$), is determined from the horizontal one, u^f ($= u_x^f$), by making different assumptions, which, in general, violate the laws of elastodynamics. First, a retrograde motion, with $|w^f(\omega)| = |u^f(\omega)|$ for all Fourier terms, is selected and denoted as "retrograde." Second, a prograde motion is introduced, again with $|w^f(\omega)| = |u^f(\omega)|$ ("prograde"). For both types of waves, the peak vertical acceleration is approximately equal to that in the horizontal direction. The waves are assumed to propagate in the positive x-direction. The apparent velocity $c_a = 500$ m/s is used in the calculations, which is reasonable when considering the dispersion curves for surface waves of this site. The corresponding dimensionless frequency $a_o = \omega a/c_a$ (in which a is the radius at the foundation level of the tower and ω is the fundamental frequency of the soil–structure system $2\pi \times 1.96$ Hz) equals approximately 1.5, for which both the translational component u_o^g and the rocking component β_o^g exhibit significant values (Fig. 7-31). Horizontally propagating wave effects should thus become important. For comparison, the horizontal and the vertical components of the motion are assumed to originate from vertically incident waves ($c_a = \infty$), while for the latter component the retrograde wave is used. In addition, the horizontal component alone is processed, with $c_a = 500$ m/s, and for $c_a = \infty$. The dynamic model of the tower is discussed in Section 4.3.3. For horizontally propagating waves, harmonics higher than the first have to be included.

Evaluating the results, it is convenient to compare the following maximum-dynamic values: the total base shear force Q, which characterizes the overall lateral response; the maximum normal column force N_{max} (tension), taken along the circumference without deadweight (the force that governs the design of the columns); and finally the total horizontal acceleration $\ddot{u}^t$ at the top of the shell at $\theta = 180°$ (this acceleration acts as an indicator of local shell motions). In Table 9-2, these three values, normalized to the corresponding ones for the horizontal and vertical motions with vertical incidence, are presented.

Studying the effect of a horizontally propagating wave, the horizontal component only is examined at the beginning. Deleting the vertical earthquake

TABLE 9-2 Ratios of Maximum Response

Wave Pattern		Total Shear Force, Q	Maximum Column Force, N_{max}	Acceleration, $\ddot{u}^t$
Horizontal/vertical:	$c_a = \infty$	1	1	1
Horizontal:	$c_a = \infty$	1.00	0.98	0.88
Horizontal:	$c_a = 500$ m/s	0.59	0.76	0.59
Retrograde		1.74	1.78	1.85
Prograde		0.56	0.81	2.24

for the case $c_a = \infty$ reduces N_{max} and, to a somewhat larger degree, $\ddot{u}^t$. When this horizontal component propagates, the resulting self-canceling effect of the translational input motion reduces all three response values. Since the reduction for $\ddot{u}^t$ is the same (0.59) as that for Q, the horizontal component of the propagating wave does not seem to excite local shell modes. Introducing the propagating vertical component, the retrograde waves lead, as expected, to larger Q's and N's, while the prograde waves lead to smaller ones. This is also apparent in Figs. 9-32 and 9-33, in which the first 4 s of the time history of Q, and of the normal forces N, in the columns with positive and negative angle of inclination α around the circumference are plotted for that time when the maximum value of Q occurs. The dominance of the motion in the fundamental frequency = 1.96 Hz for all three cases is evident in Fig. 9-32. However, the rotational component of the effective seismic input excites the local shell modes for both the retrograde and the prograde waves, as is apparent from $\ddot{u}^t$ in Table 9-2. The corresponding in-structure response spectra in the x-direction (Fig. 9-34) at the same location of the shell dramatically emphasize this fact.

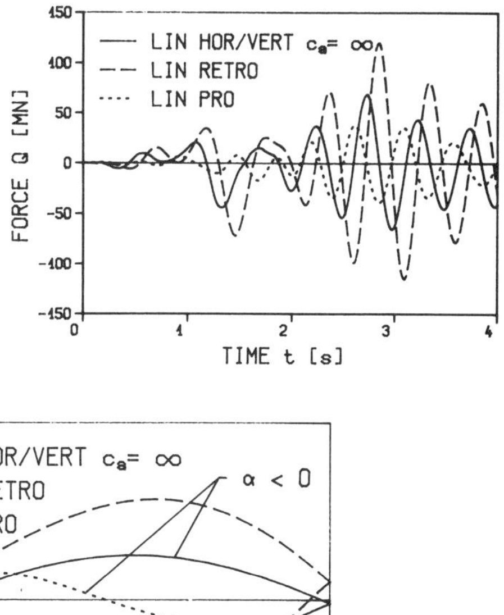

Figure 9-32 Time history of total shear force at base (Ref. [15]). LIN denotes a linear analysis. (In Ref. [15], nonlinear behavior is also examined.)

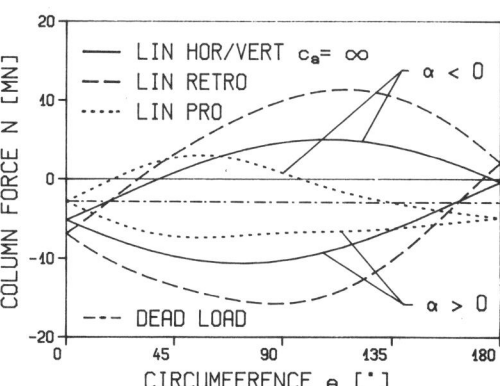

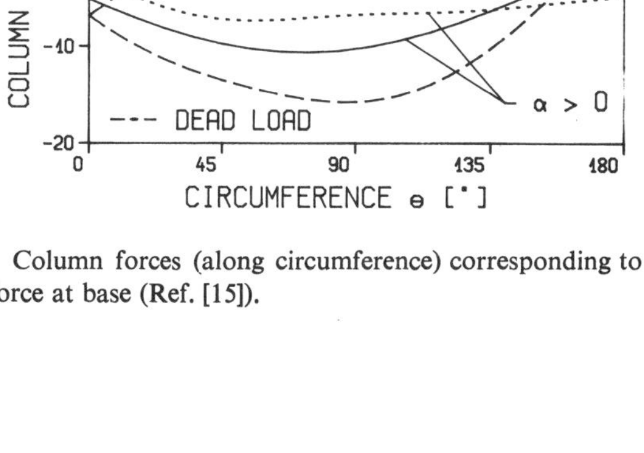

Figure 9-33 Column forces (along circumference) corresponding to maximum total shear force at base (Ref. [15]).

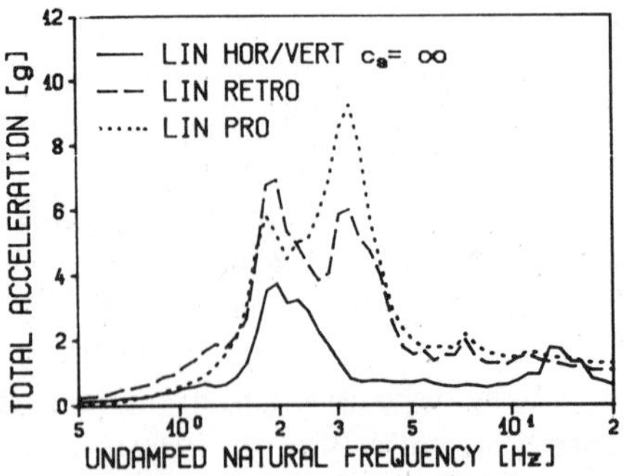

Figure 9-34 In-structure response spectra (1% damping) at top of shell, $\theta = 180°$ (Ref. [15]).

9.3.4 Horizontally Propagating Waves on Nuclear Island with Aseismic Bearings

Finally, the effect of horizontally propagating seismic waves on a nuclear island, which is isolated with aseismic bearings, is investigated. A schematic section passing through the nuclear island is shown in Fig. 9-35. All safety-related structures of the nuclear island are founded on a common upper raft.

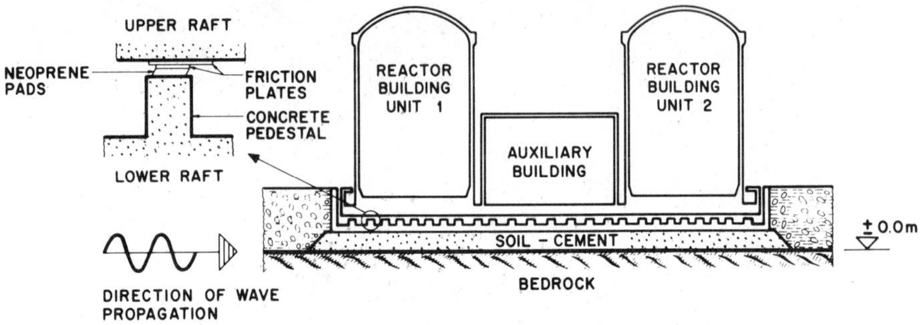

Figure 9-35 Nuclear island on aseismic bearings (Ref. [16]).

The aseismic bearings are located between the lower and upper rafts. On top of concrete pedestals resting on the lower raft, neoprene pads, which are flexible in the horizontal direction only, are placed. In addition, friction plates are used to limit the horizontal force transmitted to the structure. This results in a nonlinear isolation mechanism acting in the horizontal direction only. The lower raft rests on a layer of soil cement which is on top of the layered rock site, whose free-field response is examined in depth in Section 6.8.

Because the isolation mechanism is nonlinear, the equations of motion have to be integrated in the time domain. As mentioned in Sections 8.1 and 8.5.2, a rigorous solution involves convolution integrals. To avoid this, the dynamic-stiffness matrix of the soil is approximated as a frequency-independent spring and dashpot, as described in another context in Section 8.4.1. The equations of motion are then solved with a direct integration scheme.

As already mentioned, the nuclear island is isolated at the base in the horizontal direction only. Because the rocking component of the seismic input arising from the horizontally propagating waves is not isolated in this concept, the effects of horizontally propagating waves are, in general, on a percentage basis larger than in conventional structures. Along with the structural configuration, whereby the large dimensions of the basemat are to be mentioned, the properties of the site are of importance when evaluating wave-passage effects. For rock sites with small material damping, the surface waves hardly decay.

The out-of-plane motion of horizontally propagating SH- and Love waves is not analyzed, as the horizontal isolation mechanism acts also for the additional torsional-seismic input generated by these waves. A two-dimensional model is adopted for the analysis of the soil–structure interaction caused by the inplane motion of SV-, P-, and Rayleigh waves. The properties of the bedrock site are specified in Table 6-5. As the structure–soil interface is selected at the top of the soil cement (−6 m), the free-field soil profile of the site consists of the layer of soil cement of 6 m thickness resting on layered bedrock. The horizontal and vertical components of the control motion consist of artificial 30-s acceleration time histories, the response spectra of which follow the U.S.-NRC Regulatory Guide 1.60, normalized to 0.30*g* and 0.20*g*, respectively. The control point is located at the top of the bedrock (0 m) without the presence of the soil cement. With respect to the total site (including the layer of soil cement), the control motion thus acts at a fictitious rock outcrop. The free-field motion at the top of the soil cement (−6 m) is determined for two different body-wave assumptions. The control motion is first deconvoluted to the half-space (+120 m), resulting in the amplitudes of the incident waves and then convoluted to the top of the soil cement. In addition to vertically incident waves, inclined body waves are investigated, whose angles of incidence of the P- and SV-waves ψ_P and ψ_{SV} in the half-space are selected as 5° and 30°, resulting in apparent velocities of 7627 m/s and 5058 m/s, respectively. The corresponding amplitudes A_P and A_{SV} of the incident waves in the half-space depend on both the horizontal and vertical components of the control motion. Besides assuming these two wave patterns, denoted later as "vertical incidence" and "inclined waves," a wave train consisting mainly of Rayleigh waves that results in the same motion at the top of the soil cement as the vertically incident body waves is also investigated. For a layered system the apparent velocity depends on the frequency. The procedure is analogous to that described in detail in Section 6.8, with the site consisting in addition of the layer of soil cement. For the first two R-modes, the dispersion curves (for the rock site without the soil cement) are shown in Fig. 6-73. For frequencies above 9.1 Hz, where the second R-mode starts, the horizontal and vertical components of the motion at the top of the soil cement can be associated with

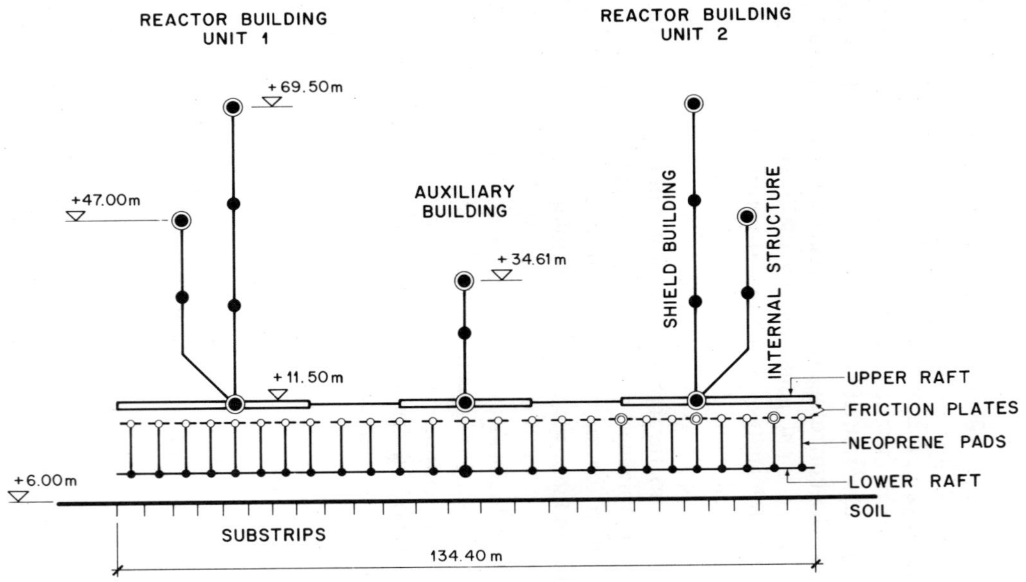

Figure 9-36 Dynamic model of nuclear island (Ref. [16]).

the R-modes. For frequencies below 9.1 Hz, a P-wave with $\psi_P = 5°$ (apparent velocity = 7627 m/s) is used in addition to the first R-mode. This wave pattern is denoted as "R- and P-waves."

The dynamic model of the nuclear island is shown schematically in Fig. 9-36. A simple model is deliberately selected which, however, can represent the characteristic effects arising from horizontally propagating waves. For this purpose only three buildings are included in the two-dimensional model. The structures are modeled with beams and lumped masses. The upper raft consists of three rigid sections connected by two flexible parts. The lower raft is assumed to be flexible over the whole length. Both rafts are rigid in their plane. Each row of the aseismic bearings in the third dimension is modeled as a discrete element representing the horizontal and vertical flexibilities of the neoprene pads and the friction plates. A total of 25 elements arise. The pads are designed to exhibit a horizontal stiffness which results in a horizontal frequency = 0.9 Hz of the total system, assuming the upper raft and the structures to be rigid and the lower raft fixed. The coefficient of friction equals 0.15. The lower raft is founded on the layer of soil cement of 6 m thickness resting on bedrock. The latter is modeled with 10 homogeneous layers which rest on a homogeneous half-space (at a depth of 120 m). The interface of the lower raft and the soil is discretized with 25 substrips analogously to the elements modeling the aseismic bearings. The (coupled) spring and dashpot coefficients of the soil are evaluated at 9 Hz. Taking the constraints imposed by the rigid connections into account, the total dynamic model has 94 dynamic degrees of freedom.

The maximum total accelerations are shown in Table 9-3 for the nuclear island with neoprene pads without and with friction plates for the various wave patterns. For vertical incidence the favorable influence of the neoprene pads on the horizontal response leads to a small amplification with height. Adding friction plates reduces the horizontal accelerations even further. The differences in the horizontal response of the two reactor buildings for vertical incidence are small. For the horizontally propagating R- and P-waves, the rotational seismic input results in an increase of the horizontal response throughout the structure. The inclined waves also lead to a larger horizontal response compared to that for vertical incidence but, in general, less than for R- and P-waves.

In-structure response spectra for the nuclear island with neoprene pads but without friction plates are presented in Fig. 9-37. The results for R- and P-waves are compared to those for vertical incidence. In the horizontal direction the peak at the frequency 0.89 Hz of the first mode is, as expected, dominant. The other peaks can also be associated with the frequencies of the higher modes. The peaks are more pronounced for the horizontally propagating wave than for vertical incidence. For the nuclear island with the complete aseismic bearings, the results are similar (not shown).

TABLE 9-3 Maximum Total Acceleration (*g*)

	Without Friction Plates		With Friction Plates		
	Vertical Incidence	R- and P-waves	Vertical Incidence	R- and P-waves	Inclined Waves
Horizontal					
Upper raft	0.304	0.338	0.186	0.189	0.198
Top shield building unit 1	0.388	0.528	0.252	0.425	0.337
Top internal structure unit 1	0.369	0.508	0.261	0.312	0.322
Top shield building unit 2	0.355	0.451	0.255	0.326	0.307
Top internal structure unit 2	0.380	0.397	0.279	0.304	0.257
Top auxiliary building	0.310	0.386	0.252	0.269	0.275
Vertical					
Center upper raft unit 1	0.301	0.303	0.301	0.303	0.334
Center upper raft unit 2	0.284	0.295	0.284	0.295	0.300
Center upper raft auxiliary building	0.302	0.284	0.302	0.285	0.310

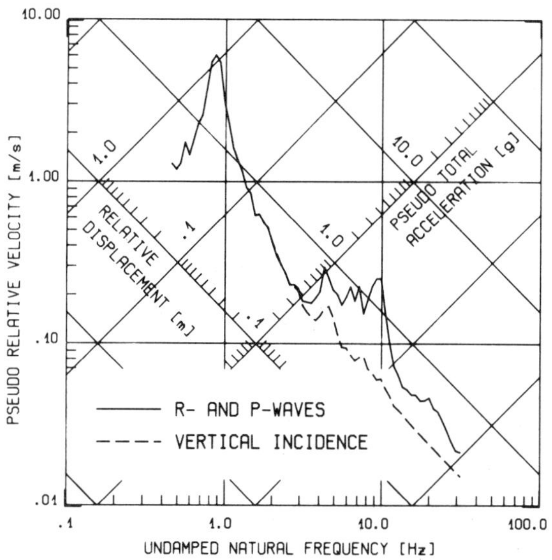

Figure 9-37 Horizontal in-structure response spectra (2% damping), nuclear island without friction plates; top of auxiliary building (Ref. [16]).

9.4 CONCLUDING REMARKS

9.4.1 Need for Adequate Consideration

Soil–structure interaction generally represents an important effect which has to be considered and evaluated properly. For loads applied directly to the structure (e.g., occurring from machine vibration), this is not disputed. For seismic excitation, the importance of the properly evaluated free-field motion (Chapter 6) is also generally acknowledged. The influence of the other part, that of the actual interaction, is more controversial. It can be small in certain cases. This applies, for example, for the bridges examined in Section 9.1.5, especially if the damping ratio of the structure is similar to that of the soil. But the seismic-design provisions applicable to building structures formulated by the Applied Technology Council ("Tentative Provisions for the Development of Seismic Regulations for Buildings," ATC-3-06 Report, Applied Technology Council, Palo Alto, California, June 1978) contain provisions for flexibly supported structures. Even in the simplest possible approach to seismic analysis, the total equivalent static-lateral force (base shear) can be reduced considering soil–structure interaction. The permissible reduction in base shear is allowed to reach 30% of the value of the base shear determined for the fixed-base structure. The concepts that lead to the equation of the reduction in base shear in these provisions are the same as those discussed in Section 3.4. The change from the fixed-base frequency ω_s to the fundamental frequency of the structure–soil system $\tilde{\omega}$ and the corresponding modification of the damping ratios from ζ to $\tilde{\zeta}$ are taken into account. The trend that soil–structure interaction affects the higher modes less in some cases is also considered. The fact that design provisions for soil–structure interaction exist demonstrates that, even for everyday building structures, the effect of the actual interaction part is recognized to be important and cannot, in general, be neglected.

For massive stiff structures founded on flexible soil, the actual interaction effects will be very large. This applies, for example, for nuclear power plants and certain off-shore structures. It must also be recognized that this type of structure contains sensitive mechanical and electronic equipment which also has to be designed for seismic excitation to remain operational during the design earthquake. In some cases, for example, for the relative displacements between two adjacent structures (which is important for the analysis of piping running between the two structures), neglecting the actual-interaction analysis can lead to a response which is too small. In most cases, however, taking soil–structure interaction into account will lead to a significant reduction of the seismic response. Neglecting this favorable effect is economically inadmissible. In certain extreme circumstances, only the use of this reduction will enable the design of the structure at all without changing its basic layout. Neglecting the effect of interaction and considering all the other (unquantified) conservatisms introduced at the other stages of seismic analysis can lead to an overly conservative design with respect to the earthquake hazard. In some cases, this will be at the expense of normal operating conditions (e.g., temperature effects) and of other extreme loading conditions arising from postulated accidents.

9.4.2 Modeling Aspects

The following features, which can be adequately modeled using the procedures discussed in this text, influence significantly the results of a soil–structure interaction analysis:

Layering of site. For a site approaching a half-space, radiation of energy will occur for all frequencies, while for a site with a significantly stiffer bedrock this will not occur for frequencies below the fundamental frequency of the layers with a finite depth.

Embedment. Even for vertically incident S-waves, the effective seismic input will consist of a translational component (reduced compared to that applicable for a surface structure) and a rocking component. The spring and damping coefficients are significantly increased. In general, the response for an embedded structure is smaller than that for the same structure resting on the surface.

Flexibility of basemat. Besides reducing the spring coefficients for the out-of-plane motion of the basemat (i.e., in the vertical and the rocking directions), the damping coefficients become smaller over the whole frequency range, indicating that a flexible basemat will radiate energy less efficiently than a rigid one does.

Through-soil coupling of neighboring structures.

Fully three-dimensional representation, if applicable.

General seismic environment, consisting of vertically incident and inclined body waves and surface waves.

The most important limitations are as follows:

Horizontal layering. An irregular soil region can, however, be selected adjacent to the structure.

(Quasi)-linear soil model. The soil's elastic moduli and damping coefficients, which can be selected to be compatible with the strains reached during the excitation, can vary with depth. The next step in the analysis of soil–structure interaction will consist of developing methods to analyze truly nonlinear material behavior in the near field (structure and surrounding soil), while the far field of the unbounded soil will be assumed to remain linearly elastic with material damping. This will allow local nonlinear effects to be modeled appropriately, such as the partial uplift of the basemat, the separation occurring between the walls of the base and the neighboring soil in the case of embedded structures, and highly nonlinear soil behavior arising adjacent to the basemat.

The many uncertainties especially associated with the modeling of the soil and the seismic environment do not destroy the confidence placed in the design of a structure and of the equipment located within, as all parameters are varied extensively. Typically, equipment is designed using an in-structure response spectrum. The location of the peaks will coincide with the first few natural frequencies of the structure–soil system and will thus shift according to the stiffness of the soil. Varying the soil properties and considering that the design-response spectrum is amplified almost uniformly throughout the range of frequencies of interest, the peaks will extend over a significant band of frequency. The exact values of the natural frequencies of the components built in at their bases are thus not so important for their designs. The values of the peaks of these broadened spectra will depend on the ratio of structural damping and of material and radiation damping of the soil, and will thus vary only within limits.

9.4.3 Recorded Field Performance

For a long time it has been known that structures founded on soft soil are affected much more by an earthquake than those resting on rock. This is a consequence of the influence of the local soil conditions on the characteristics of the seismic motion at the surface (i.e., on the free-field response).

Actual recorded motions on various soils measured during the 1957 San Francisco earthquake were used for the seismic analysis of a typical 10-story building. The maximum base shear turned out to be several times larger for the structure founded on deep soft sites than for the same structure on rock.

As another example of the large influence of the free-field motion on the response of structures, the structural damage caused in the Caracas earthquake of 1967 is cited. Figure 9-38, which is based on Ref. [17], shows the structural-damage intensity as a function of the fundamental frequencies of the soil deposits of the various sites. The fundamental frequencies depend on the depths of the sites. For structures with three to five stories, which will have a large fundamental frequency, damage was many times greater where soil depths ranged from 30 to 50 m than for soil depths over 100 m. For buildings over 10 stories high with small fundamental frequencies, the structural-damage intensity was several times higher where soil depths exceeded 160 m than for soil depths below 140 m. It is apparent that the seismic free-field motion at the surface of the soil deposit is amplified at the fundamental frequency of the site. If the fundamental frequency of the structure–soil system (which will be only somewhat smaller than that of the fixed-base structure for this type of building) is close to that of the soil deposit, a resonance condition occurs, leading to larger structural damage.

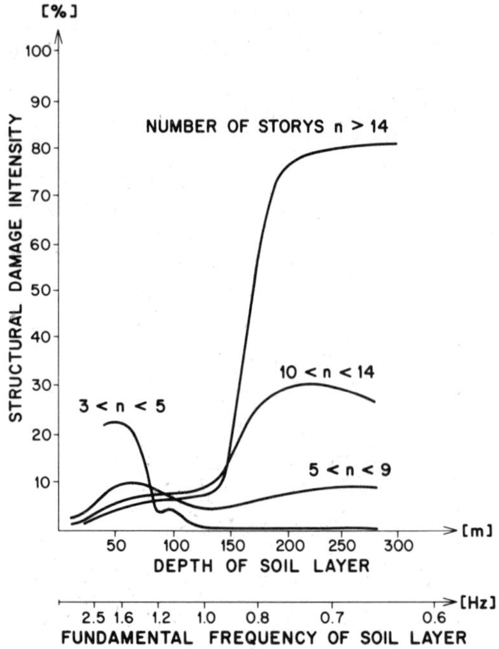

Figure 9-38 Structural damage versus influence of local soil conditions on seismic ground motions (after Ref. [17]).

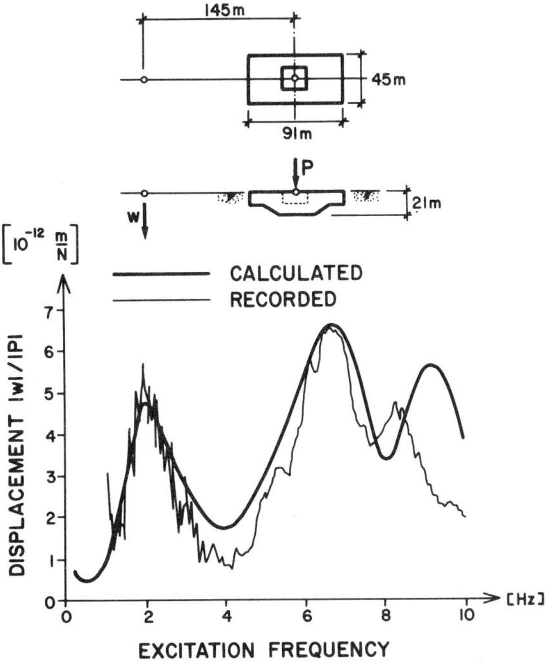

Figure 9-39 Comparison of vertical displacement of surrounding soil from vertical shaking test, large-scale shaking table foundation (after Ref. [18]).

The results of vibration tests to determine the dynamic characteristics of the foundation of a large-scale shaking table and of the surrounding soil (Nuclear-Power Engineering Test Center, Tadotsu, Japan) will be discussed next (Ref. [18]). The concrete mat foundation (Fig. 9-39) with a length = 91 m, a width = 45 m, and a thickness = 13 m to 21 m has a weight = 1.5 GN. It is embedded in a horizontally layered site of sand, gravel, and clay. The shear-wave velocity increases from 160 m/s to 640 m/s at a depth of 180 m, where weathered granite (1160 m/s) can be found. The results of the frequency-sweeping tests are shown in Fig. 9-39, where the recorded vertical displacement amplitude w of the free surface at a distance of 145 m from the center of the foundation, where the vertical load P is applied, is plotted. The agreement with the calculated values determined prior to the test is excellent. In the horizontal direction, actual earthquake records are examined. In Fig. 9-40, the amplitude ratio of the horizontal displacement u^t of the foundation to that of the free field u^f in the indicated points is plotted versus the frequency of the earthquake. Assuming the free-field motion to arise from vertically incident S-waves, the response of the foundation is calculated. The ratio $|u^t|/|u^f|$ decreases for increasing frequency until it reaches a constant value, which is caused by the

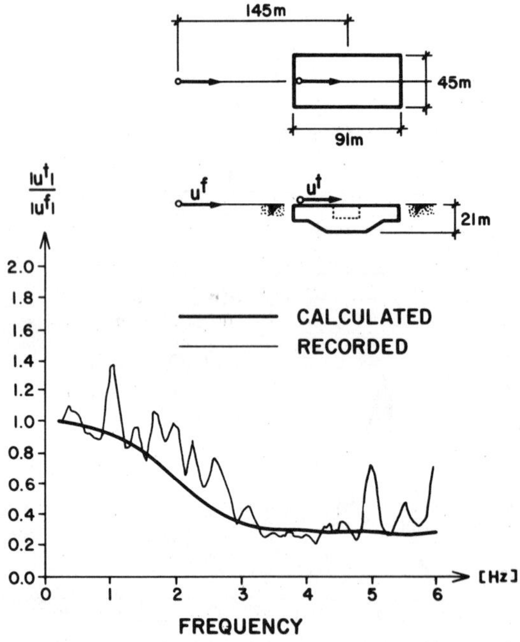

Figure 9-40 Comparison of horizontal displacement ratio from actual earthquakes, large-scale shaking table foundation (after Ref. [18]).

effect of the embedment. The agreement between the recorded and calculated values is fair.

Finally, the recorded field performance of an actual earthquake with strong motion is compared to the results of a soil–structure interaction analysis. The following comparison is based on Refs. [19] and [20], where a more complete description can be found. The buried reactor structure within the refueling building of Unit 3 of Humboldt Bay Power Plant in California consists of a massive concrete caisson embedded at a depth of about 25 m below the ground surface. The strongly layered site consists of clay and sand of various densities. During the Ferndale earthquake of June 7, 1975, a strong motion instrument at the surface recorded the free-field motion in the so-called transverse direction with a peak acceleration of 0.35*g*. This earthquake record was assumed to act at the surface of the site for the postearthquake calculation. Using the soil properties determined from the seismic analysis for the design earthquake of 0.25*g* performed prior to the recorded earthquake, the free-field response was determined. Subsequently, applying a procedure which is in essence the same as that developed in this text, the actual interaction was calculated. The response spectrum for 5% damping of the calculated horizontal motion is compared to that determined from the recorded time history within the structure at the level

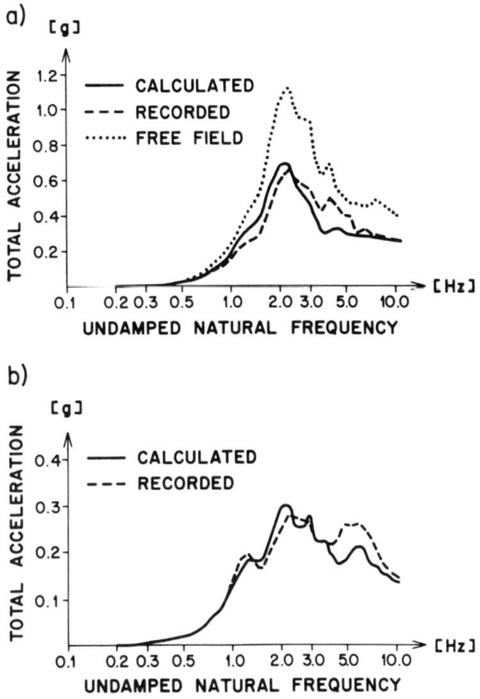

Figure 9-41 Comparison of response spectrum (5% damping), Humboldt Bay, refueling building (after Ref. [19]). (a) At ground surface; (b) at base of caisson.

of the ground surface and at the base of the caisson in Fig. 9-41. The agreement throughout the frequency range is good. The response spectrum of the recorded free-field motion is also specified in Fig. 9-41a. The large reduction of the motion at the base of the structure should also be noted. This comparison shows that, at least for this case, which exhibits quite a severe shaking, the analytical procedures developed in this text can realistically predict the seismic response of a massive structure embedded in a strongly layered site.

SUMMARY

1. The depth of the soil layer strongly affects the structural response. For a shallow layer, the fundamental frequency of the structure–soil system is smaller than that of the site. No radiation damping occurs. The equivalent damping ratio will thus be a weighted average of the material-damping ratios of the structure and of the soil. For a layer with an intermediate depth, the fundamental frequency of the structure–soil system becomes, for a sufficiently soft site, larger than that of the site, resulting in radiation damping which contributes significantly (especially for squat structure) to the equivalent damping ratio. The deeper the layer is, the less flexible the soil has to be to activate the radiation damping.

2. Adding a mass and a mass moment of inertia associated with the base to the spring–mass–dashpot system modeling the structure does not modify the characteristics of soil–structure interaction significantly.
3. Besides the stiff massive structures, for which all important interaction effects must be evaluated, types of structures exist where only certain aspects need to be examined. For instance, for continuous bridges with a large slenderness ratio, rocking dominates. The corresponding radiation damping can be neglected. Besides depending on the stiffness ratio of the structure to that of the soil and on the mass ratio, the response is thus affected by the hysteretic-damping ratios of the structure and of the soil.
4. The contribution of the second mode to the total structural response (which is significant for structures that have a small fixed-base frequency) is affected nonuniformly taking soil–structure interaction into account.
5. Turning to the structural response caused by (out-of-plane) tangential free-field motions, which are created by SH-waves, the following points are important:
   (a) For all angles of incidence, the response of the massless basemat for the horizontally propagating waves is less than that for vertical incidence. This applies to all points on the basemat, the reduction being larger in the center point.
   (b) For a structure with mass, the response for the horizontally propagating waves can be larger or smaller than for vertical incidence. For points located on the axis, horizontally propagating wave effects never dominate.
   (c) In-structure response spectra for points not located on the axis exhibit an additional substantial peak at the fundamental torsional frequency of the structure–soil system.
6. The following statements apply to the structural response arising from the radial (horizontal) and vertical (in-plane) free-field motions (P-, SV-, and R-waves):
   (a) Throughout the structure, the response caused by the radial motion of the horizontally propagating wave is always reduced, compared to that of vertically incident SV-waves.
   (b) For points on the axis, this also applies to the vertical response caused by the vertical motion of the horizontally propagating wave. The corresponding rocking input (which is about twice as large as the torsional rotation arising from an SH-wave) leads to a substantial horizontal response, especially for points located at higher elevations and to an additional vertical response for points not situated on the axis. The latter even leads to a larger vertical response in points on the boundary of the massless basemat.
   (c) For the total horizontal response the phase angle between the horizontal and vertical free-field motions is important. If the free-field motion is

retrograde (which is the case for R-waves in a half-space and for SV-waves for angles of incidence less than 45°), the horizontal response in points at higher elevations caused by the radial motion and that caused by the vertical are in phase and are thus added. For a prograde motion (SV-waves for angles of incidence between 45° and the critical angle), the opposite applies.

(d) For R-waves, whose motion is retrograde in an elastic half-space, whose vertical component is approximately 1.5 times larger than the horizontal, and whose velocity of propagation is even slightly less than the shear-wave velocity, the horizontal-structural response is magnified considerably (e.g., by a factor of 2.5 at an elevation equal to the radius for all parameters examined).

(e) Within the frequency range of interest to earthquake engineers, deviating from the vertical incidence for the P-wave does not govern the vertical response. However, the horizontal response for the SV-wave at the critical angle is significantly larger than that for vertical incidence, due to the increase of the free-field motion by a factor of approximately 1.7.

(f) When combining P- and SV-waves to generate radial and vertical free-field motions, making reasonable assumptions (vertical amplitude = 0.67 horizontal amplitude, random phase angle, etc.) will lead to a structural response similar to that obtained when vertical incidence is postulated.

7. The analytical procedures developed in this text are being applied to actual structures. The through-soil coupling of a reactor building and of a reactor-auxiliary and fuel-handling building is investigated. The seismic pile–soil–pile interaction of a reactor building is addressed. The dynamic response of a hyperbolic cooling tower and of a nuclear island with (nonlinear) aseismic bearings to horizontally propagating seismic waves is examined.

8. Soil–structure interaction generally represents an important effect which must be considered. The free-field analysis is the most important aspect. Neglecting the actual interaction analysis will lead to an overly conservative design (although certain results, such as the relative displacements between two structures, can be underestimated). Even seismic-design provisions applicable to everyday building structures allow a significant reduction of the equivalent static-lateral force (up to 30%) for soil–structure interaction effects.

9. The results are strongly influenced by the layering of the site, the embedment of the structure, the flexibility of the base, the through-soil coupling of neighboring structures, and the general seismic environment. The calculational procedure assumes a horizontal layering of the site (with an irregular soil region adjacent to the structure) and a (quasi)-linear soil model.

10. The calculated free-field response compares favorably with recorded motions of actual earthquakes. It is also demonstrated that the analytical procedures predict realistically the recorded seismic response of a reactor caisson embedded in a strongly layered site for quite severe shaking.

PROBLEMS

9.1. Comparing the dynamic-stiffness matrix of a surface foundation of a half-space to that of a finite layer, the main effect is that below the fundamental frequency of the built-in layer no radiation of energy (for the undamped case) exists. As an approximation to the radiation damping coefficients of the layer, these can be set equal to zero and to the values of the half-space for a frequency below and above the corresponding fundamental frequencies of the layer, respectively. In addition, the static-stiffness matrix increases for a finite layer. As a crude approximation, the dynamic-stiffness coefficients of a circular basemat of radius a resting on the surface of an undamped layer of depth d can be formulated as

$$S_x = \frac{8Ga}{2-\nu}\left(1 + \frac{a}{2d}\right)(k_x + ia_o c_x)$$

$$S_\phi = \frac{8Ga^3}{3(1-\nu)}\left(1 + \frac{a}{6d}\right)(k_\phi + ia_o c_\phi)$$

where k_x and k_ϕ equal 1 and

$$c_x = \begin{cases} 0 & a_o = \dfrac{\omega a}{c_s} < \dfrac{\pi}{2}\dfrac{a}{d} \\ 0.575 & a_o > \dfrac{\pi}{2}\dfrac{a}{d} \end{cases}$$

$$c_\phi = \begin{cases} 0 & a_o = \dfrac{\omega a}{c_s} < \dfrac{\pi}{2}\dfrac{a}{d}\dfrac{c_p}{c_s} \\ 0.15 & a_o > \dfrac{\pi}{2}\dfrac{a}{d}\dfrac{c_p}{c_s} \end{cases}$$

Note that c_x and c_ϕ of the half-space correspond to the expressions introduced in Eq. 3.65.

Introducing the dimensionless parameters used in Section 9.1.1 and the relation d/a, derive equations for the properties $\tilde{\omega}/\omega_s$ and $\tilde{\zeta}$ of the equivalent one-degree-of-freedom system (Fig. 9-1). The material damping of the soil can be taken into account by introducing the additional term $(1 - \tilde{\omega}^2/\omega_s^2)\zeta_g$ (analogous as in Eq. 3.66b). Plot $\tilde{\omega}/\omega_s$ and $\tilde{\zeta}$ as a function of $\bar{s}$ $(0.1 < \bar{s} < 10)$ for $d/a = 1$, 2, and for the half-space ($\bar{h} = 0.67$, $\bar{m} = 3$, $m_o = I_o = 0$, $\nu = 0.33$, $\zeta = 0.025$, $\zeta_g = 0.05$). Compare with Fig. 9-2.

Results:

$$\frac{\tilde{\omega}^2}{\omega_s^2} = \frac{1}{1 + \dfrac{\bar{m}\bar{s}^2}{8}\left[\dfrac{2-\nu}{\bar{h}^2 k_x(1 + a/2d)} + \dfrac{3(1-\nu)}{k_\phi(1 + a/6d)}\right]}$$

$$\tilde{\zeta} = \frac{\tilde{\omega}^2}{\omega_s^2}\zeta + \left(1 - \frac{\tilde{\omega}^2}{\omega_s^2}\right)\zeta_g + \left(\frac{\tilde{\omega}}{\omega_s}\right)^3 \frac{\bar{s}^3\bar{m}}{16\bar{h}}\left[\frac{2-\nu}{\bar{h}^2}\frac{c_x}{k_x^2(1 + a/2d)} + 3(1-\nu)\frac{c_\phi}{k_\phi^2(1 + a/6d)}\right]$$

$$a_o = \frac{\tilde{\omega}}{\omega_s}\frac{\bar{s}}{\bar{h}} < \frac{\pi}{2}\frac{a}{d} \longrightarrow c_x = 0$$

$$a_o < \pi\frac{a}{d} \longrightarrow c_\phi = 0$$

The equivalent parameters are plotted in Fig. P9-1. They agree well with the rigorous values shown in Fig. 9-2.

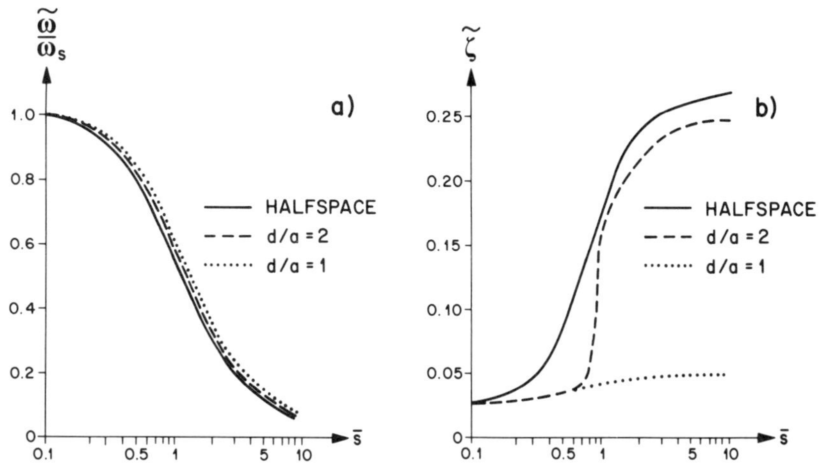

Figure P9-1 Equivalent properties. (a) Natural frequency; (b) damping.

9.2. In Section 9.1.5 the interaction effects of a continuous girder bridge with hinged columns (Fig. 3-19) are discussed. In the following, the characteristics for built-in columns (Fig. P4-6) are examined. In Problem 4.6 the equations for the natural frequency $\tilde{\omega}$ and the damping ratio $\tilde{\zeta}$ of an equivalent one-degree-of-freedom system are derived.

(a) Express $\tilde{\omega}$ and $\tilde{\zeta}$ as a function of the dimensionless parameters $\bar{s} = \omega_s h/c_s$, $\bar{h} = h/a$, $\bar{m} = m/(\rho a^3)$, ν, ζ, and ζ_g, using the frequency-independent approximations of Eq. 3.65 to evaluate the spring and damping coefficients. The radiation-damping ratios ζ_x and ζ_ϕ are defined in Eqs. 3.40 and 3.45.

(b) By plotting $\tilde{\omega}/\omega_s$ and $\tilde{\zeta}$ versus $\bar{s}$ ($0.1 < \bar{s} < 10$) for $\bar{h} = 3$, 5, and 10 using the parameters $\bar{m} = 3$, $\nu = 0.33$, $\zeta = 0.025$, and $\zeta_g = 0.05$ show that, for the statically indeterminate system (Fig. P4-6), a strong dependence on $\bar{h}$ exists, in contrast to the statically determinate one (Fig. 3-18). Verify that $\tilde{\zeta}$ for sufficiently large $\bar{s}$ will be larger than ζ_g; that is, the term of the radiation damping in the horizontal direction ζ_x is significant. This is in contrast to the girder with the hinged columns (Fig. 9-4, curve for $\bar{m} = 3$).

Results:

(a) $$\frac{\tilde{\omega}^2}{\omega_s^2} = \frac{1}{1 + \bar{s}^2\bar{m}\left[\dfrac{2-\nu}{8\bar{h}^2} + \dfrac{3(1-\nu)}{32 + (1-\nu)\bar{s}^2\bar{m}}\right]}$$

$$\tilde{\zeta} = \zeta - \frac{\tilde{\omega}^2}{\omega_s^2}\frac{(2-\nu)\bar{s}^2\bar{m}}{8\bar{h}^2}\left(\zeta - \zeta_g - 0.288\frac{\bar{s}}{\bar{h}}\frac{\tilde{\omega}}{\omega_s}\right)$$

$$- \frac{\dfrac{\tilde{\omega}^2}{\omega_s^2}\dfrac{96}{(1-\nu)\bar{s}^2\bar{m}}}{\left[1 + \dfrac{32}{(1-\nu)\bar{s}^2\bar{m}}\right]^2}\left(\zeta - \zeta_g - 0.15\frac{\bar{s}}{\bar{h}}\frac{\tilde{\omega}}{\omega_s}\right)$$

(b) See Fig. P9-2.

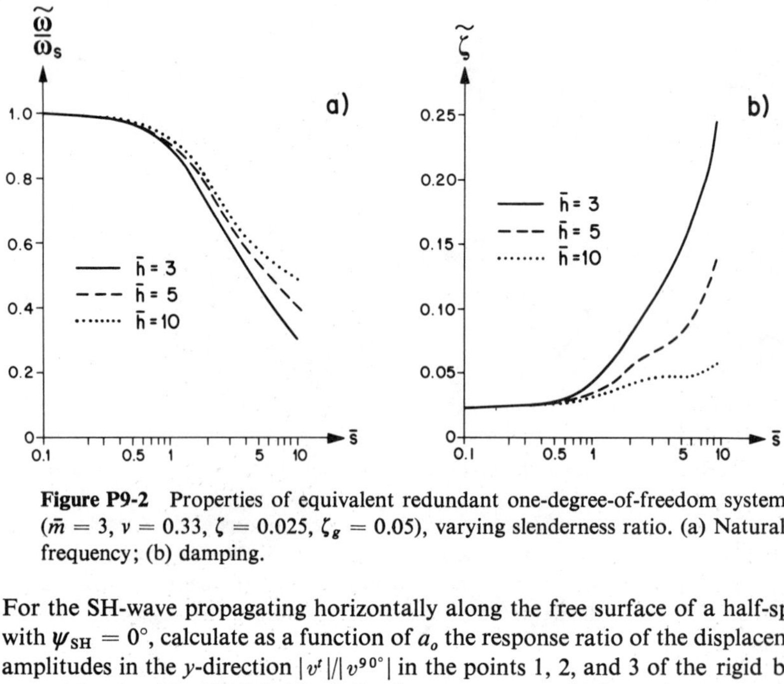

Figure P9-2 Properties of equivalent redundant one-degree-of-freedom system ($\bar{m} = 3$, $\nu = 0.33$, $\zeta = 0.025$, $\zeta_g = 0.05$), varying slenderness ratio. (a) Natural frequency; (b) damping.

9.3. For the SH-wave propagating horizontally along the free surface of a half-space with $\psi_{SH} = 0°$, calculate as a function of a_o the response ratio of the displacement amplitudes in the y-direction $|v^t|/|v^{90°}|$ in the points 1, 2, and 3 of the rigid basemat with mass (Fig. P9-3a). The following parameters which are the same as used in connection with Fig. 9-11 apply: $\bar{m} = 3$, $\nu = 0.33$, and $\zeta_g = 0.05$. Plot this ratio and compare with that one of the massless basemat ($\bar{m} = 0$). For the scattered motion v_o^g, γ_o^g, use the results of the approximate analysis of Problem 4.3. The dynamic-stiffness coefficients in the horizontal direction (subscript x) and for twisting (subscript ϕ) can be determined either by using the conical shear beam (Problems 5.5 and 5.6) or as follows [corresponds to applying for the damped case the same spring and damping coefficients as for the undamped case; see Problem 7.3, method (b)].

$$S_x = \frac{8Ga}{2-\nu}\left[k_x - \zeta_g a_o c_x + i a_o\left(c_x + \frac{2\zeta_g}{a_o}k_x\right)\right]$$

with $k_x = 1$ and $c_x = 0.575$.

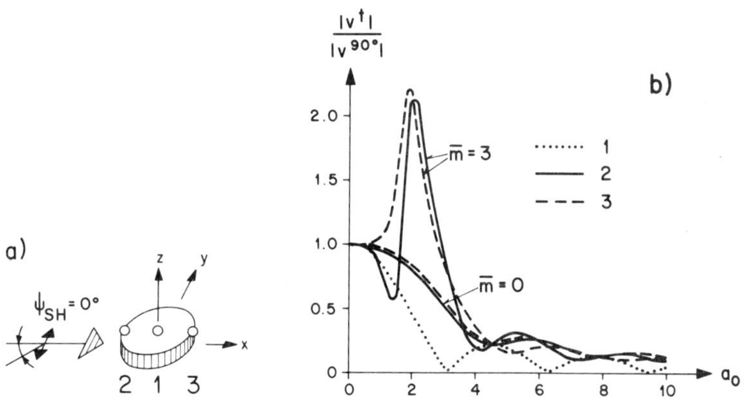

Figure P9-3 Basemat with mass excited by horizontally propagating SH-wave. (a) Dynamic system; (b) harmonic response.

$$S_\phi = \frac{16Ga^3}{3}\left[k_\phi - \zeta_g a_o c_\phi + i a_o\left(c_\phi + \frac{2\zeta_g}{a_o}k_\phi\right)\right]$$

with $k_\phi = 1$ and $c_\phi = 0.15$.

Compare the results with the exact ones shown in Fig. 9-11.

Solution:

The equations of motion (Eq. 3.20) of the rigid basemat with mass m and mass moment of inertia $I\ (= 0.5\, a^2 m)$ equal

$$(-\omega^2 m + S_x)v_o^t = S_x v_o^g$$

$$(-\omega^2 I + S_\phi)\phi^t = S_\phi \gamma_o^g$$

where (Problems 4.2 and 4.3)

$$\frac{v_o^g}{v^f} = \frac{\sin a_o}{a_o}$$

$$\frac{a\gamma_o^g}{v^f} = -i\frac{3}{2a_o}\left(\frac{\sin a_o}{a_o} - \cos a_o\right)$$

with setting $c_a = c_s$ for $\psi_{SH} = 0°$,

$$a_o = \frac{\omega a}{c_s}$$

For vertical incidence ($c_a = \infty$), $v_o^g = v^f$, $\gamma_o^g = 0$, and

$$(-\omega^2 m + S_x)v^{90°} = S_x v^f$$

This leads to (Point 1)

$$\frac{v_o^t}{v^{90°}} = \frac{\sin a_o}{a_o} \quad (\text{for any } \bar{m})$$

$$\frac{a\phi^t}{v^{90°}} = \frac{S_\phi(-\omega^2 m + S_x)}{S_x(-\omega^2 I + S_\phi)}\left[-i\frac{3}{2a_o}\left(\frac{\sin a_o}{a_o} - \cos a_o\right)\right]$$

$$\text{Point 2:}\quad \frac{v^t}{v^{90°}} = \frac{v_o^t}{v^{90°}} - \frac{a\phi^t}{v^{90°}}$$

$$\text{Point 3:}\quad \frac{v^t}{v^{90°}} = \frac{v_o^t}{v^{90°}} + \frac{a\phi^t}{v^{90°}}$$

The absolute values are plotted in Fig. P9-3b, using S_x and S_ϕ as specified in the statement of the problem. The agreement with the exact values shown in Fig. 9-11 is very good.

9.4. Determine the harmonic response of the massless structure with $\bar{h} = 1$ in points 1 and 2 (Fig. P9-4a) for the following horizontally propagating surface waves:
(a) Rayleigh wave of halfspace ($|w^f| = 1.565|u^f|$ with $c_a = 0.933c_s$ for $\nu = 0.33$)
(b) Retrograde wave with $|w^f| = |u^f|$ propagating with the same c_a
(c) Prograde wave with $|w^f| = |u^f|$ propagating with the same c_a.
For the scattered motion of the basemat, use the approximate results of Problems 4.2 and 4.4. Plot the horizontal response ratio $|u^t|/|u^f| = |u^k|/|u^f|$ versus a_o. Compare part (a) with the results shown in Fig. 9-21.

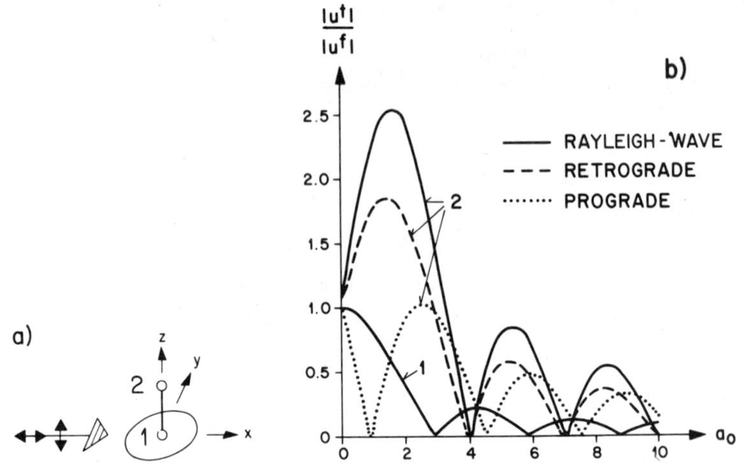

Figure P9-4 Massless structure excited by horizontally propagating surface waves. (a) Dynamic system; (b) harmonic response.

Solution:

Rayleigh wave	$w^f = -1.565iu^f$
Retrograde wave	$w^f = -iu^f$
Prograde wave	$w^f = +iu^f$

Problem 4.2:

$$\frac{u_o^g}{u^f} = \frac{c_a}{\omega a}\sin\frac{\omega a}{c_a} = \frac{0.933}{a_o}\sin\frac{a_o}{0.933} \qquad \text{with } a_o = \frac{\omega a}{c_s}$$

Problem 4.4:

$$\frac{a\beta_o^g}{w^f} = i3\frac{c_a}{\omega a}\left(\frac{c_a}{\omega a}\sin\frac{\omega a}{c_a} - \cos\frac{\omega a}{c_a}\right)$$

$$\frac{a\beta_o^g}{w^f} = i3\frac{0.933}{a_o}\left(\frac{0.933}{a_o}\sin\frac{a_o}{0.933} - \cos\frac{a_o}{0.933}\right)$$

$$\text{Point 1:}\quad \frac{u^t}{u^f} = \frac{u_o^g}{u^f}$$

$$\text{Point 2:}\quad \frac{u^t}{u^f} = \frac{u_o^g}{u^f} + \frac{h\beta_o^g}{u^f}$$

Note that for a retrograde motion the two terms on the right-hand side are positive (for small and intermediate a_o). The absolute values are plotted in Fig. P9-4b. The agreement of the approximate results for the R-wave with the corresponding exact values shown in Fig. 9-21 is very good.

REFERENCES AND ACKNOWLEDGMENTS

[1] "Strong Motion Earthquake Accelerograms," Report EERL 76-02, Earthquake Engineering Research Laboratory, California Institute of Technology, Pasadena, Calif., 1976, vol. 2.

[2] J. P. Wolf, K. M. Bucher, and P. E. Skrikerud, "Response of Equipment to Aircraft Impact," *Nuclear Engineering and Design*, *47* (1978), 169–193.

[3] J. P. Wolf and P. E. Skrikerud, "Influence of Geometry and of the Constitutive Law of the Supporting Columns on the Seismic Response of a Hyperbolic Cooling Tower," *Earthquake Engineering and Structural Dynamics*, *8* (1980), 415–437. Reprinted by permission of John Wiley & Sons, Ltd.

[4] J. P. Wolf and P. Obernhuber, "Effects of Horizontally Travelling Waves in Soil–Structure Interaction," *Nuclear Engineering and Design*, *57* (1980), 221–244.

[5] J. P. Wolf and P. Obernhuber, "Free-Field Response from Inclined SH-Waves and Love-Waves," *Earthquake Engineering and Structural Dynamics*, *10* (1982), 823–845. Reprinted by permission of John Wiley & Sons, Ltd.

[6] J. P. Wolf and P. Obernhuber, "Free-Field Response from Inclined SV- and P-Waves and Rayleigh-Waves," *Earthquake Engineering and Structural Dynamics*, *10* (1982), 847–869. Reprinted by permission of John Wiley & Sons, Ltd.

[7] J. P. Wolf and P. Obernhuber, "In-Plane Free-Field Response of Actual Sites," *Earthquake Engineering and Structural Dynamics*, *11* (1983), 121–134. Reprinted by permission of John Wiley & Sons, Ltd.

[8] J. P. Wolf and G. R. Darbre, "Dynamic-Stiffness Matrix of Surface Foundation on Layered Halfspace Based on Stiffness-Matrix Approach," in Summary Report, IAEA Specialists' Meeting on Gas-Cooled Reactor Seismic Design Problems and Solutions, IAEA IWGGCR/6 UC-77, General Atomic Company, San Diego, Calif., 1983, pp. 183–206.

[9] J. P. Wolf and G. R. Darbre, "Dynamic-Stiffness Matrix of Soil by the Boundary-Element Method: Conceptional Aspects," to appear in *Earthquake Engineering and Structural Dynamics, 12* (1984). Reprinted by permission of John Wiley & Sons, Ltd.

[10] J. P. Wolf and G. R. Darbre, "Dynamic-Stiffness Matrix of Soil by the Boundary-Element Method: Embedded Foundations," to appear in *Earthquake Engineering and Structural Dynamics, 12* (1984). Reprinted by permission of John Wiley & Sons, Ltd.

[11] J. P. Wolf and G. R. Darbre, "Dynamic-Stiffness Matrix of Embedded and Pile Foundations by Indirect Boundary-Element Method," *Transactions of the 7th International Conference on Structural Mechanics in Reactor Technology*, Chicago, 1983, vol. K(b), paper K 11/1, pp. 245–258.

[12] J. P. Wolf and G. A. von Arx, "Impedance Function of a Group of Vertical Piles," *Proceedings of the Specialty Conference on Earthquake Engineering and Soil Dynamics, ASCE,* Pasadena, Calif., 1978, vol. 2, 1024–1041.

[13] J. P. Wolf and B. Weber, "On Calculating the Dynamic-Stiffness Matrix of the Unbounded Soil by Cloning," *Proceedings of the International Symposium on Numerical Models in Geomechanics*, Zurich, A.A. Balkema, Rotterdam, 1982, pp. 486–494.

[14] J. P. Wolf, G. A. von Arx, F. C. P. de Barros, and M. Kakubo, "Seismic Analysis of the Pile Foundation of the Reactor Building of the NPP Angra 2," *Nuclear Engineering and Design, 65* (1981), 329–341.

[15] J. P. Wolf and K. M. Bucher, "Nonlinear Traveling-Wave Effects on Cooling Tower," *Journal of the Structural Division, ASCE, 108* (1982), 1424–1439.

[16] J. P. Wolf, P. Obernhuber, and B. Weber, "Response of a Nuclear-Power Plant on Aseismic Bearings to Horizontally Propagating Waves," *Earthquake Engineering and Structural Dynamics, 11* (1983), 483–499. Reprinted by permission of John Wiley & Sons, Ltd.

[17] H. B. Seed and I. M. Idriss, "Ground Motions and Soil Liquefaction during Earthquakes," Earthquake Engineering Research Institute, Berkeley, Calif., p. 18.

[18] H. Tajimi, "Recent Tendency of the Practice of Soil-Structure Interaction Design Analysis in Japan and Its Theoretical Background," Principal Division Lecture, Division K, 7th International Conference on Structural Mechanics in Reactor Technology, Chicago, 1983.

[19] J. Lysmer, M. Tabatabaie, F. Tajirian, S. Vahdani, and F. Ostadan, "SASSI, a System for Analysis of Soil-Structure Interaction," UCB/GT/81-02, University of California, Berkeley, 1981.

[20] J. E. Valera, H. B. Seed, C. F. Tsai, and J. Lysmer, "Soil-Structure Interaction Effects at the Humboldt Bay-Power Plant in the Ferndale Earthquake of June 7, 1975," UCB/EERC-77/02, University of California, Berkeley, 1977.

Certain parts of this text are based on the references mentioned above. This is indicated by including the number of the reference in the captions of the figures. The permission to include the material is gratefully acknowledged.

INDEX